T0135230

Advances in Intelligent Systems and Computing

Volume 750

Series editor

Janusz Kacprzyk, Systems Research Institute, Polish Academy of Sciences, Warsaw, Poland
e-mail: kacprzyk@ibspan.waw.pl

The series "Advances in Intelligent Systems and Computing" contains publications on theory, applications, and design methods of Intelligent Systems and Intelligent Computing. Virtually all disciplines such as engineering, natural sciences, computer and information science, ICT, economics, business, e-commerce, environment, healthcare, life science are covered. The list of topics spans all the areas of modern intelligent systems and computing such as: computational intelligence, soft computing including neural networks, fuzzy systems, evolutionary computing and the fusion of these paradigms, social intelligence, ambient intelligence, computational neuroscience, artificial life, virtual worlds and society, cognitive science and systems, Perception and Vision, DNA and immune based systems, self-organizing and adaptive systems, e-Learning and teaching, human-centered and human-centric computing, recommender systems, intelligent control, robotics and mechatronics including human-machine teaming, knowledge-based paradigms, learning paradigms, machine ethics, intelligent data analysis, knowledge management, intelligent agents, intelligent decision making and support, intelligent network security, trust management, interactive entertainment, Web intelligence and multimedia.

The publications within "Advances in Intelligent Systems and Computing" are primarily proceedings of important conferences, symposia and congresses. They cover significant recent developments in the field, both of a foundational and applicable character. An important characteristic feature of the series is the short publication time and world-wide distribution. This permits a rapid and broad dissemination of research results.

More information about this series at http://www.springer.com/series/11156

J. Dinesh Peter · Amir H. Alavi
Bahman Javadi
Editors

Advances in Big Data and Cloud Computing

Proceedings of ICBDCC18

 Springer

Editors
J. Dinesh Peter
Department of Computer Sciences
  Technology
Karunya Institute of Technology & Sciences
Coimbatore, Tamil Nadu, India

Bahman Javadi
School of Computing, Engineering and
  Mathematics
University of Western Sydney
Sydney, NSW, Australia

Amir H. Alavi
Department of Civil and Environmental
  Engineering
University of Missouri
Columbia, MO, USA

ISSN 2194-5357   ISSN 2194-5365   (electronic)
Advances in Intelligent Systems and Computing
ISBN 978-981-13-1881-8   ISBN 978-981-13-1882-5   (eBook)
https://doi.org/10.1007/978-981-13-1882-5

Library of Congress Control Number: 2017957703

This Springer imprint is published by the registered company Springer Nature Singapore Pte Ltd.
The registered company address is: 152 Beach Road, #21-01/04 Gateway East, Singapore 189721, Singapore

Preface

The boom of cloud technologies and cloud data storage has been a forerunner to the growth of big data. It has substantial advantages over conventional physical deployments. In India, the organizations that adopt big data have established the boundary between the use of private clouds, public clouds, and Internet of things (IoT), which allows better access, performance, and efficiency of analyzing the data and understanding the data analytics. The main objective of this conference is to reignite, encourage, and bring together the proficient members and professionals in the field of big data and cloud computing.

International Conference on Big data and Cloud Computing (ICBDCC18) is a joint venture of the professors in the Department of Computer Science and Engineering of Karunya University, University of Missouri (USA), and Western Sydney University (Australia). ICBDCC18 provided a unique forum for the practitioners, developers, and users to exchange ideas and present their observations, models, results, and experiences with the researchers who are involved in real-time projects in big data and cloud computing technologies. In the last decade, a number of sophisticated and new computing technologies have been evolved that strides straddle the society in every facet of it. With the introduction of new computing paradigms such as cloud computing, big data, and other innovations, ICBDCC18 provided a professional forum for the dissemination of new ideas, technology focus, research results, and discussions on the evolution of computing for the benefit of both scientific and industrial developments. ICBDCC18 has been supported by the panel of reputed advisory committee members from both India and abroad. The research tracks of ICBDCC18 received a total of 110 submissions that were full research papers on big data and cloud computing. Each research article submission was subjected to a rigorous blind review process by two eminent academicians. After the scrutiny, 51 papers were selected for presentation in the conference and are included in this volume.

This proceedings includes topics in the fields of big data, data analytics in cloud, cloud security, cloud computing, big data and cloud computing applications. The research articles featured in this proceedings provide novel ideas that contribute to the growth of the society through the recent computing technologies. The contents of this proceedings will prove to be an invaluable asset to the researchers in the areas of big data and cloud computing.

The editors appreciate the extensive time and effort put in by all the members of the organizing committee for ensuring a high standard for the articles published in this volume. We would like to record our thanks to the panel of experts who helped us to review the articles and assisted us in selecting the candidates for the best paper award. The editors would like to thank the eminent keynote speakers who have consented and shared their ideas with the audience and all the researchers and academicians who have contributed their research work, models, and ideas to ICBDCC18.

Coimbatore, India   J. Dinesh Peter
Columbia, USA   Amir H. Alavi
Sydney, Australia   Bahman Javadi

Contents

About the Editors

J. Dinesh Peter is currently working as an associate professor in the Department of Computer Sciences Technology at Karunya University, Coimbatore. Prior to this, he was a full-time research scholar at National Institute of Technology Calicut, India, from where he received his Ph.D. in computer science and engineering. His research focuses include big data, image processing, and computer vision. He has several publications in various reputed international journals and conference papers which are widely referred to. He is a member of IEEE, CSI, and IEI and has served as session chairs and delivered plenary speeches for various international conferences and workshops. He has conducted many international conferences and been as editor for Springer proceedings and many special issues in journals.

Amir H. Alavi received his Ph.D. degree in Structural Engineering with focus on Civil Infrastructure Systems from Michigan State University (MSU). He also holds a M.S. and B.S. in Civil and Geotechnical Engineering from Iran University of Science & Technology (IUST). Currently, he is serving as a senior researcher in a joint project between the University of Missouri (MU) and University of Illinois at Urbana-Champaign (UIUC), in Cooperation with the City Digital at UI+LABS in Chicago on Development of Smart Infrastructure in Chicago. The goal is to make Chicago as the Smartest City on Earth. Dr. Alavi's research interests include smart sensing systems for infrastructure/structural health monitoring (I/SHM), sustainable and resilient civil infrastructure systems, energy harvesting, and data mining/data interpretation in civil engineering. He has published 3 books and over 130 research papers in indexed journals, book chapters, and conference proceedings, along with three patents. He is on the editorial board of several journals and is serving as ad-hoc reviewer for many indexed journals. He has also edited several special issues for indexed journals such as Geoscience Frontiers, Advances in Mechanical Engineering, ASCE-ASME Journal of Risk and Uncertainty in Engineering Systems, as well as a recent special of Automation in Construction on Big Data in civil engineering. He is among the Google Scholar three hundred most cited authors within civil engineering domain (citation > 4100 times; h-index = 34). More, he is selected as the advisory board of Universal Scientific Education and Research Network (USERN), which belongs to all top 1% scientists and the Nobel laureates in the world.

Dr. Bahman Javadi is a senior lecturer in networking and cloud computing at the Western Sydney University, Australia. Prior to this appointment, he was a research fellow at the University of Melbourne, Australia. From 2008 to 2010, he was a postdoctoral fellow at the INRIA Rhone-Alpes, France. He received his MS and Ph.D. degrees in computer engineering from the Amirkabir University of Technology in 2001 and 2007, respectively. He has been a research scholar at the School of Engineering and Information Technology, Deakin University, Australia, during his Ph.D. course. He is a co-founder of the Failure Trace Archive, which serves as a public repository of failure traces and algorithms for distributed systems. He has received numerous best paper awards at IEEE/ACM conferences for his research papers. He has served as a member in the program committee of many international conferences and workshops. His research interests include cloud computing, performance evaluation of large-scale distributed computing systems, and reliability and fault tolerance. He is a member of ACM and senior member of IEEE.

Fault-Tolerant Cloud System Based on Fault Tree Analysis

Getzi Jeba Leelipushpam Paulraj, Sharmila John Francis, J. Dinesh Peter
and Immanuel John Raja Jebadurai

Abstract Cloud computing has gained its popularity as it offers services with less cost, unlimited storage, and high computation. Today's business and many emerging technologies like Internet of Things have already been integrated with cloud computing for maximum profit and less cost. Hence, high availability is expected as one of the salient features of cloud computing. In this paper, fault-tolerant system is proposed. The fault-tolerant system analyzes the health of every host using fault tree-based analysis. Virtual machines are migrated from the unhealthier host. The proposed methodology has been analyzed with various failure cases, and its throughput is proved to be the best compared with the state-of-the-art methods in literatures.

Keywords Fault tree analysis · Virtual machine migration · Fault tolerance
Cloud computing

1 Introduction

Cloud computing offers computation and storage as service in pay as you use manner. Infrastructure, software, and platform are offered on demand. The services offered should be available and reliable to the customers. Such reliability and availability of cloud service can be achieved through fault tolerance [1]. Fault-tolerant system must be able to detect the failure and also take alternate measures for uninterrupted service

G. J. L. Paulraj (✉) · S. J. Francis · J. D. Peter · I. J. R. Jebadurai
Karunya University, Coimbatore, India
e-mail: getzi@karunya.edu; getz23@gmail.com

S. J. Francis
e-mail: sharmilaanand2003@yahoo.co.in

J. D. Peter
e-mail: dineshpeter@karunya.edu

I. J. R. Jebadurai
e-mail: immanueljohnraja@gmail.com

© Springer Nature Singapore Pte Ltd. 2019
J. D. Peter et al. (eds.), *Advances in Big Data and Cloud Computing*,
Advances in Intelligent Systems and Computing 750,
https://doi.org/10.1007/978-981-13-1882-5_1

1

to the customers. Fault-tolerant system can be reactive or proactive [2]. Reactive methods analyze the system after failure and attempt to reduce its impact. It also aids in recovery of lost data. In the proactive method, the occurrence of failure is predicted and prevented. Proactive methods are more efficient methods for ensuring high availability [2].

This paper proposes a proactive fault-tolerant system. The system estimates the occurrence of failure using fault tree analysis. To analyze the failure, fault tree is constructed considering the major causes of failures at the host level. The remedial measure is handled by migrating the virtual machines from the host that is about to fail to the healthier host. The proposed methodology is simulated, and performance metrics are analyzed.

Section 2 discusses the various fault-tolerant techniques available in the literature. Section 3 presents the host level fault tree. Section 4 explains the fault-tolerant Virtual Machine (VM) migration technique. Section 5 analyzes the performance of the proposed methodology in terms of response time and throughput. Section 6 concludes the paper and suggests future research work.

2 Related Works

Fault-tolerant systems handle fault detection and fault recovery. They handle failure using two methods: reactive and proactive method. Various Techniques explain reactive fault-tolerant systems. In [3], failure of a host causes the jobs running on that host to move to a replica host. In [3], the job request is executed by primary node and backup node. The result of their execution is compared. If the results are same, it means that there is no failure. The different result shows the presence of failure. The primary node or backup node which is responsible for the failure is replaced. In [4], Byzantine architecture is involved in reactive fault detection. $3n + 1$ nodes are involved in detecting n faulty node. The fault detection capability is only 33%. In [5] a decision vector has been used between a host and its neighbor. The decision vectors are exchanged with the neighbors. The conflict is identified by deviation in updating the decision vector. However, $2n + 1$ nodes are required to identify n faulty nodes. All the techniques discussed above identify failure only after its occurrence. In above architectures, replication is used as failure recovery mechanism. In [6, 7], proactive fault detection techniques have been proposed. In [6], map reduce technique is used. The jobs are divided into smaller sections and executed in various nodes. They are combined in the reduction phase. Any failure affects only the small section of the job, and it is recovered using replication. In [7], component ranking-based approach is used. This technique ranks the host based on its reliability and host is selected for execution of its job based on its rank.

Most of the above techniques are reactive in nature. They detect failure only after its occurrence. Failure in proactive algorithms is not estimated before its occurrence. Most of the above techniques use replication as a fault recovery solution. However, replication increases the infrastructure cost. The objective of our Fault Tree-based

Fault-Tolerant (FT-FT) System is to estimate the occurrence of failure in every host. If a host is prone to failure the virtual machines are migrated to another healthier host.

3 Fault Tree Analysis

Fault Tree analysis was introduced by the U.S. Nuclear Regulatory Commission as the main instrument used in their reactor safety studies. Fault tree analysis addresses the identification and assessment of catastrophic occurrences and complete failures [8]. It is a deductive approach used to analyze the undesired state of the system. The fault tree has two nodes: events and gates. An event is an occurrence within the system. The event at the top of the tree is called the top event. The top event has to be carefully selected. Gates represents the propagation of failures through the systems. The occurrence of top events can be quantitatively estimated.

3.1 Structure of Fault Tree

The first step of the proposed fault-tolerant system is to construct fault tree [9]. To construct a fault tree, host failure has been identified as a primary event. Three major failure factors that contribute to the failure of a host have been identified. They are hardware failure (E_1), system crashes (E_2), and network outages (E_3). As failure of any one of this factor causes the host to fail, they are connected to the primary event by means of OR gate. The fault tree for primary event along with major failure factors are depicted in Fig. 1.

Reason for the occurrence of major failure factors is also identified. Power system failure (F_1), board malfunction (F_2), and driver malfunction (F_3) are identified as sub-events for hardware failure (E_1). Application crashes (G_1), operating system crashes (G_2), and virtual machine crashes (G_3) are categorized as sub-events for

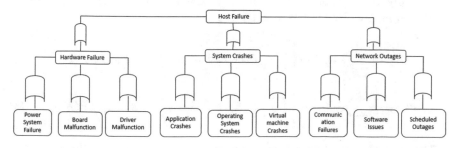

Fig. 1 Category of major risk factors

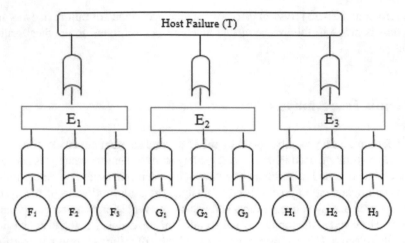

Fig. 2 Fault tree for host failure event

system crashes (E_2). The sub-events for network outages (E_3) are communication failures (H_1), software issues (H_2), and scheduled outages (H_3).

3.2 Failure Rate Estimation Using Probability

Fault tree analysis helps to identify potential failures. It also estimates the reliability of the system. The fault tree for host failure (T) is represented in Fig. 2. The reliability of the host is analyzed using probability theory [10].

The probability of occurrence of hardware failure $P(E_1)$ is given by Eq. 1

$$P(E_1) = \sum_{i=1}^{3} P(F_i) - \left[\left(\sum_{i=1}^{2} P(F_i) * P(F_{i+1}) \right) - (P(F_3) * P(F_1)) \right] \quad (1)$$

Similarly, the probability of occurrence of System crashes $P(E_2)$ and Network outages $P(E_3)$ are given in Eqs. 2 and 3

$$P(E_2) = \sum_{i=1}^{3} P(G_i) - \left[\left(\sum_{i=1}^{2} P(G_i) * P(G_{i+1}) \right) - (P(G_3) * P(G_1)) \right] \quad (2)$$

$$P(E_3) = \sum_{i=1}^{3} P(H_i) - \left[\left(\sum_{i=1}^{2} P(H_i) * P(H_{i+1}) \right) - (P(H_3) * P(H_1)) \right] \quad (3)$$

The probability is set using random and exponential distribution. The probability of occurrence of host failure $P(T)$ is the union of the probability of occurrence of hardware failure $P(E_1)$, system crashes $P(E_2)$, and network outages $P(E_3)$, and it is given by Eq. 3

$$P(T) = \sum_{i=1}^{3} P(E_i) - \left[\left(\sum_{i=1}^{2} P(E_i) * P(E_{i+1}) \right) - (P(E_3) * P(E_1)) \right] \quad (4)$$

The reliability of host S_i at time t is denoted by Eq. 5

$$R_{S_i}(t) = 1 - P(T) \quad (5)$$

where $P(T)$ is the cumulative distributive function, and it is denoted by the Eq. 6

$$P(T) = \int_0^t p(t) \, dt \quad (6)$$

Applying (6) in (5), we get

$$R_{S_i}(t) = 1 - \int_0^t p(t) dt \quad (7)$$

$\frac{p(t)}{R_{S_i}(t)}$ represent the conditional probability of failure per unit time, and it is denoted by λ. The Mean Time to Failure (MTTF) [11–13] is given by Eq. 8

$$\text{MTTF} = \int_0^{\infty} R_{S_i}(t) dt \quad (8)$$

The MTTF value is updated as the Health Fitness Value (HFV) for every host in the migration controller. The migration controller migrates the host based on its health condition using the fault-tolerant migration algorithm.

4   Fault-Tolerant VM Migration Algorithm

Based on the failure rate and MTTF value, every host is assigned a Health Fitness Value (HFV). The host updates its HFV to the migration controller. The migration controller checks for any host which is about to fail. The migration controller sorts the host on its ascending order of the HFV value. The HFV value of every host is compared with its job completion time. The job completion time of every host (Ω_{n+1}) is estimated using Eq. 9

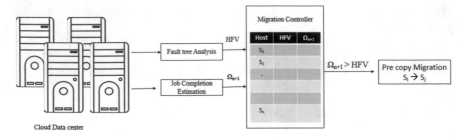

Fig. 3 FT-FT-based migration technique

$$\Omega_{n+1} = \alpha^* \, \omega_n + (1 - \alpha)^* \, \Omega_n \tag{9}$$

where α is the constant and it takes the value of 0.8. ω_n is the actual job completion time at nth time period, and Ω_n is the estimated job completion time at nth time period.

The migration is initiated in any host based on the below condition C_{migrate}

$$C_{\text{migrate}} = \begin{cases} \Omega_{n+1} > \text{HFV} \to \text{migrate (VM} \in \text{host } S_i) \\ \text{Otherwise} \quad \to \text{no migration} \end{cases}$$

As shown in Fig. 3, the condition C_{migrate} is checked in every host S. If the condition is false, the migration controller checks the condition of the next host. If the condition is true for any host S_i, the migration controller checks the resource requirement. Based on the resource requirement, it selects the destination host S_j. The resource availability is checked in host S_j. The virtual machines from host S_i is migrated to the destination host S_j using pre-copy VM migration technique [14]. The virtual machine resumes their execution from host S_j to host S_i and host S_j is submitted for maintenance.

5 Performance Analysis

The proposed work has been implemented using CloudSim. The simulation was performed using single Datacenter handling 500 hosts. The host has 4 CPU cores, 1 GB RAM, and 8 GB disk storage. The number of virtual machines is 600. The virtual machines are categorized as small instance that requires 1 CPU core, 128 MB RAM, and 2 GB disk storage; medium instance with 2 CPU cores, 256 MB RAM, and 3 GB disk storage; and large instance with 3 CPU cores, 256 MB RAM, and 5 GB disk storage. Space shared scheduling is used to initially schedule the virtual machines. The proposed scenario was tested for simulation period of 800 s. The cloudlets are modeled using random and Planet Lab workload [15].

Fig. 4 Throughput (random workload)

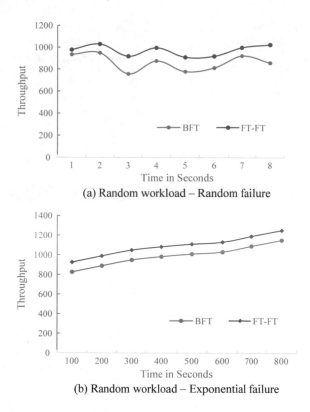

(a) Random workload – Random failure

(b) Random workload – Exponential failure

Initially, the VM is scheduled in the hosts using space shared scheduling. The simulation was performed by introducing failure with random and exponential distribution. The result was analyzed under random workload-random failure, random workload-exponential failure, Planet Lab workload-random failure, Planet Lab workload-exponential failure. The throughput of the proposed technique is compared with Byzantine fault tolerance framework [6]. The throughput is a measure of ratio of number of cloudlets completed to that of number of cloudlets assigned. Figure 4 depicts the throughput of Cloud datacenter.

Figures 4 and 5 depict the throughput of Cloud datacenter. The throughput of cloud datacenter is measured by varying the time. The throughput is measured for FT-FT method and compared with BFT Protocol. The random workload is used with 100 servers. Failure is introduced randomly during the simulation time. It is observed from Fig. 4a that the throughput of the proposed FT-FT method is increased by 13.6% when compared with BFT protocol.

Figure 4b depicts throughput of cloud datacenter for random workload with exponential failure. The failure was introduced using random distribution. The throughput of the FT-FT protocol has been improved by 12.8% when compared with the BFT protocol. The FT-FT protocol estimates MTTF using fault tree and

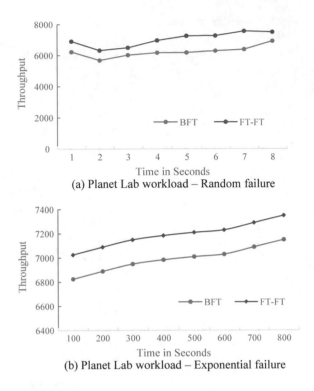

Fig. 5 Throughput (Planet Lab workload)

(a) Planet Lab workload – Random failure

(b) Planet Lab workload – Exponential failure

migrates to healthier host. This performance improvement is due to the fact that the jobs are migrated to suitable host and the number of failed job is less.

Figure 5a, b depicts the throughput of cloud datacenter with Planet Lab workload by introducing failure using random and exponential distribution. It is observed that the throughput of the proposed technique using random failure distribution has improved by 10.18 and 2.86% using exponential failure distribution.

This is because the failure is estimated well ahead and the jobs are migrated to healthier host. The failure is detected proactively and the jobs are migrated before failure. This improves the throughput of the proposed technique compared with BFT protocol.

6   Conclusion

Reliability is highly expected parameters in cloud datacenter where services are offered in demand. Our proposed method improves reliability by estimating failure in a proactive manner using fault trees. The host is ordered based on its MTTF. The VMs from the host which is about to fail are migrated to the healthier host. The proposed algorithm has been simulated and throughput has been measured using random and Planet Lab workload. The failure was introduced using random and exponential distribution. It is observed that the proposed method has improved the throughput compared with the Byzantine fault tolerance framework. In future, the proposed technique can be enhanced using more failure factors, more reliability analysis technique, and optimized VM placement technique.

References

1. Leelipushpam, P.G.J., Sharmila, J.: Live VM migration techniques in cloud environment—a survey. In: 2013 IEEE Conference on Information & Communication Technologies (ICT). IEEE, New York (2013)
2. Cheraghlou, M.N., Khadem-Zadeh, A., Haghparast, M.: A survey of fault tolerance architecture in cloud computing. J. Netw. Comput. Appl. **61**, 81–92 (2016)
3. Kaushal, V., Bala, A.: Autonomic fault tolerance using haproxy in cloud environment. Int. J. Adv. Eng. Sci. Technol. **7**(2), 222–227 (2011)
4. Zacks, S.: Introduction to Reliability Analysis: Probability Models and Statistical Methods. Springer Science & Business Media, Berlin (2012)
5. Lim, J.B., et al.: An unstructured termination detection algorithm using gossip in cloud computing environments. In: International Conference on Architecture of Computing Systems. Springer, Berlin (2013)
6. Zhang, Y., Zheng, Z., Lyu, M.R.: BFTCloud: a byzantine fault tolerance framework for voluntary-resource cloud computing. In: 2011 IEEE International Conference on Cloud Computing (CLOUD). IEEE, New York (2011)
7. Zheng, Q.: Improving MapReduce fault tolerance in the cloud. In: 2010 IEEE International Symposium on Parallel & Distributed Processing, Workshops and Phd Forum (IPDPSW). IEEE, New York (2010)
8. Clemens, P.L.: Fault tree analysis. JE Jacobs Severdurup (2002)
9. Veerajan, T.: Probability, Statistics and Random Processes, 2nd edn. Tata McGraw Hill, New Delhi (2004)
10. Xing, L., Amari, S.V.: Fault tree analysis. In: Handbook of Performability Engineering, pp. 595–620 (2008)
11. Van Renesse, R., Minsky, Y., Hayden, M.: A gossip-style failure detection service. In: Middleware'98. Springer, London (1998)
12. Yuhua, D., Datao, Y.: Estimation of failure probability of oil and gas transmission pipelines by fuzzy fault tree analysis. J. Loss Prev. Process Ind. **18**(2), 83–88 (2005)
13. Larsson, O.: Reliability analysis (2009)
14. Clark, C., et al.: Live migration of virtual machines. In: Proceedings of the 2nd Conference on Symposium on Networked Systems Design & Implementation, vol. 2. USENIX Association (2005)
15. Calheiros, R.N., et al.: CloudSim: a toolkit for modeling and simulation of cloud computing environments and evaluation of resource provisioning algorithms. Softw. Pract. Experience **41**(1), 23–50 (2011)

Major Vulnerabilities and Their Prevention Methods in Cloud Computing

Jomina John and Jasmine Norman

Abstract A single name for dynamic scalability and elasticity of resources is nothing but a cloud. Cloud computing is the latest business buzz in the corporate world. The benefits like capital cost reduction, globalization of the workforce, and remote accessibility attract people to introduce their business through the cloud. The nefarious users can scan, exploit, and identify different vulnerabilities and loopholes in the system because of the ease of accessing and acquiring cloud services. Data breaches and cloud service abuse are the top threats identified by Cloud Security Alliance. The major attacks are insider attacks, malware and worm attack, DOS attack, and DDOS attack. This paper analyzes major attacks in cloud and comparison of corresponding prevention methods, which are effective in different platforms along with DDoS attack implementation results.

Keywords Cloud computing · Worm attack · Insider attack · DDOS attack · XML DDOS attack · Forensic virtual machine

1 Introduction

Cloud computing is a new generation computing paradigm which provides scalable resources as a service through Internet. This works as a model which provides network access from a shared pool of resources which is an on-demand service (e.g., storage, network, services, applications, or even servers) [1]. Cloud can provide us with a large-scale interoperation and controlled sharing among resources which are managed and distributed by different authorities. For that all the members involved in this scenario should trust each other so that they can share their sensitive data with

J. John (✉)
School of Information Technology, VIT Vellore, Vellore, India
e-mail: jominacj@gmail.com

J. Norman
VIT University, Vellore, Tamil Nadu, India
e-mail: jasmine@vit.ac.in

© Springer Nature Singapore Pte Ltd. 2019
J. D. Peter et al. (eds.), *Advances in Big Data and Cloud Computing*,
Advances in Intelligent Systems and Computing 750,
https://doi.org/10.1007/978-981-13-1882-5_2

the service provider and use their resources as software, as a platform, as a server, etc., which are termed as software as a service, platform as a service, storage as a service, etc. [2]. There are different types of cloud like public cloud, private cloud, community cloud, and hybrid cloud.

Security is the major concern in cloud computing. A trust can be generated among the cloud service provider (CSP) and the clients. In two level, there are only CSPs and clients whereas in the case of three level, there are CSPs, clients, and customers of the client.

In all these, there is a chance of direct and indirect attacks against the data or the services that are provided by the CSPs [3].

In this paper, different types of attacks and the different models of prevention in cloud computing are analyzed. Major attacks revealed in this paper include insider attacks, worm attacks, DDoS attacks, EDoS attacks, XML DoS attack, and HTML DoS attack [4]. Also, this paper points to the implementation result of DDoS attack simulation in a real cloud environment.

2 Attacks in Cloud

2.1 Insider Attack

For a company when they are appointing anyone, they will enquire about the employee in all possible ways especially in the case of an IT specialist since the most effective attack anyone can expect is from inside the organization.

Normally the insider can be a previous employee or a less privileged employee who want to steal data or services for financial or penetrating purpose. The multi-tenant nature of cloud computing environment makes it so difficult to identify different attacks or unprivileged actions taken by an insider [5]. Insider threat can be classified into two.

1. Inside the cloud provider: It is a malicious employee working for the cloud provider.
2. Inside the cloud outsourcer: It is an employee which outsourced part or whole of its infrastructure on the cloud.

2.1.1 Detection and Prevention Methods in Insider Attack

Insider activities in cloud computing can be monitored using rule-based learning. In this method in order to monitor insider activities, there are mainly two goals are there.

1. It will be able to detect an attack at least at the time of attack perpetuation or when it starts.

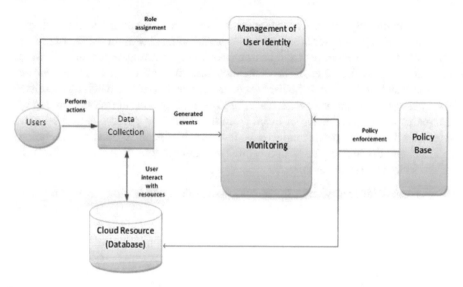

Fig. 1 .

2. This model will be able to tell customers what kind of attack happened by looking at the pattern of attack, even if cloud providers try to hide attack information from the customers [6].

In this method using any attack tools generate attack scripts and can use the information about different attack scenarios described in Internet security sites [7]. By using attack scripts we can reduce the human effort to program it with respect to the actual attack duration on multiple VMs and its timing.

Another method is detecting a malicious insider in the cloud environment using sequential rule mining. In this case, use a model that reduces attacks that are originated from the users of the CSP who are legitimate. To predict the behavior pattern of users sequential pattern mining technique using the system, i.e., event "*f*" should occur after event "ABC" with the specified minimum support and minimum confidence [8] (Fig. 1).

1. *Policy Base*: It enforces nontechnical control before and after users access the entire system.
2. *Log Events*: Capture and store all events. After sequence of event is learned, the user profile is created.
3. *Management of user identity*: It ensures the insider has valid credentials. Each user will be assigned with a role and ensure each privileged insider has access to the resources that are required for the smooth conduct of their job roles. Insiders are provided with very limited access to cloud data.
4. *Monitoring component*: It is the core component from which developed two main algorithms. This includes mainly rule learning algorithm and pattern matching algorithm.

Rule mining algorithm learns how each user performs actions in the system in order to come up with user profile and pattern matching algorithm match user profile with set of single sequences from testing data. In rule learning algorithm, it extracts raw data from log file. Then, it sets a threshold and checks the items matches with the threshold and then adds it to the list of sequential rule. In pattern matching algorithm, for a particular user, it retrieves the user profile and sequence of events from log and compares a single sequence of event from log with user profile. If that sequence triggers a rule in user profile and its satisfied think it as positive otherwise malicious.

To detect the insider attack in a healthcare environment, first ensures that some of the following factors are present:

1. A trusted third party and a trusted cloud, trusted by all healthcare organizations.
2. Secure transmission of all the keys.
3. TTP has a public key PuK known by all healthcare organization and private key PrK only known to TTP.
4. A symmetric key K shared by all clinics and another symmetric key K1 shared by doctors.

For implementing this, there is actually three modules are there.

A.   Watermarking module.

- In this embed watermark on at least significant bits of the image (LSB). The watermark should be invisible, fragile, and blind.
- Region of noninterest embedding.

1. *Watermark preparation and embedding*

Watermark divided into three parts:

a. Header: 16 bit to store the size of watermark.
b. Encrypted Hash: 160-bit data is generated by first applying SHA I hash then encrypt it.
c. Payload: Payload contains patient information.

2. *Watermark extraction*

First 16-bit header extracted. Then extract the load from the image boarder. Then find the hash and encrypt it with shared key. Compare h1 and h2, if both are same, means the image is not tampered.

3. *Modification detection*

Noise is inserted for generating tampered image for testing so that successfully detects the tampering in all cases.

B.   Logging module

Logging is required continuously in order to track when and who has done the modification. It should have the feature like,

- Constant monitoring of logged users.
- Sending the log files to TTP for generating audit trails.
- Maintaining integrity of log files.

C. Security module

It provides an interface for the secure transmission of all data between clients (Table 1).

2.2 Worm Attacks

In cloud worm injection attack, the attacker is actually trying to damage the service, the virtual machine or the application and implements its malicious code into the cloud structure [9]. Signature-based antivirus can detect very high accuracy when the signature has been known but if it changes its signature completely, it is too difficult to compute [10, 11].

Different methods are there for the detection of worm attacks which include

- Check the authenticity of all received messages,
- Using hash function store the original image files,
- It uses file allocation technique,
- Hypervisor method is utilized,
- Portable executable format file relationships,
- Map reduce job,
- Hadoop platform,
- File indexing, and
- File relaxation index [12].

2.2.1 Methods in Malware Detection

There are different methods in malware detection with VM introspection, using symptom analysis, with the help of hypervisor, etc. Virtual machines can be monitored using LIBVMI for identifying the malware presence. In this method, it is employing a virtual machine monitoring system with a hypervisor, i.e., Xenhypervisor with preexisting libraries for VM monitoring like LibVMI [13].

Distributed Detection: For security as a whole to be assured, the whole cloud should be monitored. For that, the first step is to establish a communication between individual detection instances. For that collect information and then do the analysis at each hypervisor to distribute points of failure.

For a single point of detection systems, the traffic to a single point will become an issue. Here the individual detection agents which are deployed inside the VMM of a physical machine in the cloud. Each agent can use different techniques like obtaining

Table 1 Comparison table of different insider attack detection methods

S. No.	Mail components	Method following	Algorithms used	Attackers identified	Complexity	Efficiency	No. of VMs inspected	Customer notification
1	Machine learning component	Rule-based learning	Naïve Bayer/decision tree	Both within CSP and outside	Low	Medium	Multiple	Yes
2	Monitoring component	Sequential pattern matching	Rule mining and pattern matching	Legitimate users within CSP	Medium	Medium	Multiple	No
3	Private and public keys	Monitoring by trusted third party	Watermarking method	Clients authenticity	Medium	High	…	Yes

VM memory state information, network traces, and accessing the VM disk image for file system data, etc. For this use appropriate library such as LibVMI.

Usually, single cloud composed of multiple different physical machines, each should not operate as an individual system but part of network agents. It is possible only by developing a secure protocol that allows information to be shared between different agents. By sharing information on health of cloud, it enables agents to determine the threat level to the VMM or VMs with which they are associated.

Agent Architecture: There should be some features which are to be incorporated into the design of each agent present.

1. There should be a method of collecting data from each of the VMs that belongs to a single virtual machine monitor (VMM).
2. An algorithm for determining the presence or absence of malware in a particular virtual machine and analyzing the data.
3. With the current threat level update each VM on a single machine.
4. In order to communicate with other agents in the cloud a protocol should be used.

Gathering of data is achieved through VM introspection. Next, we should have an algorithm for analyzing the data. Since malware which makes use of the features which is unique to virtualization will be undetectable in signature-based methods, there should apply some form of heuristic analysis in the agents.

Also, detection based on the behavior of the malware or detection of anomalies in VMs will be better.

Third step is the updation of VM which is very important since without VM modification mechanism if the infection is detected, it is impossible to perform a remediation phase. Also become impossible to increase any defenses if threat level increase. Tools are already available in hypervisor architecture like xm in Xen hypervisor.

Since the detection system is distributed in nature it requires a decentralized Peer-to-Peer (P2P) architecture, since a centralized system would be vulnerable at the single point so P2P protocols are available [13].

Malware detection by VM introspection is another option. In this method combined the file system clustering, Virtual Machine Introspection (VMI), malware activity recording, etc., since recording entire VM activities require considerable resources.

For classifying the files K method algorithm is used on predefined clusters. Preprocessing is done before clustering, avoiding irrelevant metadata, pronouns, prepositions, etc., and it is converted to vector space model for frequency counting. It is treated by dimension reduction techniques and team variance for attribute selection and distance between files is calculated and go for clustering [9]. So we conclude code is malware or not by using existing malware detection software or by using signature, i.e., in the code generated from entire graph is detected for all API calls jump instruction and remote references and check for any system Hook for changes in file properties then make it as a malicious one [14].

Malware detection is also possible by anomaly detection. Virtualized network of cloud can offer new chances to monitor VMs with its all internal events without any direct access or influence. So our challenge is to detect spreading computer worms as fast as possible. Through the hypervisor Virtual machine introspection (VMI) allows to get information on running VMs layer without the need to directly access the machines [15].

For that first, we have to define what an anomaly is. An anomaly is a one that creates inconsistencies like

- A new process is started which is unknown.
- A new module is loaded which is unknown.

This is identified as a threat and an ongoing threat can be identified as,

a. From a chosen VM, retrieve list of running process.
b. Check for any unknown process.
c. After a number of scans can identify inconsistencies in different VMs.
d. Take corrective action if the occurrence of an identified process exceeds a limit L1.
e. If occurrence decreases, add it to the known list.

This helps to distinguish between the malicious software spread or regular updates.

Spreading process monitor collects process lists of different hardware nodes or randomly chosen virtual guest machines. So it is easy to detect the inconsistencies but it increases their appearance in other VMs. For that network traffic for all VM is routed through a bridge, which is configured in administrative demo 0 VM so that beach guest VM can be scanned and traffic of infected VM can be isolated, so that to prevent infection on uninfected VMs [15].

Malware detection can be possible using FVM. In this method, malwares are detected by symptoms and to monitor other VMs for symptoms in real time via VM introspection, which uses a Forensic Virtual Machine (FVM) which are monitoring VMs. In VM introspection, one guest VM monitors, analyzes, and modifies the state of other guest VM by observing VM memory pages [16]. Some VMs have been given the capacity to inspect the memory pages that the number of small independent VMs is called forensic VM (FVM). Once the symptom was detected, FVM reports its findings to other FVMs and other FVMs inspect that VM and report to command and control center.

Symptom detection via FVMs:

VM introspection is done through a number of forensic virtual machines (FVMs). FVM is a VM configured by user or admin. The integrity of the FVM is very important, also make sure that an FVM is not conducting undesirable activities and inform the clients. So FVMs should follow some guidelines.

A. Guidelines for FVM

1. FVM only reads: In virtualization reading and writing into VM is possible. Here FVMs can't change the states of a VM, and thus, all operations within VM can ensure integrity.
2. FVMs are small: one symptom per FVM: FVMs are designed to be small so that manually can inspect also it deals with identifying a single symptom.

3. One VM at a time is inspected by FVM: FVM inspect one VM at a time in order to avoid leakage of information and before inspecting other VMs, it flushes its memory.
4. Secure communication: By sending message through a secure multicast channel, FVMs communicate with each other.

B. FVM implementation

1. The offset of the target guest kernel task structures is located.
2. The same physical page can be mapped into the FVM and contents examined by converting the known target guest kernel virtual address of the task structure into a machine's physical address.

C. The actual page tables and memory regions being used are identified from the located target guest OS task structure
D. Formalizing the forensic VM

Use Greek letters Φ, $\Phi1$, $\Phi2$, …. for referring the FVMs. For detecting a unique symptom, each FVM is responsible which is its own symptom. Each FVM deals with symptoms with given neighborhood $N(\Phi)$, which is a subset of all VMs. The messages from other FVMs are used by each FVM to get a picture of surrounding world.

Each FVM keeps a record of the last time that a VM is visited with the help of the local variable "last visited". "Last visited assigning the local time of the last visit to a VM in $N(\Phi)$ via an FVM of type Φ. Each time it is updated by the last visited by the messages arriving at the FVM.

E. Lifecycle of an FVM

Most time FVM inspects VM, if discovery is successful, a message discovered is sent. FVMs sent a message of absence if symptoms have disappeared. Otherwise, it will send depart message, at the end of permissible time to stay. Then chooses next VM, then a message "arrive" is sent and inspection is carried out at next.

VM MOBILITY ALGORITMS

a. Guidelines for mobility algorithms

- All VMs must be visited.
- Algorithm must make sure that when more symptoms are detected urgency of visiting a VM increases.
- Movement of VM must not follow a predefined pattern.
- Multiple inspection of VMs by multiple FVMs should be carried out.

b. Mobility algorithm

Selection of VM starts from subset of VMs called Neighborhood (NEIB). Set an upper bound $Max(vj)$ on number of FVMs associated to each VM in vj.

Suppose VM v1 find symptoms s1, s2, and s3 whereas v2 finds symptom s4. A double array variable $\{Disc(Ci, V)\}$ is included to keep the record of the total number of symptoms discovered in FVM v k [16] (Table 2).

Table 2 Comparison table of different worm attack detection method

S. No.	Name	Main component	Method following	Algorithms used	Complexity	Efficiency	Symptoms identified	Clustered
1	VM monitoring using LibVMI	Xen hypervisor with Lib VM libraries	Agents used for VM introspection	Analysis algorithm • Signature-based • Behavior-based	Low	Low	No	No
2	Malware detection by VM introspection	File system clustering malware activity recording	Record all malicious activities performed	Clustering-K Medoid algorithm Malware file detection-back track method	Medium	High	No	Yes
3	Malware detection by anomaly detection	Anomaly detection, VMI	Identifies anomalies and prevent its spreading	Anomaly detection algorithm	Medium	Medium	Yes	No
4	Malware detection using FVM	Forensic virtual machine	Using FVM VM introspection and mobility algorithm	Mobility algorithm, FVM formalizing algorithm	High	High	Yes	No

2.3 DDoS Attacks

DDOS attacks are performed by three main components.

1. Master: The attacker who launches the attack indirectly by a series of compromised systems is known as the master.
2. Slave: It is used to launch the attack and the system is compromised.
3. Victim: An attack launched by an attacker will be applied to this one.

Two Types of victims

1. Primary victim: Systems which are under attack.
2. Secondary victim: Systems which are compromised to launch an attack.

Different phases of DDOS attack include the following:

1. Interruption Phase: Less important systems are compromised by the Master to perform DDOS attack by flooding large number of request.
2. Attack Phase: To attack a target system, it installs DDOS tools.

DOS Vs DDOS: In DOS it prevents users from accessing a service from victim where in DDOS in order to cause denial of service for users of a targeted system, a multitude of compromised systems attack a single target.

XML-BASED DDOS ATTACK: Security in SOA and XML messages are major concern and messages will transit ports that are open for Internet access because the use of XML-based web services removes the network safety. In XML DDOS, the attacker needs to spend only very less processing power or bandwidth that victim needs to spend to handle the payload.

HTTP-BASED DDOS ATTACK: When an HTTP client (web browser) want to communicate to HTTP server (web server), it can be a GET request or a POST request. GET is for requesting a web content where POST is for forms, which submit data from input fields. In this case, flooding can be done when in response to a single request the server allocates a lot of resources so that a relatively complex processing needed on the server [17].

2.3.1 DDoS Attack Prevention and Detection Techniques

New framework to detect and prevent DOS in cloud is by using Covariance matrix modeling. In this method to detect DOS attack, a framework is designed in the cloud which depends on covariance matrix mathematical modeling. It includes three stages like training stage, then prevention stage, and detection stage [18].

(1) Training stage: In this stage, it monitors incoming network traffic in virtual switch using any flow traffic tool and then summarizes all packet traffic in matrix values form, after that matrix is converted into a covariance matrix. (2) Detection and prevention stage: In this covariance matrix resulted from new captured traffic is compared with profile of normal traffic and if resultant, matrix is all zero's means no

attack otherwise if anomaly degree values more than a predefined threshold means there is an attack. To find attacker's source IP address number of nodes from attacker to victim is counted by counting value of TTL (Time to Live). After determining the source all IP address by the attacker is blocked. When attack has been known legitimate traffic to victim's VM shifted to same VM but in another physical machine, since in cloud multiple copies of one cloud is available.

Another method is the usage of a novel cloud computing security model to detect and prevent DOS and DDOS attack. In DDOS, large volume of data packets will be present which can be grouped as trusted or untrusted using TTL or hop counts.

Packets can be monitored using TTL approach. Here, technique used is the hop count to detect DOS attack. In this, the data packets are monitored continuously and three parameters are executed [19], TTL, SYN flag, Source IP.

There are four possible scenarios for each packet.

1. In the IP2HC table, SOURCE IP is already available and SYN flag is high and then calculates hop count using TTL information. If hop count matches with hop count in IP2HC table, then do nothing otherwise update IP2HC table with new hop count. If it is one in IP2HC table, then packet is real; otherwise, it is spoofed.
2. If source IP exist in the table and if SYN flag is low, then calculate the hop count. The packet is real if the hop count matches with the hop count in IP2HC Table.
3. If the source IP information is not present in the table and the SYN flag is low, we can make sure that the packet is spoofed. The advantage of this method is, this algorithm uses only information of SYN, SOURCE IP and TTL and using TTL hop count is calculated. By comparing hop count with the hop count in IP2HC table authenticity can be verified. One drawback is continuous monitoring of packets is required [20].

Entropy-based anomaly detection is also possible. Entropy is the measure of randomness. For the data packets, headers of the sample data are analyzed for IP and Port then their entropy is computed. If entropy increases beyond the value which is set as threshold, that system should generate an alarm of DDOS attack [19]. For multilevel DDOS process includes the following steps:

1. First step is the user allowed to pass through the router for the first time and user is verified using detection algorithm.
2. Second time entropy is computed depending on user's authenticity and data packet size. If the value does not meet the standard range, it is considered as the intruder node and message sent to CSP.
3. Entropy for each and every packet is calculated. After that, it is compared with threshold value. If anomaly exists, message sent to CSP.

CBF packet filtering method is another option. A modified confidence-based filtering method is introduced to increase processing speed and reduce storage needs. It is deployed at cloud dB. In the optional fields of IPV4 header confidence value is stored. For enhanced CB, one which has confidence value is above a threshold is considered as the legitimate packet [21]. A packet with confidence less than threshold will be discarded otherwise accepted.

Table 3 Comparison table of different DDOS detection methods

S. No.	Type of DOS	Method following	Detection factor	Technology used	Complexity	Efficiency
1	Simple DOS	Covariance matrix modeling	TTL	Any flow control toll	Medium	Medium
2	DOS and DDOS	Reverse checking mechanism	Hop count and TTL	Hardware-based watermark-ing technology	Medium	High
3	DDOS	Packet monitoring	Hopcount	IP2HC table	Low	Medium
4	DDOS and HTML DDOS	Entropy-based anomaly detection	IP-based entropy computa-tion	Level of anomaly based on entropy	Medium	Medium
5	XML DDOS	Confidence-based filtering	Confidence value	IPV4 confidence value checking	Low	Low
6	DDOS	Cloud fusion unit	IDS alerts	Dempster Shafter THEORY and combi-nation rule	High	High

Dempstersh after theory can be used for intrusion detection. This method includes Cloud Fusion Unit (CFU) and sensor VMs. CFU collects alerts from different IDS (Intrusion Detection system). For detection of DDoS, a virtual cloud can set up with frontend and three nodes [22]. Detection was done using IDS installed in VMs. Assessment is carried out in CFU. The alerts are analyzed using Dempster Shafter theory. After obtaining probabilities of each attack packet, using fault tree method probabilities of each VM-based ID is calculated. For assessment to maximize the true DDoS attack alerts, Dempster's combination rule is used [19] (Table 3).

3 DDOS Attack Simulation in Cloud

Simulation of DDOS attacks in a real-time cloud is also possible. A cloud environment is set up using OpenStack, which is an open-source software that creates public and private cloud. Major simulation includes SYN flood attacks, TCP flood attack, UDP flood attack, and ICMP flood. After the simulation, an analytical study

Table 4 Comparison table of different DDOS methods simulation

S. No.	Attack	Method	Bandwidth consumed	No. of packets (Avg) in a second	Effect on target
1	SYN flood attack	Works in a three-way handshake method	Medium	0–5	Average
2	TCP flood attack	Connection-oriented attack	Low	0–2	Less
3	ICMP flood attack	Vulnerability-based attack	Very high	10–20	Very high
4	UDP flood attack	Connectionless random packets are sending	High	5–10	High

is done based on the IOgraphs generated using Wireshark in the OpenStack cloud environment, and the results are tabulated (Table 4).

4 Discussion and Open Problems

In this paper after the detailed study of various attacks and their prevention methods, we can identify that all these prevention methods for various attacks are not effective for all the attacks and they can be prevented up to a particular extend. For the insider attack, we should make the cloud admin works properly whereas in the case of worm and DDoS attack, between the cloud clients we should ensure security.

A protocol which can be developed for the cloud service provider and the clients, then only these prevention methods can be implemented properly and its effectiveness will be there in a secure cloud environment.

5 Conclusion

As more and more industries stepping into the cloud, the security issues and vulnerabilities become a major issue in this field. A lot of methods are there to prevent each type of this attack but still, they are not strong enough to defend against these attacks in all scenarios. A system which is not at all vulnerable to any of these major attacks is a real challenge. To achieve that, we have to build a standard with all the security policies between a CSP and the clients so that we can ensure data privacy, integrity, and authenticity in cloud environment.

References

1. Ahmed, M., Xiang Y.: Trust ticket deployment: a notion of a data owner's trust in cloud computing. In: 2011 International Joint Conference of IEEE TrustCom-11/IEEE ICESS-11/FCST-11
2. Bradai, A., Afifi, H.: Enforcing trust-based intrusion detection in cloud computing using algebraic methods. In: 2012 International Conference on Cyber-Enabled Distributed Computing and Knowledge Discover
3. Rajagopal, R., Chitra, M.: Trust based interoperability security protocol for grid and cloud computing. In: ICCCNT'12 26–28 July 2012, Coimbatore, India
4. Kanwal, A., Masood, R., Ghazia, U.E., Shibli, M.A., Abbasi, A.G.: Assessment criteria for trust models in cloud computing. In: 2013 IEEE International Conference on Green Computing and Communications and IEEE Internet of Things and IEEE Cyber, Physical and Social Computing
5. Duncan, A., Creese, S., Goldsmith, M.: Insider attacks in cloud computing. In: 2012 IEEE 11th International Conference on Trust, Security and Privacy in Computing and Communications
6. Khorshed, M.T., Shawkat Ali, A.B.M., Wasimi, S.A.: Monitoring insiders activities in cloud computing using rule based learning. In: 2011 International Joint Conference of IEEE
7. Guo, Q., Sun, D., Chang, G., Sun, L., Wang, X.: Modeling and evaluation of trust in cloud computing environments. In: 2011 3rd International Conference on Advanced Computer Control (ICACC 2011)
8. Nkosi, L., Tarwireyi, P., Adigun, M.O.: Detecting a malicious insider in the cloud environment using sequential rule mining. In: 2013 IEEE International Conference on Adaptive Science and Technology (ICAST)
9. Bisong, A., Rahman, M.: An overview of the security concerns in enterprise cloud computing. Int. J. Netw. Secur. Appl. (IJNSA) 3(1) (January 2011)
10. Yang, Z., Qin, X., Yang, Y., Yagnik, T.: A hybrid trust service architecture for cloud computing. In: 2013 International Conference on Computer Sciences and Applications
11. Habib, S.M., Hauke, S., Ries, S., Muhlhauser, M.: Trust as a facilitator in cloud computing: a survey. J. Cloud Comput. Adv. Syst. Appl. (2012)
12. Noor, T.H., Sheng, Q.Z.: Trust management of services in cloud environments: obstacles and solutions. ACM Comput. Surv. 46(1), Article 12, Publication date: October 2013
13. Watson, M.R.: Malware detection in the context of cloud computing. In: The 13th Annual Postgraduate Symposium on the Convergence of Telecommunications, Networking, and Broadcasting
14. More, A., Tapaswi, S.: Dynamic malware detection and recording using virtual machine introspection. In: Best Practices Meet, 2013 DSCI IEEE
15. Biedermann, S., Katzenbeisser, S.: Detecting computer worms in the cloud. In: iNetSec'11 Proceedings of the 2011 IFIP WG 11.4 International Conference on Open Problems in Network Security
16. Harrison, K., Bordbar, B., Ali, S.T.T., Dalton, C.I., Norman, A.: A framework for detecting malware in cloud by identifying symptoms. In: 2012 IEEE 16th International Enterprise Distributed Object Computing Conference
17. Rameshbabu, J., Sam Balaji, B., Wesley Daniel, R., Malathi, K.: A prevention of DDoS attacks in cloud using NEIF techniques. Int. J. Sci. Res. Publ. 4(4) (April 2014) ISSN 2250-3153
18. Ismail, M.N., Aborujilah, A., Musa, S., Shahzad, A.: New framework to detect and prevent denial of service attack in cloud computing environment. Int. J. Comput. Sci. Secur. (IJCSS) 6(4)
19. Sattar, I., Shahid, M., Abbas, Y.: A review of techniques to detect and prevent distributed denial of service (DDoS) attack in cloud computing environment. Int. J. Comput. Appl. 115(8), 0975–8887 (2015)
20. Syed Navaz, A.S., Sangeetha, V., Prabhadevi, C.: Entropy based anomaly detection system to prevent DDoS attacks in cloud. Int. J. Comput. Appl. 62(15), 0975–8887 (2013)
21. Goyal, U., Bhatti, G., Mehmi, S.: A dual mechanism for defeating DDoS attacks in cloud computing model. Int. J. Appl. Innov. Eng. Manage. (IJAIEM)

22. Santhi, K.: A defense mechanism to protect cloud computing against distributed denial of service attacks. Int. J. Adv. Res. Comput. Sci. Softw. Eng. 3(5) (May 2013) (ISSN: 2277 128X)
23. Khalil, I.M., Khreishah, A., Azeem, M.: Cloud computing security: a survey. ISSN 2073-431X, 3 February 2014
24. Noor, T.H., Sheng, Q.Z., Zeadally, S.: Trust management of services in cloud environments: obstacles and solutions. ACM Comput. Surv. 46(1), Article 12, Publication date: October 2013
25. Kanaker, H.M., Saudi, M.M., Marhusin, M.F.: Detecting worm attacks in cloud computing environment: proof of concept. In: 2014 IEEE 5th Control and System Graduate Research Colloquium, August 11–12, UiTM, Shah Alam, Malaysia
26. Praveen Kumar, P., Bhaskar Naik, K.: A survey on cloud based intrusion detection system. Int. J. Softw. Web Sci. (IJSWS), 98–102
27. Rahman, M., Cheung, W.M.: A novel cloud computing security model to detect and prevent DoS and DDoS attack. Int. J. Adv. Comput. Sci. Appl. 5(6) (2014)
28. Shahin, A.A.: Polymorphic worms collection in cloud computing. Int. J. Comput. Sci. Mob. Comput. 3(8), 645–652 (2014)
29. Quinton, J.S., Duncan, A., Creese, S., Goldsmith, M.: Cloud computing: insider attacks on virtual machines during migration. In: 2013 12th IEEE International Conference on Trust, Security and Privacy in Computing and Communications
30. Nicoll, A., Claycomb, W.R.: Insider threats to cloud computing: directions for new research challenges. In: 2012 IEEE 36th International Conference on Computer Software and Applications
31. Nguyen, M.-D., Chau, N.-T., Jung, S., Jung, S.: A demonstration of malicious insider attacks inside cloud IaaS Vendor. Int. J. Inf. Educ. Technol. 4(6) (December 2014)
32. Garkoti, G., Peddoju, S.K., Balasubramanian, R.: Detection of insider attacks in cloud based e-healthcare environment. In: 2014 13th International Conference on Information Technology
33. Kumar, M., Hanumanthappa, M.: Scalable intrusion detection systems log analysis using cloud computing infrastructure. In: 2013 IEEE International Conference on Computational Intelligence and Computing Research (ICCIC)
34. Praveen Kumar, P., Bhaskar Naik, K.: A survey on cloud based intrusion detection system. Int. J. Softw. Web Sci. (IJSWS), ISSN (Print) 2279-0063 ISSN (Online) 2279-0071
35. Sun, D., Chang, G., Suna, L., Wang, X.: Surveying and analyzing security, privacy and trust issues in cloud computing environments. SciVerse Sci. Direct Procedia Eng. 15, 2852–2856 (2011)
36. Oktay, U., Sahingoz, O.K.: Attack types and intrusion detection systems in cloud computing. In: Proceedings of the 6th International Information Security & Cryptology Conference, Bildiriler Kitabı
37. Sevak, B.: Security against side channel attack in cloud computing. Int. J. Eng. Adv. Technol. (IJEAT) 2(2) (December 2012) ISSN: 2249-8958
38. Siva, T., Phalguna Krishna, E.S.: Controlling various network based ADoS attacks in cloud computing environment: by using port hopping technique. Int. J. Eng. Trends Technol. (IJETT) 4(5) (May 2013)
39. Bhandari, N.H.: Survey on DDoS attacks and its detection &defence approaches. Int. J. Sci. Modern Eng. (IJISME) 1(3) (February 2013) (ISSN: 2319-6386)
40. Wong, F.F., Tan, C.X.: A survey of trends in massive DDoS attacks and cloud-based mitigations. Int. J. Netw. Secur. Appl. (IJNSA) 6(3) (May 2014)
41. Goyal, U., Bhatti, G., Mehmi, S.: A dual Mechanism for defeating DDoS attacks in cloud computing model. Int. J. Appl. Innov. Eng. Manage. (IJAIEM) 2(3) (March 2013) ISSN 2319-4847
42. Bhandari, N.H.: Survey on DDoS attacks and its detection &defence approaches. Int. J. Sci. Modern Eng. (IJISME) 1(3) February ISSN: 2319-6386
43. Santhi, K.: A defense mechanism to protect cloud computing against distributed denial of service attacks. Int. J. Adv. Res. Comput. Sci. Softw. Eng. 3(5) (May 2013) ISSN: 2277-128 X

Assessment of Solar Energy Potential of Smart Cities of Tamil Nadu Using Machine Learning with Big Data

R. Meenal and A. Immanuel Selvakumar

Abstract Global Solar Radiation (GSR) prediction is important to forecast the output power of solar PV system in case of renewable energy integration into the existing grid. GSR can be predicted using commonly measured meteorological data like relative humidity, maximum, and minimum temperature as input. The input data is collected from India Meteorological Department (IMD), Pune. In this work, Waikato Environment for Knowledge Analysis (WEKA) software is employed for GSR prediction using Machine Learning (ML) techniques integrated with Big Data. Feature selection methodology is used to reduce the input data set which improves the prediction accuracy and helps the algorithm to run fast. Predicted GSR value is compared with measured value. Out of eight ML algorithms, Random Forest (RF) has minimum errors. Hence this work attempts in predicting the GSR in Tamil Nadu using RF algorithm. The predicted GSR values are in the range of 5–6 kWh/m^2/day for various solar energy applications in Tamil Nadu.

Keywords Machine learning · Big data · Global solar radiation · Random forest

1 Introduction

Accurate knowledge of solar radiation data is necessary for various solar energy based applications and research fields including agriculture, astronomy, atmospheric science, climate change, power generation, human health and so on. However, in spite of its importance, the network of solar irradiance measuring stations is comparatively rare throughout the world. This is due to the financial costs involved in the acquisition, installation, and difficulties in measurement techniques and maintenance of these

R. Meenal (✉) · A. Immanuel Selvakumar
Department of Electrical Sciences, Karunya Institute of Technology and Sciences,
Coimbatore 641114, India
e-mail: meenasekar5@gmail.com

A. Immanuel Selvakumar
e-mail: iselvakumar@yahoo.co.in

© Springer Nature Singapore Pte Ltd. 2019
J. D. Peter et al. (eds.), *Advances in Big Data and Cloud Computing*,
Advances in Intelligent Systems and Computing 750,
https://doi.org/10.1007/978-981-13-1882-5_3

measuring equipments. Due to the lack of hourly measured solar radiation data, the prediction of solar radiation at the earth's surface is essential.

Several models have been developed to predict solar radiation for the locations where the measured radiation values are not available. Empirical correlations are the most widely used method to determine GSR using other measured meteorological parameters namely sunshine duration [1] and air temperature [2]. Hatice Citakoglu compared artificial intelligence method and empirical equations for prediction of GSR [3]. It was observed that Artificial Neural Network (ANN) has a good capability for the prediction of GSR. The major drawback of neural networks is their learning time requirement. A novel machine learning model namely support vector machine (SVM) has been widely used recently for the estimation of GSR [4]. Several studies have proved that SVM performed better than ANN and conventional empirical models [5]. An overview of forecasting methods of solar radiation using different machine learning approaches is reviewed by Voyant et al. [6]. Due to unavailability of sunshine records in all the meteorological stations in India, this work attempts in predicting the GSR using various machine learning techniques with commonly measured maximum and minimum temperature as input parameters. This work is focused on assessing the solar energy resource potential of ten smart cities of Tamil Nadu state, India which is seldom found in the literature. Ten smart cities include Cuddalore, Chennai, Coimbatore, Dindigul, Erode, Madurai, Salem, Thanjavur, Tirunelveli, and Trichy. Smart Cities Mission is an urban renewable program introduced by Government of India in the year 2015. The objective of the Smart Cities Mission is to formulate economic growth and progress the quality of life of people by enabling local area development and harnessing technology that results in smart outcomes.

The structure of this paper is organized as follows. Section 2 gives the details of the data set. Section 3 describes empirical model. Various machine learning models and performance evaluation of model is described in Sect. 4. Section 5 describes the results and discussion. Conclusions are finally presented in Sect. 6.

2 Data Set

Data are been collected from the database of Indian Meteorological Department (IMD), Pune. The training dataset includes the monthly mean maximum temperature and minimum temperature and daily GSR in $MJ/m^2/day$. Further the daily radiation data is converted into monthly by taking average. After that it is converted to .CSV (Comma Separated File) format. Values of the geographical parameters like latitude, longitude, time frame of the training database are displayed in Table 1. The machine learning models are trained using the experimental training dataset (see Table 1) from IMD, Pune and the geographical parameters namely the latitude, longitude, and the month numbers. The input testing data including the maximum and minimum temperature for the ten smart cities of Tamil Nadu state are taken from Atmospheric Science Data Centre of NASA [7].

Table 1 Geographical parameters of training and testing locations

Training locations/training period	Latitude (N)	Longitude (E)	Testing locations	Latitude (N)	Longitude (E)
Trivandrum (2005–2012)	8.48	76.94	Tirunelveli	8.71	77.76
Coimbatore (2005–2009)	11.01	76.95	Madurai	9.93	78.11
Mangalore (2004–2008)	12.91	74.85	Dindigul	10.36	77.98
Chennai (2003–2011)	13.08	80.27	Thanjavur	10.78	79.13
Hyderabad (2000–2008)	17.38	78.46	Trichy	10.79	78.70
Bhubaneswar (2003–2008)	20.27	85.83	Coimbatore	11.01	76.95
Nagpur (2004–2010)	21.14	79.08	Erode	11.34	77.71
Patna (2000–2008)	25.59	85.13	Salem	11.66	78.14
New Delhi (2003–2011)	28.65	77.22	Cuddalore	11.74	79.77

3 Empirical Model

Empirical correlations are the most commonly used conventional method for the estimation of GSR. Chen et al. [2] proposed the following empirical equation for estimating GSR as follows:

$$\frac{H}{H_0} = a\left(T_{max} - T_{min}\right)^{0.5} + b,$$ (1)

where H is the monthly mean GSR on horizontal surface, H_0 is the monthly mean daily extraterrestrial radiation in kWh/m^2/day, a, b and c are empirical constants determined by statistical regression technique. T_{min} and T_{max} are the minimum and maximum temperature. Bristow and Campbell model is also reviewed in this study [5].

4 Machine Learning Models

The Solar radiation for ten smart cities of Tamil Nadu is predicted by various machine learning approaches using WEKA. "WEKA" stands for the Waikato Environment for Knowledge Analysis, is developed by University of Waikato, New Zealand in 1993. WEKA is a collection of machine learning algorithms for solving real-world data mining jobs. It contains tools for data preprocessing, classification, regression, clustering, association rules, and visualization. Temperature-based machine learning models are applied for GSR prediction in Tamil Nadu. The best input parameters are selected using Attribute Evaluator and search method of WEKA.

Attribute selection: The role of the attributes selection in machine learning is to reduce the dimension of the input data set, to speed up the prediction process, and to improve the correlation coefficient and prediction accuracy of the Machine Learning algorithm. The input parameters are ranked according to the correlation with measured GSR value. Maximum temperature secured the first rank (0.785) followed by minimum temperature (0.445) and Relative Humidity has the least rank (0.298). So it is omitted from the input data set. The following eight methods available in WEKA are used to predict the GSR in Tamil Nadu.

Linear and Simple Linear regression (SLR): Regression analysis is a technique for examining functional relationships among variables that is articulated in the form of an equation or a model relating dependent variable and one or more predictor variables.

Gaussian Process: Gaussian process regression (GPR) is a method of interpolation. The interpolated values are modeled by a Gaussian process administered by prior covariance. In this process, new data points can be constructed inside the range of a discrete set of well-known data points. Temporal GPR approach is said to be more robust and accurate than the other Machine Learning techniques since it finds covariance between two samples based on time-series [8].

M5Base: An M5 model tree is a binary decision tree which has linear regression functions at the terminal (leaf) nodes. This can predict continuous numerical attributes. Quinlan [9] invented this algorithm. The advantage of M5 over CART method is that the model trees are much smaller than regression trees.

REP tree and Random tree: Quinlan proposed a tree model called Reduced Error Pruning (REP). Each node is replaced with its class which is more admired (starting at the leaves). If the accuracy of prediction is not affected then the change is kept. Basically REP tree is quick decision tree learning and it constructs a decision tree based on the information gain or reducing the variance. Leo Breiman and Adele Cutler introduced Random tree algorithm solves both classification and regression problems. Also Random Model Trees produce predictive performance better than the Gaussian Processes Regression [10].

SMOreg: The regression algorithm named SMOreg implements the supervised machine learning algorithm Support Vector Machine (SVM) for regression. SVM is developed by Vapnik in 1995 for solving both classification and regression problems [11]. The classifiers in WEKA are designed to be trained to predict a single 'class' attribute, which is the target for prediction. Selection of kernel to use is a key parameter in SVM. Linear kernel is the simplest kernel that separates data with a straight line or hyper plane. The default in WEKA is a Polynomial kernel that will fit the data using a curved or wiggly line.

Random Forest: Random Forest is an ensemble algorithm based on decision tree predictors developed by Breiman [12]. RF classification algorithm is most suitable for the analysis of large data sets. This algorithm is popular because of high-prediction accuracy and gives information on the importance of variables for classification. It does not require complicated or expensive training procedures like ANN or SVM. The main parameter to adjust is the number of trees. Also compared to ANN and SVM, training can be performed much faster. Random Forests are also robust to outliers and noise if enough trees are used.

4.1 Evaluation of Model Accuracy

The performance of the machine learning models is evaluated based on the statistical parameters namely Root mean square (RMSE), Mean Absolute Error (MAE), and Correlation coefficient (R) are defined as follows:

$$\text{RMSE} = \sqrt{\frac{1}{n} \sum_{i=1}^{n} (H_{im} - H_{ic})^2} \tag{2}$$

$$\text{MAE} = \frac{1}{n} \sum_{1=1}^{n} \left(\frac{H_{ic} - H_{im}}{H_{im}} \right) \tag{3}$$

where $H_{i,c}$ is the ith calculated value $H_{i,m}$ is the ith measured value of solar radiation and n is the total number of observation files. The linear correlation coefficient (R) value is used to find the relationship between the actual measured and calculated values. For better modeling R value should be closer to one. If $R = 1$, it means that there is an exact linear relationship between measured and calculated values.

5 Results and Discussion

Eight machine learning algorithms available in WEKA are used to predict the GSR in India. The machine learning models are trained using the input dataset and measured GSR collected from IMD, Pune. The best input parameters are selected using attribute selection of WEKA. Maximum temperature received the first rank and RH is omitted from dataset since its correlation with GSR is very low (0.298). Figure 1 shows the map of India and Tamil Nadu showing the training and testing locations. The training and testing results of various ML techniques are shown in Table 2 for the locations Chennai and Patna. Similar results are obtained for other locations also. Figure 2 shows the comparison between measured GSR and predicted GSR by SVM and Random Forest method for Chennai city. It also shows R value of different machine learning algorithms. The performance ranking of different approaches is

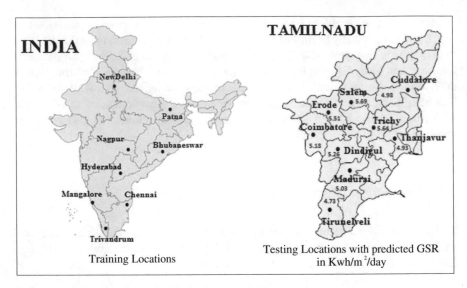

Fig. 1 Map of India and Tamil Nadu showing training and testing locations

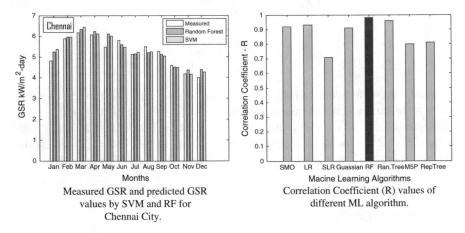

Fig. 2 Comparison between measured GSR and predicted GSR values and Correlation coefficient—*R* values of different ML algorithm

complicated because of the diversity of the data set, and statistical performance indicators. The error statistics of prediction is relatively equivalent in our results. Overall, Random forest has minimum errors and high correlation coefficient value. Also the selected RF ML model is compared with the conventional temperature-based empirical models. Table 3 shows the comparison of error statistics between empirical models and the current random forest model. It is observed that *R* value is closer to one and the error value is less (RMSE: 0.66) for RF model when compared with empirical models (RMSE: 1.5). Hence this work attempts in predicting the

Table 2 Error statistics of the various machine learning models for the prediction of monthly average daily GSR for the training sites of India (Chennai and Patna)

Algorithm/location	Training			Testing			Rank
	MAE	RMSE	R	MAE	RMSE	R	
Chennai							
SMO	0.8948	1.1436	0.888	1.47	1.7335	0.9152	4
Linear regression	1.0513	1.2262	0.8683	1.6391	1.944	0.9329	3
SLR	1.3803	1.7626	0.7011	2.0407	2.5686	0.7087	8
Gaussian	1.2824	1.4118	0.8809	1.9137	2.3217	0.9050	5
Random forest	**0.625**	**0.746**	**0.9714**	**1.5438**	**1.8296**	**0.9803**	**1**
Random tree	0	0	1	1.0958	1.4199	0.9564	2
M5P	1.2104	1.5272	0.7987	1.9112	2.4116	0.8006	7
Rep tree	1.1833	1.3883	0.8274	1.9064	2.1993	0.8129	6
Patna							
SMO	0.6930	1.0773	0.9277	1.4128	1.8569	0.8127	6
Linear regression	0.5867	0.6706	0.9724	0.919	1.2814	0.9172	2
SLR	1.4948	1.6611	0.8163	1.6137	1.8498	0.8176	5
Gaussian	1.4650	1.8112	0.8819	2.091	2.4534	0.7604	7
Random forest	**0.6611**	**0.7993**	**0.9776**	**1.1784**	**1.4261**	**0.9309**	**1**
Random tree	0.01	0.0245	1	1.3025	1.6763	0.8418	4
M5P	0.7808	0.8983	0.95	1.3045	1.6033	0.8647	3
Rep tree	1.8089	2.2964	0.6018	2.5317	2.9291	0.4282	8

Table 3 Comparison of error statistics between temperature based empirical model and RF ML model (Patna)

Author/source		R	RMSE
Chen et al. model [5]	$\frac{H}{H_0} = a\left(T_{\max} - T_{\min}\right)^{0.5} + b$	0.8876	1.5365
	$\frac{H}{H_0} = a\ln\left(T_{\max} - T_{\min}\right) + b$	0.8938	1.5000
Bristow and Campbell [5]	$\frac{H}{H_0} = a\left(1 - \exp(-b\Delta T)^c\right)$	0.8908	1.5171
Current study	RF Machine learning model	0.9776	0.6611

solar radiation of ten smart cities of Tamil Nadu using random forest method. The predicted values are in good matching with NASA- and Meteonorm-based data given by TANGEDCO—Tamil Nadu Generation and Distribution Corporation Limited [13]. The predicted GSR values are in the range of 5–6 kWh/m²/day (see Table 4).

Table 4 Predicted global solar radiation data ($kWh/m^2/day$) of smart cities of Tamil Nadu state, India using random forest method

Smart cities/month No.	Chennai	Coimbatore	Cuddalore	Dindigul	Erode	Madurai	Salem	Thanjavur	Tirunelveli	Trichy
1	5.25	5.83	4.85	5.66	5.86	4.92	5.84	4.82	4.88	5.74
2	5.97	6.22	5.29	6.32	6.25	5.76	6.39	5.26	5.62	6.41
3	6.33	6.26	5.82	6.35	6.30	6.20	6.42	5.83	5.70	6.44
4	6.23	6.02	5.77	6.05	6.34	5.67	6.48	5.62	5.02	6.47
5	6.11	5.61	5.51	5.59	5.92	4.93	6.41	5.10	4.69	5.96
6	5.60	4.43	5.05	4.74	5.21	4.84	5.68	4.99	4.29	5.65
7	5.14	4.13	4.78	4.61	4.99	4.82	5.40	4.80	4.02	5.43
8	5.20	4.19	4.94	4.71	5.16	4.94	5.49	4.92	4.56	5.54
9	5.13	4.82	4.86	5.12	5.34	4.86	5.48	4.79	4.61	5.45
10	4.52	4.82	4.51	4.62	5.06	4.56	5.06	4.57	4.52	5.00
11	4.37	4.82	4.33	4.69	4.82	4.43	4.78	4.42	4.42	4.82
12	4.40	5.05	4.04	4.70	4.87	4.48	4.83	4.10	4.46	4.81
Annual GSR	5.25	5.18	4.98	5.26	5.51	5.03	5.69	4.93	4.73	5.64
Meteonorm data		5.52	5.58	5.49	5.63	5.59	5.58	5.41	5.51	5.55
NASA data		4.97	5.14	4.99	5.11	5.10	5.19	5.18	4.91	5.20

6 Conclusion

In this work, various machine learning algorithms namely Linear Regression, simple linear regression, M5P, REP tree, Random tree, Random forest and SMOreg using WEKA are evaluated for the prediction of GSR in Tamil Nadu. The dimensionality of the input data set is reduced using feature selection method. The best parameter is selected based on the correlation with measured GSR value. The selected input parameters namely maximum and minimum temperature and geographical parameters namely month number, latitude, and longitude is used as inputs to different machine learning models in WEKA. The output of these models namely the GSR were tabulated and compared with the experimental values. The comparison was made by calculating MAE, RMSE and correlation coefficient (R). It was observed that the error statistics of prediction is relatively equivalent for all the algorithms. Overall, Random forest and SVM have minimum errors and the predicted values are closer to the actual values. A higher value of correlation coefficient for Random forest (0.9803) is found. Therefore, Random forest is the first choice of option for the GSR prediction. Hence Random forest is selected for the prediction of GSR for ten smart cities of Tamil Nadu. The predicted annual GSR varies between 5 and 6 kWh/m^2/day. This excellent solar potential in Tamil Nadu can be better utilized for a wide range of solar energy applications. Hence, RF Model developed using WEKA can be implemented to estimate solar potential of any location worldwide.

Acknowledgements Authors would like to thank IMD, Pune for data support.

References

1. Prescott, J.A.: Evaporation from a water surface in relation to solar radiation. Trans. R. Soc. S. Aust. **64**, 114–118 (1940)
2. Chen, R., Ersi, K., Yang, J., Lu, S., Zhau, W.: Validation of five global radiation models with measured daily data in China. Energ. Convers. Manage. **45**, 1759–1769 (2004)
3. Citakoglu, H.: Comparison of artificial intelligence techniques via empirical equations for prediction of solar radiation. Comput. Electron. Agric. **115**, 28–37 (2015)
4. Belaid, S., Mellit, A.: Prediction of daily and mean monthly global solar radiation using support vector machine in an arid climate. Energ. Convers. Manage. **118**, 105–118 (2016)
5. Meenal, R., Immanuel Selvakumar, A.: Assessment of SVM, Empirical and ANN based solar radiation prediction models with most influencing input parameters. Renewable Energy **121**, 324–343 (2018)
6. Voyant, C., Notton, G., Kalogirou, S., Nivet, M.-L., Paoli, C., Motte, F., Fouilloy, A.: Machine learning methods for solar radiation forecasting: a review. Renewable Energy **105**, 569–582 (2017)
7. NASA: Atmospheric Science Data Centre. https://eosweb.larc.nasa.gov/cgi-bin/sse/grid.cgi
8. Salcedo-Sanz, S., Casanova-Mateo, C., Muñoz-Marí, J., Camps-Valls, G.: Prediction of daily global solar irradiation using temporal gaussian processes. IEEE Geosci. Remote Sens. Lett. **11**(11) (November 2014). https://doi.org/10.1109/lgrs.2014.2314315
9. Quinlan, J.R.: Learning with continuous classes. In: Adams and Sterling (eds.) Proceedings AI'92, pp. 343–348. World Scientific, Singapore (1992)

10. Pfahringer, B.: Random Model Trees: An Effective and Scalable Regression method. University of Waikato, New Zealand. http://www.cs.waikato.ac.nz/~bernhard
11. Vapnik, V.: The Nature of Statistical Learning Theory. Springer, New York (1995)
12. Breiman, L.: Random forests. J. Mach. Learn. **45**(1), 5–32 (2001)
13. TANGEDCO, Solar irradiance data in Tamil Nadu (2012). http://www.tangedco.gov.in/linkpdf/solar%20irradiance%20data%20in%20Tamil%20Nadu.pdf

Taxonomy of Security Attacks and Risk Assessment of Cloud Computing

M. Swathy Akshaya and G. Padmavathi

Abstract Cloud Computing is an international collection of hardware and software from thousands of computer network. It permits digital information to be shared and distributed at very less cost and very fast to use. Cloud is attacked by viruses, worms, hackers, and cybercrimes. Attackers try to steal confidential information, interrupt services, and cause damage to the enterprise cloud computing network. The survey focuses on various attacks on cloud security and their countermeasures. Existing taxonomies have been widely documented in the literature. They provide a systematic way of understanding, identifying, and addressing security risks. This paper presents taxonomy of cloud security attacks and potential risk assessment with the aim of providing an in depth understanding of security requirements in the cloud environment. A review revealed that previous papers have not accounted for all the aspects of risk assessment and security attacks. The risk elements which are not dealt elaborately in other works are also identified, classified, quantified, and prioritized. This paper provides an overview of conceptual cloud attack and risk assessment taxonomy.

Keywords Cloud computing · Security challenges · Taxonomy · Zero-day attack · Risk assessment

1 Introduction

Cloud Computing (CC) has become popular in organizations and individual users. According to Gartner cloud adoption will continue to rise at a compound increase rate of 41.7% in the year 2016 [1, 2]. Cloud computing inherently has a number of

M. Swathy Akshaya (✉) · G. Padmavathi
Department of Computer Science, Avinashilingam Institute for Home Science and Higher Education for Women (Deemed to be University), Coimbatore, Tamil Nadu, India
e-mail: akshayakulandaivel@gmail.com

G. Padmavathi
e-mail: ganapathi.padmavathi@gmail.com

© Springer Nature Singapore Pte Ltd. 2019
J. D. Peter et al. (eds.), *Advances in Big Data and Cloud Computing*,
Advances in Intelligent Systems and Computing 750,
https://doi.org/10.1007/978-981-13-1882-5_4

operational and security challenges. The security of data outsourced to the cloud is increasingly important due to the trend of storing more data in the cloud.

Currently multi-cloud database model is an integral component in the cloud architecture. The security of multi-cloud providers is a challenging task [3]. Confidentiality and integrity of software applications are the two key issues in deploying CC. Challenges arise when multiple Virtual Machines (VMs) share the same hardware resources on the same physical host. For example, attackers can bypass the integrity measurement by reusing or duplicating the code pages of legitimate programs. Cloud Service Provider (CSP), service instance, and cloud service users are the components in the basic security architecture [4].

When the "classical" taxonomies were developed systems such as cloud computing, 3G malware, VoIP and social engineering vulnerabilities were unheard of [5]. These approaches are very general and apply to all networks and computer systems.

In brief the objective of this paper is to deal with three important aspects of cloud computing namely security attacks, taxonomy, and risk assessment.

This paper is organized as follows:

- Section 1 briefly discussed about the introduction, motivation, objective of the paper, and organization of the paper.
- Section 2 presents the different attack taxonomies.
- Section 3 describes about threats and vulnerabilities of cloud.
- Section 4 deals with most vulnerable attacks of CC.
- Section 5 explains the risk assessment.
- Section 6 accounts for conclusion and future scope.

2 Attack Taxonomies

A taxonomy is defined as the classification and categorization of different aspects of a given problem domain. It serves as a basis for a common and consistent language [6]. Based on literature survey it has been concluded that taxonomy should be purposeful because the taxonomy's purpose can significantly impact the level of detail required [7].

Attacks on cloud services have unique characteristics for developing cloud attack and risk assessment taxonomy. A cloud service's elasticity, abstract deployment, and multi-tenancy might impact the risk assessment.

2.1 Attack Levels in CC

In cloud computing the attack levels are classified into two categories namely VM to VM or Guest-to-Guest attacks which are similar to each other and Client-to-Client attacks shown in Fig. 1 [1].

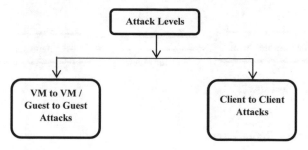

Fig. 1 Levels of attacks

2.1.1 VM to VM Attacks

A container which contains applications and guest operating systems is called virtual machines. A cloud multi-tenant environment consists of potential vulnerabilities in which cloud providers use hypervisor and VM technologies such as VMware vSphere, Microsoft Virtual PC, Xen, etc., due to these vulnerabilities in the attacks shown below in Fig. 2.

2.1.2 Guest-to-Guest Attacks

When an attacker gains administrative access to the hardware then the attacker can break into VMs because securing the host machine from attacks is an important factor. Moreover the attacker can hop from one VM to another by compromising the underlying security framework; this scenario is called guest-to-guest attacks.

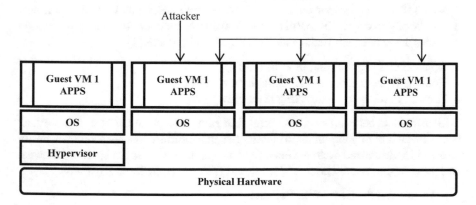

Fig. 2 VM-to-VM/guest-to-guest attacks

Fig. 3 Client-to-client attacks

2.1.3 Client-to-Client Attacks

If there is one physical server for several VMs, one malicious VM can infect all the other VMs working on the same physical machine. By gaining the benefit of vulnerabilities in client application that runs on a malicious server, the client attacks other client's machines refer Fig. 3.

In such cases, the entire virtualized environment becomes compromised and malicious clients can escape the hypervisor and access the VM environment. As a result the attacker gets the administrative privileges to access all the VMs; hence it is a major security risk to virtualized environment. The next chapter discusses the attack surfaces available in CC.

2.2 Surface Attacks in CC

In an attack surface unauthorized users can gain access to systems and cause damage to the software environment. The most critical attack vector in multi-tenant cloud environment is resource sharing. Practice and theory has differences between them, in theory large hypervisors has low attack vectors, but in practice emerges number of real-world attacks targeting hypervisors for instance covert channel calls, use of root kits [8]. A hypervisor when compromised by side-channel attack has a new threat, layer spoofing in Blue Pill root kit leaks information to the outsourced cloud becomes the new attack vector in a virtualized environment [1].

- Application Programming Interface (API)
- Hooking System Calls and Hooking Library Calls
- Firewall Ports using Redirecting Data Flows

Sharing data between systems is easy but sometimes it turns out to be an attack vector and it is highly important to identify the attack surface that are prone to security attacks in a virtualized system. Generally CC and its resources are based on Internet and these resources are classified into three types shown in Table 1 namely

- Software as a Service—Web Browser
- Platform as a Service—Web Services and APIs: Simple Object Access Protocol (SOAP), Representational State Transfer (REST) and Remote Procedure Call (RPC) protocols

Table 1 Attack surface in cloud service models

Attack surface	Attack vectors		
	SaaS	PaaS	IaaS
Application level	Input/output validation	Runtime engine that runs customer's applications	Virtual workgroups
Data segregation	Unauthorized access of data	Data service portal	Multi-tenancy and isolation
Data availability	Hosted virtual server	Network traffic	Virtual network
Secure data access	Encryption/ decryption keys	Third party components	Cloud multi-tenant architecture
Data center security	Server based data breaches	Datacenter vulnerabilities	Virtual domain environments
Authentication/ authorization	ID and password	Client API password reset attack	Poor quality credentials

- Infrastructure as a Service—VMs and Storage Services: Virtual Private Network (VPN), File Transfer Protocol (FTP).

2.2.1 Attack Surface in SaaS

In CC web applications are called software that constitutes a service and these dynamic services collects data from various sources in a distributed cloud environment. With this feature it aids hackers to insert text in the web page by using comments called scripts when executed it causes damages [9].

2.2.2 Attack Surface in PaaS

The PaaS cloud layer has the utmost responsibility of security with strong encryption techniques to provide services to customers without any disruption. So PaaS cloud layer is responsible to secure runtime engines programming framework from attackers or malicious threats which runs the customer applications. The major attack vector in PaaS cloud layer is multi-tenancy, supported by many platforms. Such as (Operating System (OS) and virtual platform) and PaaS must facilitate it to their users by providing a secure platform to run their application components. In this way, PaaS allows its customers for multiple accesses of cloud services simultaneously. It also paves path to malicious user can also have multiple ways of interfering and disrupting the normal execution of the PaaS container.

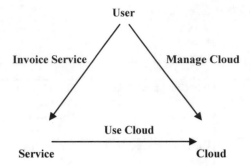

Fig. 4 Attack surface triangle

2.2.3 Attack Surface in IaaS

A technology that aids IaaS cloud to support virtualization technologies between operating system and hardware is the additional layer of hypervisor or virtual machine monitor. This technology is again used for creating APIs to perform administrative operations and causes increase in attack surface. With this new technology comes many methods such as APIs, Channels like sockets, data items such as input strings are exploited.

Table 1 depicts the nature of various attack surfaces and attack vectors in cloud service models.

The next section deals with the cloud security attacks based on service delivery models.

2.3 Security Attacks on CC

A cloud computing attack can be modeled using three different classes of participants namely: service users, service instances, and cloud provider refer Fig. 4 [10].

In recent years there have been new threats and challenges in cloud computing (CC) [1]. To meet these challenges new taxonomy and classifications are required. The new taxonomy which is based on service delivery model of CC is illustrated in Fig. 5 [11].

2.3.1 Security Attacks on SaaS

According to Gartner; SaaS is the "software that's owned, delivered and managed remotely by one or more providers" [1]. Some of the potential problems with the SaaS model includes Data-related security issues such as owners of data, data backup, data access, *data* locality, data availability, identity management and authentication,

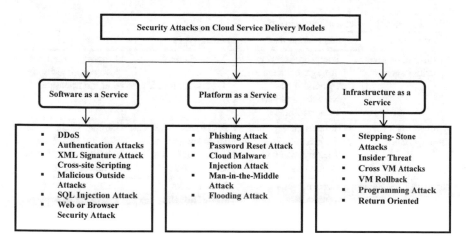

Fig. 5 Taxonomy of security attacks based on cloud service delivery models

etc. According to Forrester's research report, 70% of security breaches are caused by internal sources [12].

2.3.2 Security Attacks on PaaS

PaaS depends on the Service-Oriented Architecture (SOA) model. In this model there exist issues which result in attacks targeting PaaS cloud. These attacks include DDoS, injection and input validation related, Man-in-the-middle, Replay and eXtensible Markup Languages (XML)—related [13].

2.3.3 Security Attacks on IaaS

In this security attack the attackers first hold the operations of hosted virtual machines opening the possibility to attack on other hosted VMs or hypervisor. The attacks have two models: one changes the semantics of kernel structure and other changes the syntax [14, 15]. Table 2 deals with cloud security issues [16].

Table 2 accounts for various security issues in CC such as virtualization level, application level, network level, and physical level. It also shows the impacts of various attack types in CC [17].

It is also equally important to consider the attacks that affect the cloud systems.

Table 2 Security issues and levels in CC

Security issues	Attack vectors	Attack types	Impacts
Virtualization level security issues	Social engineering Storage, datacenter vulnerabilities Network and VM vulnerabilities	DoS and DDoS VM escape Hypervisor root kit	Programming flaws Software interruption and modification
Application level security issues	Session management Broken authentication Security misconfigurations	SQL injection attacks Cross site scripting Other application based attacks	Confidentiality Session Hijacking Modification of data at rest and in transit
Network level security issues	Firewall misconfigurations	DNS attacks Sniffer attacks Issues of reuse IP address Network sniffing, VoIP related attacks	Traffic flow analysis Exposure in network
Physical level security issues	Loss of power and environment control	Phishing attacks Malware injection attack	Limited access to data centers Hardware modification and theft

2.4 Cloud System Attacks

Attacks on cloud systems can be classified using real-world examples in the following way.

2.4.1 The Amazon EC2 Hack

The Amazon Elastic Compute Cloud (EC2) is so-called Infrastructure as a Service is one of the known commercial and publicly available CC service [18]. It offers virtual servers and allows users to deploy its own Linux, Solaris, or Windows based virtual machines. EC2 also offers SOAP interface for starting a new instances of a machine or terminating an instances [10]. A weakness was found in this control service. It was possible to modify an eavesdropped message despite of the digital signed operation by using Signature Wrapping Attack. In this way an attacker was able to execute arbitrary own machine commands in place of a legitimate cloud user and results in Denial of Service (DoS) on the users services. Therefore the attack incident is reduced by two separate actions attacking the cloud control interface and attacking the service instances.

2.4.2 Cloud Malware Injection

In the Cloud Malware Injection attack, the attacker uploads a manipulated copy of a victim's service instance so that some service request to the victim service are processed within that malicious instance. In order to carry out this, the attacker has to gain control over the victim's data in the cloud system. This attack is considered to be the major representative of exploiting the service-to-cloud attack surface. The attacker is capable of attacking security domains in this way.

2.4.3 Cloud Wars

Flooding attacks still impact the cloud resource exhaustion when the attacker uses a cloud for sending his flooding messages. Both clouds—the attacker's one and victim's one provide adequate resources for sending and receiving attack messages. This process continues until one of the both cloud systems reaches its maximum capacities. The attacker uses a hijacked cloud service for generating attack message that could trigger huge usage bills for cloud-provided services that the real user never ordered for. This is one of the side-effects in cloud wars. The next section is followed by cloud architecture based on attack categories.

2.5 Architecture of CC Attacks

The literature survey shows that there is an increase in attacks and penetrations. Email phishing, spam, Cyber Attacks, viruses and worms, blended attacks, zero-day or denial of service attacks are some of the attacks rapidly grow in malware with alarming situation. The attacks heightened risk to significant infrastructure like industrial control systems, banking, airlines, mobile and wireless network and many other infrastructures.

A general taxonomy has been broken into four categories namely.

2.5.1 Network and System Categories

The following is an indicative list of network categories. It should be noted these categories are not necessarily distinct and are often in a constant state of change and evolution [19]. Taxonomies need to address each broad category of these networks and individual implementations of each [20]. Universal taxonomy covering all types of networks and attacks is unmanageable. A taxonomy focuses on a particular network is both viable and realistic and very useful in practice.

The following categories of network infrastructures are most commonly used in all sectors (Government, industry, and individual users). The above ten sample networks interconnect systems such as devices, client/server software, applications, network processing devices or a combination of these [21]. Such devices are illustrated and each can be subject to a range of specific penetration attacks.

Selection of Network Categories

- WPAN (Bluetooth, RFID, UWB)
- WLAN (WPA/WPA2)
- Broadband Access (Wireless/Fixed)
- ISP Infrastructure
- Ad hoc, VoIP
- SCADA
- Cloud
- Messaging Networks (Email, Twitter &Face book)

Selection of Devices and Systems

- Web Browser/Client
- Web Server
- Handheld Devices (WPAN)
- GPS Location Device
- Hubs/Switches/Router/Firewall
- Industrial Device & Control System
- Cloud Client
- Operating System
- Application

2.5.2 Attack Categories

The following probable list of attack categories is not necessarily exhaustive and mutually exclusive. Each of these categories represents a class of attack and many variations of each are possible. This further indicates a general taxonomy is difficult to propose [22].

Selection of Attack Categories

- Exponential Attacks-Virus and Worms
- Hacking, Cracking, and Hijacking
- Trojans, Spyware, Spam
- Zero Day, Bots & Botnets
- Protocol Failures-Spoofing & Data Leakage
- Denial of Services (Distributed)
- Authentication Failures, Phishing, Pharming

2.5.3 Attack Techniques

The attack techniques used to create the class of attacks described in the previous section. These attack techniques are not mutually exclusive. Any one of those attack categories may utilize one or more of these techniques [23].

Selection of Attack Techniques

- Traffic Analysis (Passive/Active)
- MAC & IP Address Manipulation
- TCP Segment Falsification
- ARP & Cookie Poisoning
- Man-in-the-Middle (MITM)
- Flooding Attack & Backscatter
- Cross Site Scripting (XSS)/Forgery (XSRF)
- Input Manipulation/Falsification

2.5.4 Protection Technologies

The indicative protection technologies are used to protect against the attacks of categories described in the previous sections. A combination of these mechanisms-category of network systems, category of attacks and specific type of attacks are used for the practical configuration under study.

Selection of Protection Technologies

- Physical Security
- Encryption (Symmetric & Asymmetric)
- Authentication (1, 2 & 3 factor)
- Backup & Disaster Management
- Sandboxing
- Trace back
- Honey pots/Honey nets
- Digital Certificates

The next section deals with the review of cloud security threats and cloud attack consequences.

3 Threats and Vulnerability of CC

This Section elaborately discusses the taxonomies of cloud security attacks and security threats as shown in Table 3.

Cloud environment has large distributed resources. A CSP uses these resources for huge and rapid processing of data [24]. This exposes a user security threats. In

Table 3 Cloud scenario security threats

Nature of the threats	Security threats			
	Nomenclature	Description	Vulnerability	Prevention
Basic security	SQL injection attack	A malicious code is placed in standard SQL code	Unauthorized access to a database by the hackers	May be avoided by the use of dynamically generated SQL in the code and filtering of user input
	Cross site scripting attack	A malicious script is injected into web content	Website content may be modified by the hackers	Active content filtering, content based data leakage prevention technique, web application vulnerability detection technique
	Man-in-the-middle attack	Intruder tries to tap the conversation between sender and receiver	Important data/Information may be available to the intruder	Robust encryption tools like Dsniff, Cain, Ettercap, Wsniff and Air jack maybe used for prevention
Network layer security	DNS attack	Intruder may change the domain name request by changing the internal mapping of the users	Users may be diverted to some other evil cloud location other than the intended one	Domain name system security extensions (DNSSEC) may reduce the effect of DNS attack
	Sniffer attack	Intruder may capture the data packet flow in a network	Intruder may record, read and trace the user's vital information	ARP based sniffing detection platform and Round Trip Time (RTT) can be used to detect and prevent the sniffing attack

(continued)

Table 3 (continued)

Nature of the threats	Security threats			
	Nomenclature	Description	Vulnerability	Prevention
	IP address reuse attack	Intruder may take advantage of switchover time/cache clearing time of an IP address in DNS	Intruder may access the data of a user as the IP address is still exists in DNS cache	A fixed time lag definition of ideal time of an IP may prevent this vulnerability
	Prefix hijacking	Wrong announcement of an IP address related with a system is made	Data leakage is possible due to wrong routing of the information	Border gateway protocol with autonomous IDS may prevent it
	Fragmentation attack	Malicious insider (user) or an outsider may generate this attack	This attack use different IPdatagram fragments to mask their TCP packets from target IP filtering mechanism	A multilevel IDS and log management in the cloud may prevent these attacks
	Deep packet inspection	Malicious insider (user)	Malicious user may analyze the internal or external network and acquire the network information	
	Active and passive eavesdropping	Malicious insiders and network users	Intruder may get network information and prevent the authentic packets to reach its destination	
	Port scan	Malicious user may attempt to access the network	Malicious user may get the complete activity status of the network	Continuous observation of port scan logs by IDS may prevent this attack

(continued)

Table 3 (continued)

Nature of the threats	Security threats			
	Nomenclature	Description	Vulnerability	Prevention
Application layer attacks	Denial of service attack	The usage of cloud network may get unusable due to redundant and continuous packet flooding	Downgraded network services to the authorized user, Increases the bandwidth usage	Separate IDS for each cloud may prevent this attack
	Cookie poisoning	Changing or modifying the contents of cookies to impersonate an authorized user	Intruder may get unauthorized access to a webpage or an application of the authorized user	A regular cookie cleanup and encryption of cookie data may prevent this vulnerability
	Captcha breaking	Spammers may break the captcha	Intruder may spam and exhaust network resource	A secure speech and text encryption mechanism may prevent this attack by bots

order to prevent the threats it is important to mitigate unauthorized access to the user data over the network [25].

3.1 Attack Consequences

Cloud environment exhibits security vulnerabilities and threats that are subject to attacks. One of the most popular documents of Cloud Security Alliance's (CSA) is "The Notorious Nine Cloud Computing Top Threats" that reports on the possible security consequences [26, 27].

The use of People, Process and Technology (PPT) suits particularly for this classification and provides a fast, easy to understand and complete method for the cause of each attack consequences. Along with PPT classification, Confidentiality, Integrity and Availability (CIA) is used in computer security research [28].

- Account Hijacking
- Compromised Logs
- Data Breach
- Data Loss
- Unauthorized Elevation and Misuse of Privilege
- Interception, Injection and Redirection
- Isolation Failure

- Resource Exhaustion

Table 3 presented describes the nature of security threats-basic, network and application layer attacks. And also reveals the vulnerabilities and prevention of these cloud security attacks.

3.2 Taxonomy of Security Threats and Attacks

In cloud computing a vast data are processed. As a consequences security threats has become prone importance. The security threats cause a bottleneck in the deployment and acceptances of cloud servers [29]. There are nine cloud security threats classified based on the nature and vulnerability of attacks [30].

- Data infringe
- Loss of the important data
- Account or service traffic capturing
- Insecure interfaces and APIs
- Service denial
- Malicious insiders
- Abuse of cloud services
- Lack of due diligence
- Shared technology vulnerabilities

The next chapter deals with classification of basic and vulnerable attacks on cloud.

4 Most Vulnerable Attacks of CC

In cloud the most common attacks are Advanced Persistent Threats (APT) and Intrusion Detection (ID) attacks among them the most vulnerable is the APT attack [31]. An APT attacks become a social issue and the technical defense for this attack is needed. APT attacks threaten traditional hacking techniques to increase the success rate of attack techniques like Zero-Day vulnerability, to avoid detection techniques it uses a combination of intelligence. Malicious code based detection methods are not efficient to contain APT Zero-Day attack.

4.1 APT Attacks

An APT attack begins from the internal system intrusion. Following are four steps identified in the APT attack shown in Fig. 6.

- Malware infects specific host using network or USM vulnerability.

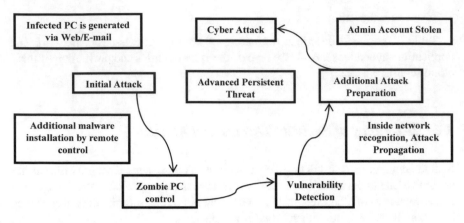

Fig. 6 APT attack process

- Download of malicious code and spread the malware through the infection server.
- The target system is found through the speed of malware.
- Important information of the target systems are leaked through malware.

4.1.1 Host-Based APT Attack Behavior

Recent misuse detection methods are found difficult to detect APT attacks because these attack uses zero-day attack or intelligent attack methods. In host system the resources for events are divided by process, thread, file system, registry, network, and service [32].

4.1.2 Malicious Code Based APT Attack Behavior

In APT attack the malicious code is performed by program specific objectives and it has the features of Zero-Day attack that is not detected by any Antivirus.

4.1.3 Network-Based APT Attack Behavior

In APT attacker uses network to detect targets and performs malicious behavior by external command.

Table 4 IDS characterization

IDS Characterization	
Parameter	Classification
Detection method	Anomaly based Specification based
Monitoring method	Network based Host based
Behavior pattern	Passive IDS Active IDS
Usage frequency	Online analysis Offline analysis

4.2   Intrusion Detection Attacks

Activities such as illegal access to network or system and malicious attacks are related intrusions that are committed via the use of Internet and electronic device. In order to prevent the threats an Intrusion Detection System (IDS) is generally preferred [33]. An IDS is an important tool to prevent and mitigate unauthorized access to the user data over the network [34].

Intrusion attacks are commonly divided into four major categories namely

- Denial of Service (DS)
- User to Root (UR)
- Remote to Local (RL)
- Probing Attack (PA)

Some of the widely used machine learning techniques for intrusion detection and prevention are Bayesian Network Learning, Genetic Algorithms, Snort, Fuzzy Theory, Information theory [35]. These techniques and approaches are still open for further research [36, 37].

Generally IDS is employed to aware the system administrator against any suspicious and possibly disturbing incident taking place in the system that is being analyzed. IDS may be classified on the basis of Table 4 [38].

IDS have parameters each of which constitutes a set of classification based on their characterization.

Depending upon the applications need a specific model is chosen for the cloud deployment. To execute real-time traffic categorization and analysis over IP networks a signature-based open source network-based IDS [SNORT] is used. Machine Learning approach is based on automatic discovery of patterns and attack classes mainly used for misuse detection [39]. The statistical methods of IDS classify the events as Normal or Intrusive based on the profile information. Intrusion Detection based on immune system concept is the latest approach and one of the difficult method to build practically [40].

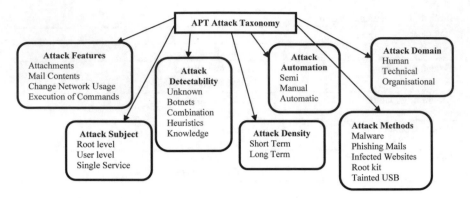

Fig. 7 APT attacks ontology

4.3 Ontology Modeling and Inference Detection for APT Attack

This section works with APT Attack Behavior and Inference Ontology.

4.3.1 APT Behavior Ontology

Ontology is a method for understanding the meaning using concepts and relations of data, information and knowledge. It can be applied to find new concepts and relations using inference and the Rules of Inference should be sound, complete and tractable refer Fig. 7 [41].

4.3.2 APT Inference Ontology

In this experiment the file system contains data set which consists of both normal files and malicious files. In such cases the normal files are excluded through general execution process and malicious files are gathered through hidden files (exe, dat, dll and lnk).

5 Risk assessmentss

A combination of likelihood of a successful attack and the impact of this incident form the basis for risk assessment [42, 43]. The five factors of the first level of taxonomy (source, vector, target, impact, and defense) are used to determine the likelihood of an attack. By identifying the dimensions of the particular attack quantification of the

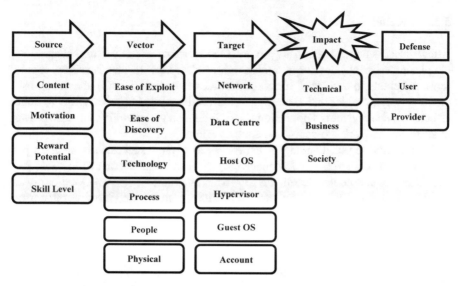

Fig. 8 Flow of cloud attack taxonomy

risk assessment could be done by rating them [7]. Using an Open Web Application Security Project (OWASP) risk assessment formula such parameter is assigned a value between 1 (low risk) and 9 (high risk) and added them to get the estimated overall impact of a given attack [44]. A risk severity of 0–3 is considered low, 3–6 medium and 6–9 high.

5.1 Flow of Cloud Attacks

The natural flow of a cloud attack comprises of five dimensions—source, vector, target, impact, and defense shown in Fig. 8. The taxonomy's second level (context, motivation, opportunities, and skill level) facilitates to identify attacker or attack source in a security incident.

5.1.1 Source

Attackers are part of a greater context with different motivations. Attackers can be part of an organized crime group, state-sponsored hacker groups, criminally motivated insiders, terrorists and hacktivists. Attacker's skill level varies from basics to sophisticate. State-sponsored attackers are better resourced and skilled as in the case of Stuxnet and Operation Shady Remote Access Tool (RAT). Generally attackers perform a cost-benefit analysis.

5.1.2   Vector

Attack success depends on many factors. One factor of attack vector is technology, people, process, physical or a combination. Other kind of attack vectors include renting Botnets to conduct Distributed Denial of Service (DDoS) attack against a cloud service provider and exploiting vulnerabilities like buffer overflow to attack affected cloud services or applications [45]. Technical vulnerability may also present in software, network protocols, and configurations.

Vectors may be challenging or expensive to exploit successfully such as performing an Structured Query Language (SQL) injection attack on a vulnerable Web application is simple and cheap, but to plant a hardware Trojan on cloud hardware would require more amount of money, bandwidth, planning and resources that is out of reach for most attackers.

5.1.3   Target

The attack target technically depends on attacker's goal and the attack vector. The target could be hypervisor in a cross-VM attack or datacenter for physical attack and also cloud service account for extortion purposes.

5.1.4   Impact

Generally the technical impact of cloud attack classified as data availability, data integrity, and data confidentiality. A cloud service's availability can be disrupted by a DDoS attack if an attacker gains unauthorized access to the data. It could breach data confidentiality and also an attacker can alter or delete user's data on cloud.

An organizations profitability and viability can have long-term effects due to financial and reputation damage. Non-compliance and privacy violation issues should be considered by Cloud Service Provider (CSP) and organizational user.

5.1.5   Defense

While considering and developing countermeasures the sequence of events is also important in cloud attacks. For the defense mechanism to be implemented the party responsible must also be determined, supported, and resourced. This varies with the deployed cloud structure. For instance, In Software as a Service [SaaS] CSP is primarily responsible for security at every layer. In Platform as a Service [PaaS] the cloud service user is responsible for securing the applications. In Infrastructure as a Service [IaaS] generally the cloud service user has greater security responsibility. In most cases the responsibility is shifted to the cloud service provider. It is also important to identify and note the levels of attacks in cloud computing.

6 Conclusion and Future Scope

Cloud computing would be gaining more importance in the near future. In utilizing CC services, the security of data and infrastructures are vulnerable to attacks and threats. In this survey paper it has been discussed at length the possible modes of security attacks such as APT, IDS, and Risk assessment. Intrusion attacks are rapidly growing cloud environment attacks which has results in higher risks for the users and organizations. A Taxonomy of attacks based on (i) Network categories and systems (ii) Attack categories (iii) Attack techniques and (iv) Protection technologies have been reviewed. This paper also focuses on eight common consequences of attacks on cloud services, affecting the services confidentiality, availability, and integrity. The security strategy while adapting CC is a multistep process. This survey leads to the conclusion there is no general or universal taxonomy is available and with the multiple layers of security and adopting security controls, it is possible to prevent cloud security threats. In future proper mitigation strategies could be developed.

References

1. Iqbal, S., Kiah, L.M., Dhaghighi, B., Hussain, M., Khan, S., Khan, M.K., Choo, K.-K.R.: On cloud security attacks: a taxonomy and intrusion detection and prevention as a service. J. Netw. Comput. Appl. **74**, 98–120 (2016)
2. Symantec, Internet Security Threat Report, vol. 17 (2011). Available http://www.symantec.com/threatreport/ (2014)
3. Singh, R.K., Bhattacharjya, A.: Security and privacy concerns in cloud computing. In: International Journal of Engineering and Innovative Technology (IJEIT) vol. 1, Issue 6, ISSN: 2277-3754 (2012)
4. Mell, P., Grance, T.: The NIST Definition of Cloud Computing, Special Publication 800-145 NIST
5. Sosinsky, B.: Cloud Computing Bible. Wiley Publishing Inc., ISBN-13: 978-0470903568
6. Simmons, C., et al.: AVOIDIT: A Cyber Attack Taxonomy. Technical Report CS-09-003, University of Memphis (2009)
7. Choo, K.-K.R., Juliadotter, N.V.: Cloud attack and risk assessment taxonomy. IEEE Cloud Comput. pp. 14–20 (2015)
8. Ab Rahman, N.H., Choo, K.K.R.: Integrating Digital Forensic Practices in Cloud Incident Handling: A Conceptual Cloud Incident Handling Model, The Cloud Security Ecosystem, Imprint of Elsevier (2015)
9. Rane, P.: Securing SaaS applications: a cloud security perspective for application providers. Inf. Syst. Secur. (2010)
10. Gruschka, N., Jensen, M.: Attack surfaces: taxonomy for attacks on cloud services. In: 3rd International Conference on Cloud Computing, pp. 276–279. IEEE, New York (2010)
11. Claycomb, W.R., Nicoll, A.: Insider threats to cloud computing: directions for new research challenges. In: 2012 IEEE 36th Annual Computer Software and Applications Conference (COMPSAC), pp. 387–394 (2012)
12. Behl, A.: Emerging security challenges in cloud computing, pp. 217–222. IEEE, New York (2011)
13. Osanaiye, O., Choo, K.-K.R., Dlodlo, M.: Distributed denial of service (DDoS) resilience in cloud: review and conceptual cloud (DDoS) mitigation framework. J. Netw. Comput. Appl. (2016)

14. Khorshed, M.T., Ali, A.B.M.S., Wasimi, S.A.: A survey on gaps, threat remediation challenges and some thoughts for proactive attack detection in cloud computing. Future Gener. Comput. Syst. **28**, 833–851 (2012)
15. Hansman, S., Hunt, R.: A taxonomy of network and computer attacks. Comput. Secur. **24**(1), 31–43 (2005)
16. Jensen, M., Schwenk, J., Gruschka, N., Lo Iacono, L.: On technical security issues in cloud computing. In: Proceedings of the IEEE International Conference on Cloud Computing (CLOUD-II) (2009)
17. Modi, C., Patel, D., Borisaniya, B., et al.: A survey on security issues and solutions at different layers of cloud computing. J. Supercomput. **63**, 561–592 (2013)
18. Deshpande, P., Sharma, S., Peddoju, S.: Implementation of a private cloud: a case study. Adv. Intell. Syst. Comp. **259**, 635–647 (2014)
19. Ab Rahman, N.H., Choo, K.K.R.: A survey of information security incident handling in the cloud. Comput. Secur. **49**, 45–69 (2015)
20. Khan, S., et al.: Network forensics: review, taxonomy, and open challenges. J. Netw. Comput. Appl. **66**, 214–235 (2016)
21. Brown, E.: NIST issues cloud computing guidelines for managing security and privacy. National Institute of Standards and Technology Special Publication, pp. 800–144 (2012)
22. Hunt, R., Slay, J.: A new approach to developing attack taxonomies for network security-including case studies, pp. 281–286. IEEE, New York (2011)
23. Asma, A.S.: Attacks on cloud computing and its countermeasures. In: International Conference on Signal Processing, Communication, Power and Embedded System (SCOPES), pp. 748–752. IEEE, New York (2016)
24. Deshpande, P., Sharma, S.C., Sateeshkumar, P.: Security threats in cloud computing. In: International Conference on Computing, Communication and Automation (ICCCA), pp. 632–636. IEEE, New York (2015)
25. Sabahi, F.: Cloud computing threats and responses, 978-1-61284-486-2/111. IEEE, New York (2011)
26. Tep, K.S., Martini, B., Hunt, R., Choo, K.-K.R.: A taxonomy of cloud attack consequences and mitigation strategies, pp. 1073–1080. IEEE, New York (2015)
27. Los, R., Gray, D., Shackleford, D., Sullivan, B.: The notorious nine cloud computing top threats in 2013. Top Threats Working Group, Cloud Security Alliance (2013)
28. Khan, S., et al.: SIDNFF: source identification network forensics framework for cloud computing. In: Proceedings of the IEEE International Conference on Consumer Electronics-Taiwan (ICCE-TW) (2015)
29. Shen, Z., Liu, S.: Security threats and security policy in wireless sensor networks. AISS **4**(10), 166–173 (2012)
30. Alva, A., Caleff, O., Elkins, G., et al.: The notorious nine cloud computing top threats in 2013. Cloud Secur. Alliance (2013)
31. Choi, J., Choi, C., Lynn, H.M., Kim, P.: Ontology based APT attack behavior analysis in cloud computing. In: 10th International Conference on Broadband and Wireless Computing, Communication and Applications, pp. 375–379. IEEE, New York (2015)
32. Baddar, S., Merlo, A., Migliardi, M.: Anomaly detection in computer networks: a state-of-the-art review. J. Wireless Mobile Netw. Ubiquit. Comput. Dependable Appl. **5**(4), 29–64 (2014)
33. Xiao, S., Hariri, T., Yousif, M.: An efficient network intrusion detection method based on information theory and genetic algorithm. In: 24th IEEE International Performance, Computing, and Communications Conference, pp. 11–17 (2005)
34. Amin, A., Anwar, S., Adnan, A.: Classification of cyber attacks based on rough set theory. IEEE, New York (2015)
35. Murtaza, S.S., Couture, M., et al.: A host-based anomaly detection approach by representing system calls as states of kernel modules. In: Proceedings of 24th International Symposium on Software Reliability Engineering (ISSRE), pp. 431–440 (2013)
36. Vieira, K., Schulter, A., Westphall, C.: Intrusion detection techniques for grid and cloud computing environment. IT Prof. **12**(4), 38–43 (2010)

37. Deshpande, P., Sharma, S., Sateeshkumar, P., Junaid, S.: HIDS: an host based intrusion detection system. Int. J. Syst. Assur. Eng. Manage. pp. 1–12 (2014)
38. Kaur, H., Gill, N.: Host based anomaly detection using fuzzy genetic approach (FGA). Int. J. Comput. Appl. **74**(20), 5–9 (2013)
39. Sommer, R., Paxson, V.: Outside the closed world: on using machine learning for network intrusion detection. In: IEEE Symposium on Security and Privacy, Oakland (2010)
40. Chen, C., Guan, D., Huang, Y., Ou, Y.: State-based attack detection for cloud. In: IEEE International Symposium on Next-Generation Electronics, Kaohsiung, pp. 177–180 (2013)
41. Khan, S., et al.: Cloud log forensics: foundations, state of the art, and future directions. ACM Comput. Surv. (CSUR) **49**(1), 7 (2016)
42. Juliadotter, N., Choo, K.K.R.: CATRA: Conceptual Cloud Attack Taxonomy and Risk Assessment Framework, The Cloud Security Ecosystem. Imprint of Elsevier (2015)
43. Peake, C.: Security in the cloud: understanding the risks of Cloud-as-a-Service. In: Proceedings of IEEE Conference on Technologies for Homeland Security (HST 12), pp. 336–340 (2012)
44. OWASP, OWASP Risk Rating Methodology, OWASP Testing Guide v4, Open Web Application Security Project. www.owasp.org/index.php/ OWASP Risk Rating Methodology (2013)
45. Bakshi, A., Dujodwala, Y.B.: Securing cloud from DDOS attacks using intrusion detection system in virtual machine. In: Proceeding ICCSN '10 Proceedings of 2010 Second International Conference on Communication Software Networks, pp. 260–264 (2010)

Execution Time Based Sufferage Algorithm for Static Task Scheduling in Cloud

H. Krishnaveni and V. Sinthu Janita Prakash

Abstract In cloud computing applications, storage of data and computing resources are rendered as a service to the clients via the Internet. In the advanced cloud computing applications, efficient task scheduling plays a significant role to enhance the resource utilization and improvise the overall performance of cloud. This scheduling is vital for attaining a high performance schedule in a heterogeneous-computing system. The existing scheduling algorithms such as Min-Min, Sufferage and Enhanced Min-Min, focused only on reducing the makespan but failed to consider the other parameters like resource utilization and load balance. This paper intends to develop an efficient algorithm namely Execution Time Based Sufferage Algorithm (ETSA) that take into account, the parameters makespan and also the resource utilization for scheduling the tasks. It is implemented in Java with Eclipse IDE and a set of ETC matrices are used in experimentation to evaluate the proposed algorithm. The ETSA delivers better makespan and resource utilization than the other existing algorithms.

Keywords Cloud computing · Scheduling · Makespan

1 Introduction

The cloud environment possesses a number of resources to provide service to the customers on payment basis [1]. Cloud consumers can utilize the cloud services provided by the cloud service provider. As the cloud is a heterogeneous environment, many issues will crop up while providing these services. The major concern is the best possible usage of the on hand resources [2].

Proper scheduling provides the better utilization of the resources. As the performance of the cloud applications is mainly affected by scheduling, makes scheduling to play a vital role in research. The requirement for scheduling arises when tasks are to be executed in parallel [3]. The parallel computing environment has undergone

H. Krishnaveni (✉) · V. Sinthu Janita Prakash
Department of Computer Science, Cauvery College for Women, Trichy, Tamil Nadu, India
e-mail: hkrish6677@gmail.com

© Springer Nature Singapore Pte Ltd. 2019
J. D. Peter et al. (eds.), *Advances in Big Data and Cloud Computing*,
Advances in Intelligent Systems and Computing 750,
https://doi.org/10.1007/978-981-13-1882-5_5

various advancements in the recent years. Different kinds of scheduling algorithms are available in the cloud computing system under the categories of static, dynamic, centralized, and distributed scheduling. Also job and task scheduling are two divisions of scheduling concept. Task scheduling is an essential component in managing the resources of cloud. It manages tasks to allocate appropriate resources by using scheduling polices. In static scheduling, during the compile time itself, the information regarding all the resources and tasks are known. Scheduling of tasks plays the key role of improvising the efficiency of the cloud environments by reducing the makespan and increasing the resource consumption [4].

Braun et al. [5] provided a simulation base for the research community to test the algorithms. They concluded that the performance of the Genetic Algorithm (GA) is good in many cases. But the functioning of the Min-Min algorithm which is in the next position to GA is not satisfactory. Opportunistic Load Balancing (OLB), Minimum Execution Time (MET), Minimum Completion Time (MCT), Min-Min and Max-Min are the simple algorithms proposed by Braun.

Most of the scheduling algorithms give importance to parameters such as energy efficiency, scalability, cost of the computing resources, and network transmission cost. Some of the existing algorithms concentrate in balancing the load of the resources in dynamic environment. This paper concentrates in minimizing the makespan and maximizing the consumption of resources with a balanced load.

This paper is structured as: Sect. 2 expresses the related works and various scheduling algorithms which form the basis of many other works. In Sect. 3, the fundamental concept of task scheduling in heterogeneous environments and the proposed scheduling algorithm are given. Section 4 shows the experimentation and the comparative result of Execution Time based Sufferage Algorithm. At last, Sect. 5 states the conclusion of the paper.

2   Related Works

Parsa et al. [6] presented a hybrid task scheduling algorithm. It is a combination of two conventional algorithms Max-min and Min-min. This algorithm utilizes the pros of Max-min and Min-min algorithms and downplays their cons. The algorithm works under the batch scheduling mode with makespan as its parameter, but not taken into account, the task deadline, arrival rate and computational cost.

A new algorithm for optimizing the task scheduling and resource allocation using Enhanced Min-Min algorithm has been proposed by Haladu et al. [7]. This algorithm includes the usefulness of Min-Min strategy and avoids its drawbacks. The main intention is to decrease the completion time with effective load balancing.

Chawda et al. [8] derived an Improved Min-Min task scheduling algorithm to execute a good number of tasks with available machines. It balances the load and increases the utilization of the machines, to diminish the total completion time. The algorithm has two stages. In the first stage, Min-Min algorithm is executed and in the second stage, it minimizes the load of the heavy loaded machine and increases the utilization of the machine that is underutilized.

3 Proposed Methodology

The mapping of meta-tasks to a resource is a NP-complete problem. Only the heuristic methods can be used to solve this problem. A lot of algorithms have been developed to schedule the independent tasks by using the heuristic approach. Different metrics such as makespan and resource utilization can be used to evaluate the efficiency of scheduling algorithms.

In static scheduling, the estimate of the expected execution time for each task on each resource is known in advance to execution and it is represented in an Expected Time to Compute (ETC) matrix. ETC (T_i, R_j) is the approximate execution time of task i on resource j.

The main aspire of the scheduling algorithm is to diminish the overall completion time (i.e.) makespan.

Any scheduling problem can be defined by using the ETC matrix as follows:

Let $T = t_1, t_2, t_3, ..., t_n$ be the group of tasks submitted to scheduler.

Let $R = r_1, r_2, r_3, ..., r_v$ be the set of resources available at the time of task arrival.

Makespan produced by any algorithm can be calculated as follows:

$$\text{makespan} = \max(\text{CT}(t_i, r_j)) \tag{1}$$

$$\text{CT}_{ij} = \text{ET}_{ij} + r_j \tag{2}$$

where

CT_{ij} completion time of machine.

ET_{ij} approximate execution time of task i on resource j.

r_j availability time or ready time of resource j after completing the previously assigned tasks.

Existing scheduling algorithms such as Min-Min, Max-Min, and sufferage are easy to implement but are considered unsuitable for large scale applications.

3.1 Execution Time Based Sufferage Algorithm (ETSA)

The waiting queue contains number of unassigned tasks. The number of task is N and the number of resource is M. For each task in waiting queue, calculate the completion time (CT) of task i in resource j.

For each task T_i find First Minimum Completion Time (FMCT$_i$) and Second Minimum Completion Time (SMCT$_i$) of task T_i. Also find First Minimum Execution Time (FMET$_i$) and Second Minimum Execution Time (SMET$_i$) of task T_i.

Calculate Sufferage Value for completion time using

$$SV_i = SMCT_i - FMCT_i \tag{3}$$

Calculate Sufferage Value for execution time EXSV$_i$ using

$$EXSV_i = SMET_i - FMET_i \tag{4}$$

Then sort the sufferage value SV$_i$ of all tasks T_i. According to the sorted SV, arrange EXSVi. For each task T_i from N, if Sufferage Value SV$_i$ is greater than EXSV$_i$, then task T_i is assigned to the machine that gives the minimum completion time of task. Remove the selected task T_i from unassigned task but update ready time of machine $R_j = C_{ij}$, otherwise, assign the T_n as selected task.

After this execution, the algorithm tries to reduce the burden of the resource having more number of tasks comparing to other resources. This will reallocate the task that has least execution time on the maximum loaded resource to the resource which is under utilized by holding the condition that its completion time in other resources is less than the makespan obtained. The pseudo code for ETSA is given in Table 1.

4 Experimental Results

The experimental result evaluates the performance of the proposed approach. The proposed work has been experimented using Java with Eclipse. The makespan and resource utilization are produced for ETC matrices of 512 tasks and 16 resources based on the benchmark model developed by Braun et al. [5].

The proposed algorithm uses the ETC matrices generated by EMGEN tool [9]. It has been evaluated in terms of the parameters such as makespan and resource utilization. The experimental result of proposed algorithm is compared with the existing Min-Min, Enhanced Min-Min, and Sufferage.

Table 2 presents the expected execution time of tasks in milliseconds for the 6×6 sub matrix of 512×16 consistent resources with high task and low resource heterogeneity. Table 3 exhibits the makespan in ms for 512×16 consistent, inconsistent and partially consistent resources. Each of these matrices has combination of low task and high resource, low task and low resource, high task and high resource, high task and low resource heterogeneity.

Figure 1 shows the window of implementation in java with eclipse IDE. The console displays the allocation of various tasks to different machines.

Figure 2 shows the makespan (in ms) produced for 512×16 consistent, inconsistent and partially consistent resources for all the combinations of low task and high resource, low task and low resource, high task and high resource, high task, and low

Table 1 Pseudo code for ETSA

While (Unassigned_Tasks ≠null)
N=number of unassigned tasks T
V=number of resource M
For i=1 to N
 For j=1 to V
//Calculate the completion time

$$CT_{ij} = ET_{ij} + r_j$$

End for
End for
For each task T_i
Find First Minimum Completion Time (FMCT$_i$) and Second Minimum Completion Time (SMCT$_i$) of task T$_i$
Find First Minimum Execution Time (FMET$_i$) and Second Minimum Execution Time(SMET$_i$) of task T$_i$

$$SV_i = SMCT_i - FMCT_i$$

$$EXSV_i = SMET_i - FMET_i$$

End For
Sort SV$_i$
Arrange ExSV according to sorted SV
Sel_Task=null //Initial Selected Task is assigned as null
For i=N to 0
If (SV$_i$>ExSV$_i$)
Sel_Task=T$_i$;
break;
Else
Sel_Task=T$_n$;
End If
End For
Assign Sel_Task (selected task) to machine j that gives the minimum completion time of task .
Remove Sel_Task from Unassigned_Tasks.
Update Ready time of resource R$_j$
Update Completion Time of all the resources
Compute makespan=max(Compeletion Time(R))
End while
Do while
Select the task with minimum ET on the resource R$_k$ that has maximum load
 Compute CT of task produced by the other resource R$_m$
 If CT<makespan
  Reassign task to resource R$_m$
 Update ready time of both R$_k$ and R$_m$
 End If
 End Do

Table 2 Expected execution time (ms) of tasks for the 6×6 sub matrix of 512×16 consistent resources with high task and low resource heterogeneity

Task list	Resources					
	R0	R1	R2	R3	R4	R5
T0	7568.17	7642.05	9203.20	9830.38	10,765.57	13,707.93
T1	2009.83	2134.44	2989.71	3082.95	3780.618	4341.14
T2	982.81	1968.64	5015.41	14,664.73	21,870.84	22,019.25
T3	6621.23	12,789.13	14,811.64	21,505.17	25,420.81	26,500.3
T4	2217.31	2973.44	2989.37	3642.86	5380.82	5471.64
T5	417.89	1418.92	1686.77	2607.07	4679.20	5922.67

Table 3 The makespan (ms) for 512 tasks and 16 consistent, inconsistent and partially consistent resources with combination of all heterogeneity

ETC	Min-Min	Enhanced Min-Min	Sufferage	ETSA
LH_C.ETC	8,428,883	1.14E+07	1.02E+07	1.01E+07
HH_C.ETC	266,164.9	357,113.9	316,891.5	314,123.9
HL_C.ETC	169,594	205,454.5	172,472.2	171,686.8
LL_C.ETC	5417.97	6818.413	5701.696	5627.88
HH_PC.ETC	3,651,382	5,757,560	3,920,190	3,549,727
HL_PC.ETC	141,116.6	184,699	132,126.5	132,199.8
LL_PC.ETC	91,898.49	145,363.8	84,931.54	84,753.93
HH_IC.ETC	2963.92	4750.84	2769.78	2746.74
HL_IC.ETC	4,089,279	4,430,988	3,094,128	3,047,772
LL_IC.ETC	131,479.3	153,864.5	114,132.8	108,157.7
LH_PC.ETC	78,816.11	129,390.5	75,690.8	72,196.73
LH_IC.ETC	2709.41	4645.26	2562.13	2529.94

resource heterogeneity. Table 3 shows that the proposed ETSA algorithm provides better makespan than the algorithms Enhanced Min-Min and Sufferage for all type of resources but gives a greater makespan than Min-Min algorithm for the consistent resources only.

It is clearly understood that the proposed ETSA provides the better results in terms of overall completion time of all tasks (i.e., makespan) than the existing algorithms Min-Min, Enhanced Min-Min, and Sufferage.

Fig. 1 ETSA implementation window

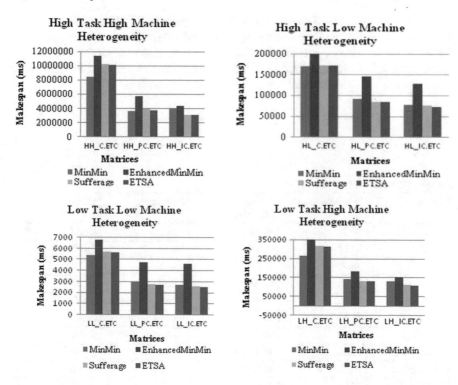

Fig. 2 Makespan (ms) produced by different resources and tasks for all heterogeneity

4.1 Resource Utilization Rate

Resource utilization rate [10] of all the resources is calculated by using the following equation:

$$\text{ru} = \left(\sum_{j=1}^{m} \text{ru}_j \right) \div m \tag{5}$$

Here, ru_j is the resource utilization rate of resource r_j. It can be calculated by the following equation:

$$\text{ru}_j = \left(\sum (\text{te}_i - \text{ts}_i) \right) \div T \tag{6}$$

where, te_i is the end time of executing task t_i on resource r_j, ts_i is the start time of executing task t_i on resource r_j.

From Table 4 and Fig. 3 it is observed that the proposed ETSA provides the better resource utilization than the existing algorithms Min-Min, Enhanced Min-Min and Sufferage for consistent, partially consistent and inconsistent resources for various heterogeneity. The better utilization of the resources implies that the idle time of the resources has been reduced and hence the load is also balanced.

Table 4 Resource utilization (%)

Matrices	Min-Min	Enhanced Min-Min	Sufferage	ETSA
HH_C.ETC	90.74	98.60	98.60	97.69
LH_C.ETC	88.18	97.74	98.49	98.65
HL_C.ETC	92.00	98.74	98.22	98.34
LL_C.ETC	95.05	98.88	98.80	98.51
HH_PC.ETC	85.76	96.26	87.66	99.71
LH_PC.ETC	81.83	96.90	96.57	99.41
HL_PC.ETC	87.72	99.47	98.08	97.99
LL_PC.ETC	89.83	98.97	98.12	98.94
HH_IC.ETC	70.72	95.82	95.71	98.96
LH_IC.ETC	73.91	96.38	92.52	99.63
HL_IC.ETC	90.30	98.91	94.59	99.68
LL_IC.ETC	91.90	98.65	97.93	99.60

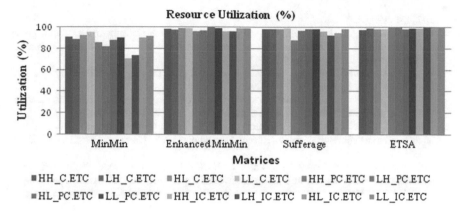

Fig. 3 Resource utilization comparison

5 Conclusion

The conventional algorithms are suitable for small-scale distributed system. All static heuristics aim to mitigate only the completion time while schedule the tasks but not distribute the tasks on all the resources evenly. Proposed ETSA, compares the SV of each task with EXSV and then take the decision to give out the tasks to the resource. It also tries to decrease the makespan with a balanced load across the resource. ETSA is used to schedule the number of tasks to the various resources based on the execution time efficiently. It gives better result in terms of makespan and resource utilization with a balanced load when compared with existing Min-Min, Enhanced Min-Min, and Sufferage. The further escalation of the opus can be carried by applying the ETSA in actual cloud computing environment (CloudSim). Since the makespan produced by the ETSA is greater than that of Min-Min algorithm for consistent resources, the reason has to be analyzed and rectified in the future. Also the proposed ETSA can be directed towards the parameters such as computational cost, storage cost, and deadline of the tasks.

References

1. Shawish, A., Salama, M.: Cloud Computing: Paradigms and Technologies, pp. 39–67. Springer, Berlin
2. Awad, A.I., El-Hefnawy, N.A., Abdel_kader, H.M.: Enhanced particle swarm optimization for task scheduling in cloud environments. In: International Conference on Communication, Management and Information Technology. Procedia Computer Science, Elsevier B.V. (2015)
3. Yang, T., Cong, F., Pyramid, S., Jose, S.: Space/time-efficient scheduling and execution of parallel irregular computations. ACM **20**, 1195–1222 (1998)
4. Maheswaran, M., Ali, S., Siegel, H.J., Hensgen, D., Freund, R.F.: Dynamic mapping of a class of independent tasks onto heterogeneous computing systems. J. Parallel Distrib. Comput. **59**,

107–131 (1999)
5. Braun, T.D., Siegel, H.J., Beck, N.: A comparison of eleven static heuristics for mapping a class of independent tasks onto heterogeneous distributed computing systems. J. Parallel Distrib. Comput. **61**, 810–837 (2001)
6. Parsa, S., Entezari-Maleki, R.: RASA: a new grid task scheduling algorithm. Int. J. Digit. Content Technol. Appl. pp. 91–99 (2009)
7. Haladu, M., Samual, J.: Optimizing task scheduling and resource allocation in cloud data center, using enhanced Min-Min algorithm. IOSR J. Comput. Eng. **18**, 18–25 (2016)
8. Chawda, P., Chakraborty, P.S.: An improved min-min task scheduling algorithm for load balancing in cloud computing. Int. J. Recent Innovation Trends Comput. Commun. **4**, 60–64 (2016)
9. Kokilavani, T., George Amalarethinam, D.I.: EMGEN—a tool to create ETC matrix with memory characteristics for meta task scheduling in grid environment. Int. J. Comput. Sci. Appl. **2**, 84–91 (2013)
10. Jinquan, Z., Lina, N., Changjun, J.: A heuristic scheduling strategy for independent tasks on grid. In: Proceedings of the Eighth International Conference on High-Performance Computing in Asia-Pacific Region (2005)

QoS-Aware Live Multimedia Streaming Using Dynamic P2P Overlay for Cloud-Based Virtual Telemedicine System (CVTS)

D. Preetha Evangeline and P. Anandhakumar

Abstract Telemedicine system is one vital application that comes under live streaming. The application hits the criticality scenario and requires high Quality of Service when compared with other live streaming applications such as video conferencing and e-learning. In order to enhance better Quality of service in this critical live streaming application, the paper proposes Cloud-based Virtual Telemedicine System (CVTS) which handles technical issues such as Playback continuity, Playback Latency, End-to-End delay, and Packet loss. Here, P2P network is used to manage scalability and dynamicity of peers but the limitation associated with peer-to-peer streaming is lack of bandwidth both at the media source as well as among the peers which leads to interrupted services. The paper deals with the proposed work Dynamic Procurement of Virtual machine (DPVm) algorithm to improve QoS and to handle the dynamic nature of the peers. The performance evaluation was carried out in both Real-time and Simulation with Flash crowd scenario, and the Quality of service was maintained without any interruptions which proves that the proposed system shows 98% of playback continuity and 99% of the chunks are delivered on time and even if it fails, the chunks are retrieved from the Storage node with a maximum delay of 0.3 s.

Keywords Cloud computing · Multimedia streaming · Real-world applications
Resource optimization · Cost efficiency

D. Preetha Evangeline (✉)
Vellore Institute of Technology, Vellore, Tamil Nadu, India
e-mail: Preethadavid4@gmail.com

P. Anandhakumar
Anna University, Chennai, Tamil Nadu, India
e-mail: anandh@annauniv.edu

© Springer Nature Singapore Pte Ltd. 2019
J. D. Peter et al. (eds.), *Advances in Big Data and Cloud Computing*,
Advances in Intelligent Systems and Computing 750,
https://doi.org/10.1007/978-981-13-1882-5_6

1 Introduction

Recent media services that are in need of enriched resources to guarantee rich media experience are Video conferencing system, E-learning, Voice over IP and among them the most critical one is Telemedicine [1]. As this critical application requires high bandwidth to support dissemination of uninterrupted content to the users and, moreover today's network are completely heterogeneous with a mixed of wired/wireless, various networking protocols and most important is the variation in devices. Users preferably use mobile devices when it comes to streaming services and one challenge involved here is the bottleneck of the available bandwidth which serves as an obstacle for live streaming. In spite of the manageable bandwidth to stream a stable quality of videos, there comes a problem with scalability [2]. Number of users cannot be restricted from participating in a live session, say for suppose, considering our Telemedicine application, there may be specialized doctors discussing on a particular case online and medical students may be interested in viewing the live discussions that are live streamed. In that case, more number of users must be allowed in participation provided with a guaranteed rate of quality in order to experience a clear picture of the case. The application is highly critical and it is sensitive to acute timings. The live streams that are transmitted during discussions must be assisted with high definition (Audio/Video synchronization), uninterrupted connectivity (Network bandwidth), and playback continuity (transmission delay) [3]. The idea behind cloud computing-based telemedicine service, where the contents are made available on the web and it can be accessed by physicians from anywhere. The paper concentrates on how cloud computing is used to improve the Qos in healthcare application.

1.1 Issues in Telemedicine Streaming Related to Quality of Service

- Infrastructural Issues: Limited availability of bandwidth in the devices and at the media source.
- Heterogeneity of Devices: Users preference of using mobile phones, laptops, tablets, etc., according to their viewing perception and on the fly transcoding according to the specification of the device.
- Implementation and Cost: The cost of building a telemedicine system with all its requirements and its operational cost is high. Hence, proper resources are not enabled which leads to poor QoS.

Figure 1 shows the system overview of cloud-assisted p2p live streaming where the video server transfers the streams to both the cloud storage and to the peers. The video chunks are disseminated through the push and pull mechanisms in the p2p environment [4]. The peer pulls the chunks from its neighboring peer and uploads the same to the peers requesting for the video chunk. The size of the swarm is calculated and the total capacity of the overlay is calculated. The computational node (Cn) in the

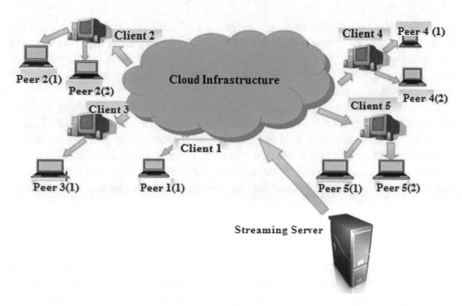

Fig. 1 Cloud-assisted P2P live streaming overview

cloud receives the chunks and pushes them to other Cns and/or peers in the swarm. Based upon the request, computational nodes (Cns) are requested from the cloud to participate in the streaming activity, thereby providing better Quality of Service.

The paper is organized as follows: Sect. 2 gives a brief literature survey on cloud-assisted p2p live streaming. Section 3 describes the overall system architecture and the entities associated with it. Section 4 explains the Proposed Dynamic procurement of Virtual machine algorithm (DPVm) Sect. 5 briefs the Implementation details and Sect. 6 illustrates the obtained experimental results and its discussions.

2 Related Works

This literature survey gives an insight to the existing techniques in cloud-assisted p2p live for streaming. The advantages and disadvantages of each mechanism have been analyzed.

2.1 Cloud-Based Live Streaming

CAC-Live, a P2P live video streaming system that uses the cloud for compensating the lack of resources with regard to cost [5]. The system is based on centralized algorithms for monitoring QoS discerned by the peers and virtual machines are managed on dynamic basis which are added or removed to/from the network. For every iteration, each peer periodically sends information about its experience in QoS from which the system decides to re-structure its overlay network by adding or removing virtual machines.

CloudMedia an on-demand cloud resource provisioning methodology to meet dynamic and intensive resource demands of VoD over the internet [6]. A novel queuing network model has been designed to characterize users viewing behaviors and two optimization problems were identified related to VM provisioning and storage rental. A dynamic cloud provisioning algorithm was proposed by which a VoD provider can effectively configure the cloud services to meet its demands. The results proved that the proposed CloudMedia efficiently handled time-varying demands and guaranteed smooth playback continuity. It was observed that the combination of cloud and the P2P technology achieved high scalability, especially for streaming applications.

Cloud-based P2P Live Video Streaming Platform (CloudPP) introduced the concept of using public cloud such as Amazon EC2 to construct an efficient and scalable video delivery platform with SVC technology [7]. It addressed the problem of satisfying video streaming requests with least possible number of cloud servers. A multi-tree P2P structure with SVC was proposed to organize and manage the cloud servers. The proposed structure reduced around 50% of the total number of cloud servers used in the previously proposed methodologies and showed 90% improvement of the total benefit cost rate.

DragonFly a hybrid Cloud-P2P architecture which was specifically devised to support, manage, and maintain different kinds of large-scale multipoint streaming applications. The DragonFly architecture used two-tier edge cloud with the user level P2P overlay to ease the streaming activity. The proposed architecture maintains the structuring of P2P overlay which distinguishes Dragonfly from the other P2P-based media streaming solutions. DragonFly enables scalability, resource-sharing, and fairness of the system. Dragonfly constructs an edge cloud to handle location-based challenges to enhance content dissemination. The source-specific application-level multicasting at the local level and the latency-aware geographic routing approach results in lower latency [9].

3   System Overview

The overall proposed architecture and the modules associated with this architecture is the Media server which acts as the source, Swarm of peers, Overlay Constructor, Data Disseminator, Starving Peer Estimator, Capacity Estimator, Cloud Server, and a Resource Allocator. The Resource allocator manages the allocation/de-allocation of computational nodes based on the proposed DPVm algorithm (Fig. 2).

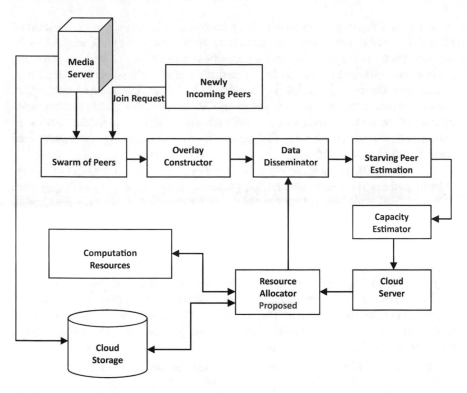

Fig. 2 Proposed cloud-based virtual telemedicine system

4 Proposed Work

The contributions made in this paper is the proposed DPVm to dynamically allocate/remove the computational node to improve the Qos considering the parameters such as Playback Continuity, Playback Latency, and End-to-End delay with Flash crowd and without flash crowd scenario. Another contribution is to estimate the number of computational nodes to be added to build a cost-effective system. The system has used one storage node from the cloud since increase in the number of storage node increases the cost, hence computational nodes can be increased alternatively thereby reducing the cost of renting resources from the cloud.

4.1 Proposed Dynamic Procurement of Virtual Machine (DPVm) Algorithm

The proposed algorithm deals with the mentioned scenario, and periodically the peers keeps updating the status of their buffer information and Qos to the capacity estimator from which a threshold is determined for both adding/removing of computational nodes. The buffer information is updated to notify the lost chunk which can be adopted from the storage node and a peer updates the status of other neighboring peers too to determine the threshold. A threshold cannot be fixed because peers may join or leave the overlay anytime and threshold keeps changing based on the number of peers active in the overlay. The proposed DPVm algorithm that is written based on the QoS of the peers and two threshold values are decided one for the addition of computational node Threshold$_{add}$ (TH$_{add}$) and another for removing of computational node Threshold$_{remove}$ (TH$_{remove}$).

The ratio of the peers receiving Low Qos (N$_{LQoS}$) to the total number of peers is estimated and if the estimated ratio is greater than the Threshold$_{add}$ (TH$_{add}$), then computational nodes are rented from the cloud and added to improve the Quality of service. Suppose the estimated ratio of the Low Qos (N$_{LQoS}$) peers is lesser than the Threshold$_{add}$ (TH$_{add}$), then computational nodes can be removed from the peers. Next, the ratio of the peers with High Qos (N$_{HQoS}$) to the total number of peers is estimated and if the estimated ratio is lesser than the Threshold$_{add}$ (TH$_{add}$), then to improve the Quality of streaming, the computational node are added to improve the upload bandwidth, likewise the ratio of peers with High QoS (N$_{HQoS}$) is estimated and if found to be higher than the Threshold$_{remove}$ (TH$_{remove}$), then the computational node is removed from the overlay. The algorithm for managing the computational nodes from the cloud is given below.

Algorithm for adding or removing virtual machine based on QoS

```
THadd // threshold for adding virtual machine
THremove // threshold for removing virtual machine
N // total number of peers in the swarm
NHQoS // number of peers having high QoS in the swarm
NLQoS // number of peers having low QoS in the swarm

NL = (N - NHQoS) / N //number of peers having low QoS
NH = (N - NLQoS) / N //number of peers having high QoS
If (NHQoS is set)
{
While (until session continues) do
If (NL ≥ THadd)
        Add a virtual machine into the swarm
Else if (NL < THremove)
        Remove a virtual machine from the swarm
Else
        Don't add or remove a virtual machine
End while
}
Else
{
While (until session continues) do
If (NH ≤ 1- THadd)
        Add a virtual machine into the swarm
Else if (NH > THremove)
        Remove a virtual machine from the swarm
Else
        Don't add or remove a virtual machine
End while
}
```

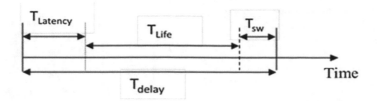

Fig. 3 Time model of a chunk

4.1.1 Estimation of Starving Peers

Starving peers are those peers that request for chunks but do not receive them within the specified time delay due to loss of packets. Increasing the upload bandwidth in the network by adding computational nodes will no way help in recovering the lost chunks, hence the lost chunks from the storage node (Sn). Initially, the video server sends the copy of the chunks to the cloud storage and another copy to the swarm of peers. Hence, when the situation arises among peers starving for chunks after its acceptable playback delay, then that particular chunk can be directly fetched from the cloud storage node (Sn). The Qos with respect to lifetime of a chunk is estimated below (Fig. 3).

4.1.2 List of Symbols

T_{delay} Maximum acceptable latency

T_{sw} If a chunk is not available in the buffer before its playback time it can be retrieved from the storage node

T_{Latency} The maximum time for a newly generated block to reach the root peers

T_{Life} Lifetime of a chunk

Cn   Computational node

Sn   Storage node

C_{vm} Cost of running a virtual machine

C_{b} Cost of transferring one block from a Sn to a peer

n   No. of blocks uploaded

C_{storage} storage cost

C_{r} Cost of retrieving a block from Sn

r   No. of blocks retrieved from Sn

λ No. of peers that is economically reasonable to serve from Sn

T_{Life} is computed as.

Fig. 4 Variation of cost with
respect to peers

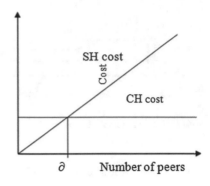

4.1.3 Impact of T_{sw} on the QoS

T_{sw} is a system parameter that has a vital impact on the quality of the media files
that is received by the end user, as well as on the total cost. Finding an appropriate
value for T_{sw} is one of the challenges that have to be concentrated. With a too small
T_{sw} peer may fail to fetch blocks from Sn in time for playback, while a too large T_{sw}
increases the number requests to Sn, thus, increases the cost. Therefore, the question
is how to choose a value for T_{sw} to achieve (i) the best QoS with minimum cost.

Each peer buffers a number of blocks ahead of its playback time, to guarantee a
given level of QoS. The number of buffered blocks corresponds to a time interval of
length T_{sw}. The length of T_{sw} should be chosen big enough, such that if a block is
not received through other peers, there is enough time to send a request to Sn and
retrieve the missing block from it in time for playback.

As stated earlier, as the number of storage node increases the cost increases,
hence based on the swarm size and the available upload bandwidth, the number of
computational node (Cn) has to be increased to minimize the economic cost. The
cost of Computational node and the cost of retrieving a chunk from the storage node
are computed below (Fig. 4).

The cost C_{cn} of one Cn (computational helper) in one round is

$$C_{cn} = C_{vm} + n \cdot C_b$$

The cost C_{sn} of pulling blocks from Sn (storage helper) per round is

$$C_{sn} = C_{storage} + r \cdot (C_b + C_r)$$

Figure 5 shows variation of cost with respect to peers

$$\lambda = C_{cn}/C_{sn}$$

Fig. 5 Average end-to-end delay using the proposed DPVm algorithm

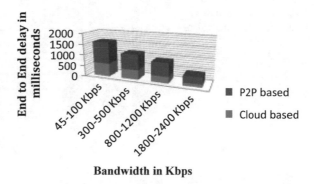

If load > λ: add Cn
If load < λ − H: remove Cn
Otherwise, do not change Cns.

5 Experimental Setup

The performance had been evaluated in both Real-time and Simulation considering QoS parameters such as Playback continuity, Playback latency, and End-to-end delay. In real-time P2P streaming, set up was integrated with the private cloud. Simulation was done using Kompics Framework. The experiment was carried out in real-time using a private cloud set up established in MaaS Laboratory at the Department of computer technology, MIT Campus. Simulation of the experiment was carried out in order to test scalability using KOMPICS framework, a test bed for performing analysis of P2P streaming. The private cloud setup was done using Openstack with the help of Ubuntu Cloud Infrastructure. The streaming rate of the video was set to 512 kbps and the chunk size is set equal to a frame.

6 Performance Evaluation

The End-to-End Delay has been considered as one of the most important parameters in live streaming. The experiment is carried out with the size of the video (1280 × 720 HD) kept constant and streamed across peers with varying bandwidth availability. Figure 5 shows the average End-to-End delay for P2P-based streaming is found to be 0.9 s and cloud-based streaming is found to be 0.3 s which proved 60% improvement in streaming live events.

The next experiment is conducted for varying the size of video as 1280 × 720, 720 × 486 and 352 × 258. The end-to-end delay is found out for varying bandwidth

Fig. 6 End to end by
varying size of the videos

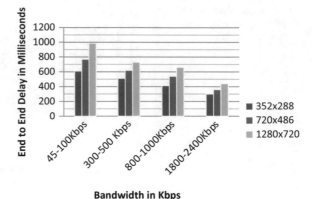

Bandwidth in Kbps

Fig. 7 Percentage of
playback continuity on
receiving the chunks on time
(flash crowd—join and
leave)

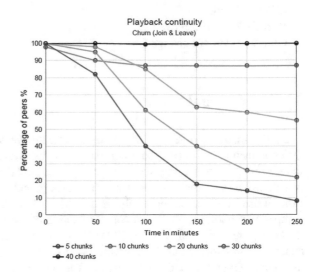

and varying video sizes. Figure 6 shows the graph plotted for varying video sizes
and varying bandwidth availability of peers.

Figure 7 illustrates churn scenario (Join and Leave) where, 100 peers enter and
leave the system with an inter arrival time of 10 ms which approximately has 0.1
and 1% of the peers leave the system per second and join back as newly joining
peers, however, the experiments were conducted with 1% churn rate to prove that
the system works in high dynamic environment with an average of 88% of playback
continuity.

7 Conclusion

The main contribution of this paper is Dynamic Procurement of Virtual machine (DPVm) algorithm which calculates the capacity of the existing p2p network and identifies the number of peers with High Qos and Low Qos to allocate a sufficient number of computational nodes to improve the upload bandwidth of the system. Another contribution is the estimation of the starving peers and how additional computational nodes can be added to minimize the cost of the chunks retrieved from the storage node. Resulting values were obtained from both Real-time and simulation environment which shows only slight variations. Flash crowd scenario was implemented and the Quality of service was maintained without any interruptions which prove that the proposed system shows 98% of playback continuity and 99% of the chunks are delivered on time and even if it fails, the chunks are retrieved from the Storage node with a maximum delay of 0.3 s.

References

1. Zhenhui, Y., Gheorghita, G., Gabriel-Miro, M.: Beyond multimedia adaptation: quality of experience-aware multi-sensorial media delivery. IEEE Trans. Multimedia **17**(1), 104–117 (2015)
2. Yuanyi, X., Beril, E., Yao, W.: A novel no-reference video quality metric for evaluating temporal jerkiness due to frame freezing. IEEE Trans. Multimedia **17**(1), 134–139 (2015)
3. Chun-Yuan, C., Cheng-Fu, C., Kwang-Cheng, C.: Content-priority-aware chunk scheduling over swarm-based p2p live streaming system: from theoretical analysis to practical design. IEEE J. Emerg. Sel. Top. Circuits Syst. **4**(1), 57–69 (2014)
4. Korpeoglu, E., Sachin, C., Agarwal, D., Abbadi, A., Hosomi, T., Seo, Y.: Dragonfly: cloud assisted peer-to-peer architecture for multipoint media streaming applications. In: Proceedings of the IEEE Sixth International Conference on Cloud Computing, pp. 269–276 (2013)
5. Mehdi, S., Seyfabad, B., Akbari, D.: CAC-Live: centralized assisted cloud p2p live streaming. In: Proceedings of the Twenty Second Iranian Conference on Electrical Engineering, pp. 908–913 (2014)
6. Nazanin, M., Reza, R., Ivica, R., Volker, H., Markus, H.: ISP-friendly live p2p streaming. IEEE/ACM Trans. Networking **22**(1), 244–256 (2014)
7. Yu, W., Chuan, W., Bo, L., Xuanjia, Q., Lau, C.M.: CloudMedia: when cloud on demand meets video on demand. In: Proceedings of the Thirtieth International Conference on Distributed Computing Systems, pp. 268–277 (2011)
8. Simone, C., Luca, D., Marco, P., Luca, V.: Performance evaluation of a sip-based constrained peer-to-peer overlay. In: Proceedings of the International Conference on High Performance Computing & Simulation, pp. 432–435 (2014)
9. Takayuki, H., Yusuke, H., Hideki, T., Koso, M.: Churn resilience of p2p system suitable for private live-streaming distribution. In: Proceedings of the Sixth International Conference on P2P, Parallel, Grid, Cloud and Internet Computing, pp. 495–500 (2014)

Multi-level Iterative Interdependency Clustering of Diabetic Data Set for Efficient Disease Prediction

B. V. Baiju and K. Rameshkumar

Abstract Clustering with diabetic data has been approached using several methods, though it suffers to achieve the required accuracy. To overcome the issue of poor clustering, multi-level iterative interdependency clustering algorithm has been presented here. The method generates initial cluster with random samples of the known classes and computes interdependency measure on different dimensions of the data point, and will be computed on the entire cluster samples for each data point identified. Then, the class with higher interdependency measure has been selected as the target class. This will be iterated for several times, until there is a movement of point. The number of classes is around the number of diseases considered and for each subspace, the interdependency measure has been estimated to identify the exact subspace of the data point. The method computes the multi-level disease dependency measure (MLDDM) on each disease class and their subspace, for prediction. A single disease class can be identified and their probability can be estimated according to the MLDDM measure. This method produces higher results in clustering and disease prediction.

Keywords Big data · High dimensional clustering · Interdependency measure
MLDDM · Subspace clustering

1 Introduction

The modern information technology has no restriction for their size or dimension of data and can be represented in various forms. But, it has limit for the type of data being used. Big data contains heterogeneous data types in the same data points, unlike homogenous data points. Big data is being used to represent modern data

B. V. Baiju (✉) · K. Rameshkumar
Hindustan Institute of Technology and Science, Chennai, India
e-mail: bvbaiju@hindustanuniv.ac.in

K. Rameshkumar
e-mail: krkumar@hindustanuniv.ac.in

© Springer Nature Singapore Pte Ltd. 2019
J. D. Peter et al. (eds.), *Advances in Big Data and Cloud Computing*,
Advances in Intelligent Systems and Computing 750,
https://doi.org/10.1007/978-981-13-1882-5_7

points and clustering such big data has various challenges. In earlier days, the data has been formed by combining numerical and alphanumeric values. Data mining has been used to uncover hidden patterns and relations to summarize the data in ways to be useful and understandable in all types of businesses to make a prediction for future perspective. Medical data is considered the most famous application to mine that data.

In today's scenario, the data has been formed with the help of images and so on. Also, the big data would contain a lot of varying form of data points, and it is not necessary that all the data points should have the same set of features. As the dimension of the data points increases without any limit, the high dimensional clustering has come to play. The high dimensional clustering should consider more number of dimensions in estimating their similarity between them. The complexity of clustering is directly proportional to the dimension of data points. While performing high dimensional clustering, the ratio of false classification and false clustering increases due to the sparse dimensions.

The data points of the class are compared with the input sample and based on the similarity between them, the clustering has been performed successfully. There will be number of subspace or subclass by which the data points can be grouped. For example, consider there exist a class c, which denotes the data points of the patients affected by fever, where, the fever can be further classified into Typhoid, Malaria, Dengue, and so on as subclasses. Hence, in-order-to produce an efficient cluster, it is more essential to identify the exact space of the data points. There are number of clustering algorithms presented to identify the cluster of the data points of big data. Most of the algorithms consider certain dimensions to compute similarity on the selective dimensions. This introduces us to the higher false classification or false indexing ratio.

To improve the performance of clustering, it is necessary to consider the maximum dimensions of the data point in computing the similarity measure. The previous methods do not compute the similarity measure for the input data point in an iterative manner. Due to this, it indexes the data point in another subspace and increases the false classification or indexing ratio. The interdependency measure represent the similarity of data points between the various class data points. By computing interdependency measure, the similarity of the data points can be measured in efficient manner, thereby producing efficient clusters.

The modern society suffers with various life-threatening diseases. Diabetic is the most common disease in humans. It has no age constraint as it happens to any age group people and it can encourage any other disease also. By using this diabetic data, the future diseases can also be identified. This will help by predicting the disease at an earlier stage. The disease prediction is the process of predicting the possible disease with the minimum information. To perform disease prediction, various measures have been used in earlier days, which include the fuzzy rules. The fuzzy rule approach performs disease prediction based on the values and by counting the number of values which falls within the range.

This method computes the similarity and predicts the possible diseases. But, in most cases, computing the range value is not essential, as the values would vary between different patients. To improve the performance of disease prediction, the multi-level disease dependency measure can be used. The multi-level disease dependency measure (MLDDM) could be computed by estimating the similarity of data points of each level in all the dimensions. It is not necessary that all the dimensions should get matched. By computing the MLDDM measure, the exact disease would be probably identified for the patient.

2 Related Works

There are number of methods which have been discussed earlier for big data clustering and disease prediction. This section discusses about various approaches towards the problem of disease prediction and big data clustering.

Improved K-means Clustering Algorithm for Prediction Analysis using Classification Technique in Data Mining [1], in this K-Means clustering vanishes off the two major drawbacks present in the algorithm. Accuracy level and calculation time consumed in clustering the data set are the two major drawbacks. The accuracy and calculation time will not matter much when the small data sets are used, but when a larger number of data sets are used, it may contain trillions of records, then little dispersion in accuracy level will matter a lot. This may lead to a disastrous situation, if not handled properly.

Prediction of Diseases Using Hadoop in Big Data—A Modified Approach [2], is made more effective by making them to converge using extrapolation technique. Moreover, MapReduce framework techniques are designed to handle larger data sets and the main objective of the proposed algorithm is to predict the diseases more accurately by doctors.

Myocardial Infarction Prediction Using K-Means Clustering Algorithm [3], is used to develop a myocardial infarction prediction using K-Means clustering technique which can discover and extract hidden information from historical heart disease data sets. Feature Selection in data mining refers to an art of minimizing the number of input attributes under evaluation. Grouping of similar data objects in the same cluster and dissimilar objects into another cluster is known as clustering. In several applications, clustering is also known as data segmentation because according to the similarities it divides large data set into groups.

Hybrid Approach for Heart Disease Detection Using Clustering and ANN [4], Heart disease prediction is the most difficult task in the field of medical sciences. Data mining can answer complicated queries for diagnosing heart disease and thus assist healthcare practitioners to make intelligent clinical decisions which the traditional decision support systems cannot. It also helps to reduce treatment costs by providing effective treatments. The major goal of this study is to develop an artificial neural networks-based diagnostic model using a complex of traditional and genetic factors for heart disease.

Intelligent Heart Disease Prediction System with MONGODB [5], focuses on the aspect of the data which is not mined. To take preventive measures to avoid the chances of heart disease, 14 attributes are used to predict the heart disease. The system is expandable, reliable, web–based, and user-friendly. It also serves a purpose of training nurses and doctors newly introduced in the field related to heart disease.

Heart Disease Prediction System using ANOVA, PCA, and SVM Classification [6], are used to develop an efficient heart disease prediction system using feature extraction and SVM classifier. This can be effectively used to predict the occurrence of heart diseases. Physicians and healthcare professionals are wisely using this heart disease prediction system as a tool for heart disease diagnosis. PCA with SVM classification technique can act as a quick and efficient prediction technique to protect the life of a patient from heart diseases. This technique is widely used to validate the accuracy of medical data. By providing the effective treatments, it also helps to reduce the treatment costs.

Prediction of Heart Disease Using Machine Learning Algorithms [7], provides a detailed description of Naïve Bayes and decision tree classifier. These are applied in our research particularly for the prediction of Heart Disease. To compare the execution of predictive data mining technique on the same data set few experiments has been conducted, and the consequence of experiments reveals that Decision Tree outperforms over Bayesian classification.

Prediction of Heart Disease Using Decision Tree Approach [8], using a data mining technique, decision tree is used as an attempt to assist in the diagnosis of the disease. Using classification techniques, a supervised machine learning algorithm has been used (Decision Tree) to predict heart disease. It has been shown that by using a decision tree, it is possible to predict heart disease vulnerability in diabetic patients with reasonable accuracy. Classifiers of this kind are used to help in early detection of the vulnerability of a diabetic patient to heart disease.

An Intelligent Heart Disease Prediction System Using K-Means Clustering and Naïve Bayes Algorithm [9], implementing a heart disease prediction system is a combination of both K-Means clustering and Naïve Bayes Algorithm. It signifies by predicting the output as in the prediction form and helps in predicting the heart disease using various attributes. K-Means algorithm is used for grouping of various attributes and Naïve Bayes algorithm is used for predicting.

Analyzing Healthcare Big Data with Prediction for Future Health Condition [10], a probabilistic data collection mechanism is designed and the correlation analysis of those collected data is performed. A stochastic prediction model is designed to foresee the future health condition based on the current health status of the patients. The performance evaluation of the proposed protocols is realized through extensive simulations in the cloud environment.

All the abovementioned methods suffer to achieve higher prediction accuracy because of considering only limited number of features and produces higher false prediction ratio.

3 Preliminaries

In this section, this paper describes the algorithms and techniques.

3.1 Multi-level Iterative Interdependency Clustering-Based Disease Prediction

The algorithm produces initial random clustering with the known class and indexes random samples to each class. Further, for each data point given, the method estimates the interdependency measure towards each class and each subspace to identify the target class. To predict the possible diseases, the method computes the multi-level disease dependency measure towards each disease class and its subspace. Finally, a single disease class has been identified as the possible disease. The detailed method is discussed in this section.

Figure 1 shows the architecture of multi-level disease dependency measure-based disease prediction and shows various steps of the proposed disease prediction algorithm.

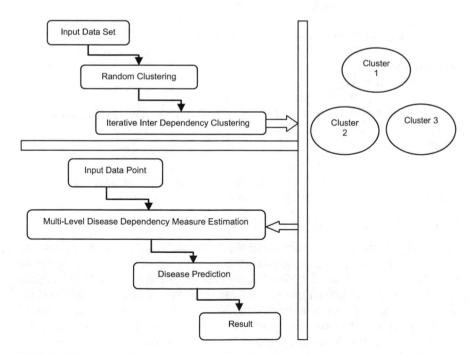

Fig. 1 Architecture of MLDDM disease prediction

Random sample clustering

The random sample clustering algorithm is produced with the known class. The method generates initial clusters and for each cluster, a random samples are added. The data set given is first preprocessed to identify the noisy records. The dimensions are used as the key to identify the noisy data. The data points are verified for the possession of all the dimensions and values. If any of the data points is identified as incomplete and with missing values, then it has been considered as noisy one and will be removed from the data set. Generated sample clusters are used to perform original clustering in the next stage.

Algorithm

Input: Diabetic Data set DDs
Output: Initial Cluster Ic
Start
    Read data set DDS.
        Initialize cluster set Cs.
            Identify all dimensions present.
Dimension Set DimSet $= \sum_{i=1}^{size(DDS)}$(Dimset $\cup$ ($\forall$(Dimension(DDS(i))$\nexists$Dimset))
For each data point Di
        Verify for the presence of all dimensions.
                If Di$\nexists\forall$(Dimensions(Dimset))
    Then
            DDS = DDS$\cap$ Di //Remove the data point.
                End   End
For each cluster Ci
        For each subspace C_{si}
            Add random samples
        Ci(CSi) $= \sum$ Datapoint$\big($Ci(Csi)$\big)$ $\cup$ RandomSample(DDs(i). class $=$ Ci)
                End   End
    Stop

The random sample clustering algorithm performs noise removal on the input data set given and generates random sample clusters.

Iterative Interdependency Clustering

The iterative interdependency clustering algorithm reads the cluster set and data set. For each of the data point, the method computes the interdependency measure on each class data points. The interdependency measure is estimated on each subspace also. The same is estimated towards various class data points also. Based on the interdependency measure, a target class is identified. The data point has been added to the target class. Whenever a data point is added to a subspace or class, then the same will be estimated for all the data points of the class and other class data points. This will be performed for number of times till there is no movement of points between the clusters.

Algorithm

Input: Cluster set Cs, Data set DDs.

Output: Cluster Set Cs, Read data set DDs, Read cluster set Cs.

    For each data point d_i

        For each class C_i

            For each data point Td_i

                For each dimension dim

Estimate inter dependency measure IDm.

$$IDM = \frac{\sum_{i=1}^{Number\ of\ Dimension} Dist\big(Tdi(dim), di(dim)\big) < Std\big(\sum Tdi(Ci)\big)}{No\ of\ dimensions}$$

If IDM>Th then

        Count +1.

    End   End   End

Compute cumulative inter dependency measure CIDM.

$$CIDM = \frac{\sum IDM}{size(C_i)} \times \frac{Count}{size(Ci)}$$

End

Choose the class with maximum CIDM.

    Assign di to selected class.

        For each cluster Ci

            For each data point di

                For each other data point of same class

                Compute inter dependency measure

$$ICIDM = \frac{\sum IDM}{size(Ci)} \times \frac{Count}{size(Ci)}$$

    End

    For each other class OCi

        Compute cumulative inter dependency measure

$$OCIDM = \frac{\sum IDM}{size(Ci)} \times \frac{Count}{size(Ci)}.$$

    End

        If ICIDM<OCIDM then

            Move data point to target class.

        End   End

        Continue till there is no movement.

    End

Stop

The interdependency clustering algorithm computes the interdependency measure on various cluster and dimensions. Based on the measure estimated, the method performs clustering in an iterative manner.

MLDDM Disease Prediction

The disease prediction algorithm first cluster the data set according to the random clustering and interdependency clustering algorithm. Then for the given input sample, the method verifies the presence of all dimensional values. If it succeeds the test, then the method computes the multi-level disease dependency measure with all the class and their subspace. The dependency measure is estimated on each dimension and based on the measure, a single disease has been selected as the possible prone.

Algorithm

Input: Cluster C_s, Data point D, Dimension Set Dimset
Output: Disease Predicted Dp.
Start
        Read cluster set Cs.
        Read data point D
            Verify the data point.
                For each dimension Dim
        If $D \in \forall \big(\text{Dimensions(Dimset)}\big)$ && $D(\text{Dim}).\text{value}! = \text{null}$ Then
        Test succeed.
        Else
            Noisy data and stop.
     End   End
For each disease class C(Cs)
        For each subspace s
            For each dimension di
Compute dependency factor

$$Df = \frac{\sum_{i=1}^{size(s)} s(di).value}{size(s)}$$

Compute multi-level disease dependency measure MLDDM.

$$MLDDM = \frac{\sum_{i=1}^{size(s)} Dist(s(di).value, D(di).value)}{Df}$$

End
Compute cumulative MLDDM.

$$CMLDDM = \frac{\sum MLDDM}{no\ of\ dimensions}$$

            End   End
            Choose the disease class with maximum CMLDDM.
Stop

The disease prediction algorithm reads the input data point and verifies for its completeness. Then, the method computes the multi-level disease dependency measure on each class with each dimension. Then, the cumulative disease dependency measure is estimated. Based on the measure estimated, a single disease class has been selected as the prediction result.

4 Results and Discussion

The multi-level interdependency based disease prediction has been implemented and evaluated for its efficiency. The evaluation of the method has been performed using diabetic data set. The method has produced efficient results in disease prediction and compared with various methods.

The details of data set being used for the evaluation of the proposed approach are shown in Table 1.

Figure 2 shows the comparative result on disease prediction accuracy produced by various methods on different number of data points. In all the case, the proposed MLDDM approach has produced higher prediction accuracy than any other method.

Figure 3 shows the false classification ratio comparisons that are generated from different methods by varying number of data points. The method has produced less false rate than other methods.

Table 1 Data set details

Parameter	Value
Data set used	UCI
No. of tuples	1 million
No. of dimensions	20
No. of diseases	10

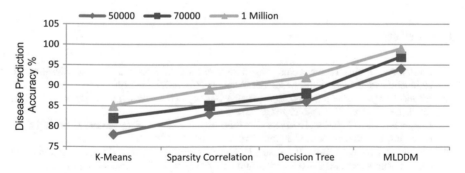

Fig. 2 Comparisons on disease prediction accuracy

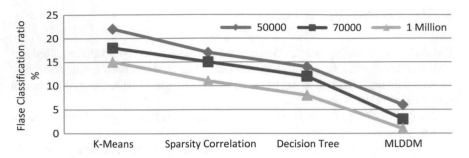

Fig. 3 False classification ratio comparisons

5 Conclusion

In this paper, an efficient multi-level interdependency measure-based disease prediction algorithm is presented. First, the method generates random sample clusters and computes interdependency measure to perform iterative clustering. Then with the input data point, the method computes multi-level disease dependency measure for each class of diseases. The method selects a disease class based on the estimated measure. UCI diabetic data set is used to evaluate the efficiency of this method. The method produces higher prediction ratio up to 99.1% and produces less false rate up to 1.2.

References

1. Bansal, A.: Improved k-mean clustering algorithm for prediction analysis using classification technique in data mining. Int. J. Comput. Appl. **157**(6), 0975–8887 (2017)
2. Jayalatchumy, D.: Prediction of diseases using Hadoop in big data–a modified approach. Artif. Intell. Trends Intell. Syst. pp. 229–238 (2017)
3. Umamaheswari, M., Isakki, P.: Myocardial infarction prediction using k-means clustering algorithm. Int. J. Innovative Res. Comput. Commun. Eng. **5**(1) (2017)
4. Chikshe, N., Dixit, T., Gore, R., Akade, P.: Hybrid approach for heart disease detection using clustering and ANN. Int. J. Recent Innovation Trends Comput. Commun. **4**(1), 119–122 (2016)
5. Jarad, A., Katkar, R., Shaikh, R.A., Salve, A.: Intelligent heart disease prediction system with MONGODB. Int. J. Emerg. Trends Technol. Comput. Sci. **4**(1), 236–239 (2015)
6. Kaur, K., Singh, M.L.: Heart disease prediction system using ANOVA, PCA and SVM classification. Int. J. Adv. Res. Ideas Innovations Technol. **2**(3), 1–6 (2016)
7. Prerana, T.H.M., Shivaprakash, N.C., Swetha, N.: Prediction of heart disease using machine learning algorithms-Naïve Bayes, introduction to PAC algorithm, comparison of algorithms and HDPS. Int. J. Sci. Eng. **3**(2), 90–99 (2015)
8. Reddy, K.V.R., Raju, P.K., Kumar, J.M., Sujatha, C.H., Prakash, R.P.: Prediction of heart disease using decision tree approach. Int. J. Adv. Res. Comput. Sci. Softw. Eng. **6**(3), 530–532 (2016)

9. Shinde, R., Arjun, S., Patil, P., Waghmare, J.: An intelligent heart disease prediction system using k-means clustering and Naïve Bayes algorithm. Int. J. Comput. Sci. Inf. Technol. **6**(1), 637–639 (2015)
10. Sahoo, P.K.: Analyzing healthcare big data with prediction for future health condition. IEEE Access, vol. 4 (2016)

Computational Offloading Paradigms in Mobile Cloud Computing Issues and Challenges

Pravneet Kaur and Gagandeep

Abstract Mobile Cloud Computing (MCC) is an excellent communication offspring obtained by blending virtues of both Mobile computing and Cloud Computing Internet technologies. Mobile Cloud Computing has found its advantages in technical and communication market in innumerable ways. Major radical advantages of Mobile Cloud Computing are the computation offloading, enhancement of the smartphone application by utilizing the computational power of resource-rich cloud and enabling the smart mobile phone (SMP) to execute resource-intensive applications. In this survey paper, a SWOT (Strengths, Weakness, Opportunities and Threats) analysis of different Computation offloading techniques, is made. Computation Offloading uses the technique to migrate heavy computational resource applications from smartphone device to the cloud. In particular, this paper laid emphasis on the similarities and differences of computation offloading algorithms and models. Moreover, some important issues in offloading mechanism are also addressed in detail. All these will provide a glimpse of how the communication between the Mobile Device and the Cloud takes place flawlessly and efficiently.

Keywords Mobile cloud computing · Smart mobile phones
Computation offloading

1 Introduction

With the ever going usage of Internet, there is a growing trend of disseminating and exchanging information via the Mobile Cloud Computing paradigm. It becomes cumbersome to transfer huge amount of data from one place to a remote, distant place electronically, i.e. through use of Internet solely. Many shortcomings come

P. Kaur (✉) · Gagandeep
Department of Computer Science, Punjabi University, Patiala 147002, Punjab, India
e-mail: Sidhu.pravneet@gmail.com

Gagandeep
e-mail: gdeep.pbi@gmail.com

© Springer Nature Singapore Pte Ltd. 2019
J. D. Peter et al. (eds.), *Advances in Big Data and Cloud Computing*,
Advances in Intelligent Systems and Computing 750,
https://doi.org/10.1007/978-981-13-1882-5_8

in way, namely the space needed to store huge amounts of data, the energy consumption, the total computation time taken by the request and the computation cost incurred. Cloud computing basically consisted of the IaaS, PaaS and SaaS vocabulary which stated the hardware and physical devices, the software environment and the software application, respectively. The deficiencies in mobile phones were limited computational power, limited batteries to be operated upon, limited space and the energy challenges. Henceforth, Mobile computing along with Cloud computing, both worked symbiotically to solve any further upcoming issues. Mobile Cloud Computing included the usage of cloud's immense resources as execution infrastructure for resource-intensive applications. The application outsourcing to the cloud enormously helped in reducing the overhead in terms of efficiency and time. In MCC, the massive huge intensive tasks of mobile devices are offloaded to cloud resources. This aids in reducing the energy, time spent and mainly the cost incurred by the transfer and computation of huge data. Now, the main focus in MCC was that which operations will be transferred to the cloud and which operations will be operated on the mobile device. This compartmentalization of applications is termed as Computation Offloading in MCC. In other words, Computation Offloading is migrating all the process or a part of the process to the Cloud for execution. Mobile Cloud Computing uses different application development models that support computation offloading. Some of the important cloud application models are [1] CloneCloud, Mahadev Satyanarayanan and colleague's model, [2] MAUI, ThinkAir and eXcloud. These models migrates heavy applications to the cloud through a process module and application that is in the cloud and helps to execute mobile phone's requests. Generally, there are two environments where the applications can be held—the smartphone and the cloud. Hence, Mobile computing assisted with cloud computing gives us the benefit to migrate some of the applications to the cloud. In MCC, during data transfer, storage of data and its processing is done at the cloud end. To deploy this advantage, the consumer and enterprise markets are increasingly adopting this mobile cloud approach to provide best services to the customers and end users. It will boost their profits by reducing the development cost incurred in developing mobile applications. We have taken into accordance the infrastructure-based architecture of MCC [3]. In this architecture, the computer infrastructure do not change positions and provides services to mobile users, via Wi-Fi or 3G.

1.1 Overview of Computation Offloading

Researchers and entrepreneurs had stated that the MCC concept had brought a technical revolution in IT industry. As it is not so costly, the user's are more benefited. Various advantages of MCC includes the resources, storage and application are always available on demand, pay as you use, scalability [3], dynamic provisioning and computation offloading. The battery lifetime is prolonged as the heavy computations are executed in the cloud. The cloud is a place where higher data storage and processing power capabilities are provided. While choosing which applications have to be migrated to the cloud and which applications are to be run locally, the application developer has to be aware of all its pros and cons. Different applications models are to be used efficiently. In other words, different application models have to be context aware because computation offloading do not always gives the best results as a trade-off [4]. The performance degradation or energy wastage may happen in certain cases. So, the model has to be wisely chosen to ensure that both these parameters add to the efficiency. Such models are termed as User-Aware application models. In MCC, the applications are divided in a user-aware manner with a consideration to different parameters like user interaction frequency, resource requirement, intensity of computation and bandwidth consumption. Context-Aware Application models are generally categorized on the basis of two perspectives: User-Aware Application Partitioning and User-Aware Computation Offloading. User-Aware Application Partitioning consists of transferring heavy applications to the cloud leading to efficient execution. The applications are compartmentalized into smaller modules. The two demarcations are static and dynamic. In static partitioning, the applications are transferred to the cloud during development. In dynamic partitioning, the applications are migrated to the cloud during runtime.

User-Aware Computation Offloading depends on the user requirements and may not be always beneficial. For instance, two applications were developed. The first application was used to find out the factorial of five-digit number. The second application was used to ethically hack seven-character password. All the two applications were executed on Samsung Quattro (Mobile) using Wi-Fi along with OpenStack (Cloud). It is a cloud operating system that controls a large pool of storage, network resources through a data center. When the first application was executed, it required more energy than when it was executed on mobile. Henceforth, offloading was not useful in this scenario. When the second application was run, the computation time and energy consumed were less than when it was executed on the mobile phone. Using above instances, it was clearly observed that the offloading decisions must be context aware in terms of objective awareness, performance awareness, energy awareness and resource awareness.

User Awareness Types: There are different types of User Awareness, namely user awareness, efficiency awareness, energy awareness and resource awareness.

User Context consists of objective context which depends upon the user's perspective and requirements. As there is a trade-off between computation performance, energy efficiency [5] and execution support, all the three parameters cannot be satisfied by using a single computational model and hence multiple computational models are used.

Performance Context uses profilers to make offloading decisions. The profilers monitor the computation time of the cloud application and that of mobile application (along with the computation offload time) and make decisions by keeping the time constraint as the main parameter.

Energy Context specifies that computation offloading is favourable when the energy required by transferring the application to the cloud is less than the local execution. During transferring the application to the cloud energy, there is certain overhead which include the energy to transfer the application, energy to execute and energy to integrate the result and sending it back to the mobile device.

Resource Context specifies that computation offloading is done when there is scarcity of resources at the smartphone end. It is very important to be aware of the resources available both at the mobile end and the cloud end for best results.

So, the above parameters are necessary for the awareness of which strategy to be used while computation offloading.

2 Structural Analysis of the Offloading Schemes

A detailed analysis of the various offloading schemes was made. Generally, the demarcation was based on static and dynamic partitioning.

Figure 1 shows the partitioning of Computation Offloading [6]. It is bifurcated into Static decision and Dynamic decision. In static as well as dynamic offloading, there is both partial and full offloading. Partial offloading means only a selected part of the application is migrated to the cloud and the rest is computed on the smart mobile device. Full offloading means that the whole process is migrated to the public cloud for processing. A combination of static and dynamic algorithms is also available so as to get the benefits of both these decisions [7]. Different Computational Offloading Mechanisms were used depending upon the nature of applications and the availability of resources. The three Computational Offloading Mechanisms are Virtual Machine Migration, Code Migration and Thread-State Offloading.

Table 1 clearly shows the difference between the three schemes of computation offloading [8]. In some cases, VM Clone Migration is useful and more efficient while in some other cases, Code Migration and Delegation or Thread Synchronization might prove to be beneficial. This is in regard to the user's perspective and is subjective in nature. The user's demand can be high speed, bandwidth, and reliable data or optimized code output. Hence, all these schemes must be carefully evaluated so as

Fig. 1 Computation offloading types

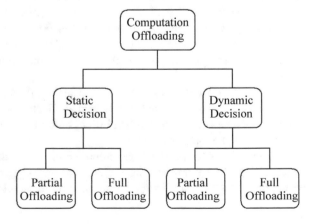

Table 1 Computation offloading mechanisms

Virtual machine transfer	Code transfer	Thread state transfer
With the help of virtualization technique, a mirror image is formed on the cloud [21]. Calculations are formed in the cloud and the result is migrated to smartphone	All or a part of the code is transferred to the cloud for further processing	Threads are transferred to the cloud
There is no modification in application	There is a noticeable modification in application [10]	There is no modification in the application [1]
High communication cost is there with the addition of incoming and outcoming requests. For example, Cloudlet and CloneCloud [22]	Static partitioning with low connectivity is there. For example, Coign	Occurrence of profiling overhead is there. For example, CloneCloud

to maximize the throughput. During analysis, single-site offloading also came into sight. Single-site offloading means that the applications are migrated to a single Public cloud only. This did not always prove to be beneficial. Another scheme came which is known as Multi-site offloading, in which applications can be offloaded to a number of sites so as to get better results in terms of energy consumption and computational time [9]. The performance of multi-site offloading is quite better than single-site offloading. Since it is NP-Hard pComputation offloading to saveroblem, so a near-optimal result is looked for. Single-site offloading techniques [1, 10] have been defined in detail using various [11] algorithms.

3 Related Work

A brief survey of the computation offloading schemes has been made depicting the static analysis and dynamic analysis. In 1999, GC Hunt discovered Coign, a technique that used static partitioning [10]. Coign is an automatic distributed partitioning system, which divides the applications into modules without accessing the source code. A graph model was constructed which depicted the inter-component communication. Coign deployed automatic partition and usage of binary applications. However, this system was too generic and is applied only for distributed programming. In 2001, Wang and R. Xu, stated an algorithm which was applicable to handheld and embedded systems, and again static partitioning was there. Profilers were used to get information of processing time, energy used and also bandwidth used. A cost graph was also made for the given process. Static partitioning divided the programme into server tasks and client tasks to minimize the energy consumption. Nevertheless, it was only applicable to personal digital assistants and embedded systems [12]. In 2004, C. Wang and Z. Li discovered a polynomial time algorithm, which was static in nature. A client-server demarcation of the application was performed using a polynomial time algorithm. Accordingly, the client code was processed at the mobile and the server code was processed on the cloud. An optimal partitioning scheme was used for a given set of input data. However, partitioning and abstraction were used mainly in distributed and parallel architectures.

In 2007, comparison methods (performing locally and on server) used to analyse offloading in wireless environments were used by S. Ou and K. Zang. These used dynamic partitioning for computation overloading. It was dependent on online statistics only and was not so accurate. In 2009, C. Xian and H. Liu used online statistics, which were used to compute optimal timeout. If the time ran out at and was more than optimal timeout, then it was offloaded to the server cloud. It was too cumbersome to work upon. A Context-Aware adapter algorithm was proposed, which could determine the gaps occurring and would use an adapter for each identified gap. This was given by H. H. La and S. D. Kim in 2010. It had large overheads in identifying the gaps [2]. MAUI (Mobile Assistance Using Infrastructure) was proposed by E. Cuervo and A. Balasubramaniam in 2010. It dynamically partitions an application at runtime in steps and then offloads. In MAUI, partitioning consisted of annotation of method as remote or local. Two versions of mobile application are made one for the cloud and the other for the smart device. It has dynamic decision and code offloading and method level granularity. Deng proposed GACO (Genetic Algorithm for computation offloading) for static partitioning in 2014. As the mobile phones are not stationery, the connectivity of mobile networks affects the offloading decision. A Genetic algorithm (GA) was modified to cater the needs for mobile offloading problem. Complex hybrid and mutation programming were involved [13]. In 2012, Kovachev discovered MACS (Mobile Augmentation Cloud Services) which minimizes energy and power consumption by using relation graphs. When highly complicated algorithms were involved it showed up energy savings up to 95% performing transparent offloading. It was mainly meant for Android development process. No built-in security features were there [14]. In 2013, P. Angin practiced Computation Offloading Framework based on Mobile Agents using application partitioner component to partition the application [15]. Liu and J. Lee proposed dynamic Programming (DP) algorithm in 2014, which included a two dimensional DP table to offload applications. A Dynamic Programming based Offloading Algorithm (DPOA) was used to find the optimal partitioning between executing subcomponents of a mobile process at the smart device and the public cloud. Like a profiler, it takes into account the smart device processor speed, the performance of the network and the type of an application. It also keeps into account the efficiency of the cloud server. The complexity of DPOA is much less than the Branch and Bound algorithm. In 2015, Zhang introduced Adaptive algorithm that reduced the energy consumption during offloading. This algorithm is used for making an energy efficient offloading scheme. The given algorithm takes energy saving and Quality of Service factors also [16]. In 2015, Gabriel Orsini, Dirk Bade and Winfried attempted to explain the context-aware computation offloading. It emphasized on the changing needs of the user. The context here meant the intermittent connectivity, scalability and heterogeneity in transferring data. The work introduced centralized offloading and opportunistic offloading. It worked well with the virtualization but was expensive to implement. In 2016, Zhang developed an heuristic algorithm for multi-site computation offloading. The algorithm basically aimed at basically three things, viz. energy consumed by the smart device, task completion time and charged price [17]. A greedy hill climbing algorithm was developed and a comparison between exhaustive search and heuristic algorithm were made to sieve out advantages of both the algorithms. It provided solution for multi-site offloading only [18]. In 2017, a hybrid of Branch and Bound Algorithm and optimized particle swarm optimization algorithm were used for making the decision for offloading data. It includes two decision algorithms to achieve the optimal solution of multi-site offloading considering the application size. It evaluates the efficiency of the proposed solution using both simulation and testbed experiments. This was more beneficial than other branch and bound algorithms.

4 Offloading Mechanism

The offloading mechanism is generally illustrated with the help of a Relation Graph [19]. The vertices and edges of a relation graph symbolize the application's units and their invocations, respectively. Object mechanism is used to extract application's units. This can be achieved either statically or dynamically. Online or offline profiling is used to determine the cost of smart device and cloud execution of each application's unit. The cost between two nearby vertices is also calculated through a weighted cost model. Different strategies like Greedy Methods, Dynamic Programming, Genetic and Optimization were used to find the optimal solution. Finally, the offloading problem was represented as an objective function minimizing the total execution cost of application. By making an analysis and cost estimation we can identify or conclude that whether the application or a part of it should be run locally (mobile) or remotely (cloud). Table 2 shows the SWOT (Strengths, Weaknesses, Opportunities and Threats) analysis of various offloading algorithms.

The table shows the static partitioning algorithms where data is transferred during compile time. All the resources and bandwidth are reserved a priori [20] (Table 3).

In Dynamic Offloading, when some partial applications are migrated to the cloud, concurrently the mobile can independently run its own applications as well.

A comparison chart is made in Fig. 2 which clearly shows that the dynamic offloading computation time is less than the static offloading computation time.

In Fig. 2, there was the pictorial representation of Static versus dynamic offloading. Computation time in microseconds was detailed in x-direction and ideal time was detailed in y-direction. As per the process, computation time of dynamic offloading is less than the static offloading [23]. Another analysis made after critical review was to use multi-site offloading rather than single site. This multi-site offloading models will also include the single-site offloading scheme's properties as well.

Table 2 SWOT analysis of various static partitioning offloading algorithms

Offloading scheme	Algorithm	Researchers	Result
Static partitioning	Coign is an automatic distributed partitioning system, which divides the applications into modules without accessing the source code [10]	G. C. Hunt and M. I. Scott, 1999	Applied only for distributed programming and was quite generic in nature
	Branch and Bound Algorithm was applied to minimize total energy consumed and other resources consumed through online profilers [11]	Li, C., Wang, R., Xu, 2001	Applied to embedded and PDA's only
	A Heuristic Polynomial was used to find an optimal programme partitioning [12]	C. Wang and Z. Li, 2004	In this, abstraction was used mainly for distributed and parallel structures
	Online statistics were used to offload the application if it crossed its threshold value	C. Xian, 2009	Results were largely depended on online statistics
	A Context-Aware adapter algorithm was proposed which could determine the gaps occurring and would use an adapter for each identified gap.	S. D. Kim, 2010	Overhead in identifying the gaps and also used in selecting the particular adapter
	CloneCloud decreases the total energy consumption using a thread-based relation graph [1].	Chun et al., 2011	It ignores runtime parameters and sometimes gives inaccurate solution
	Branch and Bound application partitioning uses the Branch and Bound algorithm which is used to find out least bandwidth required [5]	Niu, 2014	Cumbersome to work upon
	Genetic Algorithm Computation Offloading uses static analysis and online profiling to transfer the applications	Deng et al., 2014	Complex hybrid, crossing over and mutation programming involved

Table 3 SWOT analysis of the various dynamic partitioning offloading algorithms

Offloading scheme	Algorithm	Researchers	Result
Dynamic partitioning	Comparison methods (performing locally and on server) used to analyse offloading in wireless environments	S. Ou, K. Yang, A. Liotta and L. Hu, 2007	Inefficient method and time consuming
	MAUI which dynamically partition an application at runtime in steps and then offloads [2]	E. Cuervo, A. Balasubramaniam, et al., 2010	Used finely grained architecture for dynamically offloading a part of application
	A relation graph is constructed in Mobile Augmentation Cloud Services (MACS), which minimizes the energy consumption [13]	Kovachev et al., 2012	Partial offloading is done to mobile and cloud and the mobile independently executes other applications but no built-in security features
	Computation Offloading Framework based on Mobile Agents using application partitioner component to partition the application [15]	P. Angin, B. Bhargava, 2013	Security issues were perceived
	Dynamic Programming (DP) algorithm included a two-dimensional DP table to offload applications	Wu and Huang, 2014	It did not consider an execution time constraint when computing the offloading decisions
	Adaptive algorithm used to reduce the energy consumption during offloading	Zhang et al., 2015	Stochastic Wireless Channels were used
	An Energy-Efficient Multi site of Terefe et al. 2016 floating policy uses discrete-time Markov chains		It provided a solution for multi-site offloading

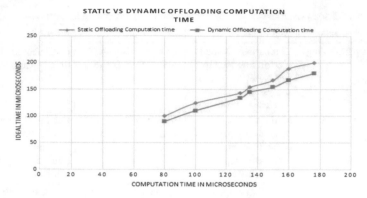

Fig. 2 Pictorial representation of static versus dynamic offloading computation time

5 Future Work

After making the analysis, we deduce that Context Awareness for offloading schemes is mandatory for proper utilization of the mobile and the cloud services. Three main challenges have come into light. Besides the objective context, performance context, resource context and energy context, another concept came up known as Emotion Awareness. Emotion Awareness could be a challenging aspect in Mobile Cloud Computing as it is quite difficult to gather and analyse human emotions electronically. But with the help of AI techniques and human sensors, it may be possible.

Second, a cost model based on energy consumption and execution time were proposed for multi-site offloading. It used a hybrid of Branch and bound algorithm and PSO algorithm to get an optimal results. It could be further implemented using Cloudlets.

Lastly, it is also a great challenge to detect and formulate a model or an efficient algorithm to solve the Computation Offloading problem along with Emotional Context as it is proved to be NP-Complete Problem which cannot be solved in polynomial time.

References

1. Chun, B.-G., et al.: CloneCloud: elastic execution between mobile device and cloud. In: Proceedings of the International Conference on Computer Systems, pp. 301–314 (2011)
2. Cuervo, E., et al.: MAUI: making smartphones last longer with code offload. In: Proceedings International Conference Mobile Systems, Applications, and Services, pp. 49–62 (2010)
3. Khan, A.R., et al.: A survey of mobile cloud computing application models. IEEE Commun. Surv. Tutorials **16**(1), 393–413 (2014)
4. Dinh, H.T., Lee, C., Niyato, D., Wang, P.: A survey of mobile cloud computing: architecture, applications, and approaches. Wireless Commun. Mobile Comput. **13**, 1587–1611 (2013)

5. Chang, Z., Ristaniemi, T., Niu, Z.: Energy efficient grouping and scheduling for content sharing based collaborative mobile Cloud. In: Proceedings of IEEE International Conference on Communications (ICC'14) (2014)
6. Chang, Z., Gong, J., Zhou, Z., Ristaniemi, T., Niu, Z.: Resource allocation and data offloading for energy efficiency in wireless power transfer enabled collaborative mobile clouds. In Proceedings of IEEE Conference on Computer Communications (INFOCOM'15) Workshop, Hong Kong, China, April 2015
7. Pederson, M.V., Fitzek, F.H.P.: Mobile clouds: the new content distribution platform. In: Proceeding of IEEE, vol. 100, no. Special Centennial Issue, pp. 1400–1403, May 2012
8. Satyanarayanan, M., et al.: Pervasive personal computing in an internet suspend/resume system. IEEE Internet Comput. **11**(2), 16–25 (2007)
9. Gu, X., Nahrstedt, K., Messer, A., Greenberg, I., Milojicic, D.: Adaptive offloading inference for delivering applications in pervasive computing environments. In: Proceedings of the First IEEE International Conference on Pervasive Computing and Communications (PerCom 2003), pp. 107–114. IEEE (2003)
10. Hunt, G.C., Scott, M.L.: The Coign automatic distributed partitioning system. In: Proceedings of the 3rd Symposium on Operating Systems Design and Implementation, February 1999
11. Wang, C., Li, Z., Xu, R.: Computation offloading to save energy on handheld devices: a partition scheme. ACM, 16–17 Nov 2001
12. Wang, C., Li, Z.: A computation offloading scheme on handheld devices. J. Parallel Distrib. Comput. **64**, 740–746 (2004)
13. Kovachev, D., Cao, Y., Klamma, R.: Mobile Cloud Computing: A Comparison of Application Models, CoRR, vol. abs/1107.4940 (2011)
14. Angin, P., Bhargava, B.: An agent-based optimization framework for mobile-cloud computing. J. Wireless Mobile Netw. Ubiquit. Comput. Dependable Appl. **4**(2) (2013)
15. Wang, Y., Chen, I., Wang, D.: A survey of mobile cloud computing applications: perspectives and challenges. Wireless Pers. Commun. **80**(4), 1607–1623 (2015)
16. Ma, R.K., Lam, K.T., Wang, C.-L.: eXCloud: transparent runtime support for scaling mobile applications in cloud. In: Proceedings of the International Conference Cloud and Service Computing (CSC), pp. 103–110 (2011)
17. Giurgiu, I., Riva, O., Juric, D., Krivulev, I., Alonso, G.: Calling the cloud: enabling mobile phones as interfaces to cloud applications. In: Proceedings of the 10th ACM/IFIP/USENIX International Conference on Middleware (Middleware '09), pp. 1–20. Springer (2009)
18. Ou, S., Yang, K., Zhang, J.: An effective offloading middleware for pervasive services on mobile devices. Pervasive Mob. Comput. **3**, 362–385 (2007)
19. Kosta, S., et al.: ThinkAir: dynamic resource allocation and parallel execution on the cloud for mobile code offloading. In: Proceedings of the IEEE INFOCOM, 2012, pp. 945–953 (2012)
20. Xing, T., et al.: MobiCloud: a geo-distributed mobile cloud computing platform. In: Proceedings of the International Conference on Network and Service Management (CNSM 12), pp. 164–168 (2012)
21. Zhang, X., Jeong, S., Kunjithapatham, A., Gibbs, S.: Towards an elastic application model for augmenting computing capabilities of mobile platforms. In: Third International ICST Conference on Mobile Wireless Middleware, Operating Systems, and Applications (2010)
22. Satyanarayanan, M., et al.: The case for VM-Based cloudlets in mobile computing. IEEE Pervasive Comput. **8**(4), 14–23 (2009)
23. Ra, M.R., Sheth, A., Mummert, L., Pillai, P., Wetherall, D., Govindan, R.: Odessa: enabling interactive perception applications on mobile devices. In: Proceedings of Mobisys, pp. 43–56. ACM (2011)

Cost Evaluation of Virtual Machine Live Migration Through Bandwidth Analysis

V. R. Anu and Elizabeth Sherly

Abstract Live migration permits to transfer constantly running Virtual Machine from source host to destination host. This is an unavoidable process in data centre in various scenarios such as load balancing, server maintenance and power management. Now, live migration performance optimization is an active area of research since performance degradation and energy overhead caused by live migration cannot be neglected in modern data centres; particularly if critical business goals and plans are to be satisfied. This work analyses the effect of proper bandwidth allocation during cost-aware live migration. Here, we design a cost function model based on network bandwidth between live migration process and service based on queuing theory. From experimental analysis, we infer that link bandwidth is a critical parameter in determining VM live migration cost.

Keywords Bandwidth · Cloud computing · Downtime · Live migration
Virtualization

1 Introduction

Virtual machines are widely used in data centres with the evolution of virtualization technology. It furnishes a steady isolation and remarkably increases the physical resources utilization. Live migration is a process of copying VM operating system from source machine to destination machine, while the OS in running mode. It is an inevitable process across physical servers and is a great help for administrators of data centres. It provides power management, fault tolerance, load balancing, server consolidation and low-level system maintenance. But live migration can result in VM

V. R. Anu (✉)
Mahatma Gandhi University, Kottayam, Kerala, India
e-mail: anuvraveendran@gmail.com

E. Sherly
IIITM-K, 17, Trivandrum, Kerala, India
e-mail: sherly@iiitmk.ac.in

© Springer Nature Singapore Pte Ltd. 2019
J. D. Peter et al. (eds.), *Advances in Big Data and Cloud Computing*,
Advances in Intelligent Systems and Computing 750,
https://doi.org/10.1007/978-981-13-1882-5_9

performance degradations during the migration period. It also causes outages and consumes energy, and as a result, it is a costly operation whose overheads should not be neglected. Due to its significant role in data centre management, it is important to preserve a non-trivial trade-off between minimization of live migration implementation cost and maintains SLA-bound service availability. In this work, we investigate affecting factors and put forward innovative strategies for cost-aware live migration.

There are many factors which affect the live migration performance and optimization. A lot of research has been taking place on these factors to improve the overall performance of live migration and strategies to reduce the degradation of performance of VMs caused by live migration. Factors such as total migration time, downtime, memory dirtying rate and amount of data transferred during live migration are well analysed in many studies. But the methods presented therein mostly neglect the migration cost of VMs. They estimate the migration cost only in terms of the number of migrations. But Akoush et al. [1] showed that link bandwidth and page dirtying rate are the primary impacting factors of live migration which influence other factors like total migration time and downtime in migration behaviour. So, the variation in these factors affects the cost of migration. But the page dirtying rate during live migration is totally dynamic in nature and purely depends on the nature of application service and kind of data handled in that service. This paper analyses the live migration cost model in terms of link bandwidth and thus help data centre administrators to decide which VM to migrate in various application scenarios in a cost-effective way.

We organize the paper as follows. Section 2 describes related work. In Sect. 3, the cost system model is explained. In Sect. 4, we show the analysis of our bandwidth model. Finally, Sect. 5 concludes the paper.

2 Related Work

In [2], the authors summarize, classify and evaluate current approaches with respect to determine cost of virtual machine live migration. This work describes in detail about the parameters which affect the cost of live migration such as page dirty rate, VM memory size and network bandwidth. Cost of migration is positively connected to the performance of migration. This survey paper gave a clear picture on the fundamental factors which affect the cost of migration. In [3], Ziyu Li discussed key parameters that affect the migration time and construct a live migration time cost model based on the hypervisor KVM (Kernel Virtual Machine). The authors put forward the model based on a feature called Total migration time. They proposed two cost-effective live migration strategies for load balance, fault tolerance and server consolidation. Zhang et al. [4, 5] theoretically analyse the effect of bandwidth allocation in total migration time and downtime of a live migration and put forward a cost-aware live migration strategy. They also discussed about varying bandwidth requirement while performing pre copy live migration. Through reciprocal-based model, they determine proper bandwidth to guarantee the total migration time and downtime requirements for a successful live migration. In all the above-mentioned related works, the authors had discussed factors other than allocated bandwidth. In [6], Xudong Xu et al. proposed a network bandwidth cost function model between migration and service based on queuing theory. Here, the authors studied about the impact of co-located interference on VMs based on a queuing theory model. They found out the trade-off of network bandwidth between the application service and live migration. As compared with all other factors mentioned in [2], the most unique factor which affects the cost of live migration is the residual bandwidth. In [7], the authors put forward a weighted sum approach to formalize the performance of live migration and also presented a bandwidth allocation algorithm to minimize the sum. They used a queuing theory model to show the relationship between a VM's performance and its residual bandwidth. In [8], David Breitgand et al. suggested a concept of cost function to evaluate the migration cost in terms of SLA violations and suggested an algorithm with some assumption to perform live migration with minimum cost possible. In [9], the authors focussed on cost-aware live migration using a component-based VM, in which VM is not considered as a monolithic image, but as a collection of various components like kernel, OS, programs and user data. Strunk et al. [10] proposed a lightweight mathematical model to quantitatively estimate the energy cost of live migration of an idle virtual machine through linear regression. Hu et al. [11] put forward a heterogeneous network bandwidth allocation algorithm. This algorithm optimizes the network bandwidth resource allocation regulates the network traffic increased the network load capacity. Mazhiku et al. [12] empirically evaluate the effect of open flow end-to-end QoS policies to reserve minimum bandwidth required for successful VM migration.

In our previous works, we discussed methods to improve the strategy of live migration through delta compression and live migration through MPTCP model [13]. The performance optimization for live migration environment is discussed in [14] by discussing the issues like interference and hotspot detection. In this work, we analyse the cost function feature of live migration through bandwidth allocation strategy based on the literature review.

3 Cost System Model

In this work, we theoretically analyse a cost function model of live migration. There exist many factors affecting the performance of live migration. As Anja Frank et al. in [2], the taxonomy of migration cost is expressed in different levels such as performance of migration, energy overload and performance loss of due to live migration in virtual machines. Performance of migration is evaluated through live migration factors like total migration time (in s) and downtime (in ms). Live migration is obviously an energy overhead process since it happened between servers simultaneously along with routine IT services. But its significant role in load balancing, server consolidation and server maintenance made it an inevitable operation in cloud data centres. Performance loss of VM is measured in terms of increase in execution time of a job running inside a VM during migration and loss in throughput of job running inside a VM during migration [15].

Akoush et al. [1] showed that link bandwidth and page dirtying rate are unique impacting factors in migration behaviour. The page dirtying rate relies on the workload running in the migrated VM which cannot be predicted or modified due to its dynamic nature. Hence, allocating the required bandwidth is the critical factor affecting the cost of live migration of VMs in cloud data centres. The cost of live migration can be expressed as the summation of cost of bandwidth used for migration and cost of bandwidth used for services of users. The overall cost calculated can be represented as

$$C_{total} = \alpha C + \beta C_{mig} \tag{1}$$

where C and C_{mig} denote average cost to execute user services and live migration of VMs, respectively. α and β are the regulatory constants to balance the cost function. Furthermore, C and C_{mig} are inversely proportional to their corresponding residual bandwidth. The increase in one's migration bandwidth may result in decrease of other's residual bandwidth. Therefore, the performance improvement of live migration and the deduction of performance degradation of VMs (routine user services) are two conflicting goals in terms of bandwidth allocation.

Assume that the live migration and user routines computes for available network bandwidth, then cost minimization on C_{total} may be

$$\begin{aligned} \text{Minimized} \quad C_{total} &= \alpha(C + \Delta C) + \beta\left(C_{mig} - \Delta C_{mig}\right) \\ &= \alpha C + \beta C_{mig} + \alpha \Delta C - \beta \Delta C_{mig} \end{aligned} \tag{2}$$

where ΔC_{mig} represents the cost advancement of live migration due to increase in residual bandwidth and ΔC denote the additional cost required due to decrease of their residual bandwidth of routine services. For a cost-effective live migration, we attempt to minimize ΔC while allocating bandwidth.

In other words [16–18], cost function $F(B_s)$ to be a function on residual bandwidth B_s, and can be expressed as the group of requests which are not satisfied by their deadline, where

$$B_s = B - B_m \tag{3}$$

B is the total available bandwidth, B_m is the bandwidth consumed by migration process. $B_m \leq B$ and B_s is the residual bandwidth utilized by running services in VMs. If S_t represent the serving time of a request and t_{SLA} is the time specified in SLA for a request to be several, then

$$F(B_s) = P[S_t > t_{SLA}]. \tag{4}$$

That is, based on the Eq. (3), if the residual bandwidth B, are sufficiently large then all requests are satisfied within their time bound of SLA. When residual bandwidth decreases, the probability of a request not to be satisfied by this time limit increases and reached into a condition where almost none of the requests are satisfied when the bandwidth is too limited.

When we consider bandwidth allocation for migration process, an unpredictable total migration time and downtime might bring much difficulty to the entire process and increased total cost [19, 20]. Sometimes in multi-tenant clouds, an effective load balancing or server consolidation may lead to long total migration time. Large downtime will affect the SLA of service and increase overall cost. Moreover, if the strategy used for live migration is pre copy, then it demands bandwidth change in each iteration. The bandwidth is likely to be varied by an interval of not more than 1 s, current network technologies seldom ensures the required bandwidth in advance. So, there should be a non-trivial trade-off between VM migration bandwidth and residual bandwidth to achieve these two conflicting goals and maintain a balanced cost model for cloud data centres. Allocation of bandwidth is decided by the page dirtying rate of workload running in the migrated VM and the nature of workload is totally unpredictable. So to implement a productive cost model, it is necessary to implement compelling bandwidth allocation model for migration.

3.1 Bandwidth Allocation Strategy for Migration

Here, we consider the pre copy method as the live migration strategy and migration bandwidth B_m are considered to be constant throughout the pre copy phase [21, 22]. Let 'P' be the total number of pages in the VM. Pi be the number of pages transferred in the ith round of pre copy phase. B_a is the bandwidth allocated by the network in the pre copy phase, B_m be maximum bandwidth and 'n' be total number of iterations during the stop and copy phase [4]. M_{total} and D_{total} denote the expected total migration time and downtime of the live migration, respectively. Actual total migration time and actual downtime are represented as M_{total}' and D_{total}'. Here as part of implementing cost-effective live migration, we need to minimize the bandwidth Ba during the pre copy phase for the given B_m to satisfy M_{total} and D_{total}.

$$\text{Minimum } B_a, \; M'_{total} \leq M_{total}; \; D'_{total} \leq D_{total} \qquad (5)$$

Based on pre copy live migration algorithm strategy, the actual total migration time and downtime be obtained as

$$M'_{total} = \frac{\sum_{i=1} P_i}{B_a} + D'_{total}; D'_{total} = \frac{P_{k+1}}{B_m} \qquad (6)$$

Algorithm 1 Bandwidth allocation strategy for cost-aware live migration
    Input: expected down time, expected total migration time, number of iterations
    Output: estimation of number of pages in each round Pi and required bandwidth allocation.

Method:

Step (1) Dirtying functions are having same nature

1. Let $Dy(t)$ be dirtying distribution function of ith page. Assume that dirtying distribution function of all pages are same then,

   Deterministic distribution function
   $$Dy(t) = 0 \quad t < T$$
   $$Dy(t) = 1 \quad t > T$$

   where T is the parameter of the distribution function.

2. $(T B_a + i)$th page is finished, then the ith transferred page will get dirty. Therefore, after transferring P pages in the first round, TB_a pages are clean, and the others $P - T B_a$ get dirtied. At the next round, once a page is transferred, a page will get dirty. Therefore, we have

   $$P_i = P \quad i = 1$$
   $$P_i = P - T B_a \quad i > 1$$

3. $D'_{\text{total}} = \frac{P - T Ba}{Bm}$

4. $M'_{\text{total}} = \frac{T Ba}{Ba} + D'_{\text{total}}$

5. To satisfy the delay guarantee

   a. $P - T B_a \le D_{\text{total}}$
   b. $T + \frac{P - T Ba}{Bm} \le M_{\text{total}}$

6. $Ba \ge \max \left\{ \frac{P - D_{\text{total} Bm}}{T}, \frac{P - (M_{\text{total}} - T) Bm}{T} \right\}$

Step (2) Dirtying functions are different (reciprocal-based model)

1. Let $P(K, i)$ be denoted as the probability of a page i transferred in round $(K - 1)$ and gets dirty at the start of round K. The probability that the first transferred page is dirty at the start of the second round

   $$P(2, 1) = \int_{x = x \min}^{\infty} g(x) X P_d(x) dx$$

   where $P_d(x)$ denotes the probability that a dirtied page with dirtying frequency x becomes dirty again after transmission. The time period from the first page being transferred to the end of the first round is $K T_1 = \frac{P - 1}{Ba}$.

2. If the dirtying frequency of a page is larger than $\frac{1}{KT_1} = \frac{Ba}{P - 1}$, then the page will become dirty again at the end of the first round. Thus, the probability is

   a. $Pr(2, 1) = \int_{x = x \min}^{Ba/P - 1} g(x) X 0 dx + \int_{Ba/N - 1}^{\infty} g(x) X 1 dx$
   b. $Pr(2, 1) = \int_{Ba/P - 1}^{\infty} \tau / x^2 dx = \tau (P - 1) / Ba$
   c. $Pr(2, j) = \tau (P - j) / Ba$

3. The expected number of dirty pages at the start of the second round is

$$P_2 = \sum_{i=1}^{P} Pr(2, i) = \frac{\tau P(P-1)}{2B_a} = \frac{\tau P1(P1-1)}{2B_a}$$

4. For a jth round, it is reversely proportional to Pj. We could represent as $\beta_j = \alpha/P_j$.

$$P_j = \frac{\alpha(P_{j-1} - 1)}{2Ba}$$

$$= \frac{\alpha^{j-1}(P-1)}{(2B_a)^{j-1}} - \left(\frac{\alpha}{2B_a}\right)^{j-2} - \left(\frac{\alpha}{2B_a}\right)^{j-3} - \cdots \left(\frac{\alpha}{2B_a}\right)$$

5. We can ignore $\left(\frac{\alpha}{2B_a}\right)$ terms since they are so small when compared with large number of memory pages.

6. The number of transferred pages in the jth round will be $P_j = (P-1)\left(\frac{\alpha}{2B_a}\right)^{j-1}$.

7. Thus, number of transferred pages at pre copy phase is

$$P_p = \sum_{j=1}^{k} P_j = (P-1)\frac{1 - \left(\frac{\alpha}{2B_a}\right)^k}{1 - \left(\frac{\alpha}{2B_a}\right)}$$

8. By satisfying the delay requirements of both the total migration time and the downtime, we represent the following:

$$(P-1)\frac{1 - \left(\frac{\alpha}{2B_a}\right)^k}{1 - \left(\frac{\alpha}{2B_a}\right)} \leq (M_{\text{total}} - D_{\text{total}})B_a$$

$$(P-1)\left(\frac{\alpha}{2B_a}\right)^k \leq D_{\text{total}}B_m$$

9. From the above equation, we can write

$$k = \frac{\log\left(\frac{D_{\text{total}}B_m}{P-1}\right)}{\log\left(\frac{\alpha}{2B_a}\right)}$$

10. From the first equation,

$$B_a \geq \frac{\alpha}{2} + \frac{P - 1 - D_{\text{total}}B_m}{M_{\text{total}} - D_{\text{total}}}$$

Since $\beta = \frac{\alpha}{N}$, then $\alpha = \beta N$; β can be furnished through obtaining the sampled dirtying frequency data. Then, the value of B_a and k can be evaluated.

3.2 Optimized Bandwidth Allocation Strategy Based on Queuing Theory

In Algorithm 1, we discussed the amount of bandwidth is required to perform live migration efficiently in both scenarios of fixed distribution function and variable distribution function. Now, we provide an optimization algorithm to determine how much bandwidth can be saved or reclaim during live migration and improve total cost for the entire process.

The bandwidth saver algorithm minimizes the performance degradation of VM during live migration based on queuing theory. Queuing theory can be used in the diversity of functional situations where it is not difficult to predict accurately the arrival rate and time) of the customers and service rate (or time) of service facilities [9]. As far as a web application in a VM is considered response time is one of the significant metrics to measure the performance. The other factors affecting performance are resource utilization and arrival rate of requests. We analyse the effect of these in performance using queuing theory.

The following equation represents the performance degradation of VMs in terms of bandwidth as [7]:

$$\Delta P_i = P(B_i) - P(B_i - \Delta B) \tag{7}$$

Here, we used a bandwidth saver algorithm to minimize performance degradation and this improves cost reduction of live migration [23, 24]. As an initial step, the performance degradation of each VM is calculated. If a VMs bandwidth is reduced by one bandwidth unit, then it will be stored into a binary search tree. It consistently saves one bandwidth unit from VMs whose degradation is minimum. Each time before storing a VM into BST, its deterioration is recalculated and assured that it is minimum. The algorithm terminates if there is no more bandwidth can be saved from VMs.

Algorithm 2 Bandwidth saver algorithm for VMs
   Input:

1. $V = \{VM_1, VM_2 .... VM_m\}$
2. Current bandwidth and least bandwidth of each VM $= \{(B1', B1)... (Bm', Bm)\}$
3. Live migration providing bandwidth
   Output: Saved bandwidth from each VM $S = \{Br_1, Br_2 .... Br_n\}$

   where $Br_1 > 0$, where $1 \leq i \leq n$
   asz

1. S=0; for each $VM_i \in V$ do
2. if $B_i - B_i' \geq \Delta B$, then calculate $\Delta P = P_i(B_i) - P_i(P_i - \Delta B)$
3. else 0
4. end if
5. if $\Delta P > 0$ then, $B_i = B_i - 1$

6. end if
7. for all $\Delta P > 0$ and $0 \le i \le n$, construct a self-balanced binary search tree and store ΔP into it.
8. let $P = 0$; and $B_r = 0$
9. for $B_r \le B$ do
10. $\Delta P_i = $ minimum element from BST
11. $(B) = \left(\frac{V_{mem}}{B}\right)(1 - (D/B)n + 1)/(1 - \left(\frac{D}{B}\right))$; $B = B_i + \Delta B$
12. $\Delta P_i \min = f(B) - f(Bi)$
13. if $\Delta P_i \min > \Delta P_i$
14. reclaim $Br_1, Br_2 \ldots \ldots \ldots \ldots \ldots Br_m$.
15. else $B_m = B_m + 1$; $P = P + \Delta Pi$
16. $B_r = B_r + 1$, $B_{ri} = B_{ri} + 1$
17. Remove the node containing ΔPi from BST
18. if $((\Delta P = P_i(B_i) - P_i(P_i - \Delta B)) > 0)$ insert it into BST
19. If BST==0, then reclaim $Br_1, Br_2 \ldots \ldots \ldots \ldots \ldots Br_n$.

4 Analysis

The analysis of algorithm can be done by selecting a client-server model of VMs. A Poisson distribution is followed while a VM client sent a request to its interrelated VM server. An average rate of flow of a request is around 500 KB. The request rate of four clients was selected as 15, 25, 35 and 45. The residual bandwidth is varied from 8 to 14 MB/s, 14 to 21 MB/s, 20 to 28 MB/s and 27 to 33 MB/s proportionally. The bandwidth increment rate was 1 MB/s.

In each network bandwidth environment, the four-client VMs are run six times repeatedly, and each run may last 10 min. After collecting the response time, we calculate the accuracy of the performance model of algorithm by comparing these values with theoretical values computed by this model. The experiment indicates that the response time is inversely proportional to the residual bandwidth. The response time will strongly increase with decrease of the residual bandwidth. The results are so close to each other shows that the error variation of performance model is very small.

In live migration, we have changed the residual bandwidth increment rate from 1 MB to 5 MB. The performance of bandwidth saver algorithm is compared with the original QEMU network bandwidth allocation method. The key evaluation metric is the performance degradation in total and can be represented as

$$
\begin{aligned}
P_{total} &= \alpha P + \beta P_{mig} \\
&= \alpha(P_1 x_1 + P_2 x_2 + P_3 x_3 + P_4 x_4) + \beta P_{mig}
\end{aligned}
\tag{8}
$$

Here, we selected $\alpha = 0.99$ and $\beta = 0.01$ since total migration time is measured in seconds and migration time measured in microseconds.

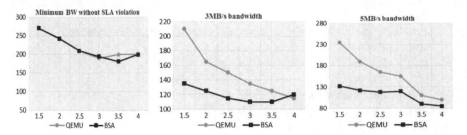

Fig. 1 Comparison of Bandwidth Saver Algorithm (BSA) with QEMU; bandwidth in x-axis and time/cost in y-axis

$x_i = 1$ if ith VM server's residual bandwidth was restored otherwise.

0   otherwise

The VM whose residual bandwidth is not reclaimed can be considered to remain unchanged during evaluation. Then, compare this model with standard QEMU model. When idle bandwidth is increased, then the reduction on P_{total} gradually decreased. This is because due to increase in residual bandwidth will reduce P_{mig} and thus P_{total}. Total time T increased sharply with decrease in residual bandwidth of VM servers (Fig. 1).

5 Conclusion

Studies focusing on performance improvement of live migration usually ignore the changes in VM due to live migration. In this paper, we presented the performance model for VMs in physical server and portray the relationship between VMs residual bandwidth and cost of liver migration. Through Algorithm 1, we theoretically decide proper bandwidth to guarantee total migration time and downtime requirement of VM live migration and its effect on the cost of live migration. The reciprocal-based method analysed dirtying frequency of memory pages and guarantee delay to reduce the performance degradation due to live migration. The concept of cost function introduced in this paper made it possible to evaluate the cost of live migration in terms of bandwidth allocation. Algorithm 2 is an optimization algorithm which helps to save bandwidth while performing migration and reduces the cost of live migration.

References

1. Akoush, S., Sohan, R., Rice, A., Moore, A.W., Hopper, A.: Predicting the performance of virtual machine migration. In: IEEE International Symposium on Modelling, Analysis & Simulation of Computer and Telecommunication Systems (MASCOTS), pp. 33–46 (2010)

2. Strunk, A.: Costs of virtual machine live migration: a survey. In: Services (SERVICES), 2012 IEEE Eighth World Congress on, pp. 323–329 (2012)
3. Li, Z., Wu, G.: Optimizing VM live migration strategy based on migration time cost modeling. In: Proceedings of the 2016 Symposium on Architectures for Networking and Communications Systems, pp. 99–109. ACM (2016)
4. Zhang, J., Ren, F., Lin, C.: Delay guaranteed live migration of virtual machines. In: INFOCOM, 2014 Proceedings IEEE, pp. 574–582 (2014)
5. Zhang, J., Ren, F., Shu, R.: Huang, T.: Liu, Y.: Guaranteeing delay of live virtual machine migration by determining and provisioning appropriate bandwidth. IEEE Trans. Comput. 65, 9, 2910–2917 (2016)
6. Xu, X., Yao, K., Wang, S., Zhou, X.: A VM migration and service network bandwidth analysis model in IaaS. In: 2012 2nd International Conference on Consumer Electronics, Communications and Networks (CECNet), pp. 123–125 (2012)
7. Zhu, C., Han, B., Zhao, Y., Liu, B.: A queueing-theory-based bandwidth allocation algorithm for live virtual machine migration. In: 2015 IEEE International Conference on Smart City/SocialCom/SustainCom (SmartCity), pp. 1065–1072 (2015)
8. Breitgand, D., Kutiel, G., Raz, D.: Cost-aware live migration of services in the cloud. In: SYSTOR (2010)
9. Gahlawat, Monica, Sharma, Priyanka: Reducing the cost of virtual machine migration in federated cloud environment using component based vm. J. Inf. Syst. Commun. 3(1), 284–288 (2012)
10. Strunk, A.: A lightweight model for estimating energy cost of live migration of virtual machines. In: 2013 IEEE Sixth International Conference on Cloud Computing (CLOUD), pp. 510–517. IEEE (2013)
11. Hu, Z., Wang, H., Zhang, H.: Bandwidth allocation algorithm of heterogeneous network. Change 14, 1–16 (2014)
12. Maziku, H., Shetty, S.: Towards a network aware VM migration: evaluating the cost of VM migration in cloud data centers. In: 2014 IEEE 3rd International Conference on Cloud Networking (CloudNet), pp. 114–119 (2014)
13. Anu, V.R., Sherly, E.: Live migration of delta compressed virtual machines using MPTCP. Int. J. Eng. Res. Comput. Sci. Eng. 4(9) (2017)
14. Anu, V.R., Sherly, E.: IALM: interference aware live migration strategy for virtual machines in cloud data centers. In: Communications (ICDMAI 2018) 2nd International Conference of Data Management, Analytics and Innovation. Springer Conference (2018)
15. Sharma, S., Chawla, M.: A technical review for efficient virtual machine migration. In: 2013 International Conference on Cloud & Ubiquitous Computing & Emerging Technologies (CUBE), pp. 20–25. IEEE (2013)
16. Wu, Q., Ishikawa, F., Zhu, Q., Xia, Y.: Energy and migration cost-aware dynamic virtual machine consolidation in heterogeneous cloud datacenters. IEEE Trans. Serv. Comput. (2016)
17. Rybina, K., Dargie, W., Strunk, A., Schill, A.: Investigation into the energy cost of live migration of virtual machines. In: Sustainable Internet and ICT for Sustainability (SustainIT), pp. 1–8. IEEE (2013)
18. Rahman, S., Gupta, A., Tomatore, M., Mukherjee, B.: Dynamic workload migration over optical backbone network to minimize data center electricity cost. In: 2017 IEEE International Conference on Communications (ICC), pp. 1–5. IEEE (2017)
19. Wang, R., Xue, M., Chen, K., Li, Z., Dong, T., Sun, Y.: BMA: bandwidth allocation management for distributed systems under cloud gaming. In: 2015 IEEE International Conference on Communication Software and Networks (ICCSN), pp. 414–418. IEEE (2015)
20. Akiyama, S., Hirofuchi, T., Honiden, S.: Evaluating impact of live migration on data center energy saving. In: IEEE 6th International Conference on Cloud Computing Technology and Science (CloudCom), pp. 759–762 (2014)
21. Ayoub, O., Musumeci, F., Tornatore, M., Pattavina, A.: Efficient routing and bandwidth assignment for inter-data-center live virtual-machine migrations. J. Opt. Commun. Networking 9(3), 12–21 (2017)

22. Mann, V., Gupta, A., Dutta, P., Vishnoi, A., Bhattacharya, P., Poddar, R., Iyer, A.: Remedy: network-aware steady state VM management for data centers. In: Networking, pp. 190–204. Springer (2012)
23. Strunk, A., Dargie, W.: Does live migration of virtual machines cost energy? In: 2013 IEEE 27th International Conference on Advanced Information Networking and Applications (AINA), pp. 514–521 (2013)
24. Voorsluys, W., Broberg, J., Venugopal, S., Buyya, R.: Cost of virtual machine live migration in clouds: a performance evaluation. In: CloudCom, vol. 9, pp. 254–265 (2009)

Star Hotel Hospitality Load Balancing Technique in Cloud Computing Environment

V. Sakthivelmurugan⊕, R. Vimala⊕ and K. R. Aravind Britto⊕

Abstract Cloud computing technology is making advancement recently. Automated service provisioning, load balancing, virtual machine task migration, algorithm complexity, resource allocation, and scheduling are used to make improvements in the quality of service in the cloud environment. Load balancing is an NP-hard problem. The main objective of the proposed work is to achieve low makespan and minimum task execution time. An experimental result proved that the proposed algorithm performs good load balancing than Firefly algorithm, Honey Bee Behavior-inspired Load Balancing (HBB-LB), and Particle Swarm Optimization (PSO) algorithm.

Keywords Cloud computing · Task migration · Load balancing · Makespan
Task execution time · Quality of service

1 Introduction

Cloud computing [1–3] is fully Internet-based computing. It connects the concepts of distributed and parallel computing. It works under "pay as you use" methodology. Cloud environment provides plenty of services [4]. It gives the platform to the customers to deploy and use their applications.

The Virtual Machines are the computational unit [5–7] in the cloud. In IT sector, the VMs should execute the task as soon as possible. It arises as a big problem in

V. Sakthivelmurugan (✉)
Department of Information Technology, PSNA College of Engineering and Technology,
Dindigul, Tamil Nadu, India
e-mail: sakthi@psnacet.edu.in

R. Vimala
Department of Electrical and Electronics Engineering, PSNA College of Engineering
and Technology, Dindigul, Tamil Nadu, India

K. R. Aravind Britto
Department of Electronics and Communication Engineering, PSNA College of Engineering
and Technology, Dindigul, Tamil Nadu, India

© Springer Nature Singapore Pte Ltd. 2019
J. D. Peter et al. (eds.), *Advances in Big Data and Cloud Computing*,
Advances in Intelligent Systems and Computing 750,
https://doi.org/10.1007/978-981-13-1882-5_10

balancing the load of the customer's or user's task within the available resources. Due to increase in the number of user requests, we use load balancing technique to reduce the response time.

The main purpose of the load balancing method is to reduce the execution time of user's application. There are three types of load balancing techniques. They are (i) Static, (ii) Dynamic, and (iii) Hybrid. Static load balancing [8] is simple which is used to manage known loads earlier. Dynamic load balancing [9] is used to manage flighty or obscure processing load. It is mostly successful among heterogeneous resources. The advantages of load balancing [10] are efficiency, reliability, and scalability.

2 Related Work

Load balancing [11] is the processes to eradicate the task from the over utilized virtual machine and allocate it to the underutilized virtual machine.

Chen et al. [12] have discussed a new load balancing technique. This method considers load and processing power. So, servers are not having the ability to handle other requirements. Priyadarsini et al. [13] have suggested min-min scheduling strategy. It acts in two stages. In the first stage, total tasks completion time is calculated. In the second stage, the least time slot is chosen among them and tasks are assigned, respectively. This methodology leads to high makespan. Pacini et al. [14] have introduced Particle Swarm Optimization. It follows the bird flocking concept. The load balancing is taken place with the help of fitness value calculation. It leads to high task execution time. Chen et al. [15] have introduced a novel load balancing methodology. The load is calculated and balanced in five levels. It leads to high makespan.

Aruna et al. [16] have suggested the firefly algorithm. It consists of scheduling index and Cartesian distance which are calculated for performing load balancing. But, it leads to high task execution time and high makespan. Dinesh Babu et al. [17] have introduced another Load Balancing algorithm. It follows the behavior of honey bee foraging methodology. It helps to reduce the response time of VMs. But, it leads to high task migration and high makespan. It works very well for homogenous as well as heterogeneous system. It works for non preemptive independent task.

3 Proposed Method: Star Hotel Load Balancing Method (SHLB)

SHLB is a dynamic method. It helps to reduce makespan and task execution time. This method is inspired by serving the food to the customers in a star hotel. There are three kinds of employees who serve the food to the customers and they are Waiters, Table Supervisors, and Kitchen Employees. All three employees play the major role for providing better hospitality to the customers who comes to the hotel. The table supervisor asks customers to get the details about the food items which they want to have. Kitchen Employee prepares the food item to the customer. The employees are grouped together and serve their food in a different direction. The employees will go back to the corresponding places after satisfying their customers.

Plenty of Virtual Machines are available in cloud computing environment and satisfy the requirements of users or customers. Servers receive more requests from the users in cloud environment. The requests are managed by cloud policies. It manages the consistency of the load of each VM during at that time.

This method is very helpful to arrange the VM in cloud environment. It can reduce makespan and execution time. The following picture Fig. 1 depicts the workflow of star hotel load balancing.

Here, we need to calculate total completion time which is called as Makespan. The task completion time is varied because of load balancing. It denotes the task completion time of $\text{Task}_p(T)$ on VM_p as CT_{pq}. It is mentioned as

Fig. 1 Flow diagram of Star Hotel Load balancing algorithm

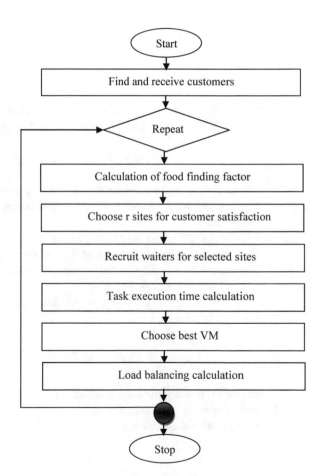

$$\text{Makespan} = \left\{ \max\left(\sum_{p=1}^{m} \sum_{q=1}^{n} \text{CT}_{pq} \right), \ p = 1, 2, \ldots m \in T, \ q = 1, 2, \ldots n \in \text{VM} \right\} \tag{1}$$

The time taken between submitting the request and the first response is called as completion time.

3.1 SHLB Algorithm

Step 1. Find and receive customers

In the starting step, we consider different types (r) of VM present in the cloud

Step 2. Calculation of food finding factor

It is calculated as

$$\text{FFF}_{pq} = \frac{\sum_{p=1}^{m} \text{TL}_p}{C} \tag{2}$$

where TL denotes task length, FFF_{pq} denotes the food finding factor of the task in VM, and C denotes capacity of VM. Capacity is calculated as

$$C = P_{\text{num}q} \times P_{\text{mis}pq} + VM_{bwq}$$

where C denotes the capacity of VM_q. $P_{\text{num}q}$ denotes number of processors. $P_{\text{mips}q}$ denotes millions of instructions per second. VM_{bwq} denotes network bandwidth.

Step 3. Choose r sites for customer satisfaction

Supervisor employees are picked SEs where every SE comprises of most astounding foot finding factor from another SE.

Step 4. Recruit waiters for selected sites

According to the supervisor's instruction, the food finding factor is calculated by Workers

$$\text{FFF}_{pq} = \frac{\sum_{p=1}^{m} \text{TL}_p + \text{FL}}{C} \tag{3}$$

where FL denotes file length before execution of task.

Step 5. Task execution time calculation

It is calculated as follows:

$$T_e = \left(\frac{1}{3600} \sum_{p=1}^{m} \left(\frac{\text{FFF}_{pq}}{P_{mipsq}} \right) \right) \tag{4}$$

where T_e denotes task execution time.

Step 6. Choose best VM
For all iterations, optimal VM has to be chosen from every group and assign the task to particular machine.

Step 7. Load balance calculation

After allocating the submitted request into VM, the VMs current workload is computed based on the load received from cloud data center controller. Standard deviation and mean for the task are calculated as follows:

$$SD = \sqrt{\frac{1}{r} \sum_{p=1}^{m} \left(T_{e_p} - T_e\right)^2} \tag{5}$$

$$\text{Mean} = \frac{\sum_{q=1}^{n} T_{e_q}}{C} \tag{6}$$

where SD stands for standard deviation, r denotes number of virtual machines. Mean denotes the execution time taken for all VMs. If SD $\leq$ mean, VM load is to be get balanced. Otherwise, imbalanced state will be reached.

4 Results and Discussion

Here, Firefly methods, Novel methods, HBB-LB, and Particle Swarm Optimization are compared. The cloudsim is a tool that provides simulation and experimentation of the infrastructure of cloud. Here, there are 100–500 tasks used for simulation.

The makespan is shown in Table 1. It can be observed that star hotel load balancing has low makespan.

The degree of imbalance is shown in Table 2. It shows that it has low degree of imbalance than other load balancing methodology.

Figure 2 clearly explains that it has low makespan than Firefly, PSO, and HBB-LB.

Figures 3 and 4 show that it has minimum task execution time than Firefly, PSO, and HBB-LB load balancing methodology.

Table 1 Makespan

Number of tasks	HBB-LB	PSO	Firefly	SHLB
100	45	40	35	30
200	102	91	84	78
300	160	147	137	119
400	217	197	185	157
500	270	241	228	191

Table 2 Degree of imbalance

Number of tasks	HBB-LB	PSO	Firefly	SHLB
100	3.9	3.5	3.3	3.1
200	3.7	3.2	3	2.9
300	4	3.5	3.3	3
400	4.1	3.6	3.3	3
500	3.6	3.4	3.3	3

Fig. 2 Makespan evaluation

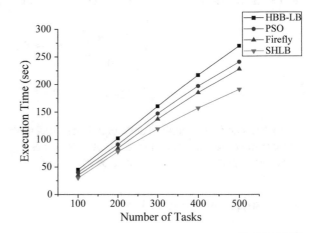

Fig. 3 Comparison of task execution time before load balancing

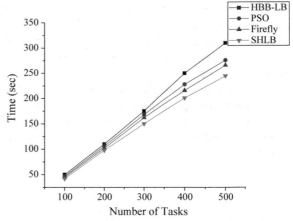

5 Conclusion

The main aim of the paper is to reduce the makespan and task execution time. The results of load balancing algorithm is computed and verified with the help of cloudsim tool. The incoming requests are tested and validated from the results. The result explains that the star hotel load balancing is an efficient and effective one when compare to HBB-LB, PSO and Firefly algorithm. It is suitable for cloud computing

Fig. 4 Comparison of task execution time after load balancing

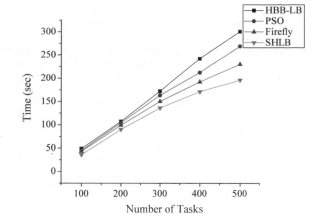

environment for reducing makespan and execution time. The performance is also improved. So, SHLB algorithm proved that it is more efficient for maintain load balancing, scheduling, and system stability effectively.

References

1. Joshi, G., Verma, S.K.: Load balancing approach in cloud computing using improvised genetic algorithm: a soft computing approach. Int. J. Comput. Appl. **122**(9), 24–28 (2015)
2. Mahmoud, M.M.E.A., Shen, X.: A cloud-based scheme for protecting source-location privacy against hotspot-locating attack in wireless sensor networks. IEEE Trans. Parallel Distrib. Syst. **23**(10), 1805–1818 (2012)
3. Tang, Q., Gupta, S.K.S., Varsamopoulos, G.: Energy-efficient thermal-aware task scheduling for homogeneous high performance computing data centers: a cyber-physical approach. IEEE Trans. Parallel Distrib. Syst. **19**(11), 1458–1472 (2008)
4. TSai, P.W., Pan, J.S., Liao, B.Y.: Enhanced artificial bee colony optimization. Int. J. Innov. Comput. Inf. Control **5**(12), 5081–5092 (2009)
5. Kashyap, D., Viradiya, J.: A survey of various load balancing algorithms in cloud computing. Int. J. Sci. Technol. **3**(11), 115–119 (2014)
6. Zhu, H., Liu, T., Zhu, D., Li, H.: Robust and simple N-Party entangled authentication cloud storage protocol based on secret sharing scheme. J. Inf. Hiding Multimedia Signal Process. **4**(2), 110–118 (2013)
7. Chang, B., Tsai, H.-F., Chen, C.-M.: Evaluation of virtual machine performance and virtualized consolidation ratio in cloud computing system. J. Inf. Hiding Multimedia Signal Process. **4**(3), 192–200 (2013)
8. Florence, A.P., Shanthi, V.: A load balancing model using firefly algorithm in cloud computing. J. Comput. Sci. **10**(7), 1156 (2014)
9. Polepally, V., Shahu Chatrapati, K.: Dragonfly optimization and constraint measure based load balancing in cloud computing. Cluster Comput. **20**(2), 1–13 (2017)
10. Mei, J., Li, K., Li, K.: Energy-aware task scheduling in heterogeneous computing environments. Cluster Comput. **17**(2), 537–550 (2014)
11. Kaur, P., Kaur, P.D.: Efficient and enhanced load balancing algorithms in cloud computing. Int. J. Grid Distrib. Comput. **8**(2), 9–14 (2015)

12. Chen, S.-L., Chen, Y.-Y., Kuo, S.-H.: CLB: a novel load balancing architecture and algorithm for cloud services. Comput. Electr. Eng. **56**(2), 154–160 (2016)
13. Priyadarsini, R.J., Arockiam, L.: Performance evaluation of min-min and max-min algorithms for job scheduling in federated cloud. Int. J. Comput. Appl. **99**(18), 47–54 (2014)
14. Pacini, E., Mateos, C., García Garino, C.: Dynamic scheduling based on particle swarm optimization for cloud-based scientific experiments. CLEI Electron. J. **17**(1), 3–13 (2014)
15. Chen, S.-L., Chen, Y.-Y., Kuo, S.-H.: CLB: a novel load balancing architecture and algorithm for cloud services. Comput. Electr. Eng. **58**(1), 154–160 (2016)
16. Aruna, M., Bhanu, D., Karthik, S.: An improved load balanced metaheuristic scheduling in cloud. Cluster Comput. **5**(7), 1107–1111 (2015)
17. Dhinesh Babu, L.D., Venkata Krishna, P.: Honey bee behavior inspired load balancing of tasks in cloud computing environments. Appl. Soft Comput. **13**(5), 2292–2303 (2013)

Prediction of Agriculture Growth and Level of Concentration in Paddy—A Stochastic Data Mining Approach

P. Rajesh and M. Karthikeyan

Abstract Data mining is the way of separating data from stunning perspectives, and along these lines, abstracting them into important information which can be used to construct the yielding and improvement possible results in cultivating. The destinations have been assented of solidness in paddy advancement and to expand the development of creation in a maintainable way to meet the nourishment prerequisite for the developing populace. In any farming fields, it for the most part, happens that at whatever point the choices in regards to different methodologies of arranging is viewed as, for example, season-wise rainfall, region, production and yield rate of principal crops, and so forth. In this paper, it is proposed to discover the forecast level of concentration in paddy improvement for different years of time series data utilizing stochastic model approach. Numerical examinations are outlined to help the proposed work.

Keywords Data mining · Time series data · Normalization · Distribution
Agriculture and stochastic model

1 Introduction

Data mining is the clever tool for predicting and the examining huge preexisting data in order to fabricate valuable information which was previously untried. Information mining field works with number of expository devices for breaking down the new methodologies in view of agriculture dataset. It enables clients to examine information from a wide range of edges and abridge the connections recognized. Farming

P. Rajesh (✉)
Department of Computer Science, Government Arts College, C.Mutlur, Chidambaram 608102, Tamil Nadu, India
e-mail: rajeshdatamining@gmail.com

P. Rajesh · M. Karthikeyan
Department of Computer and Information Science, Annamalai University, Annamalainagar, Chidambaram 608002, Tamil Nadu, India

© Springer Nature Singapore Pte Ltd. 2019
J. D. Peter et al. (eds.), *Advances in Big Data and Cloud Computing*,
Advances in Intelligent Systems and Computing 750,
https://doi.org/10.1007/978-981-13-1882-5_11

keeps on being the most driving segment of the state economy, as 70% of the tenants are occupied with agribusiness and united exercises for their job. A stochastic model is an instrument for assessing likelihood conveyances of potential results by taking into consideration arbitrary variety in at least one contribution over the time. The unpredictable assortment is by and large in light of instabilities saw in dataset for a picked period using time-series data of action frameworks. "Stochastic" signifies "relating to risk", and is utilized to portray subjects that contain some components of arbitrary or stochastic conduct. For a framework to be stochastic, at least one sections of the framework have arbitrariness related with it. Persons leave the organizations due to unsatisfactory packages or unsatisfactory work targets or both. This leads of the depletion of manpower. The expected time to recruitment due to the depletion of manpower is derived using Shock model approach [1]. Agriculture sector improvement based on rainfall, water sources and agriculture labour. A number of models are discussed by many researchers [2, 3, 12, 13]. Actually, information mining is the way toward discovering connections or examples among a great deal of fields in expansive social databases [15]. Many authors have estimated the expected to recruitment or the breakdown point of the organization from the viewpoint of the manpower loss. In doing so, the Shock models and cumulative damage process have been used [7]. It is assumed that the threshold level which indicates the time to recruitment is a random variable following Erlang2 distribution and truncated both at left and right [9]. Precipitation time arrangement for a 30-year term (1976–2006) was surveyed. The watched time-arrangement information was utilized as contribution to the stochastic model to produce another arrangement of day-by-day time-arrangement information [8]. The stochastic precipitation generator show, never connected in Malaysia, was received for downscaling and recreation of future everyday precipitation [6]. The stochastic model is utilized to examine specialized, financial, and allocate effectiveness for an example of New England dairy ranches [5]. The demonstrating structure was set up by the advancement of an interim two-stage stochastic program, with its arbitrary parameters being given by the measurable investigation of the reproduction results of an appropriated water quality approach [11]. Harvest yield models can help leaders inside any agro-industrial inventory network, even with respect to choices that are irrelevant to the product generation for demonstrating [4].

2 System Overview

In stochastic process, a standout amongst the most critical factors is Completed Length of Service (CLS), since it empowers us to anticipate turnover. The most broadly utilized distribution for completed length of service until the point when leaving is the mixed exponential distribution with respective probability density functions [2, 13, 14] and lognormal distribution [10, 16] with respective PDF given by,

$$f(t) = \rho \delta e^{-\delta t} + (1 - \rho) \gamma e^{-\gamma t} \tag{1}$$

$$f(t) = \frac{1}{\sqrt{2\pi}\sigma t} \exp\left(-\frac{1}{2}\left(\frac{\log t - \mu}{\sigma}\right)^2\right) \tag{2}$$

These parameters might be assessed utilizing the technique for greatest probability. The main shortcoming of this lognormal hypothesis, however, is that there is no satisfactory model explains its use in terms of the internal behavior. Mixed exponential distribution [2] describes the distribution of the CLS. Therefore, the PDF of CLS is $f(t) = \rho\delta e^{-\delta t} + (1-\rho)\gamma e^{-\gamma t}$, where $\delta, \gamma \geq 0$ and $0 \leq \rho \leq 1$.

Stochastic Model

A stochastic model is an instrument for assessing likelihood appropriations of potential results by taking into account the arbitrary variety in at least one contributions over the time. The random variation is usually based on fluctuations observed in historical data for a selected period using standard time series techniques. In probability theory, a stochastic procedure or now and then arbitrary process is a gathering of irregular factors; this is regularly used to speak to the development of some arbitrary esteem, or framework, after some time. In any case, it involves significance that the correct and suitable sort of stochastic demonstrating for the concerned factors ought to be defined hypothetically simply after the scrutiny of data set.

Pr (that the cumulative damage in k decisions has not crossed the threshold level)

$$= P\left[\sum_{i=1}^{k} X_i < y\right] = \int_0^\infty g_k(x)\,\overline{H}(x)\,dx \tag{3}$$

where $\overline{H}(x) = 1 - H(x)$.

Since y follows Mixed exponential distribution with parameter δ and γ as suggested by [12]. It has density function as $h(y) = \rho\delta e^{-\delta t} + (1-\rho)\gamma e^{-\gamma t}$ and the corresponding distribution function is

$$H(y) = \rho\delta \int_0^y e^{-\delta u}\,du + (1-\rho)\gamma \int_0^y e^{-\gamma u}\,du$$

$$= \rho(1 - e^{-\delta y}) + (1-\rho)(1 - e^{-\gamma y})$$

Hence, $\overline{H}(x) = p(e^{-\delta x} - e^{-\gamma x}) + e^{-\gamma x}$

Therefore, substituting this for $\overline{H}(x)$ in Eq. (3), we have

$$P\left[\sum_{i=1}^{k} X_i < Y\right] = \rho\, g_k^*(\delta) + (1-\rho)\, g_k^*(\gamma) \tag{4}$$

Now, $S(t) = P(T > t) = $ Survivor function
= Pr (that there are exactly k decisions in $(0, t]$ and the loss has not crossed the level)

$$S(t) = P[T > t] = V_k(t) P\left[\sum_{i=1}^{k} X_i < Y\right]$$

$$S(t) = \sum_{k=0}^{\infty} [F_k(t) - F_{k+1}(t)] P\left[\sum_{i=1}^{k} X_i < Y\right]$$

$$= 1 - \rho\left[1 - g^*(\delta)\right] \sum_{k=1}^{\infty} F_k(t)\left[g^*(\delta)\right]^{k-1}$$

$$+ (1 - \rho)\left[1 - g^*(\gamma)\right] \sum_{k=1}^{\infty} F_k(t)\left[g^*(\gamma)\right]^{k-1}$$

$$L(t) = \rho\left[1 - g^*(\delta)\right] \sum_{k=1}^{\infty} F_k(t)\left[g^*(\delta)\right]^{k-1}$$

$$+ (1 - \rho)\left[1 - g^*(\gamma)\right] \sum_{k=1}^{\infty} F_k(t)\left[g^*(\gamma)\right]^{k-1} \tag{5}$$

$$l(t) = \rho\left[1 - g^*(\delta)\right] \sum_{k=1}^{\infty} f_k(t)\left[g^*(\delta)\right]^{k-1}$$

$$+ (1 - \rho)\left[1 - g^*(\gamma)\right] \sum_{k=1}^{\infty} f_k(t)\left[g^*(\gamma)\right]^{k-1} \tag{6}$$

Now taking Laplace transform of $l(t)$, we get

$$l^*(s) = \frac{\rho\left[1 - g^*(\delta)\right] f^*(s)}{\left[1 - g^*(\delta)\right] f^*(s)} + \frac{(1 - \rho)\left[1 - g^*(\gamma)\right] f^*(s)}{\left[1 - g^*(\gamma)\right] f^*(s)} \tag{7}$$

Assuming that $g(x) \sim \exp(\theta)$ and $f(x) \sim \exp(c)$, which implies that the random variable $X_i \sim \exp(\theta)$ and $U_i \sim \exp(c)$. Since $f(.)$ follows $\exp(\theta)$ know

$$f * (s) = \frac{\theta}{\theta + s} \tag{8}$$

$$l^*(s) = \frac{\rho\left[1 - g^*(\delta)\right]\frac{\theta}{\theta+s}}{\left[1 - g^*(\delta)\frac{\theta}{\theta+s}\right]} + \frac{(1 - \rho)\left[1 - g^*(\gamma)\right]\frac{\theta}{\theta+s}}{\left[1 - g^*(\gamma)\frac{\theta}{\theta+s}\right]}$$

$$l^*(s) = \frac{\rho\left[\left(s + \theta\left[1 - g^*(\delta)\right]\right)(0) - \left[1 - g^*(\delta)\right]\theta\right]}{\left(s + \theta\left[1 - g^*(\delta)\right]\right)^2}$$

$$+ \frac{(1 - \rho)\left[\left(s + \theta\left[1 - g^*(\gamma)\right]\right)(0) - \left[1 - g^*(\gamma)\right]\theta\right]}{\left(s + \theta\left[1 - g^*(\gamma)\right]\right)^2}$$

We know that, taking first-order derivatives

$$E(T) = \frac{-dl^*(s)}{ds}\bigg|_{s=0}$$

$$= \frac{p[1 - g^*(\delta)]\,\theta}{(\theta[1 - g^*(\delta)])^2} + \frac{(1 - p)[1 - g^*(\gamma)]\,\theta}{(\theta[1 - g^*(\gamma)])^2} \tag{9}$$

Let $g(.)$ follow exponential distribution with parameter β yielding
Let

$$g^*(\delta) = \frac{\beta}{\beta + \delta}, \quad g^*(\gamma) = \frac{\beta}{\beta + \gamma}$$

$$= \frac{\rho}{\theta[1 - g^*(\delta)]} + \frac{(1 - \rho)}{\theta[1 - g^*(\gamma)]}$$

$$E(T) = \frac{\rho\mu\gamma + \delta(\mu - \rho\mu + \gamma)}{\theta\delta\gamma} \tag{10}$$

We know that, taking second-order derivatives

$$\frac{d^2l^*(s)}{ds^2}\bigg|_{s=0} = \frac{(s + \theta[1 - g^*(\delta)])^2(0) + (1 - \rho)[1 - g^*(\delta)]\theta\,[2(s + \theta[1 - g^*(\delta)])]}{(s + \theta[1 - g^*(\delta)])^4}$$

$$+ \frac{(s + \theta[1 - g^*(\gamma)])^2(0) + (1 - \rho)[1 - g^*(\gamma)]\theta\,[2(s + \theta[1 - g^*(\gamma)])]}{(s + \theta[1 - g^*(\gamma)])^4}$$

$$E(T^2) = \frac{\left[2\left(\frac{\rho}{\delta^2} + \frac{1-\rho}{\gamma^2}\right)(\mu + \gamma)^2\right]}{\theta^2} \tag{11}$$

$$V(T) = E(T^2) - [E(T)]^2$$

$$V(T) = \frac{\left[2\left(\frac{\rho}{\delta^2} + \frac{1-\rho}{\gamma^2}\right)(\mu + \gamma)^2\right]}{\theta^2} - \left[\frac{\rho\mu\gamma + \delta(\mu - \rho\mu + \gamma)}{\theta\delta\gamma}\right]^2 \tag{12}$$

3 Results and Discussion

All parameters ought to have a similar scale for a reasonable correlation between them. The techniques are normally outstanding for rescaling information. Normalization is used to change the scales into single numerical value between 0 and 1. The changes in Expected level of concentration (η) in paddy development and its variance (σ) are indicated by taking the following numerical examples with different inputted data values in Eqs. 10 and 12. In our perceptions in view of the following tables, as the estimation of "θ" to be specific the parameter for actual rainfall, "μ" speak to the name of sources of irrigation, "δ" to be specific said as crop-wise gross area irrigated, "γ" characterized as sources of paddy area and "ρ" to be specific production of paddy. The parameter of the stochastic model δ, γ, ρ, μ, and θ as standardized information. Take diverse esteems, given in Tables 1, 2, 3, 4, 5, and 6, and are demonstrated in Figs. 1, 2, 3, and 4 individually. In this paper, the optional information investigation and audit includes gathering and breaking down an immense range of data. The information is taken from the Department of Economics and Statistics, Government of Tamil Nadu, Chennai.

In this stochastic model approach, the level of concentration is predicted based on Tables 1, 2, 3, 4, 5, and 6. In Table 1, the data set includes different recent paddy

Table 1 Time series data for paddy includes rainfall (mm), sources of irrigation (nos), area irrigated (ha), paddy area (ha), and paddy production (tones)

Year	Actual rainfall (in mm)	Normal rainfall (in mm)	Sources of irrigation (in nos)	Crop wise gross area irrigated (in ha)	Paddy area (ha)	Paddy Prod. (tones)
2010–11	1165.1	908.6	2,912,129	3,347,557	1,905,726	5,792,415
2011–12	937.1	921.6	2,964,027	3,518,822	1,903,772	7,458,657
2012–13	743.1	921.0	2,642,700	2,991,459	1,493,276	4,050,334
2013–14	790.6	920.9	2,679,096	3,310,877	1,725,730	7,115,195
2014–15	987.9	920.9	2,725,641	3,394,295	1,794,991	7,949,437

Table 2 Normalized time series data for paddy on rainfall (mm), sources of irrigation (nos), area irrigated (ha), paddy area (ha), and paddy production (tones)

Year	Actual rainfall (in mm)	Normal rainfall (in mm)	Sources of irrigation (in nos)	Crop wise gross area irrigated (in ha)	Paddy area (ha)	Paddy Prod. (tones)
	θ		μ	λ_1	λ_2	p
2010–11	0.5561	0.4423	0.4672	0.4513	0.4812	0.3911
2011–12	0.4473	0.4487	0.4755	0.4744	0.4807	0.5035
2012–13	0.3547	0.4484	0.4240	0.4033	0.3770	0.2734
2013–14	0.3774	0.4483	0.4298	0.4464	0.4357	0.4804
2014–15	0.4716	0.4483	0.4373	0.4576	0.4532	0.5367

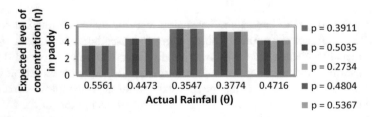

Fig. 1 Expected level of concentration (η) in paddy development

Table 3 Expected level of concentration (η) in paddy development and its variance (σ) in actual rainfall (θ)

$\mu = 0.4672, \delta = 0.4513, \gamma = 0.4812$

θ	$\rho = 0.3911$		$\rho = 0.5035$		$\rho = 0.2734$		$\rho = 0.4804$		$\rho = 0.5367$	
	η	σ	η	σ	η	σ	η	σ	η	σ
0.5561	3.5890	13.5799	3.6020	13.8729	3.5754	13.2733	3.5993	13.8125	3.6058	13.9591
0.4473	4.4622	20.9919	4.4784	21.4448	4.4453	20.5179	4.4751	21.3514	4.4831	21.5780
0.3547	5.6272	33.3833	5.6476	34.1035	5.6059	32.6295	5.6434	33.9551	5.6536	34.3154
0.3774	5.2891	29.4924	5.3082	30.1286	5.2691	28.8264	5.3043	29.9975	5.3139	30.3159
0.4716	4.2328	18.8885	4.2481	19.2960	4.2167	18.4620	4.2449	19.2120	4.2526	19.4159

Table 4 Expected level of concentration (η) in paddy development and its variance (σ) increases in rainfall (θ)

$\mu = 0.4672, \delta = 0.4513, \gamma = 0.4812$

Θ	$\rho = 0.3911$		$\rho = 0.5035$		$\rho = 0.2734$		$\rho = 0.4804$		$\rho = 0.5367$	
	η	σ	η	σ	η	σ	η	σ	η	σ
0.30	6.6535	46.6735	6.6776	47.6805	6.6283	45.6179	6.6726	47.4736	6.6847	47.9777
0.35	5.7030	34.2908	5.7237	35.0306	5.6814	33.5152	5.7194	34.8786	5.7298	35.2489
0.40	4.9901	26.2539	5.0082	26.8203	4.9712	25.6601	5.0045	26.7039	5.0135	26.9875
0.45	4.4357	20.7438	4.4517	21.1913	4.4188	20.2746	4.4484	21.0994	4.4565	21.3234
0.50	3.9921	16.8025	4.0065	17.1650	3.9769	16.4224	4.0036	17.0905	4.0108	17.2720

Table 5 Expected level of concentration (η) in paddy development and its variance (σ) increases in sources of irrigation (μ)

$\theta = 0.3547$, $\delta = 0.4513$, $\rho = 0.4812$

μ	$\rho = 0.3911$		$\rho = 0.5035$		$\rho = 0.2734$		$\rho = 0.4804$		$\rho = 0.5367$	
	η	σ	η	σ	η	σ	η	σ	η	σ
0.40	5.2235	28.8783	5.2410	29.5159	5.2052	28.2099	5.2374	29.3849	5.2461	29.7042
0.45	5.5240	32.2026	5.5437	32.9013	5.5035	31.4701	5.5396	32.7578	5.5495	33.1076
0.50	5.8246	35.708	5.8464	36.4704	5.8017	34.9086	5.8419	36.3138	5.8528	36.6954
0.55	6.1251	39.3943	6.1491	40.2231	6.1000	38.5252	6.1442	40.0529	6.1562	40.4677
0.60	6.4256	43.2617	6.4518	44.1594	6.3982	42.3201	6.4464	43.9750	6.4596	44.4244

Table 6 Expected level of concentration (η) in paddy development and its variance (σ) increases in paddy area (γ)

$\theta = 0.3547$, $\mu = 0.4672$, $\delta = 0.4513$

γ	$\rho = 0.3911$		$\rho = 0.5035$		$\rho = 0.2734$		$\rho = 0.4804$		$\rho = 0.5367$	
	η	σ	η	σ	η	σ	η	σ	η	σ
0.30	6.6341	37.2587	6.4687	32.9047	6.80742	41.7593	6.5027	33.8039	6.4198	31.6081
0.35	6.2522	34.0633	6.1573	31.3594	6.35168	36.8753	6.1768	31.9166	6.1292	30.5573
0.40	5.9658	32.8613	5.9237	31.5608	6.00987	34.2194	5.9323	31.8283	5.9113	31.1759
0.45	5.7430	32.9096	5.7420	32.8778	5.74402	32.9429	5.7422	32.8843	5.7418	32.8684
0.50	5.5648	33.8088	5.5967	34.973	5.53134	32.5875	5.5901	34.7339	5.6060	35.3160

data for the year of 2010–2011, 2011–2012, 2013–2014, and 2014–2015 and the details of time series data for paddy which includes the details of rain fall, sources of irrigation, area of irrigation, paddy area, and finally production of paddy in tones as on 2014–2015. In Table 2, data set is indicated to a normalization of Table 1 data set using Mathematica 7, this progression is critical when taking care of the parameters utilizing distinctive units and sizes of information.

The following Tables 3, 4, 5, and 6, which is utilized to show the expected level of concentration focus (η) in paddy improvement and its variance (σ). Analysts utilize the difference to see, how singular numbers identify with each other inside an informational index from the normal esteem (η and σ) (see Figs. 1, 2, 3, and 4).

The way toward examining information from Table 3, the actual rainfall (θ) values is compared with normal rainfall for the years 2010–2011, 2011–2012, 2013–2014, and 2014–2015. In these periods, the expected level of concentration in paddy development (η) and its variance (σ) also decreased. In the period 2012–2013, the estimation of θ is low, and at that point, the most extreme concentration happened and paddy advancement likewise diminished in this period, as appeared in Table 3 and Fig. 1.

The estimation of Table 2 and the year 2012–2013, the actual rainfall (θ) is 743.1 and it is equal to the normalized value as 0.3547, and for this circumstance, the expected level of concentration is increased in that period. Ultimately, the growth

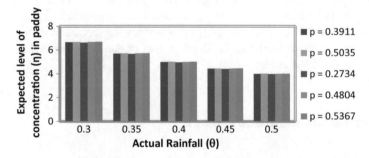

Fig. 2 Expected level of concentration (η) in paddy development for increases in θ

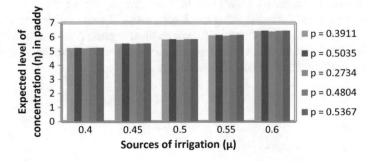

Fig. 3 Expected level of concentration (η) in paddy development for increases in μ

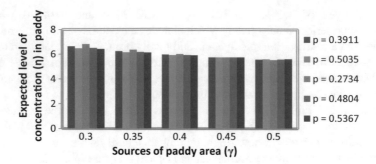

Fig. 4 Expected level of concentration (η) in paddy development for increases in γ

of paddy is also decreased. Accept that the (θ) esteem increments in other example and the development of paddy growth also increases. In these periods, the expected level of concentration in paddy development (η) and its variance (σ) also decreased, as shown in Table 4 (see Fig. 2).

In Table 5, the parameter of (μ) named as sources of irrigation. The parameter of (μ) value increases, then the expected level of concentration (η) in paddy development and its variance (σ) are likewise diminished. Eventually, the development of paddy is additionally incremented, as shown in Table 5 and Fig. 3.

With a specific end goal to think about the connections in view of Table 6, the parameter of (γ) which is named as paddy area is incremented inconsistently than the expected level of concentration (η) in paddy development and its variance (σ) are likewise diminished. At last, the development of paddy is likewise increments, as shown in Table 6 and Fig. 4.

4 Conclusion and Future Scope

The field of agriculture is the most important sector in developing country like India. The use of different data mining approaches in agribusiness can change the situation of agriculturists and can have better yield. In future, the degree to grow new stochastic model can be created for anticipating the development of conceivable outcomes and level of focus in sugarcane, wheat, beats, vegetables, organic products, and so forth. The model development is not only applicable in the field of agriculture. In a more extensive to utilize the other social impact area.

References

1. Arulpavai, R., Elangovan, R.: Determination of expected time to recruitment—a stochastic approach. Int. J. Oper. Res. Optim. **3**(2), 271–282 (2012)
2. Bartholomew, D.J.: The Stochastic Model for Social Processes, 3rd edn. Wiley, New York (1982)
3. Bartholomew, D.J., Forbes, A.F.: Statistical Techniques for Manpower Planning. Wiley, Chichester (1979)
4. Bocca, F.F., Rodrigues, L.H.A.: The effect of tuning, feature engineering, and feature selection in data mining applied to rainfed sugarcane yield modelling. Comput. Electron. Agric. **128**, 67–76 (2016)
5. Bravo-Ureta, B.E., Rieger, L.: Dairy farm efficiency measurement using stochastic frontiers and neoclassical duality. Am. J. Agr. Econ. **73**(2), 421–428 (1991)
6. Dlamini, N.S., Rowshon, M.K., Saha, U., Lai, S.H., Fikri, A. Zubaidi, J.: Simulation of future daily rainfall scenario using stochastic rainfall generator for a rice-growing irrigation scheme in Malaysia. Asian J. Appl. Sci. 3(05) (2015)
7. Esary, J.D., Marshall, A.W.: Shock models and wear processes. Ann. Probab. **1**(4), 627–649 (1973)
8. Fadhil, R.M., Rowshon, M.K., Ahmad, D., Fikri, A. Aimrun, W.: A stochastic rainfall generator model for simulation of daily rainfall events in Kurau catchment: model testing. In: III International Conference on Agricultural and Food Engineering 1152, pp. 1–10 (2016) (August)
9. Guerry, M.A., De Feyter, T.: An extended and tractable approach on the convergence problem of the mixed push–pull manpower model. Appl. Math. Comput. **217**(22), 9062–9071 (2011)
10. Lane, K.F., Andrew, J.E.: A method of labour turnover analysis. J. R. Stat. Soc. Ser. A (Gen.), 296–323 (1995)
11. Luo, B., Li, J.B., Huang, G.H., Li, H.L.: A simulation-based interval two-stage stochastic model for agricultural nonpoint source pollution control through land retirement. Sci. Total Environ. **361**(1), 38–56 (2006)
12. McClean, S.: A comparison of the lognormal and transition models of wastage. The Statistician, 281–294 (1975)
13. McClean, S.: The two-stage model of personnel behaviour. J. R. Stat. Soc. Ser. A (Gen.), 205–217 (1976)
14. McClean, S.: Manpower planning models and their estimation. Eur. J. Oper. Res. **51**(2), 179–187 (1991)
15. Rajesh, P., Karthikeyan, M.: A comparative study of data mining algorithms for decision tree approaches using WEKA tool. Adv. Nat. Appl. Sci. **11**(9), 230–243 (2017)
16. Young, A.: Demographic and ecological models for manpower planning. Aspects Manpower Plann. 75–97 (1971)

Reliable Monitoring Security System to Prevent MAC Spoofing in Ubiquitous Wireless Network

S. U. Ullas and J. Sandeep

Abstract Ubiquitous computing is a new paradigm in the world of information technology. Security plays a vital role in such networking environments. However, there are various methods available to generate different Media Access Control (MAC) addresses for the same system, which enables an attacker to spoof into the network. MAC spoofing is one of the major concerns in such an environment where MAC address can be spoofed using a wide range of tools and methods. Different methods can be prioritized to get cache table and attributes of ARP spoofing while targeting the identification of the attack. The routing trace-based technique is the predominant method to analyse MAC spoofing. In this paper, a detailed survey has been done on different methods to detect and prevent such risks. Based on the survey, a new proposal of security architecture has been proposed. This architecture makes use of Monitoring System (MS) that generates frequent network traces into MS table, server data and MS cache which ensures that the MAC spoofing is identified and blocked from the same environment.

Keywords ARP spoofing · MAC spoofing · MS · Network trace · Broadcast Man-in-the-middle attack

1 Introduction

In a networking environment, Address Resolution Protocol (ARP) matches system Internet protocol (IP) with the MAC address of the device. The client MAC ID can be disguised by the attacker as a genuine host by various breaching techniques [1]. The genuine MAC address of the host can be changed with the attacker's MAC. Consequently, the victim's system will understand the attacker's MAC address as

S. U. Ullas (✉) · J. Sandeep
Department of Computer Science, Christ University, Bengaluru 560029, India
e-mail: ullaspvm@gmail.com

J. Sandeep
e-mail: sandeep.j@christuniversity.in

© Springer Nature Singapore Pte Ltd. 2019
J. D. Peter et al. (eds.), *Advances in Big Data and Cloud Computing*,
Advances in Intelligent Systems and Computing 750,
https://doi.org/10.1007/978-981-13-1882-5_12

the genuine MAC address. In general scenario, a routing cache table is maintained, often called an ARP cache. This cache table is used to maintain a correlation between each IP address and its corresponding MAC address [2]. The ARP assigns the rules to correlate the IP with MAC. When an incoming packet is destined for a host machine on the same network, it arrives at the gateway first. The gateway is responsible to find the exact MAC address for the corresponding IP address. As soon as the address is figured, a connection will be established. Else, an ARP broadcast message will be sent as a special packet over the same network [3]. If the address is known by any one of the machines in the same network that has the IP address associated with it, then for further reference, ARP updates the ARP cache and will send a packet to the MAC address that replied. In addition, various methods of spoofing have been mentioned (Table 1) in this paper to analyse MAC spoofing.

The effects of ARP spoofing can have serious implications on the user. To ensure access to devices in any network, there is high demand for control in terms of acceptable access and response time. MAC address can be succeeded either by centralized or decentralized form or combination of both [3].

In an attacking scenario, when the attackers MAC ID is injected into a poisoned ARP table, any messages sent to the IP address will be sent to the attacker's id rather the genuine host. Once the privilege is changed, then the attacker can change the forward routing destination. Now, the genuine host cannot track or find Man-in-the-middle attack. In other words, any intruder can breach the ARP table and can act as a host machine. Packets sent through the IP address to MAC layer will not be sent through the genuine machine instead to the intruder's machine. Figure 1 represents the general ARP working scenario, where a victim broadcasts an ARP request packet across the network intended towards the router. In response, the router replies with an ARP acknowledgement. The scenario becomes vulnerable as the attacker also listens to these broadcasted messages [4]. Figure 2 represents the ARP attack scenario. In the attacking scenario, the request message is broadcasted all over the network. In this context, the attacker is also able to view the request. At this instant, the attacker tries to breach the network by spoofing the router and host (victim) by generating fake requests and reply messages [4].

2 Existing Work

Daesung Moon et al. proposed a routing trace-based network security system (RTNSS) method, wherein in the general environment, routing trace based is used to find the ARP attack [5]. A cache table is maintained to find any breach in the system. Any change in cache table, like a state where the table value will be changed from static to dynamic, is responded with the termination of attacker's connection. This model is further designed for detection and prevention techniques. On a comparison with other models, this design uses the server- and agent-based protocol. The server is used in the admin system and agent protocol is installed on the client system. Once a change in the cache table is figured, then the server will respond with all

Table 1 Comparison of the techniques used

Article	Attack method	Prevention	Solution	Solution accuracy
RTNSS	Cache table	Server and agent protocol	Only proposal model	–
ARP/NDP	Securing IPv6	New secure ARP protocol	New TPM-based technology implemented	Over heading range 7–15.4%
Attacks on IPv6	Securing IPv6 from ground	New prevention algorithm is introduced	No real-time implementation	–
DOS in WLAN	Detection and prevention on DDOS attack	ICM and EICM are the features used here	MATLAB and Wireshark analyzer outputs used to find the attack	Reduced network overhead
Anonymous arm	Prevention algorithm is used	AS-AR algorithm is used	As secure neighbour discovery, the malicious node cannot find the destination	–
ARP based MITM	Prevention technique to find genuine nodes	Voting system is introduced to identify malicious nodes	Fast reply messages from the nodes are considered as genuine	–
DS-ARP	Network routing path is traced frequently	Any changes in the routing path can identify the malicious node	Not complex algorithm it is a very simple algorithm to trace the malicious node	–
CQM mac protocol	Cycle quorum system is used as the analysis scheme	A Markhov chain model is used to get the accurate performance	channel slot allocation and cyclic quorum system. It can avoid bottle neck. Bird swarm algorithm is used here	–
Full-duplex mac layer design	Mac layer design protocol suite is used	Average throughput value is increased for both primary and secondary user	FD transmit sense reception and FD transmit sense is used for the design improvement	–

(continued)

Table 1 (continued)

Article	Attack method	Prevention	Solution	Solution accuracy
ARP deception defence	ARP virus is used in campus environment	A new proposal is introduced to share in the secure way	This system can prevent ARP virus attack in college environment	–
AH-MAC	This paper proposes an adaptive hierarchical Mac protocol for low rate and large scale with optimization of cross-layer techniques	Combination of AH-mac protocol over leech to improve the accuracy	In the proposed system, during the transfer, most of the network activities to the cluster head will reduce the node activity	Through acknowledgment, the AH consumes eight times less energy

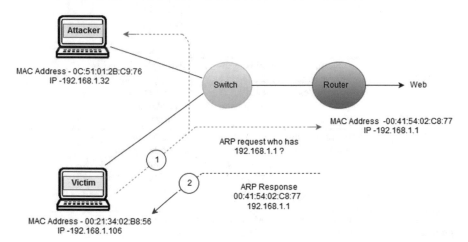

Fig. 1 General ARP working scenario

the prevention techniques. This design helps to prevent ARP attacks using routing trace-based technique.

Securing ARP/NDP from the Ground Up is proposed by Tian et al. [6]. It is mentioned clearly that ARP protocol is vulnerable due to many factors. It is perhaps because all the older proposals are based on modifying the existing protocol. The neighbour discovery protocol (NDP) has the same vulnerability and this protocol is used for IPv6 communication. A new ARP secure ARP/rARP protocol suite is developed, which does not require protocol modification, rather, it helps in continual verification of the target system. Tian et al. [6] introduced an address binding repository which follows 'a custom rules and properties'. Trusted Platform Module

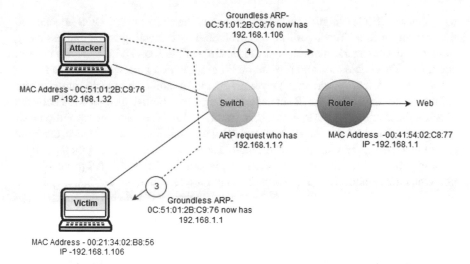

Fig. 2 ARP attack scenario

(TPM) is the latest technology which helps to prove the rules when needed. The TPM facilitated attestation has a low processing cost. It also supports IPv6 NDP as a TPM base in Linux implementation, the overhead ranging from 7 to 15.4%. This was compared with Linux NDP implementation. This work had the advantage of securing both IPv4 and IPv6 communication.

Ferdous A. et al., proposed a detection of neighbour discovery protocol [7]. To communicate, IPv6 uses NDP as ARP, by default, NDP lacks authentication and is stateless. In IPv4, the traditional way of spoofing attacks for breaching the IP to MAC resolution is relevant in NDP. A malicious host can do denial of service, Man-in-the-middle attacks, etc., using the spoofed exploited MAC address in IPv6 network. Since the IPv6 is new and not familiar, many detection prevention algorithms available for IPv4 are not implemented in IPv6 protocol. Some mechanisms which are proposed in IPv6 are not flexible and lack cryptographic key exchange. To overcome the problem, Ferdous et al., has proposed an active detection in IPv6 environment for NDP based attacks.

Detection and prevention proposal scheme are introduced by Abdallah et al. [8]. Easy installation, flexibility and portability have made wireless network more common today. The users must be aware of the security breach in the infrastructure of the networks and the vulnerability in it. In WLAN 801.11 (Denial of Service), DOS are one of the most vulnerable attacks. Wi-Fi protected access (WPA) and wired equivalent privacy (WEP) are the security features used to protect the network from intruders. Still, both the security protocol is vulnerable to DOS attack because their control frame and management are not encrypted to detect and prevent DOS attacks. The algorithm has five different tables to monitor the task done in the network. All the five tables will be processed when a client requests a connection. This can increase both latency and overhead. Abdallah et al. proposed an enhanced integrated central manager (EICM) algorithm which enhances the detection of DOS attacks and the prevention time. The algorithm was evaluated by collecting MAC address using Matlab and Wireshark analyser and used it for simulation. The results are used in successive DOS detection and prevention time by reducing network overhead.

Song G. et al., proposed an Anonymous-address-resolution model scheme where ARP protocol in the data link layer generally helps to establish a connection between MAC address and IP of the system [9]. In traditional ARP, during the resolution process, the destination address will be revealed by the malicious nodes. Hence, there is a risk of man-in-the-middle attack and denial of service attack. An anonymous address resolution protocol used here is to overcome this security threat. AS-AR protocol does not reveal the IP and MAC address of the source node. Since the destination address cannot be obtained, it cannot follow the attack. Experiment and analysis results were evident that AS-AR has good security levels, as secure neighbour discovery.

Seung Yeob Nam et al., proposed a mitigating ARP poisoning-based MIMA. Which mentions new mechanism to counteract ARP-based man-in-the-middle attack in a subnet [10]. Generally, both wired and wireless nodes coexist in subnet. New node is protected by making a vote between the good nodes and malicious nodes. The differences in the node are noted. There remains a challenge in terms of processing capability and access medium. To solve the issue, a uniform transmission capability of LAN cards with Ethernet where the access delays are of the smaller medium is proposed by Seung Yeob Nam et al. Another scheme to solve this voting issue is by filtering the votes from the responding speed of the reply messages. To figure out the fairness in the voting by analytics, nodes which respond faster with voting parameters has also been identified.

Ubiquitous system has many security threats [11]. Since it does not have much security considerations, anyone can get into the host to mitigate the security by disguising as genuine host, this can breach the genuine host to steal the private information. Min Su Song et al. proposed DS-ARP. A proposed scheme based on routing trace is modelled here. In this work, a new scheme of routing protocol is used. Where the network routing path is traced frequently for change in the network movement path. Then, the change will be determined. It does not use complex methods or heavy algorithm to work. This method can give high constancy and high stability since it does not change or alter the existing ARP protocol.

Hu, Xing et al., proposed a dynamic channel slot allocation scheme. In this architecture, high diversity node situation is compared with single-channel MAC protocol. Fewer collisions were identified for multichannel MAC protocol [12]. The research also identifies that overall performance is good for cyclic quorum-based multichannel MAC protocol. With the help of channel slot allocation and cyclic quorum, the system can avoid bottleneck. A Markov chain model is used to get the accurate performance of (CQM) MAC protocol. It helps to join the channel hopping scenario of CQM protocol and IEEE802.11 distributed coordinate function (DCF). In saturation bound situation, the optimal performance of CQM protocol was obtained. In addition, a dynamic channel slot allocation of CQM protocol was proposed by Xing et al., to improve the performance of CQM protocol in unsaturated situations. The protocol was based on wavelet neural network by using the QualNet platform wherein the performance of DCQM and CQM protocol was simulated. The results of simulations proved the performance of DCQM at these unsaturated situations.

Teddy Febrianto et al., proposed a cooperative full-duplex layer design [13]. Spectrum sensing is a challenging task. A new design was proposed by Teddy et al. which merges the full-duplex communications and cooperative spectrum sensing. The work demonstrates improvisation in terms of the average throughput of both primary user and secondary user for the given schemes. Under different full-duplex (FD) schemes, the average throughput was derived for primary users and secondary users. The FD transmit sense reception and FD transmit sense are the two types of FD schemes. For the secondary users, the FD transmit sense reception allows to transfer and receive data simultaneously. The FD transmits sense, continuously sense the channel during the transmission time. On an optimal scheme, based on cooperative FD, sensing of spectrum, this layer design displays the respective trade-offs. Under different multichannel sensing scheme and primary channel utilization for the secondary users, the average throughput value is analysed. Finally, for FD cooperative spectrum sensing, a new FD-MAC protocol is designed. It is also experimented in some applications that the proposed MAC protocol has a higher average throughput value.

Xu et al., proposed a method on college network environment [14]. It explains that in the vast development of web and Internet, the networking environment has developed a lot. In the same time, many data are exchanged and shared in the same environment on both secure and insecure way. While considering this in the fact the campus environment has various threats to the existing ARP virus and loopholes. Where in an unsecure environment, the sharing of data is happened between teachers and students in an unsecured way because of the ARP virus. A solution is proposed in this paper to prevent the ARP spoofing on a campus network in a ubiquitous way. Ismail Al-Sulaifanie et al., proposed an adaptive hierarchical MAC protocol (AH-MAC) for low-rate and large-scale optimization of cross-layer techniques [15]. The algorithm combines the strengths of IEEE 802.15.4 and LEACH. The normal nodes are battery operated while the predetermined clusters are supported with energy harvesting circuit. This protocol transfers most of the networks' activities to the cluster head by minimizing the nodes activity. Scalability, self-healing, energy efficiency are the predominant features of this protocol. In terms of throughput and energy consumption, a great improvement of the AH-MAC over LEACH protocol is found in the simulation results. While improving throughput via acknowledgement support the AH-MAC consumes eight times less energy.

3   Proposed Architecture

In this paper, a new architecture has been proposed and the section discusses the mechanism of MAC spoofing prevention in detail. The step-by-step process includes trace collection, time interval computation, monitoring system, identification and verification with IP and MAC in the table.

(i)   Trace collection

Trace collection is the process where the traces are collected by the system. Monitoring system (MS) fetch all the trace at dynamic time interval (T) [16]. A database table was created and inserted with all the collected trace information about the ARP connections which has been established from router and host. This trace information is collected under different scenarios and criteria. The computing waits time (T) based on variable traffic rate.

(ii)   Time interval (T) computation

The system computes the wait time interval (T) based on data traffic rate (1). The time interval computation is done based on three parameters. β, K and r_i (1)–(5). B is the constant time applicable for (time interval) T for any applications (2). K is the constant which has been identified for the rate at which data traffic changes (3). r_i is the different weight which has been assigned based on the variation of data traffic rate (4). The relation between the Parameter and Time interval is given below. Table 2 shows the weight parameter considered in the model.

$$T = \beta - K(1 - r_i) \tag{1}$$

$$T \propto \beta \tag{2}$$

$$T \propto \frac{1}{K} \text{ where } K \leq T \tag{3}$$

$$T \propto r \tag{4}$$

$$r_i = [1, 0.8, 0.5, 0.2, 0.0] \tag{5}$$

All the collected information will be stored in the individual database like the server database and the monitoring system database. When the system identifies low traffic, then on a given low traffic time stamp, the system will request for trace collection. The method dynamically adapts to the network's traffic. The system will trace and collect the information from the host or the agent to balance the overhead of security [17].

MS working scenario

Figure 3 presents the proposed algorithm which uses its own ID with MAC address. For example, Host B with ID 02 and BC is the corresponding MAC address

Table 2   Weight value parameter

Weight	Value	Classification
r_1	1	Very low traffic
r_2	0.8	Low traffic
r_3	0.5	Moderate traffic
r_4	0.2	High traffic
r_5	0	Very high traffic

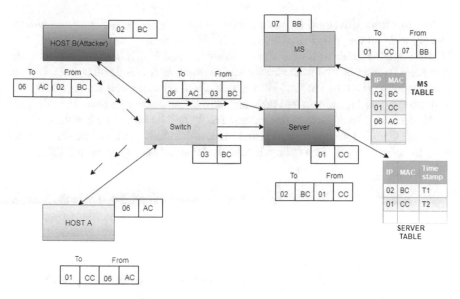

Fig. 3 Proposed monitoring system working scenario

sends information from its own address, i.e. 02 BC to host A with IP 06 and MAC address as AC. The system has monitoring system and server with cache table. The system uses checking and termination to determine the intruder. The server table has a list of ID with corresponding MAC ID and time stamp of when the request and reply is forwarded [18]. It is same with the MS table but MS acts as a monitoring admin to figure out the breach [19].

Working scenario of MS

Step 1: Trace collection at time T.

Step 2: Dynamic time allocation and routing.

Step 3: Path identification from the trace of the cache table. Then the system will start the default MS.

Step 4: If MS finds duplication in the server table with the first connected IP and MAC, then MS cross check with its own table and servers table.

Step 5: MS can identify the intruder's ID [20].

Step 6: Check for the next server's row in the table for the attack verification.

Step 7: If any attack is determined, then a message is broadcasted.

Step 8: Broadcasted message has attacker's id and data. This will be noted in all the system.

Step 9: Hosts will block the id for any communication in the network. So, within the same network, the same intruder cannot connect.

(iii) Monitoring system

The monitoring system helps to monitor the router and host data transfer with the routing table and monitoring table, where it will check duplication of IP and MAC [21]. If any MAC address is reused or duplicated, the MS will check the servers table. A Request packet is sent to the host to check whether it is a genuine host. If there is no response given by the host, then it blocks and notes it as an attacker's ID. The system, then, will terminate that table and skip to the next table data. MS will send an ARP message to the next table data and a reply will be sent since it is the Genuine host. The Genuine host is identified and the terminated IP will be blacklisted to all the ports of the switch. The system blocks the IP and denies service to it [22].

(iv)   Identification

As shown in Fig. 3, the ARP request from the host is broadcasted over the network and before the server responds to that request message, the intruder tries to pretend as a server and sends a reply message. This is how an attacking scenario occurs [23]. To check the attacker's ID, MS will check the host address of the received ARP request. In the MS table, the MS will note the IP and MAC address. In the identification part, the system will check for the current server and for the request being broadcasted from the host.

(v)   Verification with IP and MAC in the table

When an attack is figured, for the corresponding genuine IP address, the MS will check the server table for the real and duplicated MAC address. In the table, the system will check broadcasted MAC address with the requested IP address first. The MS will make the server send a ping packet to the server table with corresponding MAC address and IP address [24]. After the process, MS will wait for the reply. If the address mentioned in the server table is genuine, then immediately a response will be broadcasted to the server.

As soon as a response is broadcasted, the MS will mark the row as a genuine host and immediately block the next duplicate address in the table. If a reply message is not responded to the server, then immediately the server will block the first row of address in the table and mark the second row of address as a genuine host. At the final stage, after noticing an attacker's IP address, the MS will send an acknowledgement with deauthentication packets to all the hosts in the network. This process ensures that the attacker cannot connect to the same network. The proposed model helps to track any ARP spoofing in the network. Once an attack is traced, the proposed algorithm will help to note the MAC ID and IP address of the intruder's system.

The data will be stored in a database table. Once an attack is noted, the attacker's ID will be noted and blacklisted. Later, intruders' address will be sent towards the network [25]. So, the attackers cannot connect to any of the hosts which relate to the server, since the attacker's address is blacklisted. In addition, Wireshark tool was used to find the traces and traffic. Wireshark cache and traces of the Wi-Fi transmission is shown in Figs. 4 and 5. The change in the attributes of the cache table helps to identify the intruder's presence. In the previous paper on RTNSS by Daesung Moon et al., it was stated that, if there is any change in the ARP cache table then the system will change the value from dynamic to static. The proposed algorithm with the help

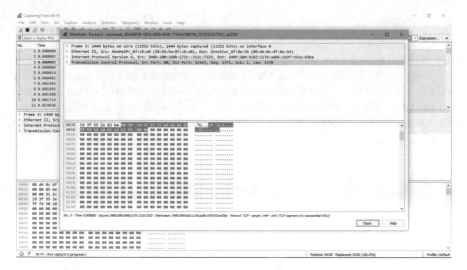

Fig. 4 Wireshark cache table attributes

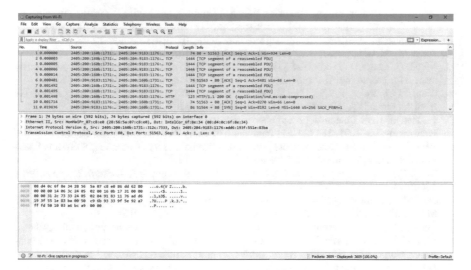

Fig. 5 Wireshark cache attribute check

of MS, cache table and time stamp identifies the attack with attacker's ID and stops the attack. In future, the attacker or the intruder cannot do the same attack on the server connected network environment.

4 Conclusion

It is essential to curb MAC spoofing. With the help of Monitoring System algorithm, an attack can be figured with attacker's address. Once an attack is figured, easily the connection to the IP can be disconnected and all the connected server hosts will get the attacker's id with additional data, such as, IP, MAC and so forth. In future, the attacker cannot spoof any of the hosts which are connected on the same server network.

References

1. Hsiao, H.-W., Lin, C.S., Chang, S.-Y.: Constructing an ARP attack detection system with SNMP traffic data mining. In: Proceedings of the 11th International Conference on Electronic Commerce (ICEC 09). ACM DL (2009)
2. Kim, J.: IP spoofing detection technique for endpoint security enhancement. J. Korean Inst. Inf. Technol. 75–83 (2017)
3. Benzaïd, C., Boulgheraif, A., Dahmane, F.Z., Al-Nemrat, A., Zeraoulia, K.: Intelligent detection of MAC spoofing attack in 802.11 network. In: '16 Proceedings of the 17th International Conference on Distributed Computing and Networking (ICDCN), Article No. 47. ACM (2016)
4. Asija, M.: MAC address. Int. J. Technol. Eng. IRA (2016). ISSN 2455-4480
5. Moon, D., Lee, J.D., Jeong, Y.-S., Park, J.H.: RTNSS: a routing trace-based network security system for preventing ARP spoofing attacks. J. Supercomput. **72**(5), 1740–1756 (2016)
6. Tian, D.(J.), Butler, K.R.B., Choi, J., McDaniel, P.D., Krishnaswamy, P.: Securing ARP/NDP from the ground up. IEEE Trans. Inf. Forensics Secur. **12**(9), 2131–2143 (2017)
7. Barbhuiya, F.A.: Detection of neighbor discovery protocol based attacks in IPv6 network. Networking Sci. **2**, 91–113 (2013) (Tsinghua University Press and Springer, Berlin)
8. Abdallah, A.E., et al., Detection and prevention of denial of service attacks (DOS) in WLANs infrastructure. J. Theor. Appl. Inf. Technol. 417–423 (2015)
9. Song, G., Ji, Z.: Anonymous-address-resolution model. Front. Inf. Technol. Electron. Eng. **17**(10), 1044–1055 (2016)
10. Nam, S.Y., Djuraeva, S., Park, M.: Collaborative approach to mitigating ARP poisoning-based Man-in-the-Middle attacks. Comput. Netw. **57**(18), 3866–3884 (2013)
11. Song, M.S., Lee, J.D., Jeong, Y.-S., Jeong, H.-Y., Park, J.H.: DS-ARP: a new detection scheme for ARP spoofing attacks based on routing trace for ubiquitous environments. Hindawi Publishing Corporation, Scientific World Journal Volume, Article ID 264654, 7 p. (2014)
12. Hu, X., Ma, L., Huang, S., Huang, T., Liu, S.: Dynamic channel slot allocation scheme and performance analysis of cyclic quorum multichannel MAC protocol. Mathematical Problems in Engineering, vol. 2017, Article ID 8580913, 16 p. (2017)
13. Febrianto, T., Hou, J., Shikh-Bahaei, M.: Cooperative full-duplex physical and MAC layer design in asynchronous cognitive networks. Hindawi Wireless Communications and Mobile Computing, vol. 2017, Article ID 8491920, 14 p. (2017)
14. Xu, Y., Sun, S.: The study on the college campus network ARP deception defense. Institute of Electrical and Electronics Engineers(IEEE) (2010)
15. Ismail Al-Sulaifanie, A., Biswas, S., Al-Sulaifanie, B.: AH-MAC: adaptive hierarchical MAC protocol for low-rate wireless sensor network applications. J. Sens. Article no 8105954 (2017)
16. Sheng, Y., et al.: Detecting 802.11 MAC Layer Spoofing Using Received Signal Strength. In: Proceedings of the 27th Conference on IEEE INFOCOM 2008 (2008)
17. Alotaibi, B., Elleithy, K.: A new MAC address spoofing detection technique based on random forests. Sensors (Basel) (2016) (March)

18. Satheeskumar, R., Periasamy, P.S.: Quality of service improvement in wireless mesh networks using time variant traffic approximation technique. J. Comput. Theor. Nanosci. 5226–5232 (2017)
19. Durairaj, M., Persia, A.: ThreV—an efficacious algorithm to Thwart MAC spoof DoS attack in wireless local area infrastructure network. Indian J. Sci. Technol. **7**(5), 39–46 (2014)
20. Assels, M.J., Paquet, J., Debbabi, M.: Toward automated MAC spoofer investigations. In: Proceedings of the 2014 International Conference on Computer Science & Software Engineering, Article No. 27. ACM DL (2014)
21. Bansal, R., Bansal, D.: Non-cryptographic methods of MAC spoof detection in wireless LAN. In: IEEE International Conference on Networks (2008)
22. Huang, I.-H., Chang, K.-C., Yang, C.-Z.: 2014, Countermeasures against MAC address spoofing in public wireless networks using lightweight agents. In: International Conference on Wireless Internet, 1–3 March 2010
23. Hou, X., Jiang, Z., Tian, X.: The detection and prevention for ARP Spoofing based on Snort. In: Proceedings of the International Conference on Computer Application and System Modeling, vol. 9 (2010)
24. Arote, P., Arya, K.V.: Detection and prevention against ARP poisoning attack using modified ICMP and voting. In: Proceedings of the International Conference on Computational Intelligence and Networks. IEEE Explore (2015)
25. Raviya Rupal, D., et al.: Detection and prevention of ARP poisoning in dynamic IP configuration. In: Proceedings of the IEEE International Conference on Recent Trends in Electronics, Information & Communication Technology. IEEE Explore (2016)

Switch Failure Detection
in Software-Defined Networks

V. Muthumanikandan, C. Valliyammai and B. Swarna Deepa

Abstract SDN networks can have both legacy and OpenFlow network elements which are managed by the SDN controller. When the entire switch fails, leading to multiple link failures, addressing link failures may become inefficient, if it is addressed at the link level. Hence, an extensive approach to address the failure at the switch level, by including the failure rerouting code at the switch prior to the faulty switch, so that minimal configuration changes can be made and those changes can be made proactively. The objective of this work is to detect switch failures by including detection logic at the switch prior to the faulty switch. The link failure is examined in terms of a single link failure or failure of a subsequent switch. Link failure if it is due to failure of a subsequent switch, corrective action is taken proactively for all links which are affected by the switch failure.

Keywords Link failure detection · Link failure protocols · Link failure recovery
Software-defined networks · OpenFlow switch

1 Introduction

Service providers can meet their end user's expectation, only when there is good network quality. To provide the best recovery mechanism with expected reliability, the failures should be detected faster. Failures in Software-Defined Networks are due to, link or switch failure or connection failure between the standardized control plane and data plane. In SDN networks, though OpenFlow switches in data plane

V. Muthumanikandan (✉) · C. Valliyammai · B. Swarna Deepa
Department of Computer Technology, Madras Institute of Technology Campus,
Anna University, Chennai, India
e-mail: muthumanikandanbe@rocketmail.com

C. Valliyammai
e-mail: cva@annauniv.edu

B. Swarna Deepa
e-mail: swarnadeepa@gmail.com

© Springer Nature Singapore Pte Ltd. 2019
J. D. Peter et al. (eds.), *Advances in Big Data and Cloud Computing*,
Advances in Intelligent Systems and Computing 750,
https://doi.org/10.1007/978-981-13-1882-5_13

can identify link failure since it does not have control plane, it has to wait for the controller to suggest any alternate routes to respond. Whenever a particular link fails in a software-defined network, it could be the result of a single link failure or could also be due to the failure of underlying switches. In the latter case, it may lead to multiple failures across the network. Hence, it will be beneficial to detect such failures separately so that they can be handled efficiently.

2 Related Works

When the primary path in the network is detected with link failure, in segment protection scheme [1], it provided a secure path for the packets via alternate backup path for each associated link. Independent transient plane design [2] overcame the performance limitation of segment protection scheme by limiting the number of configuration messages and flow table entries. To detect the complete failure of the data plane switches, conservative time-synchronization algorithm was used. The detection system was based on a conservative time-synchronization algorithm with message passing interface used to probe connections by exchanging messages between nodes in logical processes as well as messages within a logical process [3, 4]. A conservative time-synchronization algorithm was used to determine out-of-order time-stamp messages in the event of a failure in a Virtual Network Environment. The order of the message time-stamps was used for detection of a failure. In click modular routing [5], the threshold power consumption value was already known from the benchmark information. The benchmark information was compared with network processor at the operating system level. Using such comparison of power consumption information, any failure in the network processor was detected at the modular level. A high hardware availability physical machine that can quickly detect the hardware failure status was due to Advanced Telecommunications Computing Architecture which supported Intelligent Platform Management Interface. The new symmetric fault tolerant mechanism divided Advanced Telecommunications Computing Architecture physical machines and KVM into pairs, such that each machine of a pair supported fault tolerance for each other. The failed virtual machines were covered on the other pair of the physical machine immediately when failure was detected either at physical layer or at Virtualization layer [6].

Bi-directional Forwarding Detection (BFD) [7] was used to detect faults along a "path" between two networking elements at the protocol layer. Control packets were sent by each node with information on the state of the monitored link and the node which received these control packets, replied with an echo message containing the respective session status. After building up the handshake, each node sent frequent control messages to detect any link failure. SPIDER, [8] was a packet processing design which was fully programmable and was helpful in detecting the link failure. SPIDER provided mechanisms to detect the failure and instant recovery by rerouting of traffic demands using a stateful proactive approach, without the SDN controller's supervision. Similar to SPIDER, which used special tags, VLAN-tag was used for fast link failure recovery by consuming low memory [9].

The interworking between different vendors were allowed through Link Layer Discovery Protocol (LLDP)-based Link Layer Failure Detection (LLFD) mechanism [10]. LLFD used the packet-out and packet-in messages present in the OpenFlow protocol to discover configuration inconsistencies between systems and to detect link failure among switches. Fast Rerouting Technique (FRT) was based on fast reroute and the characteristics of the backup path. During failure detection, the controller evaluated the set of shortest and efficient backup paths using FRT which reduced the packet drop and the memory overhead [11]. All the packet flows affected by link failure were clustered in a new "big" flow, once the link failure was detected in Local Fast Re-route [12]. For the new clustered flow, SDN controller dynamically deployed a local fast re-route path. Operational flows between the switches and controller were greatly reduced in the local fast reroute technique. The controller heartbeat technique used the heartbeat hello messages which sent out directly from the controller instead of switches to detect the link failures [13]. Failure detection service with low mistake rates (FDLM) was based upon modified gossip protocol which utilized the network bandwidth efficiently and reduced false detections [14]. The failure consensus algorithm was modified, in order to provide strong group membership semantics. It also reduced the false detection rate using an extra state to report false failure detection.

3 Proposed System

The architecture of the system is shown in Fig. 1. The architecture consists of Open-Daylight SDN Controller and OpenFlow switches. The LLDP defines basic discovery capabilities of the network and is used for detecting link failures. Upon detection of the first link failure, the Switch Failure Detection (SFD) is triggered which takes the network topology and failed link as inputs and returns if any switch has failed. The obtained information is used to proactively update optimal routes for all the flows that involved in the failed switch in their paths.

3.1 Working Principle

The SFD technique gets the network topology of the system and failed link as its input. The failed link refers to the first link failure detected. Using network topology information, SFD finds out the source and destination of the link. In the subsequent step, it finds out if the source and destination of the failed link are in the OpenFlow switches. If either source or destination of the failed link is a switch, SFD further continues execution to verify if the entire link in the switch has failed. For each identified switch, it finds all the connected hosts. Next, it finds the reachable hosts and the packet loss ratio is calculated. If the drop rate is 100%, then SFD detects that the identified switch has failed. The process flow of switch failure detection is shown in Fig. 2.

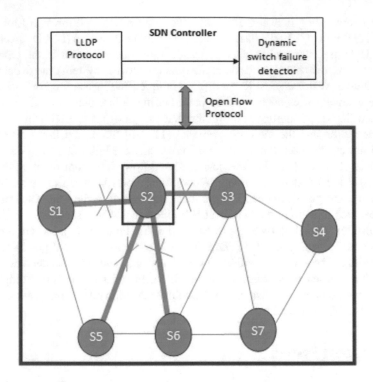

Fig. 1 Proposed SFD system architecture

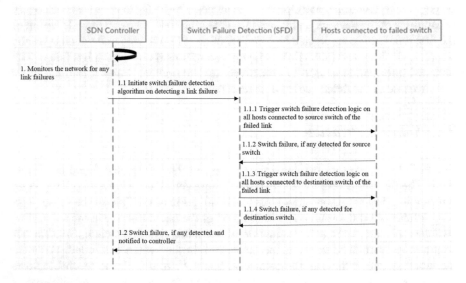

Fig. 2 Process flow of switch failure detection

3.2   SFD Algorithm

Input: The network topology G (V, L)-where V denote the switches and L denote the links, the failed link l.
Output: Failed switch, if any
1. Set SWF$_s$= SWF$_D$={Ø}
2. The controller in SDN checks the status of network topology and finds link l is disrupted.
3. $G \leftarrow G\ (V, L - l)$.
4. **if** s(l) is a switch - if s(l) is a switch, **then**
        HS←{ H$_1$,H$_2$,H$_3$,...,H$_i$}←Find all the hosts connected to s(l) on G   //HS-hosts connected to source switch
        if all the pings fail for the set of hosts in HS, **then**
            Set SWF$_s$ ←s(l) // source switch failed
        end if
        if SWF$_s$ is not empty, **then**
            controller learns that the switch SWF$_s$ has failed
        end if
5. **end if**
6. **if** d(l) is a switch, **then**
        HD←{ H$_1$,H$_2$,H$_3$,...,H$_j$}←Find all the hosts connected to d(l) on G   //HD-hosts connected to destination switch
        if all the pings fail for the set of hosts in HD, **then**
            Set SWF$_D$ ←d(l) // destination switch failed.
        end if
        if SWF$_D$ is not empty, **then**
            controller learns that the switch SWF$_D$ has failed
        end if
7. **end if**
8. Switch failure, if any is detected

4   Experimental Setup

The experimental setup is created with two Ubuntu servers over an Oracle Virtual Box. The first server is used to emulate the network using Mininet. The second server acts as the SDN controller, running OpenDaylight Beryllium version 0.4.0. Wireshark packet analyzer is installed for analyzing the packets transmitted. A custom topology is created with four hosts and seven switches in Mininet. A bandwidth of 10 Mbps is set for the links between switches and a delay of 3 ms is configured. OpenFlow protocol is used for the communication between the switches and controller. A remote open daylight controller is configured to manage the entire network and it listens on port number 6633 for any OpenFlow protocol messages. The hello packets are transmitted in the form of the ping requests for a period of 5 ms in the network setup. LLDP packets are used to discover the links between various switches.

5 Performance Evaluation

The performance of the SFD approach is evaluated when different switches fail in the network setup. The time taken for switch failure detection is shown in Fig. 3. The average detection time is 35.28571 ms.

The performance of the SFD approach is evaluated by detecting the failure in the various links associated with a particular failed switch and the time taken for failure detection is shown in Fig. 4. The average detection time is 35 ms.

The network throughput before and after switch failure is analyzed using the data obtained from Wireshark and it is shown in Fig. 5.

6 Conclusion and Future Work

The proposed SFD technique detects the switch failures in SDN. The detection approach detects the failure in the switch prior to the faulty switch. Whenever a particular link fails, SFD examines whether the link failure is due to single link failure or due to a subsequent switch failure. The proposed work minimizes the detection time that would otherwise be required for the other links connected to

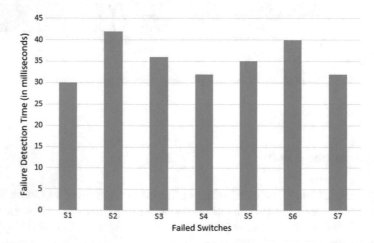

Fig. 3 Comparison of failure detection time for different switches

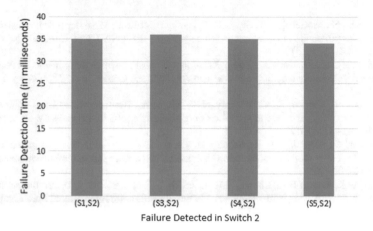

Fig. 4 Comparison of failure detection time for different links of failed switch

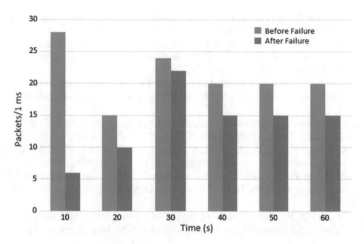

Fig. 5 Network throughput before and after a switch failure

the faulty switch. The proposed work also increases the network throughput. The proposed work can be extended to provide recovery from such switch failures during the switch downtime. It can be extended to arrive at an optimal active path for recovery immediately after the occurrence of the switch failure by avoiding the affected links.

References

1. Sgambelluri, A., Giorgetti, A., Cugini, F., Paolucci, F., Castoldi, P.: OpenFlow-based segment protection in ethernet networks. J. Opt. Commun. Netw. **9**, 1066–1075 (2013)

2. Kitsuwan, N., McGettrick, S., Slyne, F., Payne, D.B., Ruffini, M.: Independent transient plane design for protection in OpenFlow-based networks. IEEE/OSA J. Opt. Commun. Networking **7**(4), 264–275 (2015)

3. Al-Rubaiey, B., Abawajy, J.: Failure detection in virtual network environment. In: 26th International Telecommunication Networks and Applications Conference (ITNAC), pp. 149–152 (2016)

4. Shahriar, N., Ahmed, R., Chowdhury, S.R., Khan, A., Boutaba, R., Mitra, J.: Generalized recovery from node failure in virtual network embedding. IEEE Trans. Netw. Serv. Manage. **14**(2), 261–274 (2017)

5. Mansour, C., Chasaki, D.: Real-time attack and failure detection for next generation networks. In: International Conference on Computing, Networking and Communications (ICNC), pp. 189–193 (2017)

6. Wang, W.J., Huang, H.L., Chuang, S.H., Chen, S.J., Kao, C.H., Liang, D.: Virtual machines of high availability using hardware-assisted failure detection. In: International Carnahan Conference on Security Technology (ICCST), pp. 1–6 (2015)

7. Tanyingyong, V., Rathore, M.S., Hidell, M., Sjödin, P.: Resilient communication through multihoming for remote healthcare applications. In: IEEE Global Communications Conference (GLOBECOM), pp. 1335–1341 (2013)

8. Cascone, C., Sanvito, D., Pollini, L., Capone, A., Sansò, B.: Fast failure detection and recovery in SDN with stateful data plane. Int. J. Network Manage. (2017)

9. Chen, J., Chen, J., Ling, J., Zhang, W.: Failure recovery using vlan-tag in SDN: High speed with low memory requirement. In: IEEE 35th International Performance Computing and Communications Conference (IPCCC), pp. 1–9(2016)

10. Liao, L., Leung, V.C.M.: LLDP based link latency monitoring in software defined networks. In: 12th International Conference on Network and Service Management (CNSM), pp. 330–335 (2016)

11. Muthumanikandan, V., Valliyammai, C.: Link failure recovery using shortest path fast rerouting technique in SDN. Wireless Pers. Commun. **97**(2), 2475–2495 (2017)

12. Zhang, X., Cheng, Z., Lin, R., He, L., Yu, S., Luo, H.: Local fast reroute with flow aggregation in software defined networks. IEEE Commun. Lett. **21**(4), 785–788 (2017)

13. Dorsch, N., Kurtz, F., Girke, F., Wietfeld, C.: Enhanced fast failover for software-defined smart grid communication networks. In: IEEE Global Communications Conference (GLOBECOM), pp. 1–6 (2016)

14. Yang, T.W., Wang, K.: Failure detection service with low mistake rates for SDN controllers. In: 18th Asia-Pacific Network Operations and Management Symposium (APNOMS), pp. 1–6 (2016)

A Lightweight Memory-Based Protocol Authentication Using Radio Frequency Identification (RFID)

Parvathy Arulmozhi, J. B. B. Rayappan and Pethuru Raj

Abstract The maturity and stability of the widely used service paradigm have brought in a variety of benefits not only for software applications but also all kinds of connected devices. Web Services hide all kinds of device heterogeneities and complexities and present a homogeneous outlook for every kind of devices. Manufacturing machines, healthcare instruments, defence equipment, household utensils, appliances and wares, the growing array of consumer electronics, handhelds, we are able, and mobiles are being empowered to be computing, communicative, sensitive and responsive. Device services are enabling these connected devices to interact with one another in order to fulfil various business requirements. XML, JSON and other data formats come handy in formulating and transmitting data messages amongst all kinds of participating applications, devices, databases and services. In such kinds of extremely and deeply connected environments, the data security and privacy are being touted as the most challenging aspects. It should be noted that even security algorithms of steganography and cryptography provides us with the probability of 0.6 when it comes to protection in a service environment. Having understood the urgent need for technologically powerful solutions for unbreakable and impenetrable security, we have come out a security solution using the proven and promising RFID technology that has the power to reduce the probability of device-based attacks such as brute-force attack, dictionary attack and key-log-related attacks—which would make the device applications and services immune from malicious programmes.

1 Introduction

We tend towards to the cloud era with the emergence of the device/Internet infrastructure as the most affordable, open, and public communication infrastructure [1].

P. Arulmozhi (✉) · J. B. B. Rayappan
Department of Electronics & Communication Engineering, SEEE, SASTRA University,
Thanjavur, India
e-mail: parvathy@ece.sastra.edu

P. Raj
Reliance Jio Cloud Services (JCS), Bangalore, India

© Springer Nature Singapore Pte Ltd. 2019
J. D. Peter et al. (eds.), *Advances in Big Data and Cloud Computing*,
Advances in Intelligent Systems and Computing 750,
https://doi.org/10.1007/978-981-13-1882-5_14

There are a number of device-related technologies enabling the era of device service [2]. Not only applications are being device-enabled with device interfaces but also all kinds of personal and professional devices are being expressed and exposed as publicly discoverable, network-accessible, interoperable, composable and extensible services [3]. Devices and machines are being instrumented with communication modules, so that they can connect with their nearby devices as well as with remote devices and applications over any network. Thus, the concept of machine-to-machine (M2M) communication has emerged and evolved fast. With the arrival of a plenty of connected devices and machines, the security implications have acquired a special significance [4]. Devices, when interacting with other devices in the vicinity or with remote ones, tend to transmit a variety of environmental and state data. The need here is how to encapsulate the data getting transmitted over the porous Internet communication infrastructure in order to be away from hackers [5, 6]. That is, the security, privacy, integrity and confidentiality of the data getting sent over the public network have to be guaranteed through a host of technical solutions [7].

We are slowly yet steadily entering into the IoT era. Every common, casual and cheap things in our everyday environments are systematically digitized and enabled to participate in the mainstream computing [8, 9]. There are a bunch of IoT platforms enabling the realization of business-specific and agnostic IoT services and systems. Another buzzword is cyber-physical systems (CPM). Devices are not only connected to other devices but also with remotely held cloud applications, services, and databases in order to be empowered accordingly to do things adaptively [10, 11] Devices become self and situation-aware in order to exhibit adaptive behaviour. The Cloud, IoT and CPM technologies, tools and tips come handy in adequately empowering all kinds of embedded systems to be context-aware [12].

With all these advancements towards the knowledge era, one hurdle and hitch, which is very prominent, is the security aspect [13, 14]. This paper is specially prepared for conveying a new and innovative security mechanism.

The rest of the paper explains as below: Sect. 2 examines the braves and security concerns of Internet services, Sect. 3 tells the related work for Internet device services RFID, Sect. 4 explains the RFID security tokens framework, Sect. 5 provides results and discussions and finally, Sect. 6 draws some conclusions.

2 RFID Security Concerns and Threats

RFID is not a flawless system. Similar to other dump devices, it has a lot security and privacy risks which are addressed by IT, and any organizations. In spite of its widespread applications and usage, RFID posses security threats and challenges that needs to be dealt properly prior to use on the internet. It consists of three parts such as Reader, transponder and antenna. In many web applications, the readers are trusted only with the RFID tag storage. To protect the tag data, in this paper, a lightweight memory algorithm of 128 bits as a key size is used for security with large number of rounds.

2.1 An Insecure Tags Vulnerable to the Following Issues

The needs for new tags should comprise and support the new security applications on the internet. Nowadays the enhance RFID tags such as Mifare class tag with 4 k memory. First, they deal with the difficulty of authenticity. This is applicable if any company wants to defend its labelled products from stealing or if RFID tags are worked in car keys to toggle the immobilizer scheme on and off. The next problem is, to stop the tracking of clients via the RFID tags they carry. This protection of privacy is generally a big challenge in the face of the emergence of new wireless technology. Table 1 shows the attacks of RFID system on the internet.

Attackers can mark information to a basic black tag or can alter data in the tag, in order to gain access or confirm a product's authenticity. The things that an attacker can do with basic tags are:

1. They can modify the existing data in the basic tags and can make invalid tags into a valid tag and viceversa.
2. The attackers can change the tag of an object to that of another tag embedded in another object.
3. They can create their own tag using personal information attached to another tag.

Table 1 RFID attacks and causes

S. No.	Name RFID Security attack	Cause of issues
1	Sniffing	*A request sent by a fake RFID reader*
2	Traffic analysis	*A traffic analysis tools to track predictable tag responses over time*
3	Denial of service attacks	*An attacker can use the "kill" command, implemented in RFID tags*
4	Spoofing	*The software permits intruders to overwrite existing RFID tag data with spoof data*
5	RFID counterfeiting	*Depending on the computing power of RFID tags*
6	Insert attack	*An attacker tries to insert system commands to the RFID system*
7	Physical attack	*Electromagnetic interference can disrupt communication between the tags and the reader*
8	Virus attack	*Like any other information system, RFID is also prone to virus attacks*

So, make sure your tag is using some sort of security algorithms such as Tiny Encryption Algorithm, Blowfish or Extended Tiny Encryption Algorithm. Any device that dealing with sensitive objects such a passports or any identity documents make sure that has a proper security or not. For real-time application or healthcare monitoring system, the confidential data of RFID system could not be revealed. In order to protect the RFID tag's data and avoid any anomalies in that RFID system it is necessary to go for the encryption.

2.2 Adoption of RFID Tags

Industrial applications of RFID can be established today in supply chain management, automated payment systems, airline baggage management and so on. The sensitive information of RFID data is being communicated over Internet and is stored in different devices. Certain communication protocols security algorithms make the RFID system as a communication device on Internet with safe. RFID device that dealing with sensitive objects such a passports or any identity documents make sure that has a proper security or not. For real-time application or health care monitoring system, the confidential data of RFID data could not be revealed. In order to protect the RFID tag's data and avoid any anomalies in that RFID system, it is necessary to go for the encryption.

2.3 Move Towards for Tackling Security and Privacy Issues for RFID System

RFID tags are well thought-out 'dumb' devices, in that they can pay attention and react. This brings up risks of unauthorized access and alteration of tag data. A lot of clarification for undertaking the security and privacy issues surrounding RFID. They can be categorised into the following areas:

1. Protection of Tag's data.
2. Integrity of the RFID Reader.
3. Personal Privacy.

The integrity of the RFID reader and the personal privacy are related with readers. In many web applications, the readers are trusted with only the RFID tag storage. Among these areas the RFID tag storage is concentrated because the tag data should not be revealed in security environment. So the encrypted data are collected by the readers and send to backend server for decryption. Here the Tiny Encryption Algorithm is used and the results are verified in LAB View software. The following diagram shows the solution for RFID tag protection.

Fig. 1 Device Service
model

Execut

3   Design of Device Service Model

The device service is an eye-catching, powerful and hottest technology in the world for the development of device-based applications. It must be secured in a business environment with remote login. To gain the acceptance by the programmer or developer, it can be located and invoked over a network with RFID device security algorithms. The RFID Tiny Encryption Algorithms are a sprint for several platforms. In this methodology, the device programme is written in python and encrypted tokens are generated RFID system. This technology actually similar to remote login authentication and pretended by the wireless sensor networks, unique registry identifier and SOA protocol. The process consists of three segments, hunting (discovery), tying and finishing. The Device Service Model is shown in the Fig. 1.

SOAP: Simple object access protocol uses a message path and flows from the dispatcher to the receiver, consisting of envelope, Body and Header.
WSDL: Device services description language (WSDL) used to describe the well-designed model of the services.
UDDI: Universal Description Discovery and Integration used as a record for the Device Services.

4   Related Work for RFID Security Framework

RFID technology is not the flawless system as it hosts a number of security and privacy concerns, which may significantly limit its operation and diminish the potential effects. Some techniques are explained for strengthening the resistance of EPC tags against cloning attacks, using PIN-based access to achieve challenge-response authentication. This approach is used to padlock the tag without storing the access key. The following diagram shows the authentication flow for accessing the device applications as shown in Fig. 2.

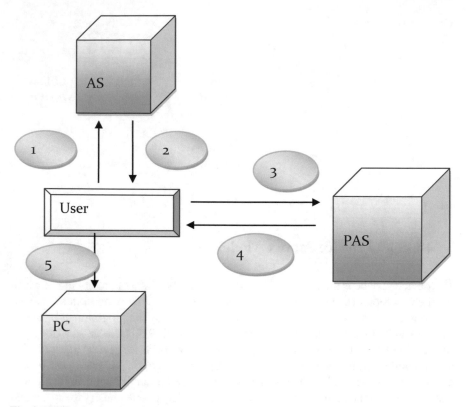

Fig. 2 RFID token generation from the authentication server

AS—Authentication Server
PAS—Privileged Attribute Server

The RFID tokens are generated and it is encrypted and decrypted with the private key. The proposed communication scheme is used to protect the privacy of RFID system in device applications.

4.1 Three Pass Mutual Authentication Procedure Between RFID Reader and Tags

The authentication is the process of identifying the entity or users. If the truthful users want to identify the authentication level should be increased with security algorithms. The key to this issue is, first the supplicant is identified with the registry process. Once the supplicant has been confirming red by using their login ID and password, there should be a route established and RFID tokens are generated for altered device services on the basis of their relationship of the hope that is called three-pass authentication (3PA). That protocol is similar to Hypertext Transfer Protocol which allows

Fig. 3 EPC based RFID token generation for device authentication

a transmission between RFID user and application layer. The user's stores a username and a password, and presently can be broadcast either in clear transcript or in a digested form. The following diagrams illustrate the steps for generating RFID tokens (Fig. 3).

4.1.1 RFID Token Algorithm Steps

1. RFID user swipe the RFID tag [sends credentials to Authentication server (AS)].
2. Authentication server sends token to privileged Attribute server (PAS).
3. The RFID user request to access resource and send token to PAS.
4. PAS creates and sends PAC to user.
5. Users sends PAC to authenticate to the resource.

The preferred Python device service implementation in IoT platform and called sec-wall which takes this to a complete innovative level by fascinating a (reverse) proxy, and this approach, security can be isolated from the service to another layer, which also keeps happy (Fig. 4).

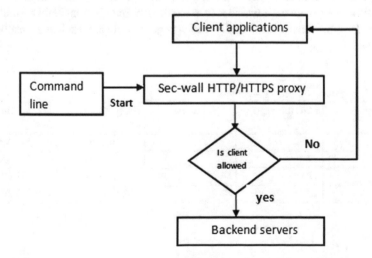

Fig. 4 RFID data transfer between server to the protocol

5 Results and Discussions

In general, the RF-TEA circuit displays a fixed latency of 128 clock cycles while doing an encryption or decryption algorithm on a solo block of 64-bit RFID data. In addition, the location and path providing processes are delivered by a clock period of 7.00 ns. Those four values of RFID data could be smooth the progress of computation. The throughput calculations are carried out by the following equation:

$$\textbf{Throughput} = (\textbf{no. of bits processed})/(\textbf{no. of clock cycles} \times \textbf{clock period})$$
$$= 96/32 \times 7 \times 10^{-9}$$
$$= 2.24 \tag{1}$$

5.1 TEA and RF-TEA Execution on Wireless Sensing Platform

Cypher	Symbol (byte)	RAM size (byte)	Execution time (ms)	CPR
TEA	870	104.002	10.0211	2.17
RF-TEA	812	104.14	9.871	2.22

The encrypted schedule, written in python and described below based on RF—encryption algorithm. Assume three times of 32-bit word size as an input and the Key is stored as $k[0...3]$ as four times of 32-bit size. The cost of performance ratio (CPR) is evaluated with minimum execution times with same RAM size of 104.

The following table code which is in a small size that gives an ideal formation of embedded applications and can easily implement in hardware, such as a RaspberryPi (Table 2).

Table 2 Comparison of security algorithm time

Language	LOC (encrypt)	LOC (decrypt)	Encrypt time (ms)	Decrypt time (ms)
C	7	7	391.001	390.009
Java	7	7	391.003	266.006
Ruby	12	13	1004.001	1034.007
Smalltalk	12	12	120,300.002	120,800.001

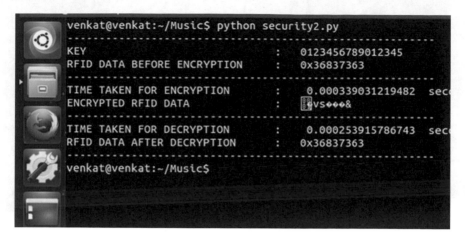

Fig. 5 RFID data with encryption and decryption for RFID tokens

Additional security can be added by increasing the number of iterations. The default number of iterations is 32, but can be configurable and should be multiples of 32. The following diagram shows the RFID data with encryption and decryption (Fig. 5).

In the above result, the RFID token encryption and decryption times are compared. Both got the same tag number which is verified by the reader.

6 Conclusions

In this paper, we have insisted on the need for RFID-enabled authentication for ensuring the utmost security for the device services. The paper carries the security features of the three pass protocols which are to substantially improve the authentication level. The RFID tokens are generated. The generated tokens are being examined by the security algorithm comprising the lightweight memory authentication protocol. This protocol is able to provide the forward and backward security. The other prominent security attacks such as high resistance to tracking, replay, DoS and man-in-the-middle attacks are also addressed and attended by this mechanism. However, this protocol holds a competitive tag-to-reader communication cost. We have therefore presented an implementation of two cyphers (TEA and RF-TEA) and compared their suitability and fitment regarding the parameters such as the memory footprint, execution time and security. Finally, we have described a successful implementation of the proposed protocol using the real world components of RFID tag with its tokens.

References

1. Chow, S.S.M., He, Y.J., Hui, L.C.K., Yiu, S.-M.: SPICE—Simple Privacy-Preserving Identity Management for Cloud Environment. In: Applied Cryptography and Network Security—ACNS 2012. LNCS, vol. 7341, pp. 526–543. Springer (2012)
2. Wang, C., Chow, S.S.M., Wang, Q., Ren, K., Lou, W.: Privacy-preserving public auditing for secure cloud storage. IEEE Trans. Comput. **62**(2), 362–375 (2013)
3. Wang, B., Chow, S.S.M., Li, M., Li, H.: Storing shared data on the cloud via security-mediator. In: International Conference on Distributed Computing Systems—ICDCS 2013. IEEE (2013)
4. Krzysztof Szczypiorski: (4 November 2003). Steganography in TCP/IP Networks. State of the Art and a Proposal of a New System—HICCUPS. Institute of Telecommunications Seminar. Retrieved 17 June 2010
5. Chu, C.-K., Chow, S.S.M., Tzeng, W.-G., Zhou, J., Deng, R.H.: Key-aggregate cryptosystem for scalable data sharing in cloud storage. IEEE Trans. Parallel Distrib. Syst. **25**(2), 468–477 (2014)
6. Tuttle, J.R.: Traditional and emerging technologies and applications in the radio frequency identification (RFID) industry. In: Radio Frequency Integrated Circuits (RFIC) Symposium, 1997, pp. 5–8. IEEE (1997)
7. Leong, K.S., Ng, M.L., Grasso, A.R., Cole, P.H.: Synchronization of RFID readers for dense RFID reader environments. In: International Symposium on Applications and the Internet Workshops, 2006. SAINT Workshops 2006, pp. 4–51 (2006)
8. Dominikus, S., Aigner, M.J., Kraxberger, S.: Passive RFID technology for the Internet of Things. In: Workshop on RFID/ USN Security and Cryptography (2010)
9. Gluhak, A., Krco, S., Nati, M., Pfisterer, D., Mitton, N., Razafindralambo, T.: A survey on facilities for experimental internet of things research. IEEE Commun. Mag. **49**(11), 58–67 (2011)
10. Chang, H., Choi, E.: User authentication in cloud computing. CCIS **120**, 338–342 (2011)
11. Kim, H., Park, C.: Cloud computing and personal authentication service. KIISC **20**, 11–19 (2010)
12. Parvathy, A., Rajasekhar, B., Nithya, C., Thenmozhi, K., Rayappan, J.B.B., Amirtharajan, R., Raj, P.: RFID in the cloud environment for Attendance monitoring system
13. Noman, A.N.M., Rahman, S.M.M., Adams, C.: Improving security and usability of low cost RFID tags. In: 2011 Ninth Annual International Conference on Privacy, Security and Trust (PST), pp. 134–141, 19–21 July 2011
14. Arulmozhi, P., Rayappan, J.B.B., Raj, P.: The design and analysis of a hybrid attendance system leveraging a two factor (2f) authentication (fingerprint-radio frequency identification). Biomed. Res. Special Issue: S217–S222 (2016)

Efficient Recommender System by Implicit Emotion Prediction

M. V. Ishwarya, G. Swetha, S. Saptha Maaleekaa and R. Anu Grahaa

Abstract Recommender systems are widely used in almost all domains to recommend products based on user's preference. However, there are several ongoing debates about increasing the efficiency with which recommendations are made to the user. So, nowadays, recommender systems not just considers user's preference, but also take into account the emotional state of the user to make recommendations. This paper aims at getting user's emotion implicitly by taking into account the time spent on different parts of the webpage. If any of these meet the predefined threshold, the user's emotion is analysed based on mouse movement in that part of the webpage. Thus, from this emotion, one gets to know whether the user is actually interested in the content of that part of the webpage. Thus, the project aims to improve the efficiency of recommendations by providing a personalized recommendation to each user.

Keywords Affective computing · Recommender system · Information overload

1 Introduction

A recommender system is used to provide a recommendation to users by predicting whether a user would be interested in the particular item or not. This is usually

M. V. Ishwarya
HITS, CSE Department, Sri Sairam Engineering College, West Tambaram, Chennai, Tamil Nadu, India
e-mail: ishwarya.cse@sairam.edu.in

G. Swetha (✉) · S. Saptha Maaleekaa · R. Anu Grahaa
Sri Sairam Engineering College, West Tambaram, Chennai, Tamil Nadu, India
e-mail: swechess97@gmail.com

S. Saptha Maaleekaa
e-mail: sapthamaaleekaa@gmail.com

R. Anu Grahaa
e-mail: sweety.grahaa3@gmail.com

© Springer Nature Singapore Pte Ltd. 2019   173
J. D. Peter et al. (eds.), *Advances in Big Data and Cloud Computing*,
Advances in Intelligent Systems and Computing 750,
https://doi.org/10.1007/978-981-13-1882-5_15

done by modelling user behaviour by analysing his past experiences. Nowadays, recommender systems are used in a variety of areas like movies, music, news, books, research articles, search queries, social tags and products in general [1]. Though these recommendations benefit the users and business owners to a great extent, sometimes users also have an annoying effect on it. This happens when the users are repeatedly recommended an item they are least interested in [2, 3]. This can be overcome by affecting computing, i.e. one can also take into account the user emotions before making recommendations. This paper extends the technique of implicit emotion prediction through user cursor movement, to recommender systems. Thus, to some extent, we can personalize and increase the efficiency of the recommender systems.

2 Existing Technology

In recent years, several approaches (Implicit and Explicit) have been made to predict users' emotions and make recommendations accordingly. In this section, we discuss a few of these approaches along with their pros and cons.

2.1 Sentiment Analysis of User Review

Today, almost all sites have a review section, where the user can provide reviews and comments about his experience with the service/product offered by the site/company. These reviews are analysed to get the user's sentiments. These sentiments are then used to provide recommendations to all the like-minded users. However, this process has a few disadvantages too. First, since the users can give their comments and reviews freely, it can sometimes be biased, which affects the whole process of recommendation. Next, online spammers can at times post unwanted content that is in no way related to the website's service or the products they sell. Another problem with this type of recommendation is that the recommendation is not personalized [4], i.e. the comments and reviews of some other person are used as a basis to recommend another like-minded person, which at times may not be that efficient.

2.2 Recommender System Using User Eye Gaze Analysis

Very recently, a company devised and patented a device that could record and maintain logs of user gaze. The main aim of this device was to obtain user emotion when he views a particular thing in the webpage by analysing his pupil dilation. For instance, using this device, one could identify whether the user is happy or angry when is viewing an advertisement. But this technology has a few disadvantages too. First, not all users would agree to wear this device due to privacy concerns. Moreover, in real time, it is not feasible to provide this device to all users of the website. However, researches are still on for tracking pupil movement through a web camera. But, even this approach poses a threat to user's privacy.

2.3 Recommender System Using Sentiment Analysis of Social Media Profiles

Social media is used by almost everyone, to express their views and to present them to the real world. Nowadays, recommender systems try to extract user's emotion and sentiments from the texts that users post in the social media. Thus, by analysing the user's sentiments extracted from his social media activity a recommender system can predict the user's interest in the product/service offered by the company and thereby make recommendations accordingly. However, though this approach eliminates several problems of recommender system like cold start problem, impersonalized suggestions, etc., it again faces the privacy breach problem.

2.4 Recommender System by Emotion Analysis of User's Keyboard Interaction

Another interesting technique of user's emotion detection is by analysing his interaction with the keyboard [5]. Several features of user's interaction with the keyboard such as keystroke verbosity (number of keys and backspaces), keystroke timing (latency measures), number of backspaces and idle times, pausing behaviour, etc., were considered. These features were later analysed and user's emotional state during interaction with the webpage was predicted by a suitable machine learning algorithm.

3 Proposed Method

The main aim of this paper is to determine user emotion through his cursor movement and use it as a basis for the recommendation. Several features of the cursor movement are analysed by the classification algorithm and based on the analysis user's emotional state is predicted. This section is divided into several subsections which gives a detailed explanation of the entire process.

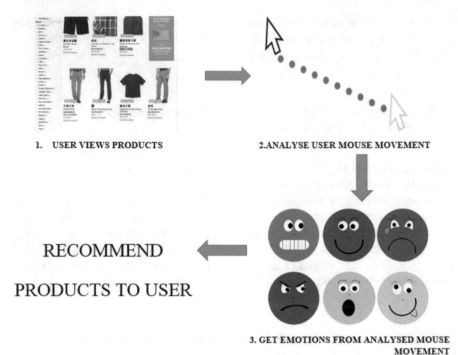

1.   USER VIEWS PRODUCTS   2.ANALYSE USER MOUSE MOVEMENT

RECOMMEND

PRODUCTS TO USER

3. GET EMOTIONS FROM ANALYSED MOUSE
MOVEMENT

3.1   Collection of Cursor Position Log

Initially, we collect a log of cursor position for the entire duration that the user spends on the webpage. We collect the cursor position, that is the x and y coordinates of the cursor continuously for a fixed time duration with interleaved intervals in between every duration. JavaScript is used for collecting x and y coordinates of the cursor. The cursor positions are maintained as a table along with the time that they were recorded.

3.2   Cursor Movement Features

The next step is to analyse several features of cursor movement from the cursor position log that we have collected. In this section, we list a few of those features. They are:

- No. of angle changes in the cursor movement of largest continuous movement.
- A list of angles by which the cursor's direction of movement changed.
- Difference between the largest and smallest angle in the above list.

- Speed of the cursor movement for continuous mouse movements (up to a certain length).

3.3 Criteria for Cursor Movement Analysis

Since it is infeasible to analyse cursor movement for the entire duration that the user spends on the webpage, criteria for cursor movement analysis is needed. Hence, we consider the scrolling speed. If the scrolling speed is zero, we start the timer. If this time meets or crosses a predefined threshold, we analyse the cursor movement. For cursor movement analysis, we take into account the features discussed in the above subsection. The results of this analysis are tabulated.

3.4 Emotion Prediction Using a Classification Algorithm

For emotion, prediction classification algorithm is used. The classification is performed by Waikato Environment for Knowledge Analysis (Weka) tool. The classifier model built here contains two target classes, i.e. here, only two emotions are considered, which are 'happiness' and 'sadness'. The table containing analysed cursor movement features is fed into the classifier, which outputs user's emotion.

3.5 Recommendation to the User

After emotion prediction, one can get to know whether the user is interested in that part (product displayed in that part) of the webpage or not. Using this result, we decide whether to recommend the product (and similar items that the user might like, found using content-based filtering algorithm) to the user or not. Content-based filtering is an existing approach used by recommender systems to recommend products by comparing the attributes of products and user profile. The user profile is generally built by analysing the attributes of the products that the user is already known to be interested in. The recommendations are displayed in the form of ads.

4 Conclusion

Since the movement of the cursor and the time spent in a particular part of the webpage is taken into account for the recommendation, personalized recommendations for each user is made more efficiently, from the emotion captured by the cursor movement. An effective content-based algorithm is used for recommending similar items. Thus, personalized recommendation is made to the user.

References

1. Madhukar, M.: Challenges & limitation in recommender systems. Int. J. Latest Trends Eng. Technol. (IJLTET) **4**(3) (2014)
2. Sree Lakshmi, Soanpet, Adi Lakshmi, T.: Recommendation systems: issues and challenges. (IJCSIT) Int. J. Comput. Sci. Inf. Technol. **5**(4), 5771–5772 (2014)
3. Sharma, L., Gera, A.: A survey of recommendation system: research challenges. Int. J. Eng. Trends Technol. (IJETT) **4**(5) (2013)
4. Jain, S., Grover, A., Thakur, P.S., Choudhary, S.K.: Trends, Problems and Solutions of Recommender System. ISBN: 978-1-4799-8890-7/15/$31.00 ©2015 IEEE
5. Shikder, R., Rahaman, S., Afroze, F., Alim Al Islam, A.B.M.: Keystroke/mouse usage based emotion detection and user identification. 978-1-5090-3260-0/17/$31.00 ©2017 IEEE

A Study on the Corda and Ripple Blockchain Platforms

Mariya Benji and M. Sindhu

Abstract Blockchain, as the name says a chain of blocks growing in forward direction contains a list of records which is linked to previous block using cryptographic methods. Typically, a block records a hash pointer, timestamp and transaction data. Inherently by design, a block cannot be modified. Blockchain acts as an open distributed ledger that records transaction between parties efficiently in a verifiable and permanent way. Blockchain technology has a wide range of application both in financial and non-financial areas working with consensus, in different platforms. This paper describes different applications of Blockchain; both in financial and non-financial fields, a study on two of its platforms—Corda and Ripple; and also a comparison between the two platforms.

Keywords Blockchain · Corda · Forking · Merkle tree · Ripple

1 Introduction

Even before industrialization, trade has been a day-to-day activity for humans. Starting from the very well-known Barter System [1] where, people exchange commodities based on their need, world trade grew tremendously down the years, facilitated by the advent of currencies, liberalization of economies and the globalization of trade. The mode of trade differs in terms of belief and trust as people were not much sure on whom to trust.

Looking at the evolution of Blockchain, where a more trustworthy and secure Triple entry book was introduced, which acted as a trusted third party alongside the participants in a transaction. All people engaged in a transaction will have a mutual

M. Benji (✉) · M. Sindhu
TIFAC-CORE in Cyber Security, Amrita School of Engineering, Amrita Vishwa Vidyapeetham, Coimbatore, India
e-mail: mariyabenji@gmail.com

M. Sindhu
e-mail: m_sindhu@cb.amrita.edu

© Springer Nature Singapore Pte Ltd. 2019
J. D. Peter et al. (eds.), *Advances in Big Data and Cloud Computing*,
Advances in Intelligent Systems and Computing 750,
https://doi.org/10.1007/978-981-13-1882-5_16

trust between each other. The trusted third party records the transaction details along with the participant's credentials, thus providing non-repudiation property for the transaction. But too much trust on third party again seems to be impractical. As an answer to these problems, cryptographic solutions emerged. To be more specific, Digital signatures [2] were introduced initially wherein a public key is broadcasted and the private key is kept secret. Here, whenever any information is sent, the message is encrypted with sender's public key and decrypted using senders public key.

This scheme brought in features like integrity, authenticity, etc., to a transaction. Although, this worked properly, still certain issues remained unsolved. For example, double-spending attack remained a problem with digital transactions in the financial arena, where the same share is transacted more than once by the sender and thus earned double profit.

Distributed network proves to be a solution to this problem, in which each successful transaction is replicated and distributed to all, thus everyone will get to know what all is transacted and also to whom all. But hurdles like geographical location, varied time-zones, etc., made this proposed system impractical. Thus, the idea of peer-to-peer network was brought in, where all the peers could get information of all transaction and a minimum of n-peers must validate them to make the transaction confirmed. But there still exists a possibility of a Sybil attack where an attacker can create n number of false peers and validate his transaction.

2   Facets of Blockchain

Distributed system is a key concept which is necessary to understand Blockchain. It is a computing environment where two or more nodes work with each other in a coordinated fashion in order to achieve a common task. A transaction is the fundamental unit of Blockchain. For a transaction we need a sender and recipient who have unique addresses to identify them. An address is usually a public key or derived from public key [3]. Blockchain can be viewed as a state transition mechanism where the transactional data is modified from one state to another as a result of a transaction execution. A node in a Blockchain is an end system that performs various functions depending on the role it takes. A node can propose and validate transaction and perform mining to facilitate consensus and thus adds security to Blockchain. This is done by consensus mechanisms. Each block is formed out of a Merkle tree of transaction. Any small change made in the data pertaining to a transaction will change the hash value of the transaction unpredictably. This makes the data, once a part of the tree, tamperproof. The hash of the root node is stored in the block as the Merkle tree root hash. This helps the transaction to verify the presence of their transaction in the block [4].

When situations like two different nodes may create a valid block simultaneously and broadcast it to update the ledger, the linear structure of the Blockchain will be changed to form a fork [5, 6]. This situation is called forking of Blockchain. Forking problem is avoided by adopting the longest chain rule to decide which fork is to be continued with. The next block is validated and formed after the forking issue is considered, and then the chain grows from there if it grows up to minimum of six nodes.

3 Brackets of Blockchain

The three broad classifications of Blockchain [7] are the public Blockchain, private Blockchain and consortium Blockchain. In public Blockchain, there is no centralized authority or party who is superior to all. All nodes are equally treated. This is also known as permission-less Blockchain. In Consortium Blockchain, not everyone has equal priority to validate a transaction, only a few are given privileges over validating a transaction.

A slightly different version of this Blockchain is private Blockchain. It has a centralized structure. A single entity has the full power to take decision and validate the process. This centralized authority should make sure that the transaction is as per the proposed consensus. Both consortium and private Blockchain are otherwise known as permissioned Blockchain. They are faster, consume less energy and are easily implemented compared to permission-less Blockchain.

4 Application of Blockchain

Blockchain technology is a revolution in the system of records. Blockchain technology can be integrated into multiple areas. Blockchain protocols facilitate businesses to use a new method of processing digital transactions [8–11]. Examples are payment system and digital currency, crowd sales, prediction markets and generic governance tools. Blockchain can be thought of as an automatically notarized ledger. Blockchain has a wide variety of applications, both in financial and non-financial fields. Major applications of Blockchain includes crypto-currencies such as Bitcoin, and many other platforms such as Factom as a distributed registry, Gems for decentralized messaging, Storj for distributed cloud storage, etc.

4.1 Currency

The very first description of the crypto-currency Bitcoin was introduced by Satoshi Nakamoto in 2008 in his whitepaper 'A Peer-To-Peer Electronic Cash System' [12]. In the beginning of 2009, the first crypto-currency became a reality with the mining of the genesis block and the confirmation of the early transactions. Today, most developed countries allow the use of Bitcoin, but consider it as private money or property. However, there are some countries, most of which are in Asia and South

America, along with Russia where Bitcoin is considered illegal [13]. You can acquire Bitcoin through mining or an exchange, receive them as payment for work or even ask donations in Bitcoin. Other notable crypto-currencies are Ripple, which was created by Ripple Labs and belongs to the category of pre-mined crypto-currencies, Litecoin, which is based on the same protocol as a Bitcoin but much user-friendly in terms of mining and transaction. Darkcoin, provides real anonymity during transaction. Primecoins, whose solution during mining procedures are prime numbers.

4.2 Voting

Voting exists still as a controversial process worldwide. Adapting Blockchain technology in a voting campaign seems to be an effective solution. The members could connect to a PC-based system through their laptop or smartphone using an open-source code that is open to editing using a kind of authentication to prove their identity. Then, the member enters their private key to access their right to vote and use their public key to select their preference and confirm it. One of the projects that promote voting through Blockchain technology is Bit Congress [14] that uses Ethereum platform [15] with an idea that every voter has access to one vote coin that gives him right to vote only one time and his vote will be recorded on the Blockchain after the system verifies it—Remotengrity, Agora Voting [16]. The transformation of the voting system from paper based to digital will increase its reliability and the convenience that it offers to the voters.

4.3 Contracts

Smart Contracts can be used to confirm a real estate transfer, thus playing the role of a notary. Also at the same time, a user can write his/her own will on the platform and contract will be executed after his death without the intervention of a thirds party to confirm it. The same technology can be used for betting purposes. The user can put their money on a digital account and creates a virtual contract that defines the conditions of winning and losing. When a result comes up in the real world, the contracts get updated from an online database and execute the terms by transferring the money to winners account.

4.4 Storage

A service used to control the data on the internet using Blockchain can be implemented. Storing the data within the Blockchain significantly increases the volume of a Blockchain, where some Blockchains adopt another means for managing this.

Storj [17] is a service used to govern various electronic files using a Blockchain. Data themselves are encrypted and stored in a scattered manner on a P2P network, and therefore cannot be accessed by third parties. It is a protocol that creates a distributed network to implement the storage of contracts between peers. It enables peers on the network to negotiate contracts, fetch data, verify the fairness and availability of remote data and retrieve data. Each peer is independent, capable of performing actions without significant human intervention.

4.5 Medical Services

The idea is to manage medical data such as electronic health records and medication records, by using Blockchain. The desired method used for protecting private records of medical data on Blockchain is by controlling the passes between the medical institutions. An existing project is Bit Health. It is a healthcare project using Blockchain technology for storing and securing health information. Data privacy, data alteration and data authenticity are biggest concerns in healthcare. Users generate public and private key and encrypts, the records using public key and store in Blockchain. Importance of using this technology solves issues like data duplication, follows CIA triad, reduce the cost of insurance and other expenditures of medical records.

4.6 IoT

Blockchain technology is used in IoT. The expected utilization method uses sensors, trackers etc. and conducts predefined tasks independently without involving a central server. ADEPT (Autonomous Decentralized Peer-to-Peer Telemetry) by IBM and Samsung uses this concept which has gained attraction. ADEPT allows the following to happen by influencing three open-source, peer-to-peer protocols that is Telehash for messaging, BitTorrent for sharing files and Ethereum for transactions.

5 Corda

Corda [18] is a distributed ledger platform made of mutually distrusting nodes which allows a single global database to record the state of deals between institutions and people. This eliminates much of the time-consuming effort currently required to keep all the ledgers synchronized with each other. This also allows a greater level of code sharing facility used in financial industry, thereby reducing the cost of financial services. The legal documents of transaction are visible only to those legitimate participants of the transaction and the hash values are used to ensure this along with the node encryption consensus. The main characteristics of Corda [19] are automated smart contracts and time-stamping of documents to ensure uniqueness.

Consensus involves acquiring the values currently available, combining it with smart contracts and producing new results or states. The two key aspects to attain consensus are transaction validity and transaction uniqueness. The validity consensus is maintained by checking the validity of the smart contract code used and also checking if it was run with appropriate signatures. Notarization, time-stamping and other constraints involved in smart contracts maintains uniqueness of the transaction.

In Corda the concept of immutable state exists and it consists of digitally signed secure transactions. The Java bytecode of Corda is also a part of the state. This runs with the help of virtual runtime environment provided by Java virtual machine (JVM). As a result, execution of consensus protocol occurs in sandbox environment making it more secure. In verification process, it calls the verification function that checks if the transaction is digitally signed by all participants which ensure a particular transaction to be executed if it is verified and validated by all participants.

5.1 Key Concepts of Corda

Corda network is semi-private. All communications are uninterrupted, where a TLS-encrypted data is sent over AMQP/1.0, which means the data is shared on a need-to-know basis. Each node has an elected doorman [20] that strictly prepares rules regarding the information that nodes must provide and the Know Your Customer (KYC) processes that they must complete before getting added to the network.

There is no single central store of data in Corda, instead, each node maintains a separate database of known facts. Each Corda identity can be represented as Legal identity and Service identity. Identities can be well-known identity or confidential which is based on whether their X.509 certificate is published or not. A state is an immutable object representing a known fact shared among different nodes at a specific point of time. State contains arbitrary data. As states are immutable, they cannot be modified directly to reflect a change in the state. Each node maintains a vault—a database which tracks all the current and historic state.

Each state is a contract and takes transaction as input and verifies it based on the contract rules. A transaction that is not contractually valid is not a valid proposal to update the ledger, and thus can never be committed to the ledger. Transaction verification must be deterministic, i.e. it should be either always accepted or always rejected. For this, contract evaluates transaction in a deterministic sandbox [20] that prepares whitelist that prevents the contract from importing unwanted libraries. It uses a UTXO (unspent transaction output) model where all states on the ledger are fixed. When creating a new transaction, the output state must be created by the proposers. The input state already exists as the outputs of previous transactions. These input state references combine all transactions overtime together and forms a chain. The assigned signers sign the transaction only if the following conditions are satisfied; transaction validity and transaction uniqueness.

Corda network use point to point messaging instead of a global broadcast. Rather than having to specify these steps manually, Corda automates the process using flows where the flow tells a node how to achieve a specific ledger update. If the proposed transaction is a valid one, ledger update involves on acquiring two types of consensus. Validity consensus—verified by authenticated signer before they sign the transaction. Uniqueness consensus—verified by notary service. The notary provides the point of finality in the system.

The core element of the architecture are a persistence layer for storing data, a network interface for interacting with other nodes, an RPC interface for interacting with the nodes owner, a service hub for allowing the nodes owners to call upon the nodes other services and plug-in registry for extending the node by installing CorDapps.

6   Ripple

Ripple is used specifically for real-time gross settlement system (RGTS) trading and allowance by Ripple also referred to as Ripple Transaction Protocol (RTXP) or Ripple Protocol, is built on distributed open-source internet protocol consensus ledger and its crypto-currency is XRP. Ripple was launched on 2012 that enables safe, quick and independent global financial transactions of any size with absolutely no chargebacks [21]. Ripple prop up tokens representing fiat currency, virtual currency or any valuable asset. Ripple is based on a shared, public ledger, which uses a consensus process that allows you in trading, payment and settlement in a distributed process.

Ripple is adopted by companies like UniCredit, UBS, Santander and specifically in bank ledgers and have numerous advantages over other virtual currencies like Bitcoin.

The open-source protocol describes the Ripple's website as a basic infrastructure technology for cross bank transactions. Both financial and non-financial companies incorporate Ripple protocol to their system. Two parties are required for a transaction to happen. Primarily, a regulated financial institution that manages and handles on behalf of customers and latterly, the market who provides liquidity in the currency which helps in trading. Ripple is based around a shared public ledger that has its contents decided by consensus.

6.1   Ripple Consensus

The consensus algorithm starts with a known set of nodes known to be participating in the consensus. This list is known as unique node list. This list is a collection of public keys of active nodes which are believed to be unique. Through the consensus algorithm, nodes on the UNL vote to determine the contents of the ledger. While the actual protocol contains a number of rounds of proposals and voting, the result can be described as basically a supermajority vote, a transaction is only approved if 80% of the UNL of a server agrees with it [21].

Initially, each server takes all valid transactions it has before initiating the consensus and makes them public in the form of a list known as the candidate set. Each server then combines each of the candidate sets of all servers on its UNL and votes on the veracity of all transaction. All transaction which meets this 80% vote is applied to the ledger and that ledger is closed becoming the new last-closed ledger [1].

7   Comparison Between Corda and Ripple

From the analysis of Corda and Ripple, it can be concluded that the security features present in both the platforms prove to be vital in ensuring security and authenticity to transactions. But, however, notarization features gives Corda platform a competitive edge over the other platform of Blockchain, which is Ripple. Uniqueness and notarization features offers Corda more reliability and stability in performance their by proving to be a more trusted and selected over Blockchain platform for conducting financial transaction. Further improvement in consensus or if Corda gets adapted to blockchain platform as a false proof, further improved security can be offered by the platform thus emerging Corda as a future platform for conducting financial transactions. The following table concludes these comparisons.

	Corda	Ripple
Governance	R3 Labs	Ripple Labs
Initial release	2016	2012
Type	Consortium Blockchain	Private Blockchain
Currency	No	XRP
Protocol	AMQP in TLS	SMTP in TLS
Participants	Only sender and receiver	The UNL list
Language	Kotlin, Java	XRP Ledger in C++
Consensus	Specific understanding of consensus (i.e. Notary nodes)	Unique Node List

8   Conclusion

Blockchain is a tool used by organizations primarily in the field of business and to be more specific, in the area of finance that allows transactions to be more efficient and secure. Data in its processed form has big monetary value these days, particularly if it pertains to the field of business. Thus, data security deserves prime importance because business organizations suffering theft or manipulation of its transmitted data becomes subject to big financial loss. In the era where a steep hike is experiencing in the area of business transactions which are of course financial in nature, a need to make transactions is more secure and authentic deserves crucial significance. Application of Blockchain serves this purpose by providing efficiency, security and authenticity to transactions. This paper discusses a comparative study between the two Blockchain platforms Corda and Ripple.

References

1. Siba, T.K., Prakash, A.: Block-chain: an evolving technology. Glob. J. Enterp. Inf. Syst. **8**(4) (2016)
2. Bozic, N., Pujolle, G., Secci, S.: A tutorial on blockchain and applications to secure network control-planes. In: Smart Cloud Networks and Systems (SCNS), pp. 1–8. IEEE (2016)
3. Andreas M. Antonopoulos, 2014. Mastering bitcoins
4. https://en.wikipedia.org/wiki/Blockchain
5. Ambili, K.N., Sindhu, M., Sethumadhavan, M.: On federated and proof of validation based consensus algorithms in Blockchain. In: IOP Conference Series: Materials Science and Engineering, vol. 225, no. 1, p. 012198. IOP Publishing (2017)
6. Sankar, L.S., Sindhu, M., Sethumadhavan, M.: Survey of consensus protocols on blockchain applications. In: 4th International Conference on Advanced Computing and Communication Systems (ICACCS), pp. 1–5. IEEE (2017)
7. Pilkington, M.: Blockchain technology: principles and applications (2015)
8. Czepluch, J.S., Lollike, N.Z., Malone, S.O.: The use of blockchain technology in different application domains. The IT University of Copenhagen, Copenhagen (2015)
9. Eze, P., Eziokwu, T., Okpara, C.: A Triplicate Smart Contract Model using Blockchain Technology (2017)
10. Foroglou, G., Tsilidou, A.L.: Further applications of the blockchain. In: 12th Student Conference on Managerial Science and Technology (2015)
11. Kuo, T.T., Kim, H.E., Ohno-Machado, L.: Blockchain distributed ledger technologies for biomedical and health care applications. J. Am. Med. Inf. Assoc. **24**(6), 1211–1220 (2017)
12. Nakamoto, S.: Bitcoin: a peer-to-peer electronic cash system (2008)
13. Bracamonte, V., Yamasaki, S., Okada, H.: A Discussion of Issues related to Electronic Voting Systems based on Blockchain Technology
14. Qi, R., Feng, C., Liu, Z., Mrad, N.: Blockchain-powered internet of things, e-governance and e-democracy. In: E-Democracy for Smart Cities, pp. 509–520. Springer Singapore (2017)
15. Wood, G.: Ethereum: a secure decentralised generalised transaction ledger. Ethereum Project Yellow Paper, 151 (2014)
16. Foroglou, G., Tsilidou, A.L.: Further applications of the blockchain. In: 12th Student Conference on Managerial Science and Technology (2015)
17. Wilkinson, S., Boshevski, T., Brando, J., Buterin, V.: Storj: a peer-to- peer cloud storage network (2014)
18. Hearn, M.: Corda-A distributed ledger. Corda Technical White Paper (2016)
19. Brown, R.G., Carlyle, J., Grigg, I., Hearn, M.: Corda: An Introduction. R3 CEV (2016)
20. https://docs.corda.net/key-concepts.html
21. Todd, P.: Ripple Protocol Consensus Algorithm Review (2015)

Survey on Sensitive Data Handling—Challenges and Solutions in Cloud Storage System

M. Sumathi and S. Sangeetha

Abstract Big data encompasses massive volume of digital data received from enormously used digital devices, social networks and real-time data sources. Due to its characteristics data storage, transfer, analysis and providing security to confidential data become a challenging task. The key objective of this survey is to investigate these challenges and possible solutions on sensitive data handling process is analysed. First, the characteristics of big data are described. Next, de-duplication, load balancing and security issues in data storage are reviewed. Third, different data transfer methods with secure transmission are analysed. Finally, different kind of sensitive data identification methods with its pros and cons in security point of view is analysed. This survey concludes with a summary of sensitive data protection issues with possible solutions and future research directions.

Keywords Big data storage · Transmission · Security and privacy issues
Sensitive data

1 Introduction

Traditionally, people had used limited number of communication channels and those channels had produced minimal quantity of data is in the form of structured data. Nowadays, people and systems are producing tremendous quantity of data with different genre as in unstructured form. Data size is reached to zettabyte (10^{21}) in size [1]. The huge volume of unstructured data is known as Big Data. Initially, it was proposed by McKinsey and Dong Lency in 2001 [2]. The sources of big data are social media, weblog, sensors, networks, telecommunications, documents, web pages, healthcare

M. Sumathi (✉) · S. Sangeetha
Department of Computer Applications, National Institute of Technology, Tiruchirappalli,
Tamil Nadu, India
e-mail: sumathishanjai.nitt@gmail.com

S. Sangeetha
e-mail: sangeetha@nitt.edu

© Springer Nature Singapore Pte Ltd. 2019
J. D. Peter et al. (eds.), *Advances in Big Data and Cloud Computing*,
Advances in Intelligent Systems and Computing 750,
https://doi.org/10.1007/978-981-13-1882-5_17

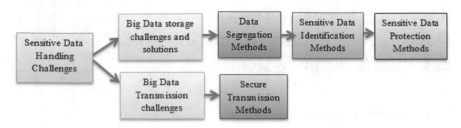

Fig. 1 Organization of the paper

records and Internet of Things. Big data is characterized by 3V's, Volume (quantity of data—ZB in size), Variety (diversity of data types—structured, semi-structured and unstructured), Velocity (data generation and processing speed—200 million emails are sent through Gmail), now fourth (Value) and fifth V's (veracity) are included in characteristic [3]. Due to its characteristics, secure data storage, communication and analysis is a complex task in traditional security techniques. Hence, alternate techniques are required to provide better security to confidential with minimal storage and communication cost.

Cloud storage system provides elasticity, reduced infrastructure cost, resource pooling, shared computing resources and broad network access. Therefore, enormous sizes of data are efficiently handled by cloud computing system with nominal cost. Data security and privacy is a serious issue of cloud storage system. The following sections focused on security issues related to data storage and transmission with its possible solutions. In Sect. 2, big data storage challenges are discussed with its solution and Sect. 3 discussed about big data transmission challenges and its solution. Figure 1 shows the organization of this paper.

2   Big Data Storage Challenges and Solutions

This section investigates secure data storage challenges and solutions in cloud storage system. Big data consists of different genre data, hence conventional storage approaches are lacked in big data storage (e.g. SQL and RDBMS). So, new storage approaches and architectures are required to handle plenty of unstructured data. The succeeding features are required to provide efficient storage systems.

- **Scalability**—When volume of data is expanded or shirked, scale-up or scale-out process is required to handle different capacities.
- **Fault tolerance**—The storage system should be able to manage faults, and provide required data to user within a precise manner and specific time period.
- **Workload and Performance**—Manage various file sizes and allow concurrent modifier operations.

- **Security**—To handle sensible data and should be able to support cryptographic techniques.
- **Privacy**—To protect private information from unauthorized users and has to provide better access control to authorized users [4].

a. **Challenges of Big Data Storage**:

- **Difficult to manage small files**—In practice, files are commonly small in size. These small files occupied more storage space. Hence, correlated files are combined into single large file for reducing storage space. But, correlation takes larger time for combine it.
- **Replica Consistency**—Replicas are required to recover original data from an unexpected resource failure. Replicas increases storage space and maintaining consistency between replicas is a critical task [5].
- **De-duplication**—Enormous sized data are stored in cloud storage system with redundant copy. The redundant copy occupies huge storage space in cloud storage. De-duplication processes uses file level, block level and hash-based techniques are used to remove redundant data in a storage place [6]. Still now de-duplication process is an on-going process, 100% de-duplication is not achieved.
- **Load Balancing**—Occasionally, plenty of data will be sending to certain nodes and minor amount of data will be sending to other nodes. The heavy storage nodes are prone to system failure or poor performance. Load balancing technique is used to balance the storage size between nodes [7].

b. **Secure Data Storage**

Big data contains different genre data. Applying identical security techniques to entire data leads to poor performance. In general user data required different security levels based on its usage. Hence, applying identical security technique to entire data will not provide efficient security system. In addition to that, entire data security increases storage cost in cloud storage system. Therefore, selective data encryption is required to provide better security with nominal storage cost. To apply selective data encryption, the data is to be classified into different categories. This part discussed about various classification techniques used for data segregation process in security point of view.

i. **Data genre-based categorization**

Big data consists of different genre data like numbers, texts, images, audio and video. Basic security mechanisms are working well for specific type of data, not suitable for big data. Applying uniform level of security to entire data leads to poor performance and provides high security to specific type of data and less security to another type of data. Hence, data is categorized into different groups based on its type. After that, separate security techniques are applied for each category. This data genre-based security mechanism provides better security to each group, but difficult to implement it. In addition to that, storage space also increased it [8].

ii. **Secret level-based segregation**

Table 1 Comparison of segregation method

Method	Storage size	Storage cost	Security level	Encryption time	Segregation time
Data genre	High	High	Low	High	High
Secret level	High	High	Low	High	High
Data access frequency	High	High	Low	High	High
Sensitive and non-sensitive	Medium	Nominal	High	Low	Medium

In big data, each data is having separate level of security. The secret levels are top secret, secret, confidential and public data. Based on secret level, security requirement is varied. The hierarchical access control mechanisms are used to provide better security to each level. In this method also encrypt entire data with different keys instead of single key. Hierarchical access control provides better security to each level but increases storage cost [9].

iii. **Data access frequency-based segregation**

To improve system performance and security level, less accessible data are moved to lower tier and frequently accessed data retained in the higher tier. Manual movement of data between different tiers is a time consuming process. Auto-tiering is required to move data between tiers. Each tier contains different kinds of sensitive data. Based on sensitivity different types of security, mechanisms are applied to it. Group key is used to access data from each tier. Each tier having a separate threshold value and it's fixed by group members within that tier [10].

iv. **Sensitive and Non-Sensitive Character-based segregation**

When a data is moving to cloud storage, data owner loss their control over their data. Nowadays, data owner are interested to provide security to their data. Present scenario, customer are willing to do their process (purchase, net banking and data transmission) through online. When a data is moving to online, inter-organization members are access the user data for improving their business, do to research work and providing better services to user. In this case, security is required for sensitive data and usability is provided to non-sensitive data. Hence, the user data can be classified as sensitive and non-sensitive. After that, security mechanisms are applied to sensitive data and visibility is provided by non-sensitive data. Here, high-end security is applied to sensitive data with nominal cost. In this technique, minimal amount of data is to be encrypted instead of entire data [11]. Table 1 shows the comparison of segregation methods.

Based on this analysis, data owner preference data segregation provides higher security than other methods. Hence, sensitive and non-sensitive character-based segregation is preferred for further process.

c. **Sensitive and Non-Sensitive Character-based segregation**

To provide security to sensitive data and visibility to non-sensitive data, sensitive data needs to be segregated from non-sensitive data. In the current scenario, data are in unstructured and semi-structured formats. To segregate sensitive data from a unstructured data is challenging task. Sensitivity differs from user to user and required different levels of security to each data. To provide better security to each data, deep learning is required to segregate the sensitive data.

Issues related to sensitive data identifications:

- **Classification technique**—Identification of sensitive term in an unstructured data using classification technique produces less accurate data. Hence, semantic and linguistic techniques are required to segregate sensitive data [12]. Sensitive data accuracy depends on classification training dataset accuracy. Particle swarm optimization, similarity measures, fuzzy ambiguous, MapReduce and Bayesian classification techniques are used for data classification process. Machine learning and information retrieval techniques are combined to gather to produce automated text classification in unstructured documents.
- **Manual Identification**—Sensitive data identification depends on the expert's semantic inference understandability. Domain experts to detect sensitive data by manually by using generated dictionary terms. The terms are semantically analysed by a tool and detect the sensitive terms. Finally, the identified sensitive terms are verified by human experts [13]. Manual identification takes more time for data classification.
- **Information Theoretic Approach**—To evaluate the cardinality of each term group from the lowest to highest informative. The higher informative terms reveals large number of terms and sensitive term identification depends on generalization threshold value. The evaluation is performed by standard measures like precision, recall and F-measures [14].

Methods to Protect Sensitive Data:

- Sanitization/Anonymization—Sanitation is a technique for preserving the confidentiality of personal data, without modifying the value of the documents. The types of anonymization are Irreversible (removing any information that can identify the individual or organization without the possibility of recovering it later) and reversible (sensitive information is cross-referenced with other information and re-identify the original information. Anonymization process is performed by human experts is a time-consuming process and will not provide added value to organization [15].
- Encryption—Used to protect data from misuse, fraud or loss in an unauthorized environment. The data is encrypted in the source or in a usage platform. The confidentiality of sensitive data depends on encryption technique and key size. Figure 2 shows security process performed in sensitive data. Sensitive data needs to protected by endpoint security, network security, physical security and access controls.

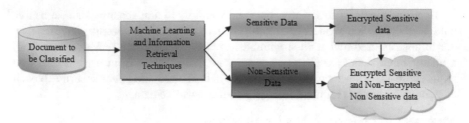

Fig. 2 Sensitive data protection method

3 Big Data Transmission Challenges and Solutions

In this section, the big data transmission challenges and its solutions are analysed. Big data transmission requires high bandwidth network channel to distribute data to different nodes in the network. Massive volume of data transfer rate depends on network bandwidth, background traffic and round trip time.

a. **Big data Transfer challenges**

- **Pipe lining, Parallelism and Concurrency-based big data transfer**—Pipe lining is used to transfer large number of small files, at full capacity of network bandwidth. In parallelism, larger files are partition into equal sized smaller files and distributed to destination. Parallelism reduces data transmission rate. When a file size and number of files are increased, concurrency mechanism is used to maximize the average throughput [16].
- **Multi-route-aware big data transfer**—Transferring data between distributed centres requires extra time and cost. To reduce these expenses, de-duplication and compression techniques are applied to transfer data. In multi-route-aware transfer, data are placed in an intermediate node when a receiver is not ready to receive the data. The stored data is forwarded to destination by multiple routes with aggregated bandwidth. This process reduces the transmission rate and cost [17].

b. **Challenges in secure data transmission**

- **Heterogeneous Ring SignCryption**—SingCryption process provides confidentiality, integrity and authentication to sensor data in a single step. Identity-Based Cryptography is used to transfer information from sensor to server in a secure manner. Ring SignCryption protects the privacy of sender. Public key is generated by group of sensor nodes through sensor ID. This group ID is used for message encryption and sent to receiver. Receiver's private key is used for decryption process and it knows the sender information through group ID [18].
- **Compression, Encryption and Compression**—Encryption techniques ensures data confidentiality during transmission, but it increases the data size and transmission cost. To reduce data size and cost, transmitted data are compressed by

Burrows–Wheeler Transform and Run-Length encoding is used for compression and reduced array based encryption technique is used for encryption. Then the encrypted data is compressed by Huffman coding technique for reducing the data size further. Multi-dictionary encoding and decoding are used to increase information retrieval speed and data size is reduced by three times [19].

4 Open Issues

1. Storage of big data with different genre is difficult to store in single storage media.
2. Difficult to analyse and extract specific data from large volume.
3. De-duplication and load balancing is not achieved 100%.
4. Providing security to entire data increases data size and storage cost with lesser security level.
5. Providing access control to specific data is a critical task.
6. Data transmission with large volume of data increases transmission cost.

5 Conclusion

Big data has become a promising technology to predict future trends. In this study, the big data characteristics are analysed along with sources. Then, big data storage challenges with de-duplication, load balancing and secure storage issues with its solutions are discussed. After that, different kind of sensitive data identification methods with its pros and cons in security point of view is analysed. Finally, big data transmission challenges with it solutions in security point of view is analysed. In future work, sensitive data are segregated from non-sensitive data in an entire dataset and security mechanisms will be applied to sensitive data instead of entire data for reducing storage and transmission cost with better security to sensitive data in cloud storage system.

References

1. Florissi, P.: EMC and Big Data. http://cloudappsnews.com
2. Minelli, M., Chambers, M., Dhiraj, A.: Text book on Big Data, Data Analytics
3. Ishwarappa, Anuradha, J.: A brief introduction of big data 5 V's characteristics and hadoop technology. In: International Conference on Intelligent Computing, Communication and Convergence, ICCC-2015, pp. 319–324
4. Alnafoosi, A.B., Steinbach, T.: An integrated framework for evaluating big-data storage solutions-IDA case study. In: Science and Information Conference (2013)
5. Zhang, X., Xu, F.: Survey of research on big data storage. IEEE, pp. 76–80 (2013)

6. Li, J., Chen, X., Xhafa, F., Barolli, L.: Secure deduplication storage systems supporting keyword search. J. Comput. Syst. Sci. (Elsevier)
7. Wang, Z., Chen, H., Ying, F., Lin, D., Ban, Y.: Workload balancing and adaptive resource management for the swift storage system on cloud. Future Gener. Comput. Syst. **51**, 120–131 (2015)
8. Basu, A., Sengupta, I., Sing, J.: Cryptosystem for secret sharing scheme with hierarchical groups. Int. J. Netw. Secur. **15**(6), 455–464 (2013)
9. Dorairaj, S.D., Kaliannan, T.: An adaptive multilevel security framework for the data stored in cloud environment. Sci. World J. **2015** (Hindawi publishing corporation)
10. Tanwar, S., Prema, K.V.: Role of public key infrastructure in big data security. CSI Communication (2014)
11. Kaur, K., Zandu, V.: A data classification model for achieving data confidentiality in cloud computing. Int. J. Mod. Comput. Sci. **4**(4) (2016)
12. Torra, V.: Towards knowledge intensive privacy. In: Proceeding of the 5th International Workshop on Data Privacy Management, pp. 1–7. Springer-Verlag
13. Perez-Lainez, R.C, Pablo-Sanchez, Iglesias, A.: Anonimytext: Anonymization of unstructured documents. Scitepress Digital Library, pp. 284–287 (2008)
14. Sanchez, D., Betat, M.: Toward sensitive document release with privacy guarantees. Eng. Appl. Artif. Intell. **59**, 23–34 (2017)
15. Domingo-Ferrer, J., Sanchez, D., Soria-Comas, J.: Database anonymization: privacy models, data utility, and micro-aggregation based inter-model connections. Morgan & Clayton, San Rafael, California USA (2016)
16. Yildirim, E., Arslan, E., Kim, J., Kosar, T.: Application level optimization of big data transfers through pipelining, parallelism and concurrency. IEEE Trans. Cloud Comput. **4**(1), 63–75 (2016)
17. Tudoran, R., Costan, A., Antoniu, G.: Overflow: multi-site aware big data management for scientific workflows on cloud. IEEE Trans. Cloud Comput. Vol. x, No. x (2014)
18. Li, F., Zheng, Z., Jin, C.: Secure and Efficient Data Transmission in the Internet of Things, vol. 62, pp. 111–122. Springer (2015)
19. Baritha Begam, M., Venkataramani, Y.: A new compression scheme for secure transmission. Int. J. Autom. Comput. **10**(6), 579–586 (2013)

Classifying Road Traffic Data Using Data Mining Classification Algorithms: A Comparative Study

J. Patricia Annie Jebamalar, Sujni Paul and D. Ponmary Pushpa Latha

Abstract People move from place to place for various purposes using different modes of transportation. This creates traffic on the roads. As population increases, number of vehicles on the road increases. This leads to a serious problem called traffic congestion. Predicting traffic congestion is a challenging task. Data Mining analyzes huge data to produce meaningful information to the end users. Classification is a function in data mining which classifies the given data into various classes. Traffic congestion on roads can be classified as free, low, medium, high, very high, and extreme. Congestion on roads is based on the attributes such as speed of the vehicle, density of vehicles on the road, occupation of the road by the vehicles, and the average waiting time of the vehicles. This paper discusses how traffic congestion is predicted using data mining classifiers with big data analytics and compares different classifiers and their accuracy.

Keywords Data mining · Classification · Traffic congestion · Accuracy
Big data analytics

1 Introduction

There are many modes of transportation in this world such as aviation, ship transport, and land transport. Land transport includes rail and road transport. People are transported using various types of vehicles. As population increases, number of vehicles on the roads also increases. Roads get congested when the numbers of vehicles grow more than the capacity of the road. This leads to a serious problem known as traffic

J. Patricia Annie Jebamalar (✉)
Karunya University, Coimbatore, Tamil Nadu, India
e-mail: patricia.benlaz@gmail.com

S. Paul
School of Engineering and Information Technology, Al Dar University College, Dubai, UAE

D. Ponmary Pushpa Latha
School of Computer Science and Technology, Karunya University, Coimbatore, Tamil Nadu, India

© Springer Nature Singapore Pte Ltd. 2019
J. D. Peter et al. (eds.), *Advances in Big Data and Cloud Computing*,
Advances in Intelligent Systems and Computing 750,
https://doi.org/10.1007/978-981-13-1882-5_18

congestion. It leads to slower speed, longer travel time and increased waiting time for the travelers. There is no standard technique to measure traffic congestion. It is possible to measure the level of traffic congestion with the advanced technologies like big data analytics and data mining.

In real world, traffic measurements can be made at every traffic junction through CCTV camera. In busy cities, the vehicle count is so enormous that a continual measurement requires big data analytics [1].

Data mining is a technique used for analyzing data in order to summarize some useful information from it. Classification is a data mining function which is used to classify the data into classes. The supervised learning method can accurately predict the target class. So, if the traffic data is supplied to the classification function, then the level of traffic congestion can be measured easily.

Many classification functions are available in data mining [2]. In this paper, the traffic data is classified using different classifiers like J48, Naive Bayes, REPTree, BFTree, Support Vector Machine, and Multi-layer Perceptron. The results were tabulated and compared to find out the best classifier for traffic data.

2 Literature Review

Data Mining is a process of acquiring knowledge from large databases. The steps involved in any data mining process are selection of data set, preprocessing, transformation, data mining, and interpretation/extraction. Classification is a data mining technique. The purpose of classification is to assign target label for each test case in the dataset. Classification develops an accurate model for the data. There are two phases in classification. They are training and testing phases [3]. Classification is used in many fields. Data mining classifiers are used for classifying educational data [4–6]. Also, classification is used in medical data analysis in order to predict the disease like diabetes [3, 7] and kidney disease [8].

Predicting the level of traffic congestion on roads is a serious problem. Traffic congestion can be predicted using neural network [9]. In [10], Search-Allocation Approach is used to manage traffic congestion. Also, Deep learning Approach can be used to predict traffic flow [2]. Congestion levels are normally grouped as free, low, medium, high, very high, and extreme [11, 12]. Since classification is a technique that can be used for assigning target class for the given data, this can also be used for classifying traffic data [13]. Since traffic data grows big in large cities, MongoDB can be used for handling traffic data [1, 2].

In this paper, the performance of various data mining classifiers is compared to predict the level of traffic congestion on roads [11, 14].

3 Data Mining Classification Algorithms

A classifier is a supervised learning technique and it is used to analyze the data. Classification models predict categorical class labels. There are two steps for any classification model. The classification algorithms build the classifiers in the first step. This step is known as learning step. The training set which has database records and their respective target class labels are used to build the classifier. In the second step, the test data is assigned a class label and the accuracy of classification rules is estimated. The classification rules can be applied to the new data sets if the accuracy is considered as acceptable.

3.1 Types of Classifier

There are different types of classifiers available for classifying data. In this research work, some of the widely used classifiers like J48, Naive Bayes, REPTree, BFTree, Support Vector Machine, and Multi-layer Perceptron are used for comparative study [4].

J48

J48 is a classification filter in data mining and it uses the C4.5 algorithm. This algorithm is used to produce a decision tree based on the entropy, gain ratio and information gain of the various attributes in the dataset. It is also a supervised learning mechanism. So, it needs a training set with desired class labels for the given data. When the test set is supplied to J48 filter, it classifies the data and assigns target label for each record.

Naive Bayes

Naive Bayes algorithm uses Bayes theorem. This technique is particularly useful when the dimensionality is very high. New raw data can be added during runtime. This is a probabilistic classifier.

REPTree

REPTree is a fast decision tree learner. It builds a decision tree using information gain. The tree is pruned using reduced error pruning (with backfitting).

BFTree

BFTree is a decision tree classifier. It uses binary split for both nominal and numeric attributes.

Support Vector Machine (SVM)

SVM is a supervised learning model. In the SVM model, the examples are represented as points in space and it is mapped. Each category is divided by a clear gap. Then, new examples are mapped into the same space and predicted to belong to a category based on which side of the gap they fall.

Multi-layer Perceptron

Multi-layer Perceptron is a feed- forward artificial neural network technique. It consists of at least three layers of nodes. It uses nonlinear activation function except the input nodes. It utilizes backpropagation for training, as this is a supervised learning technique.

4 Methodology for Measuring Level of Traffic Congestion

As discussed in the previous section, data can be classified using the data mining classifiers. The objective of this paper is to compare the accuracy of different classifiers while classifying the level of traffic congestion in an area.

4.1 Traffic Data Using Big Data Analytics

In real world, traffic measurements would be made at every traffic junction through CCTV camera and associated software system which identifies each vehicle's entry time-stamp and exit time-stamp. This provides the amount of traffic enters and exits in a particular interval. Hence, the number of vehicle present in the edge at a given point of time can be counted. This provides a measurement of traffic congestion in the edge. However, in busy cities, the vehicle count is so enormous that a continual measurement requires large data (big data) analytics. Big Data Analytics is performed using MapReduce algorithm over a distributed storage and processing elements.

One of the most prominent distributed data store available in open-source community is MongoDB, which supports very easy to program MapReduce method [1]. MongoDB Storage Cluster is explained in Fig. 1.

In MongoDB, a distributed (shredded) data store is proposed using hash code on edge ID property. This enables each edge ID is stored in a specific data node. Hence, a query to identify the congestion is computed for each edge in one of the distributed storage node. It is required to implement data mining classification algorithms in the MapReduce functions to classify the data set. The structure of MongoDB Object for traffic measurement, MapReduce function and the query to retrieve the edge record are given below.

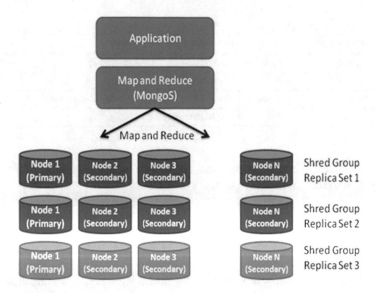

Fig. 1 MongoDB storage cluster

MongoDB Object	Map Function

{
_id:
ObjectId("50a8240b927d5d8b5891743c"),
edge_id: "tlyJn12",
time: new Date("Oct 04, 2015 11:15 IST"),
meaureAt: 'Entry',
vehicleProp:{length:"NA",type: "NA"} }
	var mapFunction1 = function() {
	for (var idx = 0; idx < this.items.length; idx++)
	{
	var key = this.items[idx].edge_id;
	var value = −1;
	if (this.items[idx].measureAt == "Entry")
	value = 1;
	emit(key, value);
	}　　};

Reduce Function	**Query**

var reduceFunction1 = function(keySKU,
countObjVals) {
reducedVal = 0;
for (var idx = 0; idx < countObjVals.length;
idx++) {
reducedVal += countObjVals[idx].value;
}
return reducedVal;
};
	db.trafficdata.mapReduce(mapFunction1,
	reduceFunction1,
	{out: {merge: "edge_num_vehicle" },
	query: {time:
	{ $gt: new Date('01/01/2017 11:00'), $and:1
	$lt: new Date('01/01/2017 11:59') } },
	})

4.2 Preparing Data Set Using Simulation

The traffic data was generated using a simulator called Simulation of Urban MObilty (SUMO) [15]. As traffic congestion depends on vehicle density on the road, occupancy of the road, average waiting time, and speed of vehicles, the attributes density,

Table 1 Traffic-related attributes

SNo.	Attribute name	Type	Description
1	Density	#veh/km	Vehicle density on the lane/edge
2	Occupancy	%	Occupancy of the edge/lane in %
3	Waiting time	Seconds	The total number of seconds vehicles stopped
4	Speed	m/s	The mean speed on the edge/lane

occupancy, waiting time, and speed are considered as high potential attributes for classifying the traffic data which is listed in Table 1.

4.3 Measuring Traffic Congestion Using WEKA

WEKA [13] is a data mining tool with a collection of various data mining algorithms. The traffic data generated using SUMO is given as input to the WEKA tool. Then the traffic data is classified using various classifiers available in WEKA. The level of traffic data is classified into free, low, medium, high, very high, and extreme.

5 Experimental Results and Comparison

Traffic congestion depends on vehicle density on the road, occupancy of the road, average waiting time and speed of vehicles. So the attributes density, occupancy, waiting time, and speed are used for classification. The level of traffic data is classified into free, low, medium, high, very high, and extreme. J48, Naive Bayes, REPTree, BFTree, Support Vector Machine, and Multi-layer Perceptron are used to classify the traffic data in this experiment. The statistical analysis is tabulated in Table 2.

The comparison of the accuracy of the classifiers based on correctly classified instances with cross validation is given in Table 3.

Figure 2 shows the comparison of various classifiers based on accuracy using a graph. From the graph, it is clear that the accuracy of J48 is better. So, J48 classifier can be used to predict the level of traffic congestion.

6 Conclusion

The experimental results show that data mining classifiers can be used to predict the level of traffic congestion on roads as free, low, medium, high, very high and extreme. Big data analytics can be used to process the data. The classifiers such as J48, Naive Bayes, and Multi-layer Perceptron are used to measure the traffic congestion on

Table 2 Statistical analysis of classifiers with cross validation

Classifier	Class	TP rate	FP rate	Precision	Recall	F-measure	ROC area
J48	Free	0.944	0.011	0.971	0.944	0.958	0.98
	Low	0.7	0.027	0.7	0.7	0.7	0.839
	Medium	0.733	0.846	0.688	0.733	0.71	0.906
	High	0.5	0.018	0.714	0.5	0.588	0.867
	Veryhigh	0.688	0.056	0.647	0.688	0.667	0.803
	Extreme	0.972	0.034	0.921	0.972	0.946	0.973
Naive Bayes	Free	0.944	0.046	0.895	0.944	0.919	0.989
	Low	0.8	0.035	0.667	0.8	0.727	0.986
	Medium	0.6	0.046	0.643	0.6	0.621	0.957
	High	0.7	0.044	0.583	0.7	0.636	0.904
	Veryhigh	0.375	0.056	0.5	0.375	0.429	0.923
	Extreme	0.861	0.046	0.886	0.861	0.873	0.989
REPTree	Free	1	0.069	0.857	1	0.923	0.956
	Low	0.5	0.035	0.556	0.5	0.526	0.838
	Medium	0.533	0.056	0.571	0.533	0.552	0.868
	High	0.5	0	1	0.5	0.667	0.892
	Veryhigh	0.563	0.028	0.75	0.563	0.643	0.744
	Extreme	0.972	0.069	0.854	0.972	0.909	0.953
BFTree	Free	0.972	0.023	0.946	0.972	0.959	0.989
	Low	0.5	0.027	0.625	0.5	0.556	0.817
	Medium	0.8	0.037	0.75	0.8	0.774	0.924
	High	0.5	0.018	0.714	0.5	0.588	0.843
	Veryhigh	0.625	0.056	0.625	0.625	0.625	0.832
	Extreme	0.944	0.057	0.872	0.944	0.907	0.962
Support Vector Machine	Free	1	0.172	0.706	1	0.828	0.914
	Low	0	0	0	0	0	0.524
	Medium	0.533	0.046	0.615	0.533	0.571	0.73
	High	0	0	0	0	0	0.799
	Veryhigh	0.5	0.131	0.364	0.5	0.421	0.713
	Extreme	0.833	0.08	0.811	0.833	0.822	0.941
Multi-layer Perceptron	Free	0.972	0.046	0.897	0.972	0.933	0.996
	Low	0.5	0.027	0.625	0.5	0.556	0.938
	Medium	0.8	0.111	0.5	0.8	0.615	0.918
	High	0	0.018	0	0	0	0.784
	Veryhigh	0.5	0.075	0.5	0.5	0.5	0.891
	Extreme	0.889	0.023	0.941	0.889	0.914	0.983

Table 3 Comparison of classifiers

Classifier	Accuracy (%)
J48	83.7398
Naive Bayes	77.2358
REPTree	79.6748
BFTree	82.1138
Support Vector Machine	66.6667
Multi-layer Perceptron	74.7967

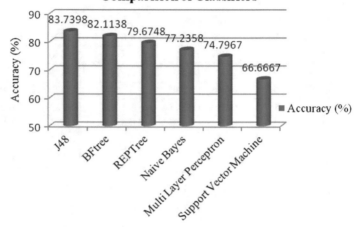

Fig. 2 Comparison of classifiers based on accuracy

roads. The accuracy of the classifiers is calculated using cross validation. From this comparative study, it is clear that J48 is performing better. So, J48 classifier can be used for predicting traffic congestion on roads.

References

1. Ananth, G.S., Raghuveer, K.: A novel approach of using MongoDB for big data analytics. Int. J. Innovative Stud. Sci. Eng. Technol. (IJISSET) **3**(8), 7 (2017). ISSN 2455-4863(Online)
2. Lv, Y., Duan, Y., Kang, W., Li, Z., Wang, F.-Y.: Traffic flow prediction with big data: a deep learning approach. IEEE Trans. Intell. Transp. Syst. **16**(2), 865 (2015)
3. Kaur, G., Chhabra, A.: Improved J48 classification algorithm for the prediction of diabetes. Int. J. Comput. Appl. **98**(22), 0975–8887 (2014)
4. Rajeshinigo, D., J. Patricia Annie Jebamalar: Educational Mining: A Comparative Study of Classification Algorithms Using Weka. Innovative Res. Comput. Commun. Eng. (2017)
5. Kaur, P., Singh, M., Josan, G.S.: Classification and prediction based data mining algorithms to predict slow learners in education sector. In: 3rd International Conference on Recent Trends in Computing 2015 (ICRTC-2015). Procedia Comput. Sci. **57**, 500–508. Elsevier (2015)

6. Adhatrao, K., Gaykar, A., Dhawan, A., Jha, R., Honrao, V.: Predicting students' performance using ID3 and C4.5 classification algorithms. Int. J. Data Min. Knowl. Manage. Process (IJDKP) **3**(5) (2013). https://doi.org/10.5121/ijdkp.2013.3504
7. Iyer, A., Jeyalatha, S., Sumbaly, R.: Diagnosis of diabetes using classification mining techniques. Int. J. Data Min. Knowl. Manage. Process (IJDKP) **5**(1) (2015). https://doi.org/10.5121/ijdkp.2015.5101
8. Vijayarani, S., Dhayanand, S.: Data mining classification algorithms for kidney disease prediction. Int. J. Cybern. Inf. (IJCI) **4**(4) (2015). https://doi.org/10.5121/ijci.2015.4402
9. Fouladgar, M., Parchami, M., Elmasri, R., Ghaderi, A.: Scalable Deep Traffic Flow Neural Networks for Urban Traffic Congestion Prediction (2017)
10. Raiyn, J., Qasemi, A., El Gharbia, B.: Road traffic congestion management based on a search-allocation approach. Transp. Telecommun. **18**(1), 25–33 (2017). https://doi.org/10.1515/ttj-2017-0003
11. Rao, A.M., Rao, K.R.: Measuring urban traffic congestion—a review. Int. J. Traffic Transp. Eng. (2012)
12. Bauza, R., Gozalvez, J., Sanchez-Soriano, J.: Road traffic congestion detection through cooperative vehicle-to-vehicle communications. In: Proceedings of the 2010 IEEE 35th Conference on Local Computer Networks, pp. 606–612 (2010)
13. Salvithal, N.N., Kulkarni, R.B.: Evaluating performance of data mining classification algorithm in weka. Int. J. Appl. Innov. Eng. Manage. (IJAIEM) **2**(10), 273–281(2013). ISSN 2319 – 4847
14. Akhila, G.S., Madhu, G.D., Madhu, M.H., Pooja, M.H.: Comparative study of classification algorithms using data mining. Discov. Sci. **9**(20), 17–21 (2014)
15. http://www.dlr.de, SUMO tutorial

D-SCAP: DDoS Attack Traffic Generation Using Scapy Framework

Guntupalli Manoj Kumar and A. R. Vasudevan

Abstract Bots are harmful processes controlled by a Command and Control (C&C) infrastructure. A group of bots is known as botnet to launch different network attacks. One of the most prominent network attacks is Distributed Denial of Service (DDoS) attack. Bots are the main source for performing the harmful DDoS attacks. In this paper, we introduce a D-SCAP (DDoS Scapy framework based) bot to generate high volumes of DDoS attack traffic. The D-SCAP bot generates and sends continuous network packets to the victim machine based on the commands received from the C&C server. The DDoS attack traffic can be generated for cloud environment. The D-SCAP bot and the C&C server are developed using Python language and Scapy framework. The D-SCAP bot is compared with the existing well-known DDoS bots.

Keywords Bots · Botnet · C&C server · D-SCAP · DDoS

1 Introduction

Bots are malicious process [1] that infects the machine to perform harmful actions such as DDoS attack, stealing credentials, email spamming, etc. [2]. Once the machine is infected with the bot, it sends its network related information to the C&C server. An attacker who sits behind the C&C server is called botmaster who controls the bots with instructions to perform harmful actions [3]. After receiving the network-related details of the infected machine, C&C server pushes commands to the infected machines to perform different attacks. Bots will execute the commands provided by the botmaster.

Botnet architecture can be centralized or de-centralized [4]. In centralized botnets, only one C&C server is used to connect to bots, whereas in decentralized approach, every machine behaves as a C&C server. Botnets are also categorized depending on

G. Manoj Kumar (✉) · A. R. Vasudevan
TIFAC-CORE in Cyber Security, Amrita School of Engineering, Amrita Vishwa Vidyapeetham, Coimbatore, India
e-mail: guntupallimanojkumarr@gmail.com

© Springer Nature Singapore Pte Ltd. 2019
J. D. Peter et al. (eds.), *Advances in Big Data and Cloud Computing*,
Advances in Intelligent Systems and Computing 750,
https://doi.org/10.1007/978-981-13-1882-5_19

the protocol used by C&C server to send commands to the bots. C&C server can use Internet Relay Chat (IRC) protocol [5], Hyper Text Transfer Protocol (HTTP), or Peer-to-Peer (P2P) protocol to send commands. IRC-based bots are the most widely used bots. HTTP bots follows the PULL approach where the attack happens at regular interval of time as instructed by the botmaster [6].

In this paper, we introduce D-SCAP bot to generate high volume of DDoS attack traffic. The D-SCAP bot is developed using Scapy framework [7] in Python language. D-SCAP bots can perform DDoS attack on victim machine. The D-SCAP bot can be installed in both Windows- and Linux-based operating systems.

The rest of the paper is organized as follows. The related work is presented in Sect. 2. In Sect. 3, the network setup, timeline of activities, and the algorithm of D-SCAP bot are presented in detail. Section 4 describes the results and analysis of D-SCAP bot and we conclude the paper in Sect. 5.

2 Related Work

The bots which are controlled by the botmaster targets the computers that are less monitored and having high bandwidth connectivity. The botmaster takes the advantage of security flaw in the network to keep the bots alive for a long duration. The understanding of bot life cycle [8] plays a major role in the development of bots. The botmaster uses different approaches to infect the bots. The bot code can be propagated through email or by means of file sharing mechanisms to infect the machines. The bot has to be developed in such a way that it should respond to the update sent by the botmaster. The update can include the new IP address for the C&C server [9]. The bots execute the commands sent by the botmaster and the results will be reported to the C&C server and wait for new instructions [10].

DDoS attacks [11] can be performed on a victim machine with the help of bots by sending continuous requests thereby crashing the victim machine [12]. DDoS attacks architecture [13] which are IRC based and Agent handler based. IRC communication channel is used by the IRC-based model [14], whereas the Agent handler model uses the handler program to perform attacks with the help of bots.

3 D-SCAP Bot: DDoS Scapy Framework Based Bot

3.1 Network Architecture

The network architecture considered for developing the D-SCAP bot is shown Fig. 1. Bot source code is distributed via email or through USB drive. After the bot installation, the bot-infected machine listens for commands from the C&C server. The proposed architecture shows the bot-infected machines are remotely controlled by

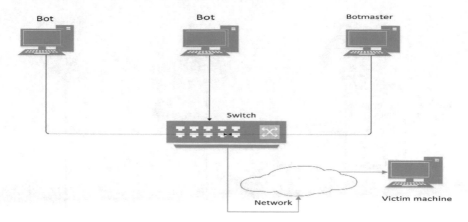

Fig. 1 Network architecture

a botmaster. The infected machines perform harmful activities such as TCP-SYN flooding attack, UDP flooding attack, and identity theft, etc., on the victim machine based on the instructions provided by the botmaster who controls the C&C server.

3.2 Timeline Activity of D-SCAP Bot

The timeline activity of the D-SCAP bot is shown in Fig. 2. After the installation of the D-SCAP bot in the machines, the infected machine sends its network-related information such as IP address and port to the C&C server controlled by the botmaster. After receiving the details from the bot machines, the C&C server sends the commands to the infected machine to generate large volume of DDoS attack traffic. The C&C server command includes the victim machine address, type of traffic to be generated, and the packet count. Once the bot machine receives the command from the C&C server, it resolves the victim address and generates (TCP/UDP/ICMP) packets to perform DDoS attack on the victim.

3.3 D-SCAP Bot Algorithms

In this section, the algorithm of C&C server and the D-SCAP bot which are developed using the Scapy framework in Python 2.7 is presented.

(a) **C&C Server Algorithm**

    Step 1:   Start

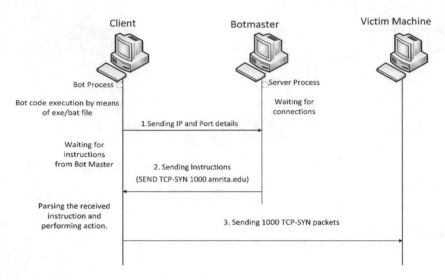

Fig. 2 D-SCAP bot communication flow

Step 2: Wait for connections from the machines which are infected with D-SCAP bot

Step 3: Receive IP and port details from infected machines

Step 4: Send instruction (SEND Attack_Type Packet_Count Victim_Domian _Name) to infected machines to perform the attack on victim

Step 5: Go to Step 2.

The Attack_Type indicates the type of attack traffic to be generated by the bot-infected machines. It includes (TCP-SYN, TCP-ACK, and RST) attacks. Packet_Count specifies the number of packets the bot has to send to the victim machine.

(b) **D-SCAP bot Algorithm**

The set of actions performed by D-SCAP bot in the infected system is as follows.

Step 1: Start

Step 2: Send the infected machine IP and port address information to C&C server

Step 3: Listen for the instruction from C&C server

Step 4: Resolve the Victim Domain name and generate the attack traffic on the victim machine

Step 5: Stop.

Table 1 Different DDoS attack bots

DDoS bots	Bot features		
	Platforms supported	Programmed language	DDoS attack types
AgoBot [15, 19]	Mostly Windows	C++	TCP-SYN, UDP, ICMP
SpyBot [16, 19]	Windows	C	TCP-SYN, UDP, ICMP
RBot [17, 19]	Windows	C++	TCP-SYN, UDP, ICMP
SDBot [18, 19]	Windows	C++	UDP, ICMP
D-SCAP bot	Windows and Linux	Python	TCP-SYN, UDP, ICMP

3.4 Comparison of Various DDoS Attack Bots

See Table 1.

DDoS attack bots are developed to support different Operating Systems (OSs). The existing DDoS attack bots are Agobot, SpyBot, RBot, and SDbot. In this work, we introduce a D-SCAP bot for performing DDoS attack which supports both Windows and Linux platforms.

4 Results

4.1 C&C Server

The C&C server is set up by the botmaster to control the botnet and listens for the connections. The C&C server can be established both in Windows and Linux environments to perform DDoS attack.

4.2 Infection Stage

The machine gets infected with the D-SCAP bot when the user clicks on the executable file (Windows)/Shell Script (Linux). The executable can be sent as an email attachment or using file sharing.

Fig. 3 Details of the D-SCAP-infected machine

Fig. 4 Attack on victim machine by the D-SCAP bot

4.3 Rallying Stage

Rallying stage refers to the first-time communication of D-SCAP bot with the C&C server. Once the D-SCAP bot successfully infects the machine, it sends the machine IP address and port details. Figure 3 shows the details of the infected machine.

4.4 Attack Performing Stage

The D-SCAP residing in the infected machine waits for the command from the C&C server. D-SCAP bot generates TCP-SYN or UDP or ICMP flooding traffic to perform attack on the victim machine based on the attack type sent by the botmaster. Figure 4 shows the network packets sent by the D-SCAP bot to the victim machine.

5 Conclusion and Future Work

DDoS attack is a serious threat to the internet. In this paper, the contributions were the construction of D-SCAP bot for DDoS attack generation using Scapy framework in Python 2.7. The D-SCAP bot is capable of producing TCP-SYN, UDP, and ICMP attack vectors destined to the victim machine. Our future work deals with the detection of the botnet using combined approach which monitors both host and network level activities.

References

1. Rajab, M., Zarfoss, J., Monrose, F., Terzis, A.: A multifaceted approach to understanding the botnet phenomenon. In: Proceedings of 6th ACM SIGCOMM Conference on Internet Measurement (IMC'06), pp. 41–52 (2006)
2. Zhang, L., Yu, S., Wu, D., Watters, P.: A survey on latest botnet attack and defense. In: 2011 IEEE 10th International Conference on Trust, Security and Privacy in Computing and Communications (TrustCom), pp. 53–60. IEEE (2011)
3. Kalpika, R., Vasudevan, A.R.: Detection of zeus bot based on host and network activities. In: International Symposium on Security in Computing and Communication, pp. 54–64. Springer, Singapore (2017)
4. Mahmoud, M., Nir, M., Matrawy, A.: A survey on botnet architectures, detection and defences. IJ Netw. Secur. **17**(3), 264–281 (2015)
5. Oikarinen, J., Reed, D.: Internet relay chat protocol. RFC1459 (1993)
6. Lee, J.-S., Jeong, H.C., Park, J.H., Kim, M., Noh, B.N.: The activity analysis of malicious http-based botnets using degree of periodic repeatability. In: SECTECH'08. International Conference on Security Technology, pp. 83–86. IEEE (2008)
7. http://www.secdev.org/projects/scapy/
8. Hachem, N., Mustapha, Y.B., Granadillo, G.G., Debar, H.: Botnets: lifecycle and taxonomy. In: Proceedings of the Conference on Network and Information Systems Security (SAR-SSI), pp. 1–8 (2011)
9. Choi, H., Lee, H., Kim, H.: BotGAD: detecting botnets by capturing group activities in network traffic. In: Proceedings of the Fourth International ICST Conference on Communication System Software and Middleware, p. 2. ACM (2009)
10. Bailey, M., Cooke, E., Jahanian, F., Yunjing, X., Karir, M.: A survey of botnet technology and defenses. In: Proceedings of the Cybersecurity Applications & Technology Conference for Homeland Security (CATCH), pp. 299–304 (2009)
11. Guri, M., Mirsky, Y., Elovici, Y.: 9-1-1 DDoS: attacks, analysis and mitigation. In: 2017 IEEE European Symposium on Security and Privacy (EuroS&P), pp. 218–232. IEEE (2017)
12. Zargar, S.T., Joshi, J., Tipper, D.: A survey of defense mechanisms against distributed denial of service (ddos) flooding attacks. IEEE Commun. Surv. Tutorials **15**(4), 2046–2069 (2013)
13. Kaur, H., Behal, S., Kumar, K.: Characterization and comparison of distributed denial of service attack tools. In: 2015 International Conference on Green Computing and Internet of Things (ICGCIoT), pp. 1139–1145. IEEE (2015)
14. Specht, S.M., Lee, R.B.: Distributed denial of service: taxonomies of attacks, tools, and countermeasures. In: ISCA PDCS, pp. 543–550 (2004)
15. https://www.sophos.com/en-us/threat-center/threat-analyses/viruses-and-spyware/W32~Agobot-NG/detailed-analysis.aspx
16. https://www.sophos.com/en-us/threat-center/threat-analyses/viruses-and-spyware/Troj~SpyBot-J/detailed-analysis.aspx
17. https://www.sophos.com/en-us/threat-center/threat-analyses/viruses-and-spyware/W32~Rbot-AKY/detailed-analysis.aspx
18. https://www.sophos.com/en-us/threat-center/threat-analyses/viruses-and-spyware/W32~Sdbot-ACG/detailed-analysis.aspx
19. Thing, V.L., Sloman, M., Dulay, N.: A survey of bots used for distributed denial of service attacks. In: IFIP International Information Security Conference, pp. 229–240. Springer, Boston, MA (2007)

Big Data-Based Image Retrieval Model Using Shape Adaptive Discreet Curvelet Transformation

J. Santhana Krishnan and P. SivaKumar

Abstract Digital India program will help in agriculture field in various ways, including a weather forecast to agriculture consultation. To find all the causing symptoms of diseased leaf, the knowledge-based Android app is proposed to refer the disease of a leaf. The user can directly capture the disease leaf image from their smartphone and upload that image into the app, and they will get all the causes and symptoms of a particular disease. Moreover, users can get information in the form of text and audio in their proffered language. This system will accept the query based on images and text format which is very useful to the farmers. In this proposed work, texture-based feature extraction using **Shape Adaptive Discreet Curvelet Transform** (SADCT) is developed using big data computing framework.

Keywords CBIR—content-based image retrieval
SADCT—shape adaptive discreet curvelet transform
HDFS—Hadoop distributed file system

1 Introduction

The era of Big data and Cloud computing became a challenge to traditional data mining algorithms. The algorithms used in traditional database framework, their processing capability, and engineering are not adaptable for big data analysis. Big Data is presently quickly developing in all fields of science and engineering, including biological, biomedical, and disaster management. The qualities of complexity plan an extreme challenge for finding useful information from the big data [1]. A large measure of agriculture data prompts meaning of complex relationship, which makes complexities and challenges in today data mining research. Current advancements

J. Santhana Krishnan (✉)
University College of Engineering Kancheepuram, Kancheepuram, India
e-mail: csesaki@gmail.com

P. SivaKumar
Karpakam College of Engineering, Coimbatore, India

© Springer Nature Singapore Pte Ltd. 2019
J. D. Peter et al. (eds.), *Advances in Big Data and Cloud Computing*,
Advances in Intelligent Systems and Computing 750,
https://doi.org/10.1007/978-981-13-1882-5_20

have prompted a surge of data from particular areas, for example, user-generated data, agriculture environment, healthcare and scientific sensor data, and Internet. Big data will be the data that surpasses the handling limit of ordinary database frameworks. The size of the data is too big, grows quickly, or does not fit the strictures of your database structures.

2 Related Work

Bravo et al. [2] find the spectral reflectance difference between in healthy wheat plants and diseased wheat plants affected by yellow rust [3]. Miller et al. [4] initiated their work in 2005. Their prime goal was the development of local capacity diagnostics in both data sharing networks and communication networks. Then perform training in classical method and modern diagnostics method and find novel diagnostics methods for evaluation. Klatt et al. [5] introduced Smart DDS, a 3-year study financed by the German Ministry of Agriculture. The ambition of this project was to develop a mobile application that can identify plant diseases with minimum computational cost. Prince et al. [6], developed a machine learning system that detects plants infected by tomato powdery mildew fungus Odium neolycopersici remotely. For developing the system, they combined both infrared and RGB image data with depth information.

The Indian Government launched "Digital India" project on July 1, 2015. The aim of this project is to enhance e-access capabilities in citizens and empower them for both government services- and livelihood-related services. The mAgriculture is the part of digital India project covered under mServices, direct impact with agriculture extension. Dr. Pratibha Sharma, (2009–12), do a project funded by ICAR (Indian Council of Agriculture Research). In this project, they identified molecular markers in relation to disease resistance and developed Disease prediction models and decision support systems. Majumdar et al. (2017), suggested data mining techniques to find the parameters which increase the production of the cost. They use DBSCAN, PAM, and CLARA algorithms to find the parameters and find a conclusion that DBSCAN is better than other techniques [7]. Rouached et al. (2017), do a review of how plants respond to nutritional limitations. This study will help to improve the yield of the crop. They integrate big data techniques to construct gene regulatory networks [8]. Xie et al. (2017) perform a study on natural social and economic factors affect the ecology of land in China. Farmers population change, GDP, per capita income are the major factors change the ecology [9]. Kamilaris et al. (2017) did a study on the topic "effect of big data in agriculture". They reviewed 34 research papers, examined the problem, methods used, the algorithms used, tools used, etc. They conclude that big data have a large role in the field of agriculture [10].

3 Visual Feature Extraction with Big Data Computing Framework

3.1 Big Data Computing Framework

In a distributed computing environment, the expansive data sets are treated by an open-source structure named Hadoop [1]. Hadoop composed of a MapReduce module, a Hadoop file distribution system (HDFS) and a number of associated projects like Zookeeper, HBase, and Apache Hive. The Hadoop Appropriated Document Frameworks (HDFS) [11] (Fig. 1).

The main node of HDFS is the NameNode that handles metadata. DataNode is slave node, which piles the data into blocks. Another important node of Hadoop is JobTracker which splits the task and assign the task to slave nodes. In slave nodes, Map and Reduce are performed, so they are called Tasktrackers. Most datasets processing is done by MapReduce programming module in both parallel and distributed environment [12]. There are two fundamental strategies in MapReduce: Map and Reduce. In general, the input and outcome are both in the form of key/value sets. Figure 2 demonstrates MapReduce programming model design. The input data is split into different blocks of 68 MB or 128 MB. The key/value pairs are provided to the mapper as information and it delivers the relative yield as key/value pairs. Partitioner and combiner are utilized as a part of amongst mapper and reducer to accomplish arranging and rearranging. The Reducer redoes through the qualities that are related to a particular key and delivers zero or more outcomes.

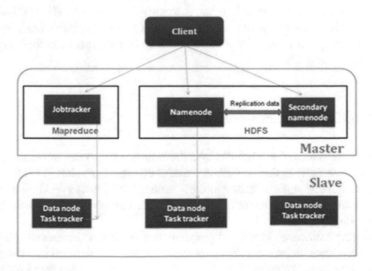

Fig. 1 HDFS architecture

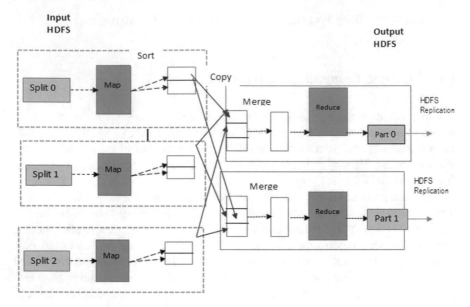

Fig. 2 MapReduce architecture

3.2 Visual Feature Extraction

Figure 3 shows the proposed architecture. The user can give a query in the form of an image or text or both. If the input is in the form of an image, then it will go to the feature extraction system. After that, the similarity is measured by Mahalanobis Distance method between input image features and feature database. The output will be the most relevant result with a corresponding text document in their preferred language and audio file.

3.3 SADCT-Based Feature Extraction

In the first section, we have familiarized the outline of curvelet transform and the relevance of curvelet in this work than wavelet change. Curvelets [13–15] are best for bent singularity approximation and are suited for takeout edge-based features from agricultural image data more effectively than that of contrasted with wavelet transform [16] (Fig. 4).

For feature extraction, we use curvelet transform on distinct images and clarify the results of recovery. In general, image recovery strategy is alienated into two phases a training phase and a classification phase. In the training phase, a set of recognized agricultural images are used to make relevant feature sets or templates. In the second phase, a comparison is done based on the features between an unknown agricultural

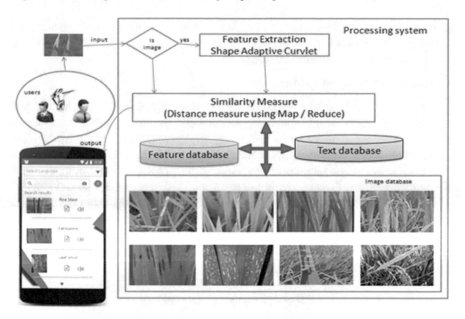

Fig. 3 The architecture of agricultural information retrieval system

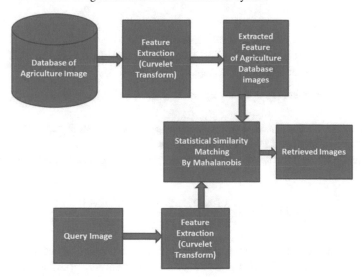

Fig. 4 SADCT-based feature extraction

image and the earlier seen images. Here, we take a query image from database and the image is transformed using curvelet transform in various scales and orientations. Use separate bands for feature extraction.

Statistical Similarity Matching

We used the Mahalanobis [17] standard for similarity measure, which considers different magnitudes for different components. Here, we input query image and find the likeness between query image and images in the database. This measure gives an idea about the similarity between these images. Mahalanobis Distance is better than Euclidean Distance measure. The retrieved images are ordered on the basis of similarity distance with respect to the query image. The likeness measure is calculated using the vectors of the query image and that of database image using the equation given below. Mahalanobis Distance (D) measurement expression is given below. The computed distances are arranged in ascending order. If computed similarity measure is less than a given threshold, then the corresponding image in the database is relevant to the input image.

$$D_2 = (x - m)^{\mathrm{T}} C^{-1} (x - m)$$

where

x Vector of data
m Vector of mean values of independent variables
C^{-1} Inverse covariance matrix of independent variables
T   Indicates vector should be transposed

Here, multi-rate vector $x = (x_1, x_2, x_3, ..., x_N)^{\mathrm{T}}$ and Mean $m = (m_1, m_2, m_3, ..., m_N)^{\mathrm{T}}$

Precision (P) is the fraction of the number of relevant instances retrieved r to the total number of retrieved instances n, i.e., $P = r/n$.

The accuracy of a retrieval model is measured using precision and it is expressed as

$$\text{Precision} = \frac{\text{No. of relevant images retrieved}}{\text{Total no. of images retrieved}} = \frac{r}{n}$$

Recall (R) is defined as the fraction of the number of relevant instances retrieved r to the total number of relevant instances m in the entire database, i.e., $R = r/m$. Robustness of retrieval model can be evaluated by the recall and it is expressed as:

$$\text{Recall} = \frac{\text{No. of relevant images retrieved}}{\text{Total no. of relevant images in DB}} = \frac{r}{m}$$

4 Experimental Results

We evaluated the execution of Agricultural image retrieval using curvelet transform using Mahalanobis Distance measure in Big data computing Framework. First, we show the result of Agricultural image retrieval on a data set using curvelet method to calculate the robustness and precision. The graphical user interface (GUI) is also created for the evaluation of the impact of the proposed technique (Fig. 5).

Figures 6, 7, 8, and 9 show the image retrieval result for different types of crops using proposed big data-based image retrieval system. Table 1 shows the retrieval system performance using precision value for wheat, paddy, sugarcane, and cotton (Table 2).

Fig. 5 Result of paddy images

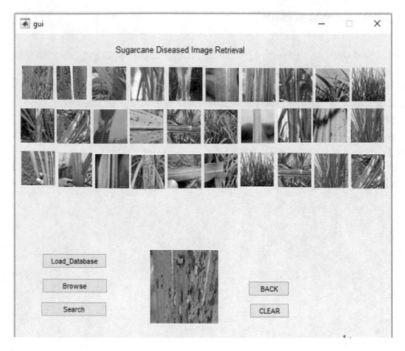

Fig. 6 Result of sugarcane images

Table 1 Precision

No. of retrieved images	Wheat	Cotton	Sugarcane	Paddy
Top 10	1	1	1	1
Top 20	0.95	0.95	0.95	0.92
Top 30	0.90	0.90	0.90	0.90
Top 40	0.88	0.88	0.88	0.90
Top 50	0.86	0.88	0.86	0.88
Top 60	0.84	0.86	0.86	0.86
Top 70	0.80	0.86	0.84	0.84
Top 80	0.75	0.84	0.82	0.82
Top 90	0.70	0.76	0.78	0.70
Top 100	0.69	0.66	0.74	0.68

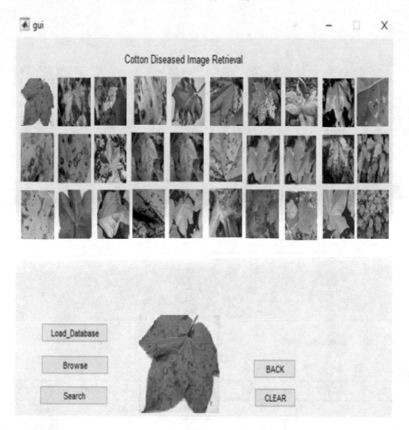

Fig. 7 Result of cotton images

The performance of the proposed model is validated with a huge number of datasets. Figure 9 represents the performance of the model using precision and recall. The proposed model is compared with existing image retrieval algorithm and it performs better than other models.

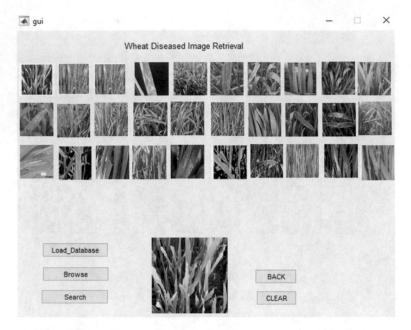

Fig. 8 Result of wheat images

Table 2 Recall

No. of retrieved images	Wheat	Cotton	Sugarcane	Paddy
Top 10	0.08	0.08	0.08	0.08
Top 20	0.15	0.16	0.15	0.18
Top 30	0.21	0.24	0.24	0.26
Top 40	0.28	0.30	0.28	0.32
Top 50	0.35	0.38	0.26	0.40
Top 60	0.41	0.44	0.40	0.46
Top 70	0.48	0.52	0.46	0.54
Top 80	0.54	0.58	0.52	0.60
Top 90	0.64	0.68	0.66	0.70
Top 100	0.70	0.78	0.72	0.80

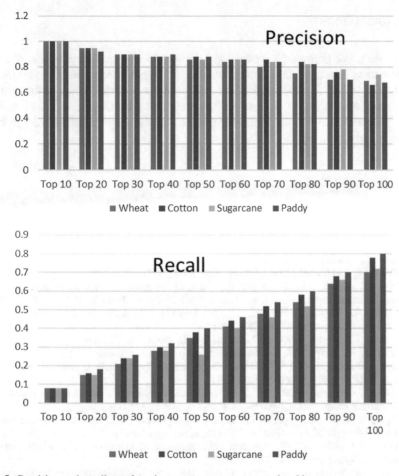

Fig. 9 Precision and recall rate for wheat, cotton, sugarcane, and paddy

5 Conclusions

In this work, we proposed a model for Agricultural image retrieval using discrete curvelet transform and Mahalanobis Distance in Big data computing environment. The main goal of this novel method is to increase the retrieval accuracy for texture-based Agricultural image retrieval in Big data computing framework. From this work, we concluded that curvelet features beats the existing texture features in both accuracy and efficiency.

References

1. Rajkumar, K., Sudheer, D.: A review of visual information retrieval on massive image data using hadoop. Int. J. Control Theor. Appl. **9**, 425–430 (2016)
2. Bravo, C., Moshou, D., West, J., McCartney, A., Ramon, H.: Early disease detection in wheat fields using spectral reflectance. Biosyst. Eng. **84**(2), 137–145 (2003)
3. Carson, C., Belongie, S., Greenspan, H., Malik, J.: Recognition of images in large databases using color and texture. IEEE Trans. Pattern Anal. Mach. Intell. **24**(8), 1026–1028 (2002)
4. Miller, S.A., Beed, F.D., Harmon, C.L.: Plant disease diagnostic capabilities and networks. Annu. Rev. Phytopathol. **47**, 15–38 (2009)
5. Klatt, B., Kleinhenz, B., Kuhn, C., Bauckhage, C., Neumann, M., Kersting, K., Oerke, E.C., Hallau, L., Mahlein, A.K., Steiner-Stenzel, U., Röhrig, M.: SmartDDS-Plant disease setection via smartphone. EFITA-WCCA-CIGR Conference "Sustainable Agriculture through ICT Innovation", Turin, Italy, 24–27 June (2013)
6. Prince, G., Clarkson, J.P., Rajpoot, N.M.: Automatic detection of diseased tomato plants using thermal and stereo visible light images. PloS One, **10**(4), e0123262 (2015)
7. Majumdar, J., Naraseeyappa, S., Ankalaki, S.: Analysis of agriculture data using data mining techniques: application of big data. J. Big Data, Springer (2017)
8. Rouached, H., Rhee, S.Y.: System-level understanding of plant mineral nutrition in the big data era. Curr. Opin. Syst. Biol. **4**, 71–77 (2017)
9. Xie, H., He, Y., Xie, X.: Exploring the factors influencing ecological land change for China's Beijinge-Tianjine-Hebei Region using big data. J. Cleaner Prod. **142**, 677e687 (2017)
10. Kamilaris, A., Kartakoullis, A., Prenafeta-Boldú, F.X.: A review on the practice of big data analysis in agriculture. Comput. Electron. Agric. **143**, 23–37 (2017)
11. Manjunath, B.S.: Texture features for browsing and retrieval of image data. IEEE Trans. Pattern Anal. Mach. Intell. **18**(8), 837–842 (1996)
12. Iakovidis, D.K., Pelekis, N., Kotsifakos, E.E., Kopanakis, I., Karanikas, H., Theodoridis, Y.: A pattern similarity scheme for medical image retrieval. IEEE Trans. Inf. Technol. Biomed. **13**(4), 442–450 (2009)
13. Rajakumar, K., Revathi, S.: An efficient face recognition system using curvelet with PCA. ARPN J. Eng. Appl. Sci. **10**, 4915–4920 (2015)
14. Rajakumar, K., Muttan, S.: Texture-based MRI image retrieval using curvelet with statistical similarity matching. Int. J. Comput. Sci. Issues **10**, 483–487 (2013)
15. Manipoonchelvi, P., Muneeswaran, K.: Significant region-based image retrieval using curvelet transform. In: IEEE Conference Publications, pp. 291–294 (2011)
16. Quellec, G., Lamard, M., Cazuguel, G., Cochener, B., Roux, C.: Fast wavelet-based image characterization for highly adaptive image retrieval. IEEE Trans. Image Process. **21**(4), 1613–1623 (2012)
17. Rajakumar, K., Muttan, S.: MRI image retrieval using wavelet with mahalanobis distance measurement. J. Electr. Eng. Technol. **8**, 1188–1193 (2013)

18. Zhang, L., Wang, L., Lin, W.: Generalized biased discriminant analysis for content-based image retrieval systems. IEEE Trans. Man Cybern. Part B Cybern. **42**(1), 282–290 (2012)
19. Zajić, G., Kojić, N., Reljin, B.: Searching image database based on content. In: IEEE Conference Publications, pp. 1203–1206 (2011)
20. Akakin, H.Ç., Gürcan, M.N.: Content-based microscopic image retrieval system for multi-image queries. IEEE Trans. Inf. Technol. Biomed. **16**(4), 758–769 (2012)
21. Li, Y., Gong, H., Feng, D., Zhang, Y.: An adaptive method of speckle reduction and feature enhancement for SAR images based on curvelet transform and particle swarm optimization. IEEE Trans. Geosci. Remote Sens. **49**(8), 3105–3116 (2011)
22. Liu, S., Cai, W., Wen, L., Eberl, S., Fulham, M.J., Feng, D.: Localized functional neuroimaging retrieval using 3D discrete curvelet transform. In: IEEE Conference Publications, pp. 1877–1880 (2011)
23. Minakshi, Banerjee, Sanghamitra, Yopadhyay, Sankar, K.P.: Rough Sets and Intelligent Systems, vol. 2, Springer link, pp. 391–395
24. Prasad, B.G., Krishna, A.N.: Statistical texture feature-based retrieval and performance evaluation of CT brain images. In: IEEE Conference Publications, pp. 289–293 (2011)

Region-Wise Rainfall Prediction Using MapReduce-Based Exponential Smoothing Techniques

S. Dhamodharavadhani and R. Rathipriya

Abstract Weather acts an important role in agriculture. Rainfall is the primary source of water that agriculturist depends on to cultivate their crops. Analyzing the historical data and predicting the future. As the size of the dataset becomes tremendous, the process of extracting useful information by analyzing these data has also become repetitive. To defeat this trouble of extracting information, parallel programming models can be used. Parallel Programming model achieves this by partitioning these large data. MapReduce is one of the parallel programming models. In general, Exponential Smoothing is one of the methods used for forecasting a time series data. Here, data is the sum of truth and error where truth can be "approximated" by averaging out previous data. It is used to forecast time series data when there is Level, Trend, Season, and Irregularity (error). In this paper, Simple Exponential Smoothing, Holt's Linear, and Holt-Winter's Exponential Smoothing methods are proposed with MapReduce computing model to predict region-wise rainfall. The experimental study is conducted on two different datasets. The first one is Indian Rainfall dataset which comprises of the year, state, and monthly rainfall in mm. The second is Tamil Nadu state rainfall dataset which consists of the year, districts, and monthly rainfall in mm. To validate these methods, MSE accuracy measure is calculated. From the results, Holt-Winter's Exponential Smoothing shows the better accuracy for rainfall prediction.

Keywords Rainfall · Prediction · MapReduce
Simple exponential smoothing method · Holt's linear method
Holt-Winter's method

S. Dhamodharavadhani (✉) · R. Rathipriya
Department of Computer Science, Periyar University, Salem, India
e-mail: vadhanimca2011@gmail.com

R. Rathipriya
e-mail: rathipriyar@gmail.com

© Springer Nature Singapore Pte Ltd. 2019
J. D. Peter et al. (eds.), *Advances in Big Data and Cloud Computing*,
Advances in Intelligent Systems and Computing 750,
https://doi.org/10.1007/978-981-13-1882-5_21

229

1   Introduction

Rainfall predictions associated with climate change may be variable across even one region to other, not speaking about larger areas on a global scale. The rainfall analysis/prediction is a mathematical approach to model the data and uses it for rainfall prediction for each region like states or district. There can be many methods used for analysis climate data, time series analysis is used in the current paper. This method is used to analyze the time series rainfall data for the period 1992–2002. The time series analysis is done to find the level and trend in rainfall, so it can be further used to forecast it for the future event.

The main intention of the time series analysis of the rainfall data is to find out any level, trend, and seasonality in the data and use this information for forecasting the region-wise rainfall for the future. In this work, forecasting the state-wise and district-wise rainfall using exponential smoothing methods. These smoothing methods are used to deliver more recent observations with larger weights and the weights reduction aggressively as the observations converted more reserved. When the parameters relating the time series are varying gradually above time, thus the exponential smoothing methods are most effective.

Exponential Smoothing is one of the traditional methods used to forecasting a leveled time series data. While the past observations moving averages are weighted equally, Exponential Smoothing allocates rapidly reducing weights as the observations become huge. In other words, current observations are assumed rather more weight in forecasting than the past observations. The observations are having trends to implements the Double Exponential Smoothing methods. Triple Exponential Smoothing is used, while the observations are having parabola trends.

In this paper, these three smoothing techniques are implemented using MapReduce framework for effective iterative and initialization process. An extensive study is conducted on Indian Rainfall dataset with different smoothing parameters (like alpha, beta, and gamma) values. The result shows that these parameters have influence in forecasting the future value.

This paper is well organized as follows: Sect. 1 discusses introduction. In Sect. 2, lists the various existing work. Section 3 discusses about methods and materials required for this research work. In Sect. 4, the proposed methodology is discussed with experimental results and discussion and finally, Conclusion is given in Sect. 5.

2   Related Work

This section lists the various existing work related to this study. They are:

Nazim and Afthanorhan [1] in this study, three exponential smoothing methods are examined such as single, double, and triple exponential smoothing methods and study their capacity to managing excessive river water level time series data.

Arputhamary and Arockiam [2] the triple exponential smoothing method (Holt-Winters), transformation function, and Z-score model are combined to develop with processing the preceding rainfall data. The rainfall data resulted from the proposed forecasting model can be used for climatic classification using the Oldeman method in the research area.

Dielman [3] single exponential smoothing, double exponential smoothing, Holt-Winter's method, and adaptive response rate exponential smoothing methods are implemented to forecast the Malaysia population. From this work, Holt's method was delivered the lowest error value. Hence, Holt's method is best forecast compared to other exponential smoothing methods.

Din [4] in this article, choosing smoothing parameters in exponential smoothing based on the further reducing the sum of squared one-step-ahead forecast error values or sum of absolute one-step-ahead forecast error values are reduced. It is concluded that these two accuracy measures are used for best forecast value.

Kalekar [5] Arima and Holt-Winter's method are traditional methods of exponential smoothing. All the methods are verified to be acceptable. Hence, these methods are useful for decision makers to lunch schemes for the agriculturist, drainage schemes, and also water resource schemes so on.

Hartomo et al. [6] in the exponential smoothing method are having two types such as multiplicative and additive seasonal methods. This paper focuses on the seasonal time series data analysis using Holt-Winter's exponential smoothing methods based on both methods.

Mick Smith [7] nowadays, dataset are rapidly growing called as Big data. MapReduce techniques are used to analyze the huge datasets. Many algorithms are difficult to handle the MapReduce framework such as traditional algorithms.

From the above study, it is known that there is no effective or benchmark method for smoothing parameter initialization. Therefore, in the present study, the exponential smoothing methods are executed with different smoothing parameter values in MapReduce environment for paralleling the forecasting process.

3 Materials and Methods

In this paper, three models of exponential smoothing are applied to predict the rainfall. They are having different varieties of models, each models delivers effects of forecasts that are weighted averages of past observations among current observations assumed comparatively more weight than past observations. The "exponential smoothing"

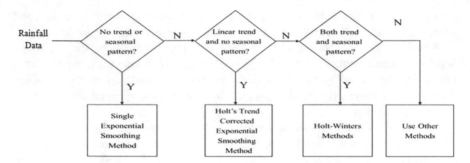

Fig. 1 Rainfall data versus exponential smoothing techniques

means follows the element that the weights reduced rapidly as the past observations. The following Fig. 1 shows rainfall data implemented using Exponential Smoothing Techniques.

Simple Exponential Smoothing (SES)

The Simple Exponential Smoothing method is applied for forecasting a time series when there is no trend or seasonal pattern, but the mean (or level) of the time series y_t is slowly altering over time. The present values are more weights than the past weighted averages values. This method is mostly applied for short-term forecasting that means commonly for periods not longer than 1 month [8].

Double exponential smoothing (DES)

This method is applied when the observation data is having a trend. Double exponential smoothing method is similar to simple exponential smoothing method but it contains two components (level and trend) that is needed for each time period. To estimate the leveled value of the observation data at the end of the period is called as level. The trend is a leveled estimate of average growth at the end of each period [8]. The method is commonly known as Holt's Linear method. It involves rapidly moving (or adjusting) the level and trend of the sequence at the end of each period. The level (L_t) is predictable by the leveled data value at the end of each period. The trend (T_t) is predictable by the leveled average increase at the end of the period. The equations for the level and trend of the series are shown in table.

Triple Exponential Smoothing

Triple exponential smoothing is advanced the double exponential smoothing to model time series with seasonality. The method is also known as the Holt-Winter's method in respects of the term of the discoverers. Holts-Winter's method is developed by the Holt's Linear method with adding a third parameter to deal with seasonality [8].

Therefore, this method consents for leveled time series while the level, trend, and seasonality are different. There are two main differences in the triple exponential method: trend and seasonality and they mostly depend on the type of seasonality. To grip seasonality, a third parameter is added in this model. The resultant set of equations is called the "Holt-Winters" [9] (HW) method after the terms of the discoverers.

	Simple Exponential Smoothing	Holt Linear Smoothing	Holt Winter Smoothing
Model of Data	Y-data, L-Level F-Forecast, t- Timepoint e- Irregular Parameter- alpha Predicted Value-predY	Y-data, L-Level T-Trend, F-Forecast t- Timepoint , e- Irregular Parameters- alpha,beta Predicted Value-predY	Y-data, L-Level T-Trend, F-Forecast t- Timepoint , e- Irregular s-seasonal value 4-quarterly, 12-monthly, 7-weekly Parameters- alpha, beta, gamma Predicted Value-predY
Initialization	$L(1:s)=mean(Y(1:s));$ $predY(1:s)=L(1:s);$	$T(1:s)=0;$ $L(1:s)=mean(Y(1:s));$ $predY(1:s)=L(1:s)+T(1:s);$	$T(1:4)=0;$ $L(1:4)=mean(Y(1:4));$ $F(1:4)=Y(1:4)-L(1:4);$ $predY(1:s)=L(1:s)+T(1:s)+F(1:s)$
Equations	$L(t)=alpha*(Y(t))+ alpha*(1-alpha)*(Y(t-1));$ $e(t)=Y(t)-L(t-1);$ $predY(t)=L(t-1)+alpha*e(t);$	$L(t)=alpha*(Y(t))+ (1-alpha)*(L(t-1)+T(t-1));$ $T(t)=beta*(L(t)-L(t-1))+(1-beta)*T(t-1);$ $predY(t)=L(t)+T(t-1);$	$L(t)=alpha*(Y(t)-F(t-s))+ (1-alpha)*(L(t-1)+T(t-1));$ $T(t)=beta*(L(t)-L(t-1))+(1-beta)*T(t-1);$ $F(t)=gamma*(Y(t)-L(t))+(1-gamma)*F(t-s);$ $e(t)=Y(t)-(L(t-1)+T(t-1)+F(t-s));$ $predY(t)=L(t-1)+T(t-1)+F(t-s);$
One Step Ahead Forecast	$L(t+1)=L(t);$ $predY(t+1)=L(t+1);$	$predY(t+1)=L(t)+h*T(t);$ where h=1	$L(t+1)=(L(t)+T(t))+alpha*e(t);$ $T(t+1)=T(t)+alpha*beta*e(t);$ $F(t+1)=F(t-s)+gamma*(1-alpha)*e(t);$ $predY(t+1)=L(t+1)+T(t+1)+F(t+1-s);$

4 Proposed Methodology

This paper proposes an approach with better communication properties for solving strongly overdetermined smoothing problems to predict rainfall using MapReduce framework for parallel implementation. There are several applications for MapReduce algorithms are present. In this research, it is adopted for Effective iterative solving purpose [10]. The proposed approach proceeds as Retrieving the Rainfall Data; Applying MapReduce_Exp_Smoothing; Visualizing the results.

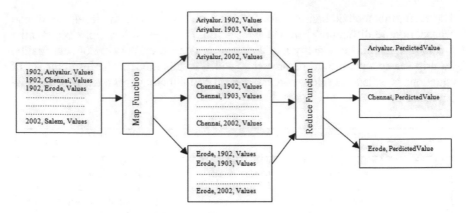

Fig. 2 Overview of the proposed approach

The overview of the proposed approach is illustrated in Fig. 2.

The pseudocode of the proposed MapReduce-based Exponential Smoothing is shown in the figure and it is implemented in MATLAB 2016. MapReduce is an algorithmic technique to "divide and conquer" big data problems. Here, it is used to predict the rainfall from time series rainfall data. In MATLAB, MapReduce needs three input arguments: Datastore, Map function, and Reduce function.

- A datastore is used to read and store rainfall dataset.
- A Map function is used to split the rainfall dataset into region/monthly rainfall pairs. It performs on given a subset of the data called chunks. It takes datastore object as its input.

In the map function, the given data is converted into key-value pairs. For each record, the district is in the Key and other details are in the record are in Value part.

Intermediate Key/Value pairs contains the output of the map function. Generally, the MapReduce calls the mapper function one time for each chunk in the datastore, with each call working independently and parallelly.

- A Reduce function is used to forecast the region-wise monthly rainfall using exponential smoothing algorithms. It operates on the given aggregate outputs from the map function. It also takes the following arguments: intermediate key-value pair and output key-value pair.

The reduce function combines the values of each district together and applies the appropriate exponential smoothing technique as illustrated in the pseudocode.

Function MapReduce_exp_smoothing(data_files)
Key: District
Value: Predict Values

data = datastore (data_files)   //Read Actual Rainfall Data (District Wise)
outds = mapreduce (data)   // Call Map Function and Reduce Function
return outds   // Get Key/Value Pairs

Function Map (data)
Key: District
Value: MonthlyRainfall for each district

add (mkey, mval)   // Generate Key Value Pairs (one for each record)

Function Reduce ({mkey, mval})
Key: District
Value: Predicted values of the district

Combine the values {mval1, mval2,.....} of each unique key (mkey) together

If data has No Trend and No Season
          Apply the Simple Exponential Smoothing
Else if data has No Season
          Apply the Holt's Linear Trend Exponential Smoothing
Else
          Apply the Holt's Winter Seasonal Exponential Smoothing
Then
          Predicted values of all months in each district

add(district, {predict_Y and measures})   // Generate Key Value Pairs (one for each district)

Pseudocode for Mapreduce_Exponential Smoothing Approach

Generally, the reduce function is called one time for each unique key in the intermediate key-value pair. It qualities the computation begun by the map function, and outputs the absolute response also in Key/Value pair format.

4.1 Experimental Results

Overview of Dataset

In this research, two Indian datasets are taken for attempting the proposed approach.

Dataset1: The historical rainfall dataset is for Tamil Nadu District province. The data was collected from www.data.gov.in. It consists of 11 years of monthly data from 1992 to 2002 in 29 Districts of Tamil Nadu. A snapshot of the dataset is shown in Fig. 3.

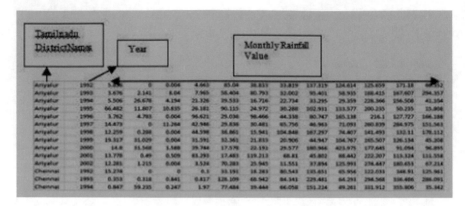

Fig. 3 A snapshot of dataset 1

SD NO.	SD_Name YEAR	JAN	FEB	MAR	APR	MAY	JUN	JUL	AUG	SEP	OCT	NOV	DEC
1 ANDAMAN	1951	82.7	7.2	0	45.4	259	619.9	665.3	101.3	360.9	489	209.6	434.8
2 ARUNACH	1951	31.8	65.3	309.9	441	344.5	757.8	554.7	278.5	383.7	101.8	46.2	39.1
3 ASSAM &	1951	4.5	2.9	54.6	230	270.6	688.4	665.9	444.4	241.5	302.2	46.3	5.4
4 NAGALAN	1951	3.9	0.2	62.4	360.1	233.7	495	436.2	440.5	191.1	310.6	54.9	20.8
5 SUB-HIMA	1951	1	0.5	22.7	69.3	299.8	546.5	778.3	503.7	126.1	158	37.3	0
6 GANGETIC	1951	0	0.4	62.1	13.1	67.4	180.7	253.5	260.3	200.1	140	36	0.4
7 ORISSA	1951	2.2	0	89.7	37.1	48	172.5	339.1	349.5	165.5	155.6	37.1	0

Fig. 4 A snapshot of dataset 2

Dataset2: The historical rainfall dataset is for Indian State province. The data was collected from www.data.gov.in and it consists of 63 years of monthly data from 1951 to 2014 in 37 States of India. A snapshot of the dataset is shown in Fig. 4.

5 Result and Discussion

This section discusses about the results obtained by the proposed MapReduce-based exponential smoothing approach. The comparative analysis of MapReduce-based Simple Exponential Smoothing, Holt's Linear Trend Exponential Smoothing, and Holts-Winter's Exponential Smoothing are shown for some districts and states like Salem, Tamil Nadu, and Pondicherry in the Figs. 5 and 6.

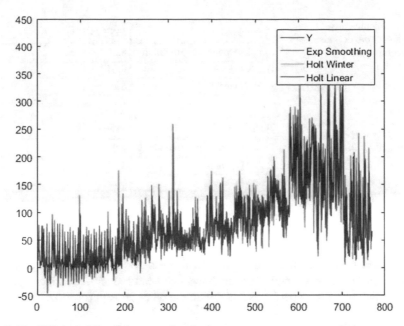

Fig. 5 Tamil Nadu and Pondicherry predicted values

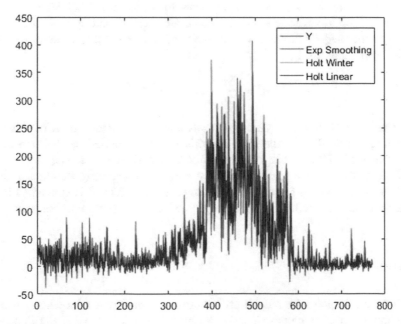

Fig. 6 Salem district predicted values

Table 1 Comparative
analysis for district-wise
prediction

Techniques	MSE ($\times 10^2$)
Simple exponential smoothing	**8.38**
Holt's Linear	4.01
Holt-Winter's	3.12

Table 2 Comparative
analysis for state-wise
prediction

Techniques	MSE ($\times 10^2$)
Simple exponential smoothing	2.6
Holt's Linear	3.98
Holt-Winter's	2.13

The graphical representation of the result shows that the three exponential smoothing techniques are performing similar to the real values. So, the standard MSE measure is used to analyze the results. The MSE measure for these three approaches are listed in the Tables 1 and 2 which shows that MapReduce-based Holt-Winter's Exponential Smoothing performs better than the other two approaches.

Error Measure:

In this work to measure the error using Mean Square Error (MSE). MSE is one of the standard criterions/error measures which are mostly used by practitioners for evaluating the model's fitness to extract series of data. The MSE accuracy measure is calculated using Eq. 1

$$\text{MSE} = \frac{\text{SSE}}{T-1}, \quad s = \sqrt{\text{MSE}} \tag{1}$$

Tables 1 and 2 tabulate the generated result obtained using the Simple Exponential Smoothing, Holt's Linear and Holt-Winter's which is applied on the time series rainfall data. The Holt-Winter's produced the lowest MSE values. The best method is defined as the one that gives the smallest error. For these three techniques, Holt-Winter's shows a better performance to compare. Therefore, it is concluded Holt-Winter's smoothing method is best for region-wise monthly rainfall forecasting.

6 Conclusion

In this paper, Region-wise rainfall prediction using MapReduce-based exponential smoothing methods is proposed, and it performs better than the existing Exponential Smoothing models. The Exponential Smoothing Techniques is an important method in modeling and forecasting rainfall. Three Exponential Smoothing methods were applied in this work, the Simple Exponential Smoothing, Holt's Linear, and Holt-Winter's. MapReduce is a framework for executing highly parallelizable and distributable algorithms across huge data sets. It is used to analyze the given data and predicts the future which showed essential runtime improvements compared to serial implementations. Therefore, it is concluded that MapReduce-based Holt-Winter's smoothing method is best for region-wise monthly rainfall forecasting. In future, smoothing methods can be optimized using better initialization procedure.

References

1. Nazim, A., Afthanorhan, A.: A comparison between single exponential smoothing (SES), double exponential smoothing (DES), holt's (brown) and adaptive response rate exponential smoothing (ARRES) techniques in forecasting Malaysia population. Glob. J. Math. Anal. **2**(4), 276–280 (2014)
2. Arputhamary, B., Arockiam, L.: Performance improved holt-winter's (PIHW) prediction algorithm for big data. Int. J. Intell. Electron. Syst. **10** (2) (2016)
3. Dielman, T.E.: Choosing smoothing parameters for exponential exponential sums of absolute errors. J. Mod. Appl. Stat. Methods **5**(1) (2006)
4. Din, N.S.: Exponential smoothing techniques on time series river water level data. In: Proceedings of the 5th International Conference on Computing and Informatics, ICOCI 2015, p. 196 (2015)
5. Kalekar, P.S.: Time Series Forecasting Using Holt-Winters Exponential Smoothing (2004)
6. Kristoko DWI Hartomo, Subanar, Winarko, E.D.I.: Winters exponential smoothing and z-score. J. Theoret. Appl. Inf. Technol. **73**(1), 119–129 (2015)
7. Mick Smith, R.A.: A Comparison of Time Series Model Forecasting Methods on Patent Groups (2015)
8. Ravinder, H.V.: Determining the optimal values of exponential smoothing constants—does solver really work? Am. J. Bus. Educ. **6**(3), 347–360 (2013)
9. Sopipan, N.: Forecasting rainfall in Thailand: a case study. Int. J. Environ. Chem. Ecol. Geol. Geophys. Eng. **8**(11), 717–721 (2014)
10. Meng, X., Mahoney, M.: Robust regression on MapReduce. In: Proceedings of the 30th International Conference on Machine Learning (2013)

Association Rule Construction from Crime Pattern Through Novelty Approach

D. Usha, K. Rameshkumar and B. V. Baiju

Abstract The objective of association rule mining is to mine interesting relationships, frequent patterns, associations between set of objects in the transaction database. In this paper, association rule is constructed from the proposed rule mining algorithm. Efficiency-based association rule mining algorithm is used to generate patterns and Rule Construction algorithm is used to form association among the generated patterns. This paper aims at applying crime dataset, from which frequent items are generated and association made among the frequent item set. It also compares the performance with other existing rule mining algorithm. The algorithm proposed in this paper overcomes the drawbacks of the existing algorithm and proves the efficiency in minimizing the execution time. Synthetic and real datasets are applied with the rule mining algorithm to check the efficiency and it proves the results through experimental analysis.

Keywords ARM · IRM · Rule construct · Crime dataset · Information gain

1 Introduction

Association rule mining is one of the important techniques in data mining [1]. These rules are made by exploring data from frequent patterns and to detect the most substantial relationships, the measures support and confidence is used. Support outputs the frequently arising items in the database based on the threshold. Confidence outputs the number of times the statements have been found to be true [2]. The existing rule mining algorithms are unproductive due to so many scans of database and also in case of large data set, it takes too much time to scan the database. The proposed rule mining algorithm proves efficiency by reducing the execution time. The aim of

D. Usha (✉)
Dr. M.G.R.Educational and Research Institute, Chennai, India
e-mail: ushahits@gmail.com

K. Rameshkumar · B. V. Baiju
Hindustan Institute of Technology and Science, Chennai, India

© Springer Nature Singapore Pte Ltd. 2019
J. D. Peter et al. (eds.), *Advances in Big Data and Cloud Computing*,
Advances in Intelligent Systems and Computing 750,
https://doi.org/10.1007/978-981-13-1882-5_22

this paper is application of crime pattern [3] and generates association rules from the mined frequent pattern with the help of rule construct algorithm and find the suitable measures to validate the rule.

2 Problem Statement

ARM [4] discovers association rules that satisfy the predefined minimum support and confidence from the given database. The main drawback of the existing Apriori algorithm is generation of large number of candidate itemset [5], which requires more space and efforts. Because of the above facts the algorithm needs too many passes and multiple scans over the whole database, so that it becomes waste and useless. The existing frequent pattern based classification also has some drawbacks. Since the frequent pattern growth approach lies on tree structure, it constructs the tree to store the data but when the data are large it may not be fit in main memory. During the computation process, the results become infeasible and over-fit the classifier. Also the existing FPM algorithm is not scalable for all types of data.

The proposed algorithm retrieves the frequently ensued patterns from large dataset with the assistance of newly built data structure. On the basis of frequent patterns, association rules are created and it discovers the rules that fulfill the predefined minimum support and confidence from a given database. The present rule mining algorithm produces wide number of association rules which comprises of non-interesting rules also. While generating rule mining algorithm [6], it reflects all the discovered rules and hence the performance becomes low. It is also incredible for the end users to know or check the validity of the huge amount of composite association rules and thereby limits the efficacy of the data mining outcomes. The generation of huge number of rules [6] also led to heavy computational cost and waste of time.

Several means have been expressed to reduce the amount of association rules like producing only rules with no repetition, producing only interesting rules, generating rules that satisfy some higher level criteria's, etc.

3 Related Work

3.1 Constraint-Based Association Rule Mining Algorithm [7] (CBARM)

In the above algorithm limitations were applied through the process of mining to produce only interested association rules instead of producing the entire association rules [7]. Generally constraints are provided by users. It can be knowledge based constraints, data constraints, dimensional constraints, interestingness constraints or rule formation constraints [7]. The task of CBARM is to discover all rules which come across all the user-specified constraints. The present apriori-based algorithms employ two basic constraints, support, and confidence. They used to produce rules which may be worthful or not informative to individual users. Also with the limit of minimal support and confidence the algorithms may miss some interesting information which dissatisfies them.

3.2 Rule-Based Association Rule Mining Algorithm (RBARM)

The RBARM algorithm constructs a rule-based classifier from association rules. It overwhelms some existing methodologies like tree based structure used to take decisions, sequential order based algorithms [8] which reflects only one feature at a time. Rule-based association rule mining algorithm mines any amount of attributes in the resultant. In another method namely class association rules resultant to be the class label. The disadvantage of this algorithm is since the number of attributes is more, efficiency cannot be achieved completely.

3.3 Classification Based on Multiple Association Rule Mining Algorithms [8] (CMAR)

The CMAR algorithm works on the basis of FP-growth algorithm to notice the class-based association rules [8]. It works with the tree structure to competently store and recover rules. It applied rule pruning methodology every time insertion takes place in the tree. The main drawback of this algorithm is that the rules pruned are sometimes with the negatively correlated classes. It reflects numerous rules when categorizing an occurrence and uses weighted measure to discover the strongest class.

4 Improved Rule Mining Algorithm (IRM)

The improved rule mining algorithm increases the efficiency through the process of reducing the computational time as well as cost. It can be succeeded by reducing the number of passes over the database, by adding additional constraints on the pattern. In legal applications, some rules will have less weightage and inefficient and some rules will have more weightage. Generation of the entire constructed rule will lead to waste of time. So, researcher likes to validate the rule to find the most efficient rule. The measure Information Gain [9] is a statistical property that measures the validity of the generated rules. It calculates lift value and with the input of the lift value it calculates information gain by taking all the possible combination of rules that

Procedure IRM_AR (k-itemset)
begin
Step 1 : for all large k-itemset l_k , $k \geq 2$ do begin
Step 2 : H_1 = { consequents of rules derived from l_k with one item in the consequent };

Step 3 : call rue_construct (l_k, H_1);

      end procedure

Procedure rule_construct (l_k: large k-itemset, H_m: set of m-item consequents)

begin

Step 1 : if ($k > m+1$) then begin

Step 2 : H_{m+1}= Apriori_gen(H_m)

Step 3 : for all $h_{m+1} \in H_{m+1}$ do begin

Step 4 : conf =support(l_k)/support($l_k - h_{m+1}$)

Step 5 : if (conf $\geq$ MinConf) and (gain $>$- MinGain) then
Step 6 : output the rule (l_k -h_{m+1})$\rightarrow$ h_{m+1} with confidence=conf

Step 7 : information gain = gain and

Step 8 : support = support(l_k)

      else

Step 9 : delete h_{m+1} from H_{m+1}

      end if

Step 10 : call rule_construct (l_k, H_{m+1})

        end for

          end if

            end procedure

Fig. 1 Rule construct procedure to construct association rules

is generated from the proposed pattern mining algorithm. The proposed Improved Rule Mining algorithm consists of association rule generation phase.

The drawback of the above-mentioned CBARM, RBARM, and CMAR algorithm is that the number of rules mined can be tremendously large. So, it takes more time to execute the process. The proposed IRM algorithm minimize the number of rules mined but proves the efficiency because of its limited and exactly predictable space overhead and is faster than other existing methods. It also increases the efficiency through the process of reducing the computational time as well as cost. It can be succeed by reducing the number of passes over the database, by adding additional constraints on the pattern. Association rule construction is a straight forward method. The rule construct algorithm uses the above procedure to construct association rule [1] (Fig. 1).

Information Gain

Information Gain is an arithmetic property that processes the validity of a given attribute which divides the training samples according to their target classification. The degree of purity is called the information. The average purity of the subsets which is produced by an attribute is increased by the measure Information Gain. In order to discerning among classes to be cultured [10], it decides the most useful training feature vectors based on its attribute. Information Gain feature ranking is derived from shanon entropy as a measure of correlation between the feature and the label to rank the features. Information Gain is a famous method and used in many papers [11]. The other method is CFS which uses the standardized conditional Information Gain to catch the most correlated features to the label and excludes those whose correlation is less than a user defined threshold [12]. It is used to decide the arrangement of attributes in the right order in the nodes of a decision tree.

Attribute Selection:

Select the attribute with the highest information gain

- Let p_i be the probability that an arbitrary tuple in D belongs to class C_i, estimated by $|C_{i,D}|/|D|$
- Predictable information (entropy) needed to classify a tuple in D:

$$\text{Info}(D) = -\sum_{i=1}^{m} p_i \log_2(p_i)$$

- Information needed (after using A to split D into v partitions) to classify D:

$$\text{Info}_A(D) = \sum_{j=1}^{v} \frac{|D_j|}{|D|} \times I(D_j)$$

- Information gained by branching on attribute A

$$\text{Gain}(A) = \text{Info}(D) - \text{Info}_A(D)$$

5 Dataset Description

The real and synthetic dataset are collected from UIML Database Repository [13] and FIMI [14] repository. The below-mentioned Table 1 shows the number of transactions and attributes of real and synthetic datasets used in this evaluation. Typically, these real datasets are very heavy and therefore they produce many long frequent itemset even for high values of support threshold. Usually the synthetic datasets are sparse when compared to the real sets.

Table 1 Synthetic and real datasets

Dataset	No. of transactions	No. of attributes
T40I10D100K	100,000	942
Mushroom	8124	119
Gazella	59,601	497
Crime dataset	5000	83

6 Experimental Study and Analysis

The Constraint-Based Association Rule Mining algorithm is compared with the proposed Improved Rule Mining algorithm. Figures 2, 3, 4 and 5 give the comparative study of execution time between IRM and CBARM varying minimum support using Gazella [15], T40I10D100K and Mushroom dataset resp. The MinSupp is set as 70%, the MinConf is set as 75% and Information Gain is set as 60%. Most of the transactions are replaced by the earlier TID which are effectively used for support calculations and easy to construct frequent patterns.

In Figs. 2, 3, 4 and 5, the performance of IRM algorithm is better than CBARM algorithm in case of small dataset and also in high density dataset. In TTC dataset, in the comparison of execution time, IRM is better than CBARM varying levels 10–20%. The execution time of IRM is reduced in case of Gazella and Mushroom dataset varying ratio between 10 and 14% respectively. In T40I10D100K, the IRM algorithm performs better than CBARM. When compared with CBARM algorithm, the IRM algorithm has reduced for about 30–35% of execution time. Hence it is proved that the proposed IRM algorithm performs better in high density dataset when compared with small dataset.

Fig. 2 Comparison—execution time between IRM and CBARM algorithm with varying support in Gazella dataset

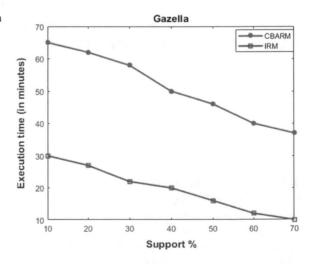

Fig. 3 Comparison—execution time between IBM and CBARM algorithm with varying support in Mushroom dataset

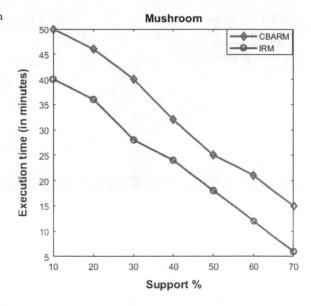

Fig. 4 Comparison—execution time between IRM and CBARM algorithm with varying support in T40I10D100 K dataset

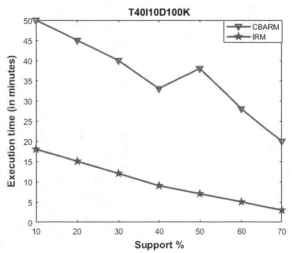

Figures 6, 7, 8 and 9 show the comparison between IRM and CBARM algorithm in terms of execution time with varying Information Gain ranges from 20 to 60%. The performance of IRM increases as the information gain increases. In TTC, Gazella and Mushroom dataset, the performance decreased by 8, 11, and 12% respectively when compared with the performance of CBARM. The proposed IRM algorithm performed in ratio varying from 30 to 40% with T40I1D100K dataset when compared with CBARM algorithm. The above-mentioned figures shows the comparative study

Fig. 5 Comparison—execution time between IRM and CBARM algorithm with varying support in Theft dataset

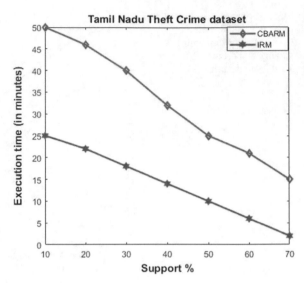

Fig. 6 Comparison —execution time between IRM and CBARM algorithm with varying information gain in Gazelle dataset

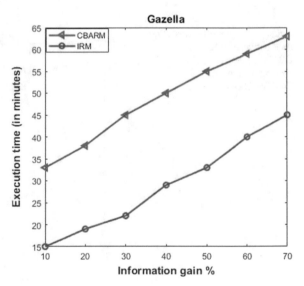

of execution time between IRM algorithm and CBARM algorithm with values using T40I10D100K, Mushroom, Gazella and TTC dataset respectively.

Fig. 7 Comparison—execution time between IRM and CBARM algorithm with varying information gain in Mushroom dataset

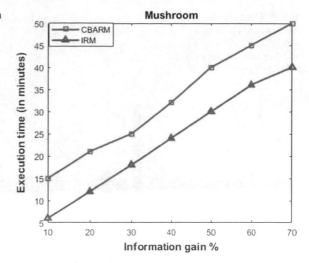

Fig. 8 Comparison—execution time between IRM and CBARM algorithm with varying Information Gain in T40I10D100 K dataset

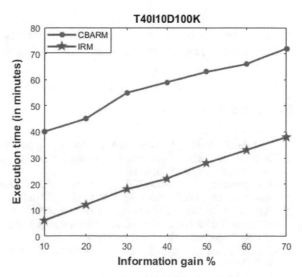

7 Conclusion

This paper detailed about association rule mining algorithm which is used to mine the rule from frequently occurred patterns from large amount of database. The projected frequent itemset acts as an input to generate association rules by using rule construct algorithm and validated the results using interesting and effective measures. The experimental study is conducted through T40I10D100K, Mushroom, Gazella, and Crime dataset. The proposed IRM algorithm is compared with above mentioned four

Fig. 9 Comparison—execution time between IRM and CBARM algorithm with varying Information gain in TTC dataset

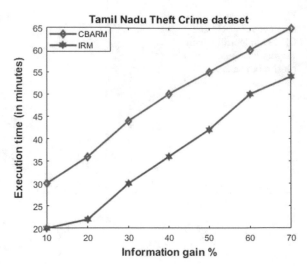

datasets. The experimental results prove that IRM algorithm is performed better than existing CBARM algorithm with varies values of measures.

References

1. Agrawal, R., Srikant, R.: Fast algorithms for mining association rules. In: Proceedings of the 20th International Conference on Very Large Data Bases, pp. 487–499 (1994)
2. Association Rule. www.ijirset.com
3. Krishnamurthy, R., Satheesh Kumar, J.: Survey of data mining techniques on crime data analysis. Int. J. Data Mining Tech. Appl. **1**(2), 117–120 (2012)
4. Agrawal, R., Imielinski, T., Swami, A.N.: Mining association rules between sets of items in large databases. In: Proceedings of the ACM SIGMOD International Conference on Management of Data, Washington, pp. 207–216 (1993)
5. Bhandari, P., et al.: Improved apriori algorithms—A survey. Int. J. Adv. Comput. Eng. Netw. **1**(2) (2013)
6. Rule Mining Algorithms. www.math.upatras.gr
7. Constraint based association rule mining algoritnm. www.lsi.upc.edu
8. Rule Based Association Rule Mining algorithm [online] Available at www.docplayer.net
9. Attribute Selection. www.research.ijcaonline.org
10. Information Gain. www.hwsamuel.com
11. Azhagusundari, B., Thanamani, A.S.: Feature selection based on information gain. Int. J. Innovative Technol. Exploring Eng. (IJITEE) **2**(2), 18–21 (2013). ISSN 2278-3075
12. Hall, M.A.: Correlation-based feature selection for discrete and numeric class machine learning. In: Proceedings of the Seventeenth International Conference on Machine Learning, pp. 359–366 (2000)
13. Gazella dataset (2000). www.gbif.org/species/5220149/datasets
14. Dataset. www.fimirepository.com
15. Biesiada, J., Duch, W., Duch, G.: Feature selection for high-dimensional data: a Kolmogorov-Smirnov correlation-based filter. In: Proceedings of the International Conference on Computer Recognition Systems (2005)

Tweet Analysis Based on Distinct Opinion of Social Media Users'

S. Geetha and Kaliappan Vishnu Kumar

Abstract The state of mind gets expressed via Emojis' and Text Messages for the huge population. Micro-blogging and social networking sites emerged as a popular communication channels among the Internet users. Supervised text classifiers are used for sentimental analysis in both general and specific emotions detection with more accuracy. The main objective is to include intensity for predicting the different texts formats from twitter, by considering a text context associated with the emoticons and punctuations. The novel Future Prediction Architecture Based On Efficient Classification (FPAEC) is designed with various classification algorithms such as, Fisher's Linear Discriminant Classifier (FLDC), Support Vector Machine (SVM), Naïve Bayes Classifier (NBC), and Artificial Neural Network (ANN) Algorithm along with the BIRCH (Balanced Iterative Reducing and Clustering using Hierarchies) clustering algorithm. The preliminary stage is to analyze the distinct classification algorithm's efficiency, during the prediction process and then the classified data will be clustered to extract the required information from the trained dataset using BIRCH method, for predicting the future. Finally, the performance of text analysis can get improved by using efficient classification algorithm.

Keywords Text classifiers · Emoticons · Twitter · Social networking

1 Introduction

High quality information get derived in the form of text is known as text mining [1]. Statistical pattern learning helps to derive the high quality information from text. A search using text mining helps to identify facts, relationships, and assertions that remain in mass of big data.

S. Geetha (✉) · K. Vishnu Kumar
CSE Department, KPR Institute of Engineering and Technology, Coimbatore, India
e-mail: geethacse14@gmail.com

K. Vishnu Kumar
e-mail: vishnudms@gmail.com

© Springer Nature Singapore Pte Ltd. 2019
J. D. Peter et al. (eds.), *Advances in Big Data and Cloud Computing*,
Advances in Intelligent Systems and Computing 750,
https://doi.org/10.1007/978-981-13-1882-5_23

The actual purpose of networking, socialization, and personalization has been completely changed over the past few years due to rise of social media among common people. The sentiment of people can be expressed to others while communicating with Internet-based social media. The Twitter is one of the well-known micro-blogging social media platforms in recent years. Various analysis tools are available for the collection of twitter data. The text mining process will help to discover the right document and extract the knowledge automatically from various unstructured data content.

The actual sentiment conveyed on the particular tweet can be analyzed exactly when dealing with group of words context including the emoticons. Twitter data is collected via the Twitter Streaming API which is a quantitative approach. A deep insight can be gained with the public views understanding.

The unstructured data requires a lion's share in digital space (i.e.) about 80% of volume. Computational study of people's opinion, sentiments, evaluation, emotions and attitudes are performed by sentimental analysis. Social Media data growth rate coincides with the growing importance of sentiment analyses. The new tool helps to "train" for sentiment analysis in the upgraded platform which includes by combining the company's natural language processing technology relatively with a straight forward machine language system.

Twitter creates an outstanding opportunity in the perspective of software engineering to track and monitor the large population end-user's opinion on various topics by establishing a unique environment. The opinions are expressed about distinct products, services, events, political parties, organizations, etc., as an enormous amount of data from users. Due to informal and unstructured data format of Twitter, there is a difficulty in reacting to the feedback quickly for government or any organizations.

The favourable and non-favourable reaction in text can be determined in the research field of sentiment analysis. The text emotions can discover opinion from the users' tweets. The major text classification based on two sentiments such as positive (☺) and negative (☹). The text multidimensional emotions are computed based on the following emotions: anger, fear, anticipation, trust, surprise, love, sad, joy, disgust. The positive or negative feelings are conveyed by sentimental analysis techniques, which completely rely on emotion evoking words and opinion lexicons for detecting feelings in the corresponding text.

While monitoring the public opinion and evaluating the former product design during the product feedback monitoring, there are lots of information need to get located with apparent emotiveness.

The sections of this paper are organized as: Literature Survey in Sect. 2. Section 3 contains related work. Current proposed system of this paper is represented in Sect. 4. Section 5 contains the module description and its Conclusion is presented in Sect. 6. Future enhancement is given at Sect. 7.

2 Literature Survey

From Twitter Data's Sentiment Analysis in Text Mining [2], the precious information source is obtained from the text messages collection to understand the actual mind-set of the large population. Opinion analysis has been performed on iPhone and Microsoft based on the tweets. Weka1 software with a word set of positive and negative used for the data mining process to compare the word set obtained from the twitter. The required features get listed to represent the tweets for assigning sentiment label while generating the Classifier using data mining tools. The collection of word frequency distribution is normal, which can be indicated by the analyzing process.

Some of the methodologies applied for Twitter data's sentiment classification in the process of text mining are: Data Collection, Data Pre-processing, Feature Determination and Sentiment Labelling.

In order to label and also represent the tweets for training data, list of sentiment words are used along with the emoticons. The document filtering and document indexing techniques are integrated here for developing an effective approach for tweet analysis in the process of text mining.

From [3] tweet sentiment analysis models are generated by machine learning strategies, which do not generalize across multiple languages. Cross-language sentiment analysis usually performed through machine translation approaches that translate a given source language into the target language of choice. Neural networks learning is accomplished by the weight changes in network connection, while an input instances set gets repeatedly passed through the network. After training process, an unknown instance passing through the network gets classified according to the values seen at the output layer. In the feed forward neural networks, the signal flows from input to output (forward direction). Machine translation is expensive and the results that are provided by thesis strategies are limited by the quality of the translation that is performed. Deep convolutional neural networks (CNN) with character-level embeddings made for owning pointing to the proper polarity of tweets, that may be written in distinct (or multiple) languages. Only the linearly separable problems get solved by a single layer neural network.

In [4], the data has been collected from the Twitter and Microblogs to pre-process it for analyzing and visualizing given data to do sentiment analysis and text mining by using the open source tools. Customers' view can be understood from the comparative value change on their products and its services, to provide the future marketing strategies and decision making policies. The business performance can be monitored based on the perspective of customer survey report.

The uncovered knowledge in the inter-relationships pattern can be detected based on data mining algorithms like clustering, classification and association rules for discovering the text pattern of new information's relationships in the textual sources. These patterns can be visualized using the word-cloud or tag cloud at the end of the text mining process.

In order to perform mining in the Twitter Microblogs, methodologies used here are: Data Access (keyword search using Twitter package), Data Cleaning (get the text and clean data using additional text mining package), Data Analysis (sentiment analysis performed for the structure representation of tweets using lexicon approach along with scoring function to assign tweet's score) and Visualization (frequency of words for customer tweets can be shown by word-cloud package and bar plots). The consumer's opinions are tracked and analyzed from social media data by utilizing the products marketing plans and its business intelligence.

From [5] computer-based tweet analysis, the address should have at least two specific issues: one, users use the electronic devices namely cell phones and tablets to post a tweet, by developing their own specific culture of vocabulary leads to increase misspellings frequency and different slang in tweets. Next, messages are posted on a variety of topics are tailored to specific topics, unlike blogs, news, and other sites, by Twitter users.

Documents get split into sentences for more than one sentences in the document, which act as the input of the system and design a strategy to represent the overall sentiment of a document. With the four different machine learning models: decision tree, neural network, logistic regression, and random forest, the experiment on Facebook comments and twitter tweets get implemented with the concept of "mixture".

In [6], the Feature set gets enriched by Emoticons available in the tweets. The number of positive and negative emotions can get complement for the bag of words and the feature hashing information. Moreover, each message has been computed with a number of positive and negative lexicons.

From [7], visualizing and understanding character-level networks has to be reasonable and interpretable even when dealing with sentences whose words come from multiple distinct languages. The multilingual sentiment analysis has the most common approach called Cross-Language Sentiment Classification (CLSC), which focuses on the use of machine translation techniques in order to translate a given source language to the desired target language. The polarity gets identified within the sentences with multiple languages together. It has a reasonable and interpretable to deal with words come from multiple languages.

The tree performs well only for training data which can be indicated by an over fitted decision tree, when it is built from the training set. For unseen data, the performance will not be that good. Hence the nodes of decision tree can "Shave off" branches easily.

From [8] WEAN, word emotion computation algorithm used to obtain the initial words emotion, which are further refined through the standard emotion thesaurus. With the words emotion every sentence's sentiment can get computed. The news event's sentiment computing task is split into two procedures: word emotion computation through word emotion association network and word emotion refinement through standard sentiment thesaurus. After constructing WEAN, compute word emotion for the different scale and intension of word circumstance that can affect word emotion. The larger of scale and stronger of the intension results to more intense of the word emotion. The main two factors that derive the sentiment values are, The Positive and Negative word frequency count and existence of a Positive and Negative Emoticon.

The sentiment classification has its effect on the various pre-processing methods in [9], includes the removing of URLs, negation replacing, inverting repeated letters, stop words removal, removing numbers, and expanding acronyms. The tweet sentiment polarity on Twitter datasets get identified using the two feature models and four classifiers. The acronyms expansion and negation replacement can help to improve the performance of sentiment classification, but changes were done barely while removing URLs, stop words and numbers.

In [10], multidimensional structure challenges address a preliminary analysis aimed at detecting and interpreting emotions present in software-relevant tweets. The most effective techniques in detecting emotions and collective mood states in software-relevant tweets get identified and then investigated based on emotions that are correlated with specific software-related events.

In [11] short text system gets processed and filtered them precisely based on the semantics information contained in it. The system analyzes the tweets for finding an interesting result and then integrates each individual keyword with sentiments. In order to find keywords and sentiments from tweets, the feature extraction is applied and the users get expressed by the keywords. For specific category, the semantic based filtering can be performed using seed list (domain specific) to reduce the information loss. Maximize the information gain by filtering on entities, verbs, keywords and their synonyms extracted from tweet.

Categorization of sentiment polarity is the fundamental problem in sentiment analysis. The piece of written text is given as a problem is to categorize the text into positive or negative (or neutral) polarity with the help of three levels of sentiment polarities namely the document level, the sentence level, and the entity and aspect level for the scope of the text are categorization. The given document gets expressed as a either negative or positive sentiment in the document level, whereas each sentence's sentiment categorization done at the sentence level. Accurate like or unlike of people from their opinions are targeted at the aspect and entity level. The improvements in classification results can be quantified by an analytical framework due to the combination of multiple models. Training data learners constructs a set of learners and combines them in the ensemble methods. The unseen data can gets classified by classifiers and are known as a target dataset. The status updates with hashtags helps to identify the required content easily from the large dataset.

From [12], the text gets automatically converted to its emotion, which is a challenge as it minimizes the misunderstanding that occurs while conveying by the internal state of the users. The two modules of the frameworks are Training Module and Emotion Extraction Module. Besides internal corporate data in training set, exploratory Business Intelligence (BI) approach gets included for the external data. The user's feedback, suggestions and the data from web-based social media such as Twitter can be obtained with the help of extracted training data from TEA database.

From [13] Joint Sentiment Topic (JST) simultaneously extracts sentiment topics and also mixture of topics. In Topic Sentiment Mixture (TSM) models mixture of topics along with the sentiment predictions. The effective four-step JST model has Sentiment labels associated with Documents, Topics, Words and then Words got associated with Topics.

The extracted statuses get classified into semantic classes [14] as useful, for people who aim to know things and also to understand better about their electorate for political decision makers. The statuses get represented typically by a feature vector with the Bag-of-Words Model (BOW) and the Vector Space Model (VSM), which are two main approaches of text representations.

From [15], the phrase carries more semantics in-order to perform better in phrase based approach and also less ambiguous for the Term-based approach which suffering from polysemy and synonymy. In statistical analysis two terms can have same frequency and it can be solved by finding term which gets contributed more meaning for concept based approach. The low frequency problem and misinterpretation problem can be solved by pattern-based approach taxonomy.

The sentiment identification algorithms represents the state of the art for solving problems in real-time applications such as customer review's summarizing, products ranking and then finding features of product that imply opinions for analyzing tweet sentiments about movies and its attempt in box office revenue for predicting.

In [16] a framework has developed to predict complex image sentiment using visual content for an annotated social event dataset. The event concept features get mapped effectively to its sentiments. The performance of the event sentiment detector can be unseen dataset of images spanning events which is not considered in model training.

3 Related Work

Various methodologies and algorithms are proposed to perform the text mining in the social media based on distinct opinions. The table represents the efficiency and performance of the classification and clustering algorithms based on the survey.

Table 1 gets predicted with the help of the survey paper analysis on general characteristics of the considered algorithms. In Table 1, the algorithms quality gets analyzed based on the efficiency (E), accuracy (A) and performance (P) factors representing as Low (L), Medium (M) and High (H) values.

Table 1 Algorithms comparison

S. No.	Algorithm	Purpose	E	P	A
1	Navie Bayes	Classify	M	H	H
2	Support vector machine	Classify (T + E)	H	M	M
3	Fisher's linear discriminant classifier	Classify	L	H	H
4	Neural network	Classify	M	M	H
5	Convolutional neural network	Classify (T + E)	M	H	M
6	Balanced iterative reducing and clustering using hierarchies	Clustering	H	H	M

Fig. 1 Classification model

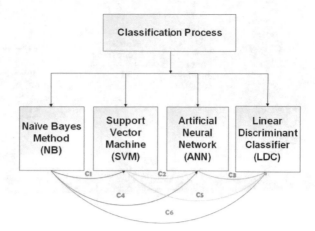

4 Proposed System

While computing the text emotion with a particular word can cause performance problem during the analysis. The classification of text is performed without considering the efficient classification algorithm, which will affect the quality factors of the text analysis. Figure 1 represents the classification model for the process to find the efficient classifier in order to improve classified data speed and accuracy. The Twitter data gets classified along with the specified emotions and punctuations with two different classifier algorithms, which are considered in parallel to compare and find the efficient one. The classifier combinations considered here are as follows C1:NB::SVM, C2:SVM::ANN, C3:ANN::LDC, C4:NB::ANN, C5:SVM::LDC, C6:NB::LDC.

A novel architecture, called FPAEC is proposed for word context associated with the emoticons and punctuations to predict the intensity for the piece of different texts, instead of using the particular word and emotion alone. The complete word context with emotion is considered here for exact analysis of the tweet dealing (Fig. 2).

The architecture description contains the information as follows: Collect the bag of words from twitter and sent to each classifiers separately to find the classification efficiency of each algorithm in the pair as specified. Secondly, combine the classifiers output using BIRCH clustering method by establishing clustering features (CF = (N, LS, SS), where N denotes a number of cluster objects, LS implies Linear Summation and SS implies Square Summation), to predict the future of the data with improved accuracy and speed of the analysis. The result of the each classifier after clustering is summed up to obtain the exact value for the feature extraction and future prediction. The processed data will get stored in the refined database for the further analysis in future with an improved performance of the model. The comparative study is made for the different classification algorithms to predict the classification algorithms quality factors in the applications. Finally the classified data is clustered to extract the required information from the trained dataset.

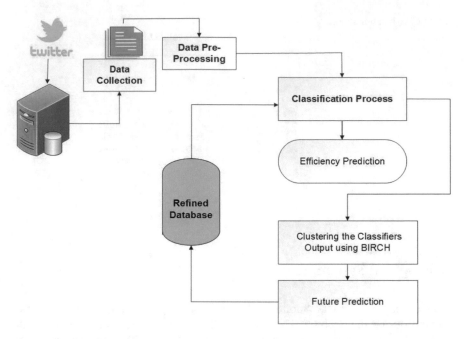

Fig. 2 FPAEC architecture

In current scenario, many people are having controversy ideas to support and oppose the education system for conducting an entrance test, called NEET in order to join the medical studies. The dataset collected based on this NEET problem is considered here as an application, in order to predict the necessity of NEET and other entrance exams. Every day, the natural language text available in electronic form gets staggered and increase accesses of the information in that text. The prediction of analysis helps to know the necessity of it in near future. The complete analysis will provide an efficient visualization of the NEET problem in the society.

5 Module Description

This section contains the riffle information regarding the process involved in the tweet analysis with a specified flow of algorithms used in this analysis. It includes Collection of Data, Text Pre-Processing, Comparison of Classification Algorithms, Clustering the classified outputs and Information Extraction.

Collection of Data The information on variables of interest get processed by gathering and measuring it in an established systematic fashion by enabling one to answer research questions that has been stated by following the test hypotheses, and then evaluate its outcomes.

Text Pre-Processing The input document is processed for removing redundancies, inconsistencies, separate words, stemming and documents are prepared for next step.

Comparison of Classification Algorithms The classification algorithms such as Support Vector Machine, Fisher's Linear Discriminant Classifier, Naïve Bayes Classifier, and Artificial Neural Network algorithm are used to perform the classification of text. The quality factors such as efficiency, speed, and accuracy are analyzed among these classification algorithms and clustering algorithm to improve the overall implementation of tweet analysis for the applications.

Clustering BIRCH clustering algorithm is an efficient and scalable method. In order to produce the best quality clustering for a given set of resources that has ability to increment and dynamically cluster the incoming multidimensional metric data point in a single attempt. The performance can extensively get denoted in terms of memory requirements like clustering quality, running time, stability, and scalability.

Information Extraction Overall analysis is performed to predict the future of the analysis with the help of values evaluated from the clustered datasets. At the final stage of the analysis, the future of the analyzed specific area can be predicted with more accuracy than the previous analysis.

6 Conclusion

Enormous amount of data are available in twitter for the opinion analysis to share and exchange the information. On-going research on mining tweets is developed to classify and analysis unstructured twitter data based on their extracted features with emotions. The twitter data can be classified accurately into three classes: positive, negative, and neutral for proposed sentiment classifier along with extracted features from word context emotions. The classifiers use the dataset containing texts along with emoticons, which will improve the intensity of the future prediction. The performance efficiency among the classifiers is analyzed and represented using graph, to use that result for predicting the future in further analysis with minimal time of I/O process. The value of emotion included to particular information gets described clearly in our proposed paper.

7 Future Enhancement

In future, neutral tweets are studied with the datasets which has got fortified with analogue domain in different form of features extraction. Automatic interpreting of various emotions in different application domains need to develop based on public's moods. Towards this direction, the more effort will be devoted in future.

References

1. www.irphouse.com
2. Wakade, S., Shekar, C., Liszka, K.J., Chan, C.-C.: Text mining for sentiment analysis of twitter data. The University of Akron (2012)
3. Younis, E.M.G.: Sentiment analysis and text mining for social media microblogs using open source tools: an empirical study. IEEE Access (2015)
4. Tang, D., Wei, F., Yang, N., Zhou, M., Liu, T., Qin, B.: Learning sentiment-specific word embedding for twitter sentiment classification. IEEE Access (2014)
5. da Silva, N.F.F., Hruschka, E.R., Hruschka, E.R.: Tweet sentiment analysis with classifier ensembles. DECSUP-12515. Federal University, Brazil (2014)
6. Omar, M.S., Njeru, A., Paracha, S., Wannous, M., Yi, S.: Mining tweets for education reforms (2017). ISBN 978-1-5090-4897-7
7. Wehrmann, J., Becker, W., Cagnini, H.E.L., Barros, R.C.: A character-based convolutional neural network for language-agnostic twitter sentiment analysis. IEEE Access (2017)
8. Jiang, D., Luo, X., Xuan, J., Xu, Z.: Sentiment computing for the news event based on the social media big data. IEEE Access (2016)
9. Jianqiang, Z., Xiaolin, G.: Comparison research on text pre-processing methods on twitter sentiment analysis. IEEE Access (2017)
10. Williams, G., Mahmoud, A.: Analyzing, classifying, and interpreting emotions in software users' tweets. IEEE Access (2017)
11. Batool, R., Khattak, A.M., Maqbool, J., Lee, S.: Precise tweet classification and sentiment analysis. IEEE Access (2013)
12. Afroz, N., Asad, M.-U., Dey, L.: An intelligent framework for text-to-emotion analyzer. IEEE Access (2015)
13. Sowmiya, J.S., Chandrakala, S.: Joint sentiment/topic extraction from text. In: IEEE International Conference on Advanced Communication Control and Computing Technologies (ICAC-CCT) (2014)
14. Akaichi, J.: Social networks' facebook' statutes updates mining for sentiment classification. Computer Science Department, IEEE (2013). doi: 10.1109
15. Chaugule, A., Gaikwad, S.V.: Text mining methods and techniques. Int. J. Comput. Appl. **85**(17) (2014)
16. De Choudhury, M., Ahsan, U.: Towards using visual attributes to infer image sentiment of social events. IEEE Access (2017)

Geetha S. is pursuing, Master of Engineering in the discipline of Computer Science and Engineering in KPR Institute of Engineering and Technology, Coimbatore, under Anna University Chennai, India. She has presented few papers in her interested areas like cloud and data analysis. She has published her papers in the journals such as IJARBEST and IJRASET.

Dr. K. Vishnu Kumar is the Head of Computer Science and Engineering Department has 11.3 years of Teaching and Research Experience. He received his Ph.D. in Computer and Information Communication Engineering from Konkuk University, Seoul, South Korea during 2011 and received M.Tech in Communication Engineering from VIT University, Vellore, India. He is an Editorial Manager at ISIUS (International Society of intelligent Unmanned System), Korea. He has published more than 40 International conference (EI), 7 SCIE Journals, 18 SCOPUS indexed Journals, 1 Book Chapter and holds one international patent.

Optimal Band Selection Using Generalized Covering-Based Rough Sets on Hyperspectral Remote Sensing Big Data

Harika Kelam and M. Venkatesan

Abstract Hyperspectral remote sensing has been gaining attention from the past few decades. Due to the diverse and high dimensionality nature of the remote sensing data, it is called as remote sensing Big Data. Hyperspectral images have high dimensionality due to number of spectral bands and pixels having continuous spectrum. These images provide us with more details than other images but still, it suffers from 'curse of dimensionality'. Band selection is the conventional method to reduce the dimensionality and remove the redundant bands. Many methods have been developed in the past years to find the optimal set of bands. Generalized covering-based rough set is an extended method of rough sets in which indiscernibility relations of rough sets are replaced by coverings. Recently, this method is used for attribute reduction in pattern recognition and data mining. In this paper, we will discuss the implementation of covering-based rough sets for optimal band selection of hyperspectral images and compare these results with the existing methods like PCA, SVD and rough sets.

Keywords Big Data · Remote sensing · Hyperspectral images · Covering-based rough sets · Rough sets · Rough set and fuzzy C-mean (RS-FCM) · Singular value decomposition (SVD) · Principal component analysis (PCA)

1 Introduction

Hyperspectral remote sensing is a standout among the most critical leaps forward in remote sensing. The current advancement of hyperspectral sensors is a promising innovation in remote sensing for examining earth surface materials. These images

H. Kelam (✉)
Department of Computer Science and Engineering, National Institute
of Technology Karnataka, Surathkal, India
e-mail: kharry551@gmail.com

M. Venkatesan
Faculty of Computer Science and Engineering Department,
National Institute of Technology Karnataka, Surathkal, India

© Springer Nature Singapore Pte Ltd. 2019
J. D. Peter et al. (eds.), *Advances in Big Data and Cloud Computing*,
Advances in Intelligent Systems and Computing 750,
https://doi.org/10.1007/978-981-13-1882-5_24

263

can be visualized as a three-dimensional datacube where the first and second dimensions represent the spatial coordinates and the last dimension represents the number of bands. In a simpler way, we can also view it as a set of images combined together, where each image is termed as 'spectral band'. Hyperspectral images are acquired in nearby spectral regions so they have highly correlated bands which results 'high dimensional feature sets' containing redundant information. Removing bands with high correlation reduce the redundancy without affecting the useful data. While portraying Big Data, it is prevalent to allude to the 3Vs, i.e. astounding developments in Volume, Velocity and Variety of information. In simple terms, Big Data can briefly be alluded to information from various sources, for example, web-based data, medicinal data, remote sensing data, and so on. For remote sensing Big Data, the 3Vs could be all the more solidly reached out to qualities of multiscale, high dimensional, multisource, dynamic state and nonstraight attributes. One of the primary difficulties experienced in calculations with Big Data originates from the basic rule that scale, or volume, includes computational complexity. Thus, as the data turns out to be huge, even minor operations can turn out to be expensive. Generally, increase in the dimensions will definitely affect the time and memory and may even turn out to be computationally infeasible on vast datasets. Figure 1 shows the dimensions of Big Data alongside their related challenges. In this paper, we deal with the curse of dimensionality, one of the challenges faced by Big Data volume dimension.

Unlike multispectral images, hyperspectral images provide much more data and can contain 10–250 contiguous spectral bands. Besides the benefits of having more bands, it suffers from issues like 'the curse of dimensionality', computationally complex, redundant bands. A more straightforward approach to overcome these issues is 'Reduction of dimensionality', in this scenario reduction of dimensionality meant to

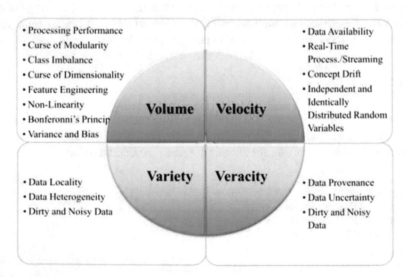

Fig. 1 [1] Big Data characteristics with associated challenges

be selecting optimal set of bands. The band selection methods are classified into two different sets, namely, supervised and unsupervised [2]. The supervised models need training set (labelled patterns) while unsupervised models don't need any training set. Equivalence relations are the statistical premise for rough set theory, in view of these relations universe of samples can be partitioned into exclusive equivalence classes. The essential objective of rough sets is to make proper classification by removing the redundant bands. However, classical rough set theory deals only with discrete data; this method showed its limitations when it comes to continuous data. Many authors have developed various extensions of rough set model to deal with continuous data that is numerously generated in real-life scenarios, few examples of the feature selection models that successfully dealt with continuous data are rough set-based feature selection with maximum relevance, maximum significance [3], fuzzy rough reduct algorithm based on max dependency criterion [4]. One more major generalization of rough sets is covering-based rough sets, in which indiscernibility relations are replaced by coverings. Many authors employed covering-based rough sets for attribute reduction. The generalized covering-based rough set is an advancement of the traditional model to tackle more complex practical issues which the traditional model cannot do.

2 Overview of Previous Work

Optimal band selection can be done in two different ways, namely, feature extraction and feature selection (in this context features are bands). In feature extraction, we build a new set of features from the existing features set, principal component analysis (PCA) [5] or K-L transform adopt feature extraction method. In PCA [5], the spectral bands (termed as principal components) are examined using covariance matrix, eigenvectors, eigenvalues and found that the first few bands contain useful information, using images of first few principal components results classification rate of 70%. On the other side, in feature selection, we select a subset of already existing features, few examples of this method are classification using mutual information (MI) [6], using factor analysis [7], continuous genetic algorithm (CGA) [8], rough sets [2], Rough set and Fuzzy C-Means (RS-FCM) [9].

Mutual information (MI) finds the statistical dependence of two random variables and can be used to calculate the relative utility of each band to classification [6]. In SVD, the data is divided into three matrices, namely, u, s, v based on the number of components and the scores values the bands are selected. Factor analysis [7] is one of the feature selection methods that use decorrelation method to remove correlated spectral bands and acquire maximum variance image bands. In CGA [8], a multiclass Support vector machine (SVM) was used as a classifier which extracts the proper subset of hyperspectral images band. In [2], a rough set method is developed to select optimal bands by computing the relevance followed by finding the significance of rest bands. In RS-FCM, the clustering is done based on fuzzy c-means algorithm which divides the bands into groups, using the attribute dependency concept in rough sets. In this paper, we propose a new band selection method for hyperspectral images using covering-based generalized rough sets, which finds the optimal spectral bands and also we compare these results with one feature extraction algorithm, namely, PCA [5] and one feature selection algorithm, namely, rough sets [2].

3 Proposed Solution

3.1 Basic Definitions of Covering-Based Rough Set

Zdzislaw I. Pawlak has proposed the concept of rough set in which indiscernibility relation partitions the universe. Many authors have generalized this concept to Covering-based rough sets where the indiscernibility relation is replaced by covering of the universe. By this generalization, we find a few subsets of the universe which are not really pairwise disjoint and this approach suites to many real-life scenarios. Zakowski used a covering of the domain and extended Pawlaks rough set theory, rather than a partition [10].

Definition 1 [10] Let a universe of discourse 'U' and family of subsets of U be $U = \{x_1, x_2, \ldots, x_n\}$ and $C = \{c_1, c_2, \ldots, c_m\}$. If no subset in C is empty, and $\cup_{i=1}^{m} C = U$ then C is called a covering of U.

Definition 2 [10] Let U be a non-empty set, C a covering of U. We call the ordered pair (U, C) a covering approximation space.

3.2 Discretization of Data

Let $U = \{x_1, x_2, \ldots, x_n\}$ be the 'n' available patterns and $B = \{b_1, b_2, \ldots, b_m\}$ be the 'm' available bands. As the values are continuous, to perform covering-based rough set theory for optimal band selection the band values of the pixels should be discretized. Unsupervised binning methods do the process of discretization without using the class information. There are 2 types of unsupervised binning methods, namely, equal width binning and equal frequency binning, we employed equal width interval binning approach to divide the continuous values into several discrete values [11]. To perform this binning, the algorithm needs to divide the data into 'k' intervals. The width of the interval and boundaries of the interval are defined as follows:

$$\text{width}(w) = (\max - \min)/k$$
$$\text{intervals} = \min + w, \min + 2w, \ldots, \min + (k - 1)w \tag{1}$$

Here max and min indicate the maximum and minimum values in the data. Using Eq. (1), the data will be discretized. The Coverings are then found for all the bands $bc = \{b_{c1}, b_{c2}, \ldots, b_{cm}\}$, in turn each b_{ci} is the set of all pattern coverings of that particular band. These coverings are used to build discernibility matrix.

3.3   Discernibility Matrix

Definition 3 [12] Let (U, C) be a covering information system. Suppose $U = \{x_1, x_2, \ldots, x_n\}$, we denote discernibility matrix by M(U, C), an $n \times n$ matrix(c_{ij}) and defined as $c_{ij} = \{c \in C : x_j \notin c(x_i)\}$ for $x_i, x_j \in U$. Clearly, we have $c_{ii} = \phi$ for any $x_i \in U$.

The discernibility matrix provides a proper description about all subsets of attributes that can differentiate any two objects [12]. This concept is consistent with the perspective in classical rough sets. Alongside, Classical rough sets and Coverings have some formal similarity among discernibility matrices. This is why the covering-based rough set is a generalization of classical rough sets. But the challenge faced here is if the dataset is small enough, construction of matrix does not take much space. But if it is Big Data (considering hyperspectral remote sensing data), the space complexity increases drastically as $n \times n$ matrix has to be stored in memory to carry out the computations. In current scenario Big Data dimensionality plays a crucial role: the time and space complexity of Machine learning algorithms is closely related to data dimensionality [1]. Taking both the parameters into consideration we proposed the algorithm for band selection.

Algorithm 1 Proposed algorithm for band selection

1: Consider $selected\_bands = \phi$ and $B = \{b_1, b_2, \ldots b_m\}$
2: Finding Coverings of the bands
3: for each pattern x_i in 'U'
4: $c_j = \{c \in C : x_j \notin c(x_i)\}$ where j = 1,2,...n and if i=j $c_j = \phi$
   if $c_j = \phi$ **then**
       do nothing
   else if $|c_j| = 1$ and $selected\_bands \cap c_j = \phi$ **then**
       $selected\_bands \leftarrow c_j, B = B - c_j$
   else if $|c_j| \neq 1$ and $B \cap c_j \neq \phi$ **then**
       Store the bands set 'bs' and its count, if same set already exists increment the count
   end if
3: for each bands set 'x' in 'bs'
   if $B \cap x \neq \phi$ and $selected_bands \cap x = \phi$ **then**
       find the band 'b' in B that is mostly repeated in complete 'bs' and if its not in selected_bands
       then $selected\_bands \cup b, B = B - b$
   end if

---

4 Experimental Results

Experiments are carried on the dataset acquired by the AVIRIS sensor over a spectral range of 400–500 nm. The dataset [13] describes the agricultural land of Indian Pine, Indiana which is taken in the early growing season of 1992. The original image consists of 220 spectral bands (0–219) (shown in Fig. 2) in which 35 bands which are noisy bands (15 bands) and water absorption bands (20 bands) were removed (0, 1, 2, 102–111, 147–164, 216–219). The resulting denoised image consists of 185 bands (shown in Fig. 3). Along with the proposed algorithm, few other methods are also implemented to find the optimal bands. All the experiments are carried out on High performance computer (HPC) in National Institute of Technology Karnataka, Surathkal. The outputs of algorithms (PCA, SVD, RS-FCM, rough sets) are shown in Figs. 4, 5, 6 and 7. In the proposed algorithm, top 10 bands are selected as shown in Fig. 8.

In Hyperspectral images, data is highly correlated in higher dimension. PCA sometimes is not a good option for multiclass data because it tries to discriminate data in lower dimension. SVD sometimes can discard useful information. Problem with RS-FCM is the number of clusters is predefined. Classical rough sets shows limitations while it comes to continuous data. The rough set algorithm that I used for comparison is an extended rough set algorithm which finds optimal bands using maximum relevance and significance values.

Fig. 2 Original image

Fig. 3 Denoised image

Fig. 4 Output of PCA

Fig. 5 Output of SVD

Fig. 6 Output of rough set

Fig. 7 Output of RS-FCM

Fig. 8 Output of covering-based rough set

Algorithm	Selected bands	Time complexity
PCA	Top 5 bands (PCA components = 5)	$O(p^2n + p^3)$ – 'n' data points and 'p' features
SVD	Top 10 bands (components = 10)	$O(mn^2)$ – 'n' data points and 'm' features
RS-FCM	Top 10 bands (9, 17, 23, 28, 56, 80, 97, 121, 167,183)	$O(mnc^2)$ – 'c' clusters and 'n' data points
Rough sets	Top 10 bands (30, 44, 80, 100, 132, 176, 184, 189, 198, 213)	O(mn + m + dm) – 'm' bands and 'n' patterns
Covering-based rough sets	Top 10 bands (56, 83, 112, 121, 173, 179, 182, 187, 209, 215)	O(mn) – 'm' bands and 'n' patterns

5 Conclusion

Deep learning cannot be directly used in many Remote Sensing tasks. Some Remote sensing images, especially hyperspectral images, contain hundreds of bands in which a small patch also resembles large datacube. For this reason, preprocessing step should be done on data to find the optimal band set. This preprocessing step will be useful while performing deep learning algorithms for crop discrimination also. There are many techniques to find the optimal bands. In this paper, we proposed a new band selection method called covering-based rough sets with time complexity o(mn) and space complexity o(n), in which equivalence relation of rough sets is replaced by coverings. This method suites many real time scenarios as coverings may not be pairwise disjoint.

References

1. L'heureux, A., et al.: Machine learning with big data: challenges and approaches. IEEE Access **5**, 7776–7797 (2017)
2. Patra, S., Bruzzone, L.: A rough set based band selection technique for the analysis of hyperspectral images. In: 2015 IEEE International Geoscience and Remote Sensing Symposium (IGARSS) (2015)
3. Maji, P., Paul, S.: Rough set based maximum relevance-maximum significance criterion and gene selection from microarray data. Int. J. Approx. Reason. (2010) (Elsevier)
4. Jensen, K., Shen, Q.: Semantics-preserving dimentionality reduction: rough and fuzzy rough based approches. IEEE Trans. Knowl. Data Eng. **16**, 1457–1471 (2004)
5. Rodarmel, C., Shan, J.: Principal component analysis for hyperspectral image classification. Surveying Land Inf. Syst. **62**(2), 115–000 (2002)
6. Guo, B., et al.: Band selection for hyperspectral image classification using mutual information. IEEE Geosci. Remote Sens. Lett. **3**(4), 522–526 (2006)
7. Lavanya, A., Sanjeevi, S.: An improved band selection technique for hyperspectral data using factor analysis. J. Indian Soc. Remote Sens. **41**(2), 199–211 (2013)

8. Nahr, S., Talebi, P., et al.: Different optimal band selection of hyperspectral images using a continuous genetic algorithm. In: ISPRS—International Archives of the Photogrammetry, Remote Sensing and Spatial Information Sciences, XL-2/W3, pp. 249–253 (2014)

9. Shi, H., Shen, Y., Liu, Z.: Hyperspectral bands reduction based on rough sets and fuzzy C-means clustering. IEEE Instrumentation and Measurement and Technology Conference **2**, 1053–1056 (2003)

10. Zakowski, W.: Approximations in the space (u, π). Demonstratio Math. **16**, 761–769 (1983)

11. Kotsiantis, S., Kanellopoulos, D.: Discretization techniques: a recent survey. GESTS Int. Trans. Comput. Sci. Eng. **32**(1), 47–58 (2006)

12. Wang, C., Shao, M., et al.: An improved attribute reduction scheme with covering based rough sets. Appl. Soft Comput. (2014) (Elsevier)

13. https://americaview.org/program-areas/research/accuracy-assessment-resources/indian-pine/

Improved Feature-Specific Collaborative Filtering Model for the Aspect-Opinion Based Product Recommendation

J. Sangeetha and V. Sinthu Janita Prakash

Abstract Utilizing the benefits of Internet services for the online purchase and online advertising has increased tremendously in the recent year. Therefore, the customer reviews of the product play a major role in the product sale and effectively describe the features quality. Thus, the large size of words and phrases in an unstructured data is converted into numerical values based on the opinion prediction rule. This paper proposes the Novel Product Recommendation Framework (NPRF) for the prediction of overall opinion and estimates the rating of the product based on the user reviews. Initially, preprocessing the set of large size customer reviews to extract the relevant keywords with the help of stop word removal, PoS tagger, Slicing, and the normalization processes. SentiWordNet library database is applied to categorize the keywords which are in the form of positive and negative based polarity. After extracting the related keywords, the Inclusive Similarity-based Clustering (ISC) method is performed to cluster the user reviews based on the positive and negative polarity. The proposed Improved Feature-Specific Collaborative Filtering (IFSCF) model for the feature-specific clusters is used to evaluate the product strength and weakness and predict the corresponding aspects and its opinions. If the user query is matched with the cache memory then shows the opinion or else extract from the knowledge database. This optimal memory access process is termed as the Memory Management Model (MMM). Then, the overall opinion of the products is determined based on the Novel Product Feature-based Opinion Score Estimation (NPF-OSE) process. Finally, the top quality query result and the recommended solution are retrieved. Thus the devised NPRF method enriches its capability in outperforming other prevailing methodologies in terms of precision, recall, F-measure, RMSE, and the MAE.

Keywords Customer reviews · SentiWordNet · Collaborative filtering · Aspects Opinions · Clustering · Memory management model

J. Sangeetha (✉)
Cauvery College for Women, Trichy, Tamil Nadu, India
e-mail: sangeetharesearch@gmail.com

V. Sinthu Janita Prakash
Department of Computer Science, Cauvery College for Women, Trichy, Tamil Nadu, India
e-mail: sinthujanitaprakash@yahoo.com

© Springer Nature Singapore Pte Ltd. 2019
J. D. Peter et al. (eds.), *Advances in Big Data and Cloud Computing*,
Advances in Intelligent Systems and Computing 750,
https://doi.org/10.1007/978-981-13-1882-5_25

275

1 Introduction

An inevitable portion of life in the current scenario is the Internet for online purchase and online advertising that attracts human for buying and selling the products to the user. A recommender framework [1] is to furnish clients with customized online item or administration proposals to deal with on the web data over-burden issue and enhance the client's relationship with the administration. The inputs of any items can be prescribed [2] in light of the socioeconomics of the client, in light of the best general vendors on a webpage, or in view of an investigation of the past purchasing conduct of the client as a forecast for future purchasing conduct.

Recommender Systems (RSs) [3] gather data on the inclinations of its clients for an arrangement of. Few reviews may be speaking particularly about the product feature like efficiency, accuracy, display, etc.

Therefore the hybrid product based recommendation system [4] is utilized to extract the relevant query of the product. The efficiency of the obtained product review based recommendation is a major drawback of the large database. Hence, a scalable product recommendation systems based on the collaborative filtering method [5] is applied to develop the competence and the scalability of the recommendation framework.

The novel technical contributions of the proposed approach are listed as follows:

- To estimate the product's strength and weakness score of the product and categorize the product aspect and its opinions by using the Improved Feature-Specific collaborative Filtering Model
- To evaluate the average score of product's strength and weakness and to show the better recommendation based on a Novel Product Feature-based Opinion Score Estimation process.

This paper is organized as follows: The detailed description of the related works regarding recommendation system, filtering techniques, and the rating prediction technique is discussed in Sect. 2. The implementation process of the Novel Product Recommendation Framework (NPRF) mechanism for proficient opinion prediction and rating estimation is pronounced in Sect. 3. The qualified investigation of the proposed NPRF model is compared with prevailing approaches is provided in Sect. 4. Lastly, Sect. 5 concludes the proficiency of the proposed NPRF approach.

2 Related Work

In this section, the review of the various recommendation system, filtering technique, and the score estimation methods are presented with their merits and demerits. Yang et al. [6] suggested the hybrid collaborative technique with the content-based filtering technique. This algorithm is capable of handling different costs for false positives and false negatives making it extremely attractive for deploying within many kinds of recommendation systems. Yang et al. [7] surveyed the collaborative filtering (CF) method of the social recommender system. Matrix Factorization (MF) based social RS and the Neighborhood-based social RS approaches. Bao et al. [8] surveyed the recommendations in the location-based social networks (LBSNs). They provided a widespread investigation of the utilized data sources, employed methodology to generate a recommendation and the objective of the recommendation system.

Hence, each recommender system explained an individual performance result with available data sets. Lei et al. [9] In this RPS, three major works were performed to tackle the recommended problems. First of all, mined the sentiment words and sentiment degree words from the user reviews by a social user sentimental measurement method, secondly, considered the interpersonal sentimental influence with own sentimental attributes that displayed the product reputation words. The rating prediction accuracy was the major issues in this recommendation system. Salehan and Kim [10] offered the sentiment mining method for analyzing the big data of the online consumer reviews (OCR). In this method, classified the consumer reviews into positive sentiment and the neutral polarity of the text. Pourgholamali et al. [11] presented the embedding unstructured side information for the product recommendation. Both the users and the products based representation were exploited in the word embedding technique. The selection of optimal reviews was the major drawbacks in the recommendation.

3   NPRF: Novel Product Recommendation Framework

This section discusses the implementation details of the proposed Novel Product Recommendation Framework (NPRF) for accomplishing a good recommendation about the product reviews through the implication of collaborative filtering with the opinion score estimation technique. Initially, the customer product reviews which are given by the users are extracted in the native form and are loaded for processing, reviews are pre-processed by applying stop word removal, POS tagger, slicing and the normalization preprocessing techniques. The keywords from the each and every user review are extracted. These keywords may represent the products polarity or the features of the products based on the SentiWord Net database. Then the extracted keywords are clustered based on the Inclusive Similarity-based Clustering (ISC) method by which the user reviews are clustered as groups based on polarity. Furthermore, the Improved Feature-Specific Collaborative Filtering (IFSCF) model is implemented to group the user reviews as a feature-specific clusters. Then the polarity detection is done and the overall opinion of the products is determined based on the Novel Product Feature-based Opinion Score Estimation (NPF-OSE) algorithm. The overall opinion about the product and the feature-specific opinion of the product are given as the query result. Here the query answer retrieval is done based on the Optimal Memory Access Algorithm (Fig. 1).

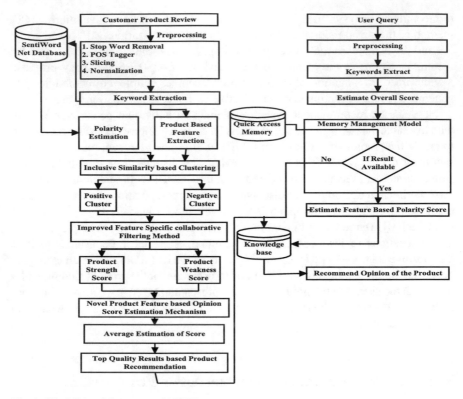

Fig. 1 Workflow of the proposed NPRF

3.1 Dataset

In this research work, the customer product reviews collected in the area of opinion mining, sentiment analysis, and opinion spam detection dataset from the social media [12] is used for evaluating the product opinions effectively. In the Customer Review Dataset (CRD) is comprised of five types of product reviews such as digital camera 1, digital camera 2, phone, Mp3 player, and the DVD player. The feature-based or aspect-based opinion mining model is preferred in this machine learning technique. Review text may contain complete sentences, short comments, or both. This dataset is taken for the purpose of both training and testing processes. The proposed algorithm is trained with the training dataset and then verified by using the testing dataset.

3.2 Preprocessing Based Document Clustering

Preprocessing is a very important step since it can improve the accuracy result of a Clustering based Inclusive Similarity (ISC) algorithm. First of all, remove the superfluous usage words in the English language. Words like is, are, was etc., such type of words are called as stop words which are removed by using the stop word removal process. It is used to remove unwanted words in each review sentence and stored in a text file. The Part-of-Speech (PoS) tagger is a computer readable program to extract the noun, verb, adverb, adjective, pronoun, preposition, conjunction, and the interjection. The main reason for using PoS tagger is to extract the relevant keywords. After, extracting the keywords the slicing technique is performed to reduce the repetitive keywords and extract only the descriptive words based on the horizontal and vertical partitioning. After, slicing the keyword to implement the normalization technique. The attribute data is scaled to fit in a specific range called Min-Max normalization.

$$Nor' = \left(\frac{Nor - Nor_{min}}{Nor_{max} - Nor_{min}} \right) * (q - p) + p, \tag{1}$$

where, Nor' represents min-max normalized data one, (Nor_{max}, Nor_{min}) represents the pre-defined boundary (p, q) represents the new interval.

After that, validate the extracted keywords with the SentiWordNet database which shows two numerical scores ranging from 0 to 1. Therefore, the positive and negative polarity based keywords Sn_{wd} are split. Then, the polarity and the product features of user reviews U_R are given to the Inclusive Similarity-based Clustering algorithm. The mathematical derivation of cumulative distance measure as,

$$CD = dist(i) * dist(j), \tag{2}$$

where i and j represent 1 to n of each review. Then, estimate the inclusive similarity

$$IS = cum.dis(i, j) * n/dist(i) * dist(j) \tag{3}$$

After that, construct an adjacency matrix of $k \times k$ matrices whose vertex denoted as (i, j) represents the set of reviews. If similar review presence in i and j then shows 1 or dissimilar reviews shows 0. Then, the ISC algorithm is applied to cluster the most similar keywords based on the adjacency matrix and merge the similar index words.

3.3 Proposed IFSCF Model

The proposed IFSCF model is utilized to extract the relevant aspects based opinions of the corresponding product and estimate the product strength and weakness percentage effectively. Initializing the cluster indices, clustered keywords, and the SentiWordNet dictionary words with the set of words in terms of polarity score for each word. Then, applying the condition for each reviews i in the clustered keywords that must satisfy the following cases in step 3. If the clustered keyword represent the positive word of SentiWordNet then assign as P_{cw}, else N_{cw}. Secondarily, applying the condition in the j of the user reviews denotes in step 9. Then categorize the aspects and its corresponding opinion of the product based on the following cases.

First of all, initialize the M number of aspects represent A_M from $U_R(i)$ and the opinion O_p be the set of opinions that corresponds to the A_M. Afterwards, the aspect–opinion pairs are termed as the AO_p. If the opinion is said to be positive then update the aspect-opinion term as strength category or it is called as weak category as,

$$A O_{\text{StrengList}} \leftarrow \left(A_M(k), O_p(k)\right) \tag{4}$$

$$A O_{\text{WeakList}} \leftarrow \left(A_M(k), O_p(k)\right) \tag{5}$$

Then, continue this process until reach the overall user reviews. After calculating the strength score value as,

$$\text{Score}_{\text{Strength}} = \frac{A O_{\text{StrengList}}\left(\text{Count}\left(A_M\right) + \text{Count}\left(O_p\right)\right)}{\text{Size}\left(A O_{\text{StrengList}}\right) + \text{Size}(A O_{\text{WeakList}})} \tag{6}$$

$$SS_p = \text{Score}_{\text{Strength}} * 100; \tag{7}$$

$$\text{Score}_{\text{weakness}} = \frac{A O_{\text{WeakList}}\left(\text{Count}\left(A_M\right) + \text{Count}\left(O_p\right)\right)}{\text{Size}\left(A O_{\text{StrengList}}\right) + \text{Size}(A O_{\text{WeakList}})} \tag{8}$$

$$WS_p = \text{Score}_{\text{weakness}} * 100; \tag{9}$$

Improved Feature specific Collaborative filtering Model
Input: cluster Words, User Reviews (U_R)
Output: Aspect-Opinion pairs, Strength and Weakness Score
Step 1: Initialize C_{id} and C_W
Step 2: Initialize Sn_{wd} with polarity score value
Step 3: For i=1 to Size (C_W)
Step 4:   If ($C_W(i)$ in Sn_{wd} (pos))
Step 5:   Add $C_W \rightarrow P_{cw}$
Else
Step 6:   Add $C_W \rightarrow N_{cw}$
Step 7:   End if
Step 8: End for i
Step 9: For j=1 to N
Step 10: Initialize A_M, O_p
Step 11: Assign $AO_p \leftarrow$ Aspect-Opinion Pair (A_M, O_p)
Step 12: For k= 1 to size (A_M)
Step 13:   If (O_p is positive)
Step 14: Update AO_p as strength by using the equation (4)
Else
Step 15:   Update AO_p as weak by using the equation (5)
Step 16:   End if
Step 17: End for K;
Step 18: End for j
Step 19: Compute Strength Score by using the equation (6)
Step 20: Estimate the product strength percentage value by using equation (7)
Step 21: Compute Weakness Score by using the equation(8)
Step 22: Estimate the product weakness percentage value by using equation (9)

3.4   Memory Management Model

The memory management model (MMM) [13] is used to match the user query with the cache memory. The execution time and the process time are reduced based on the optimal memory access. The multiple user reviews and the clustered words are given to the preprocessing model to extract the relevant key features. The query of product and the user reviews are collected in parallel to manage the cache memory effectively. The removal of stop words and the relevant keywords extraction are the successive steps in the user query processing. The results from the query processing are matched with the contents of the cache memory. Then, check whether the relevant results (services) are available in the cache memory or not. If they are available means the results are extracted from the cache memory otherwise the results are extracted from the database effectively.

3.5 Proposed NPF-OSE

The proposed NPF-OSE method shows the average score of the positive and negative reviews and estimates the better recommendation of the product. The obtained percentage score of product strength and weakness is sorted with an ascending arrangement to show the top quality ranking of the recommended reviews. Then estimate the average value of a product strength and weakness. This score shows the top quality review results and finally predicted the product recommendations "positive feedback" or "negative feedback". The proposed NPF-OSE algorithm is shown below.

Novel Product Feature based Opinion Score Estimation Algorithm
Input: Aspect-Opinion pairs, User Query
Output: Overall Opinion Score and Recommendation
Step 1: Initialize U_Q
Step 2: $U_k \leftarrow$ pre-process (U_Q) // where U_k is the keywords in the user Query
Step 3: $Asp_Q \leftarrow$ Extract $(Aspects, U_k)$
Step 4: if $(Size\,(Asp_Q) > 0)$
Step 5:   Initiate aspect specific Recommendation
Step 6:   Let AO_p be the Aspect –opinion pair
Step 7:   Sub_$AO_p \leftarrow$ Similar (AO_p, Asp_Q)
Step 8: $F_{pos} \leftarrow$ positive features (Sub_AO_p)
Step 9: $F_{neg} \leftarrow$ Negative features (Sub_AO_p)
Step 10: $Score_{strengthFeat}$ = Polarity Score (F_{pos})
Step 11: $Score_{WeakFeat}$ = Polarity Score (F_{neg})
Step 12: apply Score Based ranking to $Score_{strengthFeat}$ and $Score_{WeakFeat}$
Step 13: $Strength_{Score} = \dfrac{\sum_{i=1}^{N} Score_{strengthFeat}}{Size\,(Score_{strengthFeat})}$
Step 14: $Weakness_{Score} = \dfrac{\sum_{i=1}^{N} Score_{strengthFeat}}{Size\,(Score_{strengthFeat})}$
Step 15: $Results = \begin{cases} Recommended & if\ (Strength_{Score} > Weakness_{Score}) \\ Not\ Recommended & Other\ Wise \end{cases}$
Step 16: else
Step 17: Initialize Overall Product Recommendation
Step 18: $F_{pos} \leftarrow$ positive features (AO_p)
Step 19: $F_{neg} \leftarrow$ Negative features (AO_p)
Step 20: Follow Step 10 to 15, End if

4 Performance Analysis

This section illustrates the experimental posture for the proposed NPRF deployment and the proficiency assessment of the varied prevailing methodologies like collaborative filtering methods and the explainable recommendation methods. Therefore, the proficiency is considered by analyzing various parameters given as MSE, RMSE, MAE, precision, recall, and F-measure.

4.1 Evaluation Metrics

The Root Mean Square Error (RMSE), Mean Absolute Error (MAE), precision, recall, F-measure are used to evaluate the proposed recommendation system performance.

4.1.1 RMSE

The MAE is used to measures the average of the squares of the errors or deviations and the root based error calculation is the RMSE is also called as the root mean square deviation (RMSD). These individual differences are also called residuals, and the RMSE serves to aggregate them into a single measure of predictive power.

$$\text{RMSE} = \sqrt{\frac{\sum_{i=1}^{n} \left(X_{\exp,i} - X_{\text{est},i}\right)^2}{n}} \tag{10}$$

4.1.2 MAE

The MAE measures the normal size of the mistakes in an arrangement of expectations, without considering their bearing. It's normal for the test of the total contrasts amongst expectation and genuine perception where every individual distinction have the level of weight. Where $X_{\exp}$ represents the experimental results and X_{est} represents the estimated variable at time i.

$$\text{MSE} = \frac{\sum_{i=1}^{i=1} \left|X_{\exp,i} - X_{\text{est},i}\right|}{n} \tag{11}$$

4.1.3 Precision

The precision which is also called as the positive predictive rate is the fraction of retrieved reviews that are relevant to the user query. It is calculated as,

$$\text{Pre} = \left\{ \frac{\{\text{relevant reviews} \cap \text{retrieved opinions}\}}{\text{retreived opinions}} \right\} \tag{12}$$

Table 1 RMSE and MAE values

Customer review		
Dataset	RMSE	MAE
Digital camera1	0.6	0.75
Digital camera2	1.12	0.65
Phone	0.98	0.74
Mp3 player	0.78	0.51
Dvd player	0.51	0.62
Average error rate	0.798	0.654

4.1.4 Recall

A recall in information retrieval is the fraction of the user reviews that are relevant to the user query that are successfully retrieved.

$$R = \left\{ \frac{\{\text{relevant reviews} \cap \text{retrieved user query}\}}{\text{relevant user query}} \right\} \tag{13}$$

4.1.5 F-measure

The F-measure is defined as the weighted harmonic mean of the precision and recall of the opinion prediction based on the user query.

$$\text{F-measure} = \frac{2 \times \text{Pre} \times R}{\text{Pre} + R} \tag{14}$$

4.2 Experimental Results and Discussions

In the proposed NPRF, the RMSE and the MAE value of the CRD dataset. Both the RMSE and MAE parameter measures the error between the true ratings and the predicted rating. Table 1 shows the proposed dataset RMSE and MAE values.

The RMSE and the MAE value prediction is used to show the effectiveness of the proposed NPRF. The average error rate is very much lower.

4.2.1 Number of Training Reviews Versus Reduction in MAE

The number of training reviews is compared with the MAE rate [14]. Figure 2 shows the MAE rate of the training reviews.

Fig. 2 No. of train reviews versus reduction in MAE

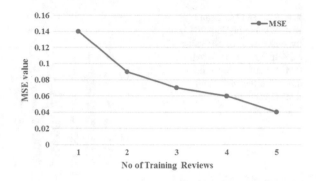

Figure 2 illustrates that the comparison of the MAE rate with the increasing number of the training reviews. If the number of training reviews increases than the MAE rate will decrease.

4.2.2 MAE and RMSE Values for Rating Prediction

The performance of all collaborative filtering methods [15] in terms of MAE and the RMSE values are validated with the existing three datasets. Figure 3a shows the MAE and RMSE value of Yelp 2013, (b) represents the MAE and RMSE value of Yelp 2014, (c) represents the MAE and RMSE value of Epinions dataset, and (d) represent the MAE and RMSE value of the proposed CR dataset. The proposed IFSCF model increases the efficiency while reduce the error rate successfully.

4.2.3 Average Precision

The average prediction is the estimation of overall product opinion among the different user queries [9]. Figure 4 represents the average precision rate.

The proposed NPF-OSE method is compared with the existing Sentiment based Rating prediction (RPS) method. This opinion prediction technique efficiency is computed with the help of different dataset such as movie dataset, Simon Fraser University (SFU) dataset, Yelp, and the CRD to estimate the average precision rate. Hence, the proposed NPF-OSE method provides the better performance than the existing RPS method.

4.2.4 Sparsity Estimation

The different experimental analysis is performed in the proposed IFSCF model to tackle the data sparsity [16]. Figure 5 shows the different experimental models of (a) precision measure, (b) recall measure, and the (c) F-measure.

Fig. 3 **a** The MAE and
RMSE value of Yelp 2013, **b**
the MAE and RMSE value
of Yelp 2014, **c** the MAE and
RMSE value of Epinions
dataset, **d** the MAE and
RMSE value of the proposed
CR dataset

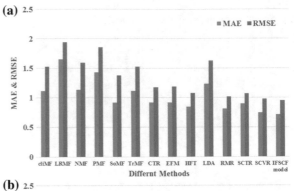

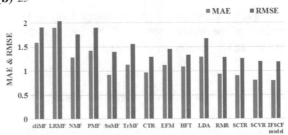

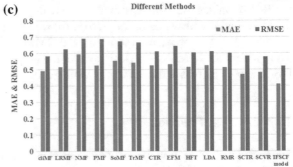

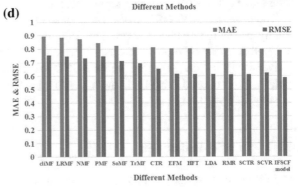

Fig. 4 Average precision range

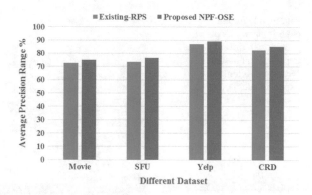

The graph 5 shows that the proposed IFSCF model is compared with the existing collaborative filtering (CF) model between the different increasing sparsity measures. It is demonstrated that the proposed IFSCF technique precision, recall, and the F-measure value is higher than the basic CF in all sparsity measures. Hence, the proposed IFSCF method attains greater performance.

5 Conclusion

The issues prevailing in the existing collaborative filtering and the recommendation systems among the customer reviews in the text mining is exhibited and the way out of obtaining a relevant opinion about the product is dealt through the recommended NPRF approach. Initially, preprocessing the large size customer reviews data and extract the relevant keyword. Then, performing ISC method to cluster the keywords based on the positive and negative polarity values. Thus, applying IFSCF technique to efficiently retrieve the product strength and weakness score and also predict the aspect and its relevant opinions. After that, applying the NPF-OSE model to estimate the average score of positive and negative features and extract the relevant query results. Finally, obtain the overall opinion score and its high recommendation solution about the product. Thus NPRF develops its capability in outperforming other prevailing methodologies in terms of RMSE, MAE, precision, recall, and the F-measure. Hence, the proposed framework achieves better recommendation than the other techniques. As a scope, recommendation can be enhanced to display a specific features about the products are arranged based on the recent time period.

Fig. 5 a Different
experimental models of
precision, **b** different
experimental models of
recall, **c** different
experimental models of
F-measure

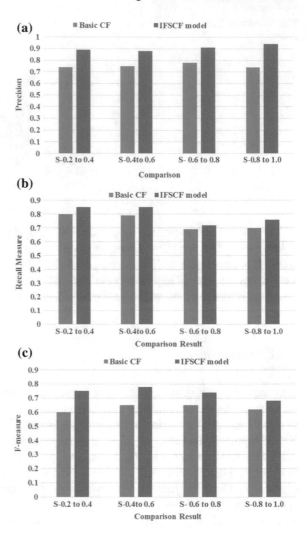

References

1. Lu, J., Wu, D., Mao, M., Wang, W., Zhang, G.: Recommender system application developments: a survey. Decis. Support Syst. **74**, 12–32 (2015)
2. Sivapalan, S., Sadeghian, A., Rahnama, H., Madni, A.M.: Recommender systems in e-commerce. In: World Automation Congress (WAC), pp. 179–184. IEEE (2014)
3. Bobadilla, J., Ortega, F., Hernando, A., Gutiérrez, A.: Recommender systems survey. Knowl.-Based Syst. **46**, 109–132 (2013)
4. Rodrigues, F., Ferreira, B.: Product recommendation based on shared customer's behaviour. Procedia Comput. Sci. **100**, 136–146 (2016)
5. Riyaz, P., Varghese, S.M.: A scalable product recommendations using collaborative filtering in hadoop for bigdata. Procedia Technology **24**, 1393–1399 (2016)

6. Yang, S., Korayem, M., AlJadda, K., Grainger, T., Natarajan, S.: Combining content-based and collaborative filtering for job recommendation system: a cost-sensitive Statistical Relational Learning approach. Knowl.-Based Syst. **136**, 37–45 (2017)
7. Yang, X., Guo, Y., Liu, Y., Steck, H.: A survey of collaborative filtering based social recommender systems. Comput. Commun. **41**, 1–10 (2014)
8. Bao, J., Zheng, Y., Wilkie, D., Mokbel, M.: Recommendations in location-based social networks: a survey. GeoInformatica **19**(3), 525–565 (2015)
9. Lei, X., Qian, X., Zhao, G.: Rating prediction based on social sentiment from textual reviews. IEEE Trans. Multimedia **18**(9), 1910–1921 (2016)
10. Salehan, M., Kim, D.J.: Predicting the performance of online consumer reviews: a sentiment mining approach to big data analytics. Decis. Support Syst. **81**, 30–40 (2016)
11. Pourgholamali, F., Kahani, M., Bagheri, E., Noorian, Z.: Embedding unstructured side information in product recommendation. Electron. Commer. Res. Appl. **25**, 70–85 (2017)
12. Opinion Mining, Sentiment Analysis, and Opinion Spam Detection (May 15, 2004). https://www.cs.uic.edu/~liub/FBS/sentiment-analysis.html
13. Sangeetha, J., Prakash, V.S.J., Bhuvaneswari, A.: Dual access cache memory management recommendation model based on user reviews. IJCSRCSEIT 169–179 (2017)
14. Zheng, L., Noroozi, V., Yu, P.S.: Joint deep modeling of users and items using reviews for recommendation. In: Proceedings of the Tenth ACM International Conference on Web Search and Data Mining, pp. 425–434. ACM (2017)
15. Ren, Z., Liang, S., Li, P., Wang, S., de Rijke, M.: Social collaborative viewpoint regression with explainable recommendations. In: Proceedings of the Tenth ACM International Conference on Web Search and Data Mining, pp 485–494. ACM (2017)
16. Najafabadi, M.K., Mahrin, MNr, Chuprat, S., Sarkan, H.M.: Improving the accuracy of collaborative filtering recommendations using clustering and association rules mining on implicit data. Comput. Hum. Behav. **67**, 113–128 (2017)

Social Interaction and Stress-Based Recommendations for Elderly Healthcare Support System—A Survey

M. Janani and N. Yuvaraj

Abstract Healthcare awareness is being increased due to an advancement made on technology and medical science field for providing consciousness about nutrition, environmental, and personal hygiene. Aging population increase life expectancy globally and cause danger to socio-economic structure in terms of cost related to wellbeing and healthcare of elderly people. Migration of people to cities and urban areas affects healthcare services in great extent. Nowadays, cities present in the world invest heavily in digital transformation for providing healthy environment to elderly people. Healthcare application is merely based on activity, social interactions, and physiological signs of elderly people for the recommendation system. Measurement of physiological signs may include wearable or ambient sensors to gather information related to elderly people health conditions. Better recommendations can be provided to elderly people merely based on three terms. First, recommendations through personal details of elderly people collected in day-to-day life. Second, measure of health conditions such as pulse rate, blood pressure and heart beat. Third, social interactions based stress of elderly people in social media is determined by collecting elderly people posts and updates. Depending on the unruffled information recommendations are generated.

Keywords Elderly assistance · HCA · Social interaction
Recommendation system

1 Introduction

In data mining, patterns are discovered from large datasets for extracting the hidden information and discovering a new knowledge. Goals include analyzing and

M. Janani (✉) · N. Yuvaraj
Department of CSE, KPR Institute of Engineering & Technology, Coimbatore, Tamil Nadu, India
e-mail: jananemuthusamy@gmail.com

N. Yuvaraj
e-mail: drnyuvaraj@gmail.com

© Springer Nature Singapore Pte Ltd. 2019
J. D. Peter et al. (eds.), *Advances in Big Data and Cloud Computing*,
Advances in Intelligent Systems and Computing 750,
https://doi.org/10.1007/978-981-13-1882-5_26

extracting the data for knowledge by using various methods and techniques of data mining [1]. Information industry consists of large data, data has to be structured to obtain meaningful information or else it becomes useless. Hence, data is converted for further analysis. Recommender system is subordinate classes derived for item recommendation to the user in information filtering system.

Consciousness about nutrition, personal hygiene and environment has increased due to the improvements made in technology and medical field. Advancements made increases life anticipation globally by increasing aging population affecting the economic structure of country based on cost [2]. It is estimated that between 2015 and 2050 the proportion of adults over 60 years old will increase from 12 to 22%. The major problems faced by elderly people are poor eyesight, isolation, forgetting their needs, unable to keep up with daily chores and housekeeping and poor nutrition.

The main challenge is to maintain a healthy social condition, which avoids loneliness and social isolation. While addressing the challenge there is a demand on cities to adapt dynamically to the needs of citizens and to make urban ambiences friendly to the elderly people. In that context, the advancement of electronic devices, wireless technologies and communication networks makes the environment to interact. Meanwhile cities are leading us to the concept called "smart cities" from smart home [3].

In our day-to-day life, recommendation systems are becoming popular in recent times. Recommender system areas are used in health, music, books, events news and social tags. Recommendation engines like twitter, Facebook, Instagram provides friend suggestions, suggestions to like similar pages and Youtube recommendation engine suggests similar videos from previously watched history. Overall recommendations engines are mainly based on user and item based collaborative filtering [4]. But the proposed recommendation system is based on hybrid approach, the combination of collaborative and content-based filtering. The approach collects personal details of elders along with their social interaction and physiological measures for providing recommendations to elderly people in the way, what they should do and what not.

Profile of elderly people is created by getting details such as name, age and gender, a password is created for them. The personal details include location, health issues, social accounts and interests. Social interactions of the elderly people are mined from twitter posts, where text to emotions is explored for mapping it with stress parameter. The physiological measurements involve measuring the blood pressure, heart rate and blood viscosity of elderly people. The data is collected from various sources but requires a common medium to relate all information and such environment is provided with smart home.

The remainder section of the paper is organized as follows. Section 2, literature survey followed by profile setup and social interaction. Section 3 describes about emotion and stress value analysis. In Sect. 4 recommendation system is provided and in Sect. 5, machine learning algorithms comparison is tabulated. Finally, Sect. 6 ends up with conclusion.

2   Literature Survey

Elderly people all over the world require support for daily living and taking care of health in regular, this is supposed to be given by the family members, friends circle or volunteers [2]. Elderly people care is offered by many paid services but seems costly and remains unreached to most of aged population having constrained financial plan conditions. Thereby, it is a need for creating awareness among elderly people about their health conditions. In order to provide awareness, recommendations can be given to elderly people who could not get proper healthcare issues solved.

Sensors are being used in the form of mobile phones, computers and wearable devices for digital era. Sensor initially collects data and then analyses the data for identifying human characterization, feelings, and thoughts. The number of elderly people in the world is growing rapidly and vital signs such as heart beat and blood pressure becomes the most basic parameters to be monitored regularly for good health. Physiological parameter is monitored based on activities of the elderly people in a continuous manner for generating alerts in emergency cases [2].

2.1   Profile Setup

Profile setup for elderly people should be simple not anonymous with much of colors and moving on next pages for further filling up details. The profile basically should carry some identity for elderly people for reference.

2.1.1   Profile Building

Sporadic social network is developed to provoke the social interaction among the elderly people by setting a profile and collecting their preferences [3]. The profile built for elderly people includes name, age, and location with a password to login, these details gets stored in database. Preferences are the events related to the cultural, sports, and music. For simulating social interaction among elderly people through sporadic social network it has four layers. First layer is communication and sensing layer and second layer is mobile layer and third layer is knowledge layer and last smart assistance layer.

2.1.2   Notifications

Social media is becoming an important platform where people find news and information and used for sharing their experiences and connect with friends and family. Nearly 41% of the elderly people aged around 75 are using social media till 2017 [5]. All over the world 15% of social media users are elderly people. When it still continues, in and around 2050, 50% of elderly people will use social media all over the world. Mobile phone users receive notifications from social media such as twitter, Facebook and Whatsapp which may annoy sometimes [6].

The proposed system collects name, age, and gender initially in login page and generates an Id for elderly people. Then health issues and their interests are collected by a survey containing few set of questions. Profile for elderly people is built and stored in a database then social accounts are linked with the profile.

2.2 Social Interaction

Social interaction exchange data between two or more individuals act as building block of society. Interactions can be studied between groups of two, three, or more larger social groups.

2.2.1 Text Collection

Nowadays people start expressing their personal feelings in social networks due to an expansion of internet using textual media communication. Emotion from text cannot be automatically derived [7]. Human affects or emotions are well understood by affective computing. Module is trained by extracting data not only from text to emotion analyzer also from the suggestions and feedbacks obtained in twitter. Data is trained using classification algorithms [8]. Sports monitoring and sport sharing platform is provided by Facebook API and friends get suggested [9].

2.2.2 Exploring Emotions from Text

Emotions in social media are explored by mining the vast amount of data and emotional state. WeFeel is a tool used for analyzing the emotion in twitter along with the gender and location. It is capable of capturing and processing nearly 45,000 tweets per minute. Investigation is made whether emotion related information obtained from twitter can be processed in real time, in research of mental health. WeFeel along with Amazon web service (AWS) obtains data from garden hose and emo hose, live streams of social media twitter [10].

2.2.3 Text Mining

Twitter plays a central role in dissemination, utilization and creation of news. The survey says nearly 74% of people daily depend on twitter for news. News from twitter and social media plays a predominant role in which neither domain's cannot be grasped or reached [11]. Even news editor completely depends on the social media

Table 1 Methodology, purpose and prediction

Methodology	Dataset	Prediction	Purposes
Factor graph model using CNN [15, 23]	Twitter – Tweet level – User level	Based on emotions – Angry – Disgust – Sad – Joy	Health care analysis for diseases prediction
Ontology model [3, 7, 11]	Sporadic social network	Based on personal information – Location – Health details – Preferences	Providing health alerts
Support vector machine with ANN [13, 27, 35]	Notifications from – Twitter – Facebook – Whatsapp	Emotion analysis	Stress prediction
Random forest [7, 12]	Twitter – Hash tags	News articles	Health care analysis
Rule based method [8, 25]	Twitter	– Stress prediction based on emotions	Sentimental analysis

for audience attention. Hashtag maps news article to the topics of content as keyword based tags, in which tweets are labeled for news stories [12].

Table 1 describes about social interactions of elderly people for predicting stress from their updates. Methodologies are compared with their data collected from the social media only for healthcare analysis. Healthcare analysis is done for predicting disease, stress based on emotions and monitoring their regular sport.

3 Emotion and Stress Value Analysis

Emotion is conscious experience characterized by intense mental activity and a certain degree of pleasure or displeasure. Stress affects body, thoughts, and feelings even behavior too. When stress is unchecked for long time, leads to health issues causing high blood pressure, obesity, heart diseases, and diabetes.

3.1 Emotion Analysis

Stress is detected based on social interactions using factor graph model in convolutional neural network (CNN). User stress state is found by relating their social interactions and employing it into the real world platforms. Stress is defined based on social media attributes relating text and visual. Social media data reflects the states of life and emotions in well-timed manner [13, 14]. Moodlens is used to perform emotion analysis for emotions such as angry, disgust, sad, and joy [15]. Rapid miner software along with machine learning algorithm is used to predict the state of people emotion by the way they are notified [6]. To detect the emotion from text, an intelligent framework is used known as TEA (Text to Emotion Analyzer) [14]. Rule based method is used in extraction module for detecting emotion. Nine varieties of emotions are listed as anger, fear, guilt, interest, joy, sad, disgust, surprise, and shame.

3.2 Mapping Emotion to Stress

Feelings of citizen, perception and well-beings are known using smart cities given rise for the emotion aware city [5]. Smart city is an intelligent city, digital city, open city, and live city for social and information infrastructure, open governance and adaptive urban living respectively. Cities all over the world cover only 2% of earth's inhabited land area. City to be truly smart needs to determine, "what people are doing" along with "Why they are behaving in certain way". Mapping the emotion with stress builds cognitive, evaluative mapping and environmental preference and affects [16]. Mobile technology is used as a tool for collecting affective data for emotion aware city where communication group is created based on topic and location. By this stress is mapped with emotions [17, 18].

3.3 Stress Analysis

Well known stress reliever is listening to music and it is very much important for patients too when they are unable to communicate. Recommendation systems based on music uses collaborative method. Hence music selection is mainly done based on the physiological measurements of people [19]. Even heart rate measurements are linked with stress calculation. Heart rate measurement is noninvasive yet easy method for stress determination. Stress level can be relieved during music listening using wireless or portable music recommendation platform [4].

3.4 Healthcare Prediction (HCA)

The major problems faced by elderly people are poor eyesight, isolation, forgetting their needs, unable to keep up with daily chores, housekeeping, and poor nutrition.

Table 2 Physiological sign measurement and recommendations

	Blood pressure [4]		Heart rate [4, 19, 20]	
	Low	High	Normal	High
Range	80–89	120–139	72 times	>76 times
Recommendation	– Drink more water – Avoid coffee at night – Eat food with high salt	– Exercise regularly – Sleep well – Healthy diet – Do house hold chores to avoid isolation	– Maintain your current health style – Provide music playlist	– Chances of Heart attack – Breathe in for 5–8 s and hold for 3-5 s, exhale slowly – Creates music playlist

3.4.1 Disease Prediction

Stress detected from tweet content is used for disease prediction [15]. To improve the health of elderly people, regular, and suitable sports are necessary. To overcome problems like body functions decline and incidence of diseases [9]. According to the survey, more than 80% of elderly people aged 65 or above suffer from at least one chronic disease. Music therapy has been successfully applied for receiving scientific consideration [19]. Based on heartbeat rhythm disruption occurs and causes health problems like wooziness and fainting [20].

3.4.2 Sign Measurement

Elderly people forget minor things and fall sick easily. Wireless sensor network (WSN) is used to monitor health of elderly people and provide safe and secure living [4]. Sensor is placed on the waist of person to identify temperature, heartbeat and pressure sensor.

In Table 2, heart rate and blood pressure of elderly people are measured in continuous fashion. Based on their physiological signs, general recommendations are provided to keep their blood pressure and heart rate within the normal range. When the blood pressure and heart rate are very high and are not in stable state in such emergency cases alerts are send to care takers.

4 Recommendation System

Information filtering system predicts the "rating" or "preference" for an item user wish to give or provide feedback for further suggestions collectively a recommendation system.

4.1 Recommendation Based on Events

Recommendation is provided regarding events with exact place, data and time. Sporadic group is mainly for avoiding loneliness and social isolation among elderly people [3]. The recommendation provided is to do at least 150 min of exercising per week [9]. Hashtagger+ is the approach used based on information and opinion type of news article for providing recommendations to news articles from twitter. Learning to Rank (L2R) algorithm combines social streams with news in real time and works well in dynamic environment with accuracy [12].

4.2 Recommendation Based on HCA

CARE is a Context Aware Recommender system for Elderly that combines functionality of digital image frame with active recommender system. The suggestion may be to do physical, mental and social activities [21]. Motivate system maps the health related data for recommendations to be provided at right location and time. This method is more flexible and efficient. Some activities along with recommendations are "Do exercise" to avoid disease, "Do gardening or painting" to create positive attitude and so on [22].

4.3 Book Recommendation

Search engine is a common tool for information, nowadays. Data mining is used to extract the hidden information or to discover new knowledge [23]. Recommendation of books is provided based on collaborative filtering algorithm based on K-Means clustering and K-Nearest neighbor [24]. Lexical and syntactic analysis is used for extracting pages in the document text and natural language technology to filter the useful information about content [25, 26]. In book recommendation system, user can tag them and can record what they read or want to read and finally rate the book [13].

4.4 Content-Based Recommendation

Natural language processing is used for providing recommendations in precise and descriptive semantic way for enhancement in book recommendation engine (BRE). Contents in book are analyzed using Content Analyzer Module [27], where similar items are recommended to the user based on the likes in the past [28]. User based collaborative filtering method collects data and analyses large amount of information [29, 30]. Latent Dirichlet Allocation is a probabilistic model designed to discover

Table 3 Recommendation based on interest

Interest	Recommendation
Reading books [1, 30, 35]	– New books arrived – Similar books you read – Give feedback – Proceed books from page number
Hearing music [19, 20]	– Recommends playlist based on heart rate – Creates playlist
Exercise [21, 36]	– To avoid age related diseases
Gardening or painting [21, 37]	– Create positive influence – To refresh them
Watching TV [5, 21, 36]	– Remainder of shows

hidden themes in collection of document [31]. LDA includes three types of entity such as documents, words and topics or concepts [14, 32].

4.5 Music Recommendation

Heartbeat of people is maintained in normal range using new linear biofeedback web-based music system [33]. Music system recommends a generated playlist based on the user's preference [1, 17]. Markov decision process (MDP) is used to generate playlist when heartbeat is higher than normal range [20]. When heartbeat is normal, preferred music playlist of user is generated to maintain normal range [34].

Based on the interest collected for personal details, is the main parameter used for providing recommendations to the elderly people. Recommendations are based on interest which may include daily routines, past time and hobbies. Recommendations based on elderly people interest are tabulated in Table 3 with their hobbies and suggestions.

5 Machine Learning Algorithms Comparison

From the literature review discussed in the previous sections, various methodologies are identified and activity prediction is performed for detecting their stress level from their social interactions. Methodologies are compared and efficient method for providing recommendations to elderly people is obtained. The algorithms are compared and tabulated in Table 4 where C—Content, Am—Ambiguity, E—Efficiency, Ac—Accuracy, R—Recommendation, F—Feedback.

Table 4 Comparison of machine learning algorithms

Method	C	Am	E	Ac	R	F
Random forest [7, 12, 36]	Yes	No	High	High	Yes	No
SVM and ANN [13, 27, 35]	Yes	No	High	High	No	No
K-means, KNN [4, 23, 24, 35]	Yes	Yes	High	High	Yes	Yes
Factor graph model [13, 15, 27, 38]	Yes	Yes	High	High	No	Yes
Content and collabora-tive filtering [16, 21, 30, 31]	Yes	Yes	High	High	Yes	No
Active recommen-dation [11, 14, 21, 28, 29]	Yes	No	Low	High	Yes	Yes
MDP [26, 29, 33, 34, 39]	Yes	Yes	Low	Low	Yes	No
Hybrid approach [4, 30]	Yes	No	High	High	Yes	Yes

6 Conclusion

The work is to provide better recommendations for elderly people based on three major terms. First, personal details and interest of elderly people is gathered and trained. Second, health conditions such as blood pressure and heart beat is measured. Third, the social interactions are mined to determine elderly people's state of mind. Finally, depending on the unruffled information obtained, recommendations are suggested to elders. At last feedback is obtained from elders for the suggestions made and emergency case, an alert is send to care takers.

References

1. Praveena Mathew, M.S., Bincy Kuriakose, M.S.: Book recommendation system through content based and collaborative filtering method. IEEE Access (2016)
2. Manjumder, S., Aghayi, E.: Smart homes for elderly healthcare-recent advances and research challenges. IEEE Access (2017)
3. Ordonez, J.O., Bravo-Tores, J.F.: Stimulating social interaction among elderly people through sporadic social networks. In: International Conference on Device, Circuits and Systems, June 2017
4. Yuvaraj, N., Sabari A.: Twitter sentiment classification using binary Shuffled Frog algorithm. Auto Soft—Intell. Autom. Soft Comput. 1(1), 1–9 (2016) (Taylor and Francis (SCI)). https://doi.org/10.1080/10798587.2016.1231479(IF=0.77)
5. Choudhury, B., Choudhury, T.S., Pramanik, A., Arif, W., Mehedi, J.: Design and implementation of an SMS based home security system. In: IEEE International Conference on Electrical Computer and Communication Technologies, pp. 1–7. 5–7 Mar 2015
6. Kanjo, E., Kuss, D.J.: Notimind: utilizing responses to smart phone notifications as affective sensors. IEEE Access (2017)
7. Yuvaraj, N., Sripreethaa, K.R., Sabari, A.: High end sentiment analysis in social media using hash tags. J. Appl. Sci. Eng. Methodol. 01(01) (2015)
8. Afroz, N., Asad, M.-U.I., Dey, L.: An intelligent framework for text-to-emotion analyzer. In: International Conference on Computer and Information Technology, June 2016
9. Wu, H.-K., Yu, N.-C.: The development of a sport management and feedback system for the healthcare of the elderly. In: IEEE International Conference on Consumer Electronics, July 2017
10. Paris, C., Christensen, H., Exploring emotions in social media. In: IEEE Conference on Collaboration and Internet Computing, Mar 2016
11. Yuvaraj, N., Sripreethaa, K.R.: Improvising the accuracy of sentimental analysis of social media with hashtags and emoticons. Int. J. Adv. Eng. Sci. Technol. 4(3) (2015). ISSN 2319-1120
12. Shi, B., Poghosyan, G. Hashtagger+: efficient high-coverage social tagging of streaming news. IEEE Tran. Knowl. Eng. 30(1) 43–58 (2018)
13. Chen, M., Hao, Y., Hwang, K., Wang, L., Wang, L.: Disease prediction by machine learning over big data from healthcare communities. IEEE Access (2017)
14. Shamim Hossain, M., Rahman, M.A., Muhammad, G.: Cyber physical cloud-oriented multi-sensory smart home framework for elderly people: an energy efficiency perspective. Parallel Distributing Comput. J. (2017)
15. Lin, H., Jia, J.: Detecting stress based on social interactions in social networks. IEEE Trans. Data Sc. Eng. (2017)
16. Moreira, T.H., De Oliveira, M.: Emotion and stress mapping. IEEE Access (2015)
17. Van Hoof, J., Demiris, G., Wouters, E.J.M.: Handbook of Smart Homes, Health Care and Well-Being. Springer, Switzerland (2017)
18. Yuvaraj, N., Sabari, A.: Performance analysis of supervised machine learning algorithms for opinion mining in e-commerce websites. Middle-East J. Sci. Res. 1(1), 341–345 (2016)
19. Shin II, H., Cha, J., Automatic stress-relieving music recommendation system based on photoplethysmography-derived heart rate variability analysis. IEEE Access (2014)
20. Liu, H., Hu, J.: Music playlist recommendation based on user heart beat and music preference. In: International Conference on Computer Technology and Development, Dec 2009
21. Rist, T., Seiderer, A.: CARE-extending a digital picture frame with a recommender mode to enhance well-being of elderly people. In: International Conference on Pervasive Computing Technologies for Healthcare, Dec 2015
22. Yuvaraj, N., Gowdham, S.: Search engine optimization for effective ranking of educational website. Middle-East J. Sci. Res. 65–71 (2016)
23. Yassine, A., Singh, S., Alamri, A.: Mining human activity patterns from smart home big data for health care applications. IEEE Access (2017)

24. Venkatesh, J., Aksanli, B., Chan, C.S.: Modular and personalized smart health application design in a smart city environment. IEEE Internet Things J. (2017)
25. Pramanik, M.I., Lau, R.Y.K., Demirkan, H., Azad, M.A.K.: Smart health: big data enabled health paradigm within smart cities. Expert Syst. Appl. (2017)
26. Perego, P, Tarabini M., Bocciolone, M., Andreoni G.: SMARTA: smart ambiente and wearable home monitoring for elderly in internet of things. In: IoT Infrastructures, pp. 502–507. Springer, Cham, Switzerland (2016)
27. Van Hoof, E., Demiris, G., Wouters, E.J.M.: Handbook of Smart Homes, Health Care and Well-Being. Springer, Basel, Switzerland (2017)
28. Yuvaraj, N., Menaka K.: A survey on crop yield prediction models. Int. J. Innov. Dev. **05**(12) (2016). ISSN 2277-5390
29. Yuvaraj, N., Shripriya R.: An effective threshold selection algorithm for image segmentation of leaf disease using digital image processing. Int. J. Sci. Eng. Technol. Res. **06**(05) (2017). ISSN 2278-7798
30. Melean, N., Davis, J.: Utilising semantically rich big data to enhance book recommendations engines. In: IEEE International Conference on Data Science and Systems, January 2017
31. Yuvaraj, N., Menaka K.: A survey on crop yield prediction models. Int. J. Innov. Dev. **05**(12) (2016) ISSN:2277-5390
32. Yuvaraj, N., Gowdham, S., Dinesh Kumar, V.M., Mohammed Aslam Batcha, S.: Effective on-page optimization for better ranking. Int. J. Comput. Sci. Inf. Technol. **8**(02), 266–270 (2017)
33. Yuvaraj, N., Dharchana, K., Abhenayaa, B.: Performance analysis of different text classification algorithms. In. J. Sci. Eng. Res. **8**(5), 18–22 (2017)
34. Gnanavel, R., Anjana, P.: Smart home system using a wireless sensor network for elderly care. In: International Conference on Science, Technology Engineering and Management, Sept 2016
35. Zhu, Y.: A book recommendation algorithm based on collaborative filtering. In: 5th International Conference on Computer Science and Network Technology, Oct 2017
36. Yuvaraj, N., Sabari, A.: An extensive survey on information retrieval and information recommendation algorithms implemented in user personalization. Aust. J. Basic Appl. Sci. **9**(31), 571–575 (2015)
37. Yuvaraj, N., Nithya, J.: E-learning activity analysis for effective learning. Int. J. of Appl. Eng. Res. **10**(6), 14479–14487 (2015)
38. Yuvaraj, N., Sugirtha, D., John, J.: Fraud ranking detection in ecommerce application using opinion mining. Int. J. Eng. Technol. Sci. **3**(3), 117–119 (2016)
39. Yuvaraj, N., Emerentia, M.: Best customer services among the ecommerce websites—A predictive analysis. Int. J. Eng. Comput. Sci. **5**(6), 17088–17095 (2016)

M. Janani ME Student, Department of Computer Science and Engineering, KPR Institute of Engineering and Technology, Coimbatore, India, carrying out a project in Data analysis under the guidance of Dr. N. Yuvaraj and published papers in the field of networks, participated in workshops and international conferences. Area of interest in data analysis, peer networks and distributed computing.

N. Yuvaraj Associate Professor, Department of Computer Science and Engineering, KPR Institute of Engineering and Technology, Coimbatore, India. He has completed his ME in Software Engineering. He has completed his Doctorate in the area of Data Science. He has 2 years of Industrial Experience and 8 years of teaching experience. He has published over 21 technical papers in various International Journals and conferences. His areas of interest include data science, sentimental analysis, data mining, data analytics and information retrieval, distributed computing frameworks.

Performance Optimization of Hypervisor's Network Bridge by Reducing Latency in Virtual Layers

Ponnamanda China Venkanna Varma, V. Valli Kumari
and S. Viswanadha Raju

Abstract Multi tenant enabled cloud computing is the de facto standard to rent computing resources in data centers. The basic enabling technology for cloud computing is hardware virtualization (Hypervisor), software defined networking (SDN) and software defined storage (SDS). SDN and SDS provision virtual network and storage elements to attach to a virtual machine (VM) in a computer network. Every byte processed in the VM has to travel in the network, hence storage throughput is proportional to network throughput. There is a high demand to optimize the network throughput to improve storage and overall system throughput in big data environments. Provisioning VMs on top of a hypervisor is a better model for high resource utilization. We observed that, as more VMs share the same virtual resources, there is a negative impact on the compute, network, and storage throughput of the system because the CPU is busy in context switching (Popescu et al. https://www.cl.cam.ac.uk/techreports/UCAM-CL-TR-914.pdf [1]). We studied KVM (Hirt, KVM—The kernel-based virtual machine [2]) hypervisor's network bridge and measured throughput of the system using benchmarks such as Iometer (Iometer, http://www.iometer.org/ [3]), (Netperf, http://www.netperf.org/netperf/NetperfPage.html [4]) against varying number of VMs. We observed a bottleneck in the network and storage due to increased round trip time (RTT) of the data packets caused by both virtual network layers and CPU context switches (https://en.wikipedia.org/wiki/Contextswitch [5]). We have enhanced virtual network bridge to optimize RTT of data packets by reducing wait time in the network bridge and measured 8, 12% throughput improvement for network and storage respectively. This enhanced network bridge can be used in production with explicit configurations.

P. China Venkanna Varma (✉)
OpsRamp Inc., Kondapur, Hyderabad, Telangana, India
e-mail: pc.varma@gmail.com

V. Valli Kumari
AP State Council of Higher Education, Tadepalli, Andhra Pradesh, India
e-mail: vallikumari@gmail.com

S. Viswanadha Raju
JNTUH College of Engineering, Jagitial, India
e-mail: svraju.jntu@gmail.com

© Springer Nature Singapore Pte Ltd. 2019
J. D. Peter et al. (eds.), *Advances in Big Data and Cloud Computing*,
Advances in Intelligent Systems and Computing 750,
https://doi.org/10.1007/978-981-13-1882-5_27

Keywords Hypervisor · TCP round trip time · Netperf · Iometer
Context switching · Software defined storage · Software defined network

1 Introduction

Virtualization technology became de facto standard for all data center requirements. Virtualization technology provided hardware equivalent components for compute, storage, and network to build the VMs with desired configurations. Multiplexing VMs on a shared cluster of physical hardware using hypervisor is a better resource utilization model adopted in all data centers. Hypervisor introduced some virtual layers to inter-operable with existing guest operating systems (OSs), such virtual layers add a significant overhead and effects the throughput of the system [6]. All the storage is provisioned from a separate hardware called "storage array" which is connected to hypervisor in a network. Every disk input/output (IO) operation on a byte associated to persistent storage should be traversed in the network to read/write the byte stream from/to the storage array. So, the storage throughput is also proportional to the network throughput. Our research is limited to optimization of network throughput.

Virtual network bridge [7] in a hypervisor is a component that provides a networking gear to all VMs. We evaluated the network and storage throughput using Netperf and Iometer benchmarks with respect to the number of VMs sharing the same hypervisor. We found a bottleneck in the network bridge due to increased data packet RTT caused by more CPU context switches [8]. We have studied KVM's network bridge operating model and derived a method to enhance the network throughput by eliminating some of the delays caused by the modules developed to emulate hardware components. We found that the Transmission Control Protocol (TCP) acknowledgement (ACK) message can be sent in advance to improve the throughout without effecting the guest OSs running in the VMs. We have extended the KVM's network bridge with a module named in-advance TCP ACK packet (iTAP) to manage in-advance ACK of in-order packets. With iTAP implementation we have measured 8, 12% throughput improvement for network and storage respectively and can be used in production with explicit configurations.

The organization of this paper is as follows: (1) Explore KVM's network bridge functionality (2) Testbed and benchmarks (3) Bottlenecks in network and storage throughput (4) Enhanced network bridge with iTAP (5) Future work and limitations (6) Conclusion.

2 KVM's Network Bridge Functionality

Red hat kernel virtual machine (KVM) hypervisor uses the Linux network bridge as networking component which implements a subset of the ANSI/IEEE 802.1d [9] standard and integrated into 2.6 kernel series. The major components of a network

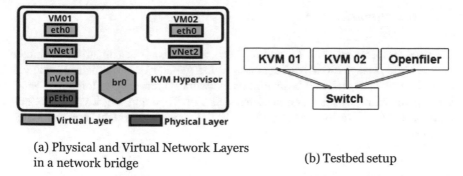

(a) Physical and Virtual Network Layers
in a network bridge

(b) Testbed setup

Fig. 1 **a** Physical and virtual network layers in a network bridge. **b** Testbed setup

bridge are a set of ports used to forward traffic between end-hosts and other switches, the control plane of a switch is typically used to run the Spanning Tree Protocol (STP), the forwarding plane is responsible for processing and forwarding input frames.

KVM network bridge uses two virtual layers, a back-end-pipe element and a frontend-pipe element for each VM. Figure 1a presents an overview of virtual layers in the network bridge. These two elements communicate each other using shared memory and ring buffers. Figure 1a describes virtual layers in a network bridge. A ring buffer holds all IO operations of a specific VM. As soon as a packet arrived at the hypervisors physical NIC, the network bridge analyzes the same packet, determines the receiver VM and send the packet to respective back-end-pipe for processing. Once all incoming packets are processed and placed the response packets, the frontend-pipe will be notified by back-end-pipe to receive packets.

3 Experiment Setup

3.1 Hardware

Installed Red hat KVM hypervisor as part of CentOS 7×64 on two Dell Power Edge servers with Intel Core i7 7700 K, 192 GB RAM, a Gigabit Ethernet Network Interface (NIC) and 1 TB 7.2 K RPM Hard disk drive (HDD). Installed Openfiler version 2.99 on a Dell Power Edge servers with Intel Core i7 7700 K, 192 GB RAM, a Gigabit Ethernet Network Interface (NIC) and 1 TB 7.2 K RPM Hard disk drive (HDD). All three hardware boxes are connected to a 10 GBps local LAN using a network switch "Dell Switch 2808". We used CentsOS $7\text{-}1611 \times 64$ operating system with linux kernel 3.10.0 as a VM (referred as CentOS7 VM). Each VM provisioned with 2 vCPUs, 3 GB of RAM, 128 GB HDD, 1 vNIC (e1000) and 1 vCD-ROM. We used Vagrant as VM orchestration and Ansible as configuration tools to configure the VMs for different benchmarks. Figure 1b describes the testbed setup.

3.2 Benchmarks

We used Netperf version 2.7.0 and Iometer version 1.1.0 benchmarks to evaluate network and storage throughput against varying number of VMs sharing same virtual resources. Netperf used to analyze the network IO throughput. Iometer is an IO subsystem characterization and measurement tool. Tcpdump tool was used to analyze the TCP RTT ready and wait time (RWT) to compute the CPU context switch overhead.

3.2.1 Netperf

All the tests runs for 6 min. Results present the average across five runs. We considered TCP STREAM, TCP RRP (Request and Response Performance) key performance indicators (KPIs). For tests TCP STREAM, TCP RRP (Request and Response Performance) netserver is running on the target VM, netperf is running on a collector VM. System firewall should be configured to allow TCP ports 12,865 and 12,866. Used n¨etserver¨ command to run the netserver on the collector. Used netperf H Collector IP address to run netperf on the target VM. Figure 2a describes the benchmark results.

3.2.2 Iometer

We considered Maximum throughput latency (ms) KPI and measured for 100% Read and 50% Read IO transactions. Figure 3a describes the benchmark results.

3.2.3 Tcpdump

Tcpdump [10] tool used to measure the TCP RTT and TCP RWT of each TCP flow in the experiment. Tcpdump command run on the VMs marked as targets and measured

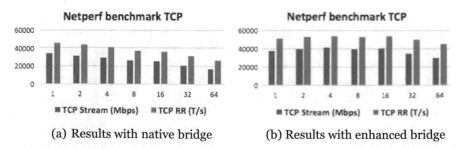

(a) Results with native bridge   (b) Results with enhanced bridge

Fig. 2 Netperf TCP Stream and RRP benchmark results

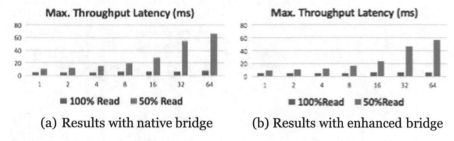

(a) Results with native bridge   (b) Results with enhanced bridge

Fig. 3 Iometer—Latency (ms) benchmark results

from the local VM. We used Wireshark [11] tool to analyze the report the RTT and RWT values. All the tests runs for 5 min and took the average across five runs. Figure 5a describes the TCP RTT and RWT test results.

4 Identified Bottlenecks in Network and Storage Throughput

All the benchmarks were executed at well-utilized system performance between 55 and 85%. We configured the system with no packet loss no firewall. Each test case sampled 3–5 times and took average value. All benchmarks were evaluated with varying number of VMs from 1, 2, 4, 8, 16, 32, and 64 having same configuration.

5 Enhanced Network Bridge to Improve the Network and Storage Throughout

We have implemented a small module named in-advance TCP ACK packet (iTAP) resides in the virtual network bridge and performs in advance TCP ACK for the VMs. The architecture and implementation of the iTAP enables no changes to the operating system in the VM. iTAP listens to all the packets (incoming and outgoing) and determines a state for the in-advance ACK decision flow. iTAP persists the state of TCP connection for each flow to decide whether to send in-advance ACK for packets or not. iTAP has intelligence to make sure no ACK for a packet that never reached the receiver VM.

State machine of the iTAP described in Fig. 4a. The iTAP algorithm persists the state of each TCP flow and decides whether to send the TCP ACK or not. iTAP uses a small data structure to hold the information of the each TCP flow and current state of the flow. The iTAP state machine is a light weight program consumes less memory and CPU. iTAP performs in-advance ACK for in-order packets if the mode is Active. iTAP discards all the void ACKs from target VM to avoid duplicate ACKs and loops

Fig. 4 iTAP state machine

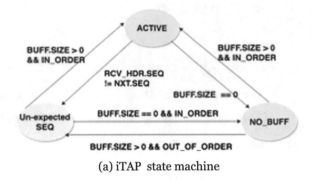

(a) iTAP state machine

in the state machine. If there are no slots in the TCP window, then the mode of the iTAP will be in either no-buffer or out of order. In this state sender never send any packets, till it gets an ACK from target. If the TCP window is full and network bridge gets its next CPU slot then all the packets in the TCP window will be processed in one shot, then the mode will be changed to Active. The sequence numbers will adjusted to match next packet and start sending in-advance ACK for all next in-order packets. If the network bridge detects out-of-order packet, it be processed out side of the iTAP, in such cases iTAP mode will be changed to out of order for that TCP flow. iTAP always tries to utilize the maximum TCP window size to improve the throughput of the network transactions.

To follow TCP rules, iTAP make sure no packet loss between virtual layers in the VM and network bridge based on the following conditions: (1) Transportation of TCP packet between virtual layer in VM and network bridge is a guaranteed memory copy operation. (2) iTAP never cross the TCP window buffer size, it will send ACK packets if and only if there is at least one slot in the window. (3) iTAP does not requires extra memory as it uses the maximum TCP window buffer size and wait till the buffer availability before processing next set of packets. All these conditions ensure guaranteed delivery of the TCP packets from network bridge to VM.

We measured same KPIs again with the same benchmarks with iTAP implementation. Figures 2b and 3b describes the network throughput is improved along the number VMs on the same hypervisor. Figure 5b describes that TCP RTT has been reduced with iTAP implementation. At the same time RWT is not changed because the CPU scheduling is same for all the experiments. We have measured 8, 12% throughput improvement for network and storage respectively.

6 Conclusion

Hypervisor is the key driver of the cloud computing. Too many abstract layers in the Hypervisor adds significant overhead to the overall throughput of the system. Based on the experiment results, we observed a bottleneck in the network bridge.

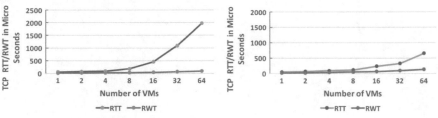

(a) Results with native bridge   (b) Results with enhanced bridge

Fig. 5 TCP ACK RTT and RWT results

Even though the data was ready in the back-end-pipe and/or frontend-pipe (i.e., kernel/application) in the VM were not ready to read the data due to wait for the CPU context switch. Most of the time round trip time (RTT) of the TCP packet was increasing due to CPU context switching, which leads to low throughput.

We have implemented a small module named in-advance TCP ACK packet (iTAP) resides in the virtual network bridge and performs in-advance TCP ACK on behalf of the VMs. iTAP listens to all the packets (incoming and outgoing) to maintain a state and send in-advance ACKs for all in-order packets. We have measured 8, 12% throughput improvement for network and storage respectively. This enhanced network bridge can be used in production with explicit configurations. This solution can not handle "out of order" packets, we can enhance the algorithm with pre-configured extra memory buffers to hold and acknowledge the "out of order" packets as a future research work.

References

1. Popescu, D.A., Zilberman, N., Moore, A.W.: https://www.cl.cam.ac.uk/techreports/UCAM-CL-TR-914.pdf
2. Hirt, T: KVM—The kernel-based virtual machine
3. Iometer: http://www.iometer.org/
4. Netperf: http://www.netperf.org/netperf/NetperfPage.html
5. https://en.wikipedia.org/wiki/Context switch
6. China Venkanna Varma, P., Venkata Kalyan Chakravarthy, K., Valli Kumari, V., Viswanadha Raju, S.: Analysis of a network IO bottleneck in big data environments based on docker containers
7. Nuutti V.: Anatomy of a linux bridge
8. Wu, Z.Z., Chen, H.C.: Design and implementation of TCP/IP offload engine system over gigabit ethernet
9. http://standards.ieee.org/about/get/802/802.3.html
10. Gupta, A.: A research study on packet sniffing tool TCPDUMP. Suresh Gyan Vihar, University, India:
11. Wireshark tool: https://www.wireshark.org

A Study on Usability and Security of Mid-Air Gesture-Based Locking System

BoYu Gao, HyungSeok Kim and J. Divya Udayan

Abstract To balance usability and security is an important aspect to be considered in any authentication systems including locking systems. Conventional authentication methods such as text and PINs passwords sacrifice security over usability, while freeform gesture passwords have been introduced as an alternative method, which sacrifices usability over security. In this work, the mid-air-based gesture authentication method for locking system is proposed, and the several criteria on discussion of its advantages over existing ones (PINs and freeform gesture-based methods) through the survey questionnaire was designed. We adopted the Multi-Criteria Satisfaction Analysis (MUSA) to analyze the user's satisfaction according to the proposed criteria. In addition, the correlation between participants' satisfaction and three aspects, age difference, gender difference, and education levels, were analyzed. The result revealed the better satisfaction on dimensions of security, use frequency and friendly experience in mid-air gesture authentication.

Keywords Usability · Security · Mid-air gesture-based authentication
Evaluation criteria

1 Introduction

Usability and security aspects of authentication method are contradictory terms yet complementary in human computer interaction (HCI) applications. Improvements in one end may negatively affect the other end. Even though, Kainda et al. [1], states that HCI researches evolved since 1975. Usability aspect is not balanced in secure

B. Gao
College of Cyber Security, Jinan University, Guangzhou, China

H. Kim
Department of Software, Konkuk University, Seoul, Republic of Korea

J. Divya Udayan (✉)
School of Information Technology and Engineering, VIT University, Vellore, India
e-mail: divya.udayan@vit.ac.in

© Springer Nature Singapore Pte Ltd. 2019
J. D. Peter et al. (eds.), *Advances in Big Data and Cloud Computing*,
Advances in Intelligent Systems and Computing 750,
https://doi.org/10.1007/978-981-13-1882-5_28

systems was stated in 2004 by Balfanz et al. [2]. From the literature, it is evident that security is aimed to prevent actions which are undesirable while usability is aimed to ease actions which are desirable. Users are the key focus for any HCI applications. Users themselves prove to be the greatest security threat to such systems such as phishing attacks pre-texting attacks, etc. This type of security breaches and human hacking are due to unsecure information sharing that are unsafe from detrimental effects when handled by frauds and hackers.

Whitten and Tygar [3] raised certain questions that need to be satisfied for usable secure systems

(i) *"users are aware of the security tasks needed to operate the system;"*
(ii) *"users know to perform the tasks successfully;"*
(iii) *"users do not attempt for unsafe errors;"*
(iv) *"users are comfortable with the interface and recommend using it."*

Usable authentication has become an important research domain due to the increasing usage of authentication methods in secure systems in our day-to-day activities. User-authentication using passwords is the most deeply discussed topics in information security. There are many solutions suggested so far to overcome issues in user-authentication. Some of them are, passphrases, pass-algorithm, cognitive passwords, graphical passwords, and so on. The limitations of the mentioned techniques are high rate of predictability and forgotten passwords.

In this work, the mid-air gesture authentication method is proposed and discussed with survey questionnaire, in which the several criteria on the aspect of security and usability are designed. We found that better satisfaction on dimensions of security, use frequency and user friendliness in mid-air gesture authentication than conventional PIN number and freeform gestures on touch-screens, through Multi-Criteria Satisfaction Analysis (MUSA) method [20].

The rest of the article consists of literature survey in Sect. 2 that summarizes the observations of current authentication methods. Section 3 shows the proposed method on mid-air authentication method in locking system. The result analysis and discussion of user study is presented in Sect. 4. Section 5 concludes the summary of this work.

2 Literature Survey

User-authentication methods can be categorized as object-based-, graphics-based-, and ID-based. Object-based authentication systems are characterized by physical possessions. This is the most traditional methods in which the user has his own physical lock and key method to lock and unlock secure items. With technology advancement, object-based authentication systems use a token which can be a pin number instead of physical lock and key [4, 5]. The main drawback of this system is that users must remember the token whenever they want to unlock a secure system.

Fig. 1 Free-form gesture
based authentication

Forgotten passwords and stolen tokens are also major security limitations object-based systems [6].

Text-based password authentication systems are widely used nowadays. The limitation of such type of systems is predictability and easy to crack [6, 7]. As we make user passwords more user-friendly, security vulnerability is higher [8]. In [9] a graphical password method to overcome usability and security issues was introduced. Figure 1 shows such a system. But recent research [10, 11] shows that even graphical password authentication methods have certain security limitations.

ID based authentication methods are unique to a particular user. Biometric methods like fingerprint, signature, facial recognition, voice recognition and so on. Also, driving license, Aadhar card can also be used as ID-based authentication methods. The major limitations of these authentication methods are cost and the replace difficulties [12]. Recently, touch ID has been used by mobile companies for unlocking mobile data introduced touch id based authentication method for mobile phones to secure user-sensitive data.

Another method has been motivated from the most natural way of expression and communication - using gestures. [13]. Recent research has widely used gestures has a secure authentication method [14, 15]. FAST [16] (Finger gesture authentication system using touch screen), has introduced a novel touch screen based authentication system for mobile devices. This system uses a digital sensor glove. Microsoft Kinect depth sensor is a recent method used to implement gesture based authentication systems for various other platforms. This system has used hand gesture depth and shape information to authenticate a user to access secure data. The authors suggested that hand motions is a renewable component of in-air hand gesture and therefore be easily compromised.

3 Proposed Methodology

According to the above literature review, we consider mid-air gesture-based authentication [17, 18] as the suitable methodology for balancing usability and security in locking applications. In order to apply mid-air gesture based authentication, guidelines are required for finding out the correct matching technologies to integrate and the most appropriate manner of authenticating using finger and hand gestures to provide higher usability. Understanding and experimenting usability and security aspects will be considered as the key foundation to develop such systems. As the major objective of this research, comparing the usability and security aspects of the researched method with other available usable authentication ones, using qualitative approaches, is required.

We propose mid-air gesture methodology as a solution to study the aspects of usability and security in locking systems. The authentication of the user is done by mid-air gestures in the room where the user can be sensed actively. There are some basic restrictions while designing the gestures such as, all designed gestures takes place in front of upper body as this space is perceived well, gesture should not be too large, fast or hectic, and gesture should not contain edges so that they are pleasant by spectators and the user, because the edges require sharp changes in the direction of the movement. With regard to these basic restrictions, possible movements of the hand and arm were evaluated, which are the basis for the designed gestures. The designed gestures are shown in Fig. 2.

The geometrical gestures in Fig. 2a–d are edgeless. The above gestures are designed to be performed in front of the user and parallel to the upper body. Some geometrical gestures for example in Fig. 2e are designed differently where they perform parallel to the floor in front of the user and they start in opposite direction of the other gestures. Such gestures are interesting for study.

The proposed method uses Leap Motion controller to detect the data required for user's hand gesture. Figure 3 shows the transfer of data (finger and hand movements) from the Leap motion to Raspberry Pi via PubNub. The information of the each of the hands and all fingers are generated by Leap Motion software so that the real time mirroring of the user's hand is recreated. The gesture of the hand enables the door in the prototype to be locked and unlocked. The test bed is shown in Fig. 4.

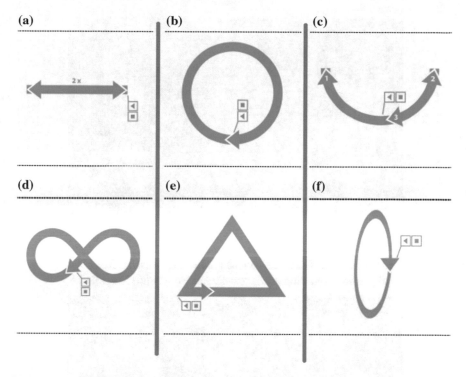

Fig. 2 Concept visualization of mid-air gesture methodology. **a** Left-right, **b** circle, **c** left-right-arc, **d** infinity, **e** triangle, **f** right-hand-rotation

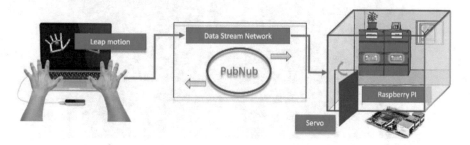

Fig. 3 Concept of simulating mid-air gesture based authentication method using leap motion and raspberry pi via PubNub communication

4 Method

In this work, we choose the Multi-Criteria Satisfaction Analysis (MUSA) method [20] to evaluate the user satisfaction and subjective experience. Basically, the MUSA model [20] was developed to measure customer's satisfaction from a specific product or service, however, the same principles can be used to measure global satisfaction

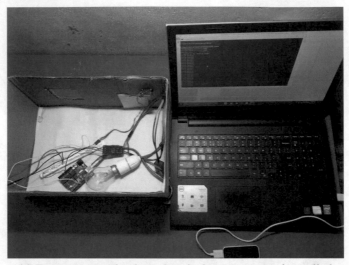

(a) Demonstrates the door closed when no gesture is applied.

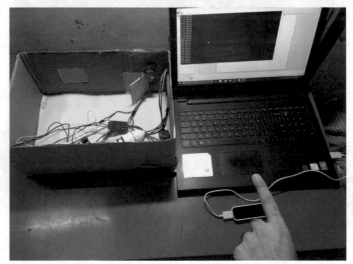

(b) Demonstrates the door open when user applies a hand gesture.

Fig. 4 Test bed of mid-air gesture methodology

of a group of individuals regarding a specific operation they interact with [19], for example, the authentication based gestures.

Based on our primary goal to study the aspects of security and usability in mid-air gesture based authentication for locking system, we developed a set of measures for this work.

Table 1 Overall participants

		Percent (%)
Gender	Male	60
	Female	40
Age	<18	10
	18–24	35
	25–34	10
	35–44	18
	45–54	8
	55–64	8
	65–74	6
	75+	5
Education	Doing high school	10
	High school	15
	Bachelor	35
	Master	15
	Ph.D.	25

4.1 Participants

The 70 participants from friends, family, and colleagues from the local campus were invited to participate in this study. The distribution of participant was shown in Table 1. Nearly half of the participants have background of computer science.

4.2 Questionnaires and Data Analysis

The evaluation questionnaires comprise of the aspects of two parts, security and usability, as shown in Table 2. Six dimensions on these two aspects were designed. The rating on aspects of usability and security were analyzed according to their beliefs.

4.3 Task Procedure

To analyze the performance of the proposed methodology, the online user studies were conducted. Three cases of locking systems were considered: (i) Pin number (ii) Freeform gestures (iii) Mid-air gesture. Three videos for the procedure of using each authentication method were captured for the three cases and uploaded online. Then, participants were asked to watch the three videos one by one and to answer the questionnaire given in Table 2. They were required to rate each criterion. A scale

Table 2 User study questionnaire

Serial No.	Questions
1. Memorability	*How easy will it be for user to remember the gestures?* *1—No effort; 2—Little effort; 3—Some effort; 4—Moderate effort; 5—Great effort*
2. Use frequency	*How often will you use it?* *1—No use; 2—Little use; 3—Some use; 4—Moderate use; 5—Great use*
3. Required training	*How much training does each case require?* *1—No training; 2—Little training; 3—Some training; 4—Moderate training; 5—Great training*
4. Security	*How secure is the method?* *1—Not secure; 2—Little secure; 3—Some secure; 4—Moderately secure; 5—Highly secure*
5. Friendly	*How user friendly is the experience?* *1—Not friendly; 2—Little friendly; 3—Some friendly; 4—Moderately friendly; 5—Very friendly*
6. Easy to learn	*How easy is it to learn?* *1—Not easy; 2—Little easy; 3—Some easy; 4—Moderately easy; 5—Very easy*

of 1–5 was given as the scoring criteria. In addition, the multiple responses from a single participant were not available with the help of recording the IP addresses of participants.

4.4 Result Analysis

The average scores on each question across 70 participants were shown in Fig. 5. Participants reported that the security of proposed method has the highest score of 4.72, while the complexity of these gestures made it difficult for participants to memorize, as average score of Q1 was the lowest at 3.48.

4.4.1 Satisfaction Analysis

The satisfaction depends on a set of criteria representing the features of usability and security in locking system. Satisfaction indices, in the range 0–100%, the level of partial satisfaction of the responders for the specific criterion, the calculation of satisfaction score was based on MUSA method [20]. Figure 6 summarizes the

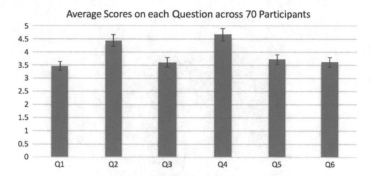

Fig. 5　Q4 represents the higher security, Q5 and Q6 give reasonable usability. Q2 shows that much more common use of this method

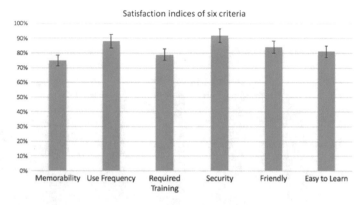

Fig. 6　Satisfaction indices of the six criteria

satisfaction indices of all criteria for the proposed method. From this graph, we notice that the three dimensions, security, use frequency, and friendly achieve relatively very high satisfaction indices, compared to the other three indices.

4.4.2　Criteria Weights

We calculated the weights of the criteria to further analyze the significance of each criterion in the overall satisfaction through MUSA. Figure 7 illustrates the weights for the six criteria. We found that there was no significant difference among these criteria.

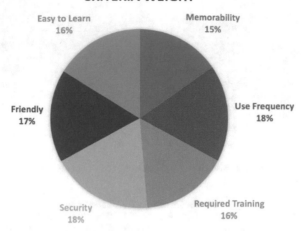

Fig. 7 The importance of each criteria

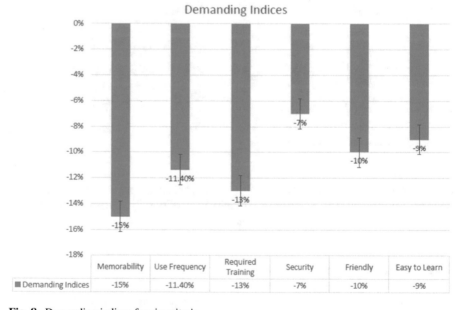

Fig. 8 Demanding indices for six criteria

4.4.3   Demanding Index

The demanding level regarding improvement effort is illustrated in Fig. 8. In particular, memorability is the dimension which is required demanding for improvement. Security is less demanding regarding the criterion with the highest level of importance.

Table 3 Chi-square test for gender

	Criteria	χ^2	df	Significance
Gender	Memorability	0.54	2	0.81
	Use frequency	1.14	2	0.76
	Required training	0.84	2	0.65
	Security	1.43	2	0.58
	Friendly	2.3	2	0.45
	Easy to learn	0.81	2	0.74

Table 4 Spearman's correlation between satisfaction and age difference

	Criteria	Spearman's rho	Significance
Age	Memorability	0.15	**0.024**[a]
	Use frequency	0.02	**0.003**[a]
	Required training	0.162	**0.03**[a]
	Security	0.076	**0.015**[a]
	Friendly	0.204	**0.02**[a]
	Easy to learn	0.762	0.127

[a]Bold font represents the significant difference

Table 5 Spearman's correlation between satisfaction and education level

	Criteria	Spearman's rho	Significance
Education	Memorability	0.12	0.76
	Use frequency	0.032	**0.023**[a]
	Required training	0.03	**0.03**[a]
	Security	0.43	0.26
	Friendly	0.325	0.47
	Easy to learn	0.42	0.25

[a]Bold font represents the significant difference

4.4.4 Correlation Between Satisfaction and Personal Demographic

The results of three aspects, different age group, gender difference, and education level on these questions, were analyzed. Complete demographics and the results are shown in Tables 3, 4 and 5.

(a) *Chi-square test for nominal demographic variables*
(b) *Spearman's correlation coefficient for demographic ordinal variables*

The demographic variables were: gender, age, education level. In terms of gender, a significant difference between each criteria and gender difference was not found, as shown in Table 3.

Regarding the correlation between age difference and criteria performance, the significant difference was found on all criteria with the exception of "Easy to Learn", as shown in Table 4.

For the effects of education on criteria evaluation, a significant difference was found on both use frequency and required training, as Table 5. While there was no significant difference on other four criteria.

4.4.5 Discussion

Regarding the easiness for user to remember the pin number, majority have voted that it is easy with little effort. Like the pin number or passwords, gesture methodology requires to remember exact gesture movements. Regarding the usage, field employees might need to use it more often than developers. With respect to training, like the biometrics methodology which requires multiple trials, 3d gesture might also require more trials. Regarding the security aspect, pin code methodology can be decoded with brute force. Even gestures, might be decoded but with little chance. The learning curve varies from user to user and with methods. It might not be possible to teach a complex use case to every employee. For example, it might be easier to teach 3D gestures and biometric to Peons and Janitor instead of complex passwords.

5   Conclusion

The findings from the thorough literature review along with user study were used to comprehend the current status of the proposed research work. After concluding the analysis, further investigation is required to make improvements to the process with the goal of identifying fundamental principles applicable to the development of mid-air gesture based authentication methods. Furthermore, the user study conducted does not provide feedback about the similarity. One of limitations is that the self-perception of similarity to estimate their ability to repeat the gesture for the participants using the methodology for the first time. And the objective evaluation that asks participants to perform the mid-air gesture base authentication will be conducted.

In future work, the refinement of the mid-air gesture-based authentication mechanism will be considered in multiple dimensions before it may be implemented successfully in daily life. Also, push-to-gesture-button approach can be another gesture direction that can be thought of by making small variations in the implemented approach so that more security can be provided while balancing usability.

Acknowledgements This work was partially supported by the Bio-Synergy Research Project (NRF-2013M3A9C4078140) of the Ministry of Science, ICT and Future Planning through the National Research Foundation and Cooperative R&D between Industry, Academy, and Research Institute funded through Korea Small and Medium Business Administration (Grants No.C0453564).

References

1. Kainda, R., Flechais, I., Roscoe, A.: Security and usability: analysis and evaluation. In: International Conference on Availability, Reliability, and Security, 2010. ARES'10, pp. 275–282 IEEE (2010)
2. Balfanz, D., Durfee, G., Grinter, R.E., Smetters, D.K.: In search of usable security: five lessons from the field. IEEE Secur. Priv. **5**, 19–24 (2004)
3. Whitten, A., Tygar, J.: Usability of security: a case study. DTIC Document, Tech. Report (1998)
4. Merchant, K., Cai, J., Maurya, S.: System and method for authenticating a smart card using an authentication token transmitted to a smart card reader., US Patent 8,522,326, 27 Aug 2013
5. Blakley III, G.R., Hinton, H.M.: Method and system for proof-of-possession operations associated with authentication assertions in a heterogeneous federated environment. US Patent 8,554,930, 8 Oct 2013
6. Gorman, L.O.: Comparing passwords, tokens, and biometrics for user authentication. Proc. IEEE **91**(12), 2021–2040 (2003)
7. Payne, B.D., Edwards, W.K.: A brief introduction to usable security. Internet Comput. IEEE **12**(3), 13–21 (2008)
8. Vu, K.P.L., Proctor, R.W., Bhargav-Spantzel, A., Tai, B.L.B., Cook, J., Schultz, E.E.: Improving password security and memorability to protect personal and organizational information. Int. J. Hum. Comput. Stud. **65**(8), 744–757 (2007)
9. Brostoff, S., Sasse, M.A., Are passfaces more usable than passwords? A field trial investigation. In: People and Computers XIV Usability or Else!, pp. 405–424. Springer (2000)
10. Davis, D., Monrose, F., Reiter, M.K.: On user choice in graphical password schemes. In: USENIX Security Symposium, vol. 13, pp. 11–11 (2004)
11. Thorpe, J., van Oorschot, P.C.: Graphical dictionaries and the memorable space of graphical passwords. In: USENIX Security Symposium, pp. 135–150 (2004)
12. Liu, J., Zhong, L., Wickramasuriya, J., Vasudevan, V.: uWave: accelerometer-based personalized gesture recognition and its applications. Pervasive Mobile Comput. **5**(6), 657–675 (2009)
13. Luff, P., Frohlich, D., Gilbert, N.G.: Computers and Conversation. Elsevier, Burlington (2014)
14. Liu, J., Zhong, L., Wickramasuriya, J., Vasudevan, V.: User evaluation of lightweight user authentication with a single triaxis accelerometer. In: Proceedings of the 11th International Conference on Human-Computer Interaction with Mobile Devices and Services, p. 15. ACM (2009)
15. Wu, J., Christianson, J, Konrad, J., Ishwar, P.: Leveraging shape and depth in user authentication from in-air hand gestures. In: 2015 IEEE International Conference on Image Processing (ICIP), pp. 3195–3199. IEEE (2015)
16. Feng, T., Liu, Z., Kwon, K.-A., Shi, W., Carbunar, B., Jiang, Y., Nguyen, N.K.: Continuous mobile authentication using touchscreen gestures. In: 2012 IEEE Conference on Technologies for Homeland Security (HST), pp. 451–456. IEEE (2012)
17. Aslan, I., Uhl, A., Meschtscherjakov, A., Tscheligi, M., Design and exploration of mid-air authentication gestures. ACM Trans. Interact. Intell. Syst. **6**(3) (2016) (Article 23)
18. Khamis, M., Alt, F., Hassib, M., von Zezschwitz, E., Hasholzner, R., Bulling, A.: GazeTouchPass: multimodal authentication using gaze and touch on mobile devices. In: Proceedings of the 2016 CHI Conference Extended Abstracts on Human Factors in Computing Systems, pp. 2156–2164 (2016)
19. Ipsilandis, P., Samaras, G., Mplanas, N.: A multicriteria satisfaction analysis approach in the assessment of operational programmes. Int. J. Project Manage. **26**(6), 601–611 (2008)
20. Muhtaseb, R., et al.: Applying a multicriteria satisfaction analysis approach based on user preferences to rank usability attributes in e-tourism websites. J. Theor. Appl. Electron. Commer. Res. **7**(3), 28–43 (2012)

Comparative Study of Classification Algorithm for Diabetics Data

D. Tamil Priya and J. Divya Udayan

Abstract Data mining techniques play a major role in healthcare centers to solve large volume of datasets. For diabetes patients if the blood glucose level diverges from typical range leads to serious complications. So, they must be monitored regularly to determine any critical variations. Implementing a predictive model for monitoring the glucose level would enable the patients to take preventive measures. This paper describes a solution for early detection of diabetes by applying various data mining techniques to generate informative structures to train on specific data. The main goal of the research is to generate clear and understandable pattern description in order to extract data knowledge and information stored in the dataset. We investigate the relative performance of various classifiers such as Naive Bayes, SMO-Support Vector Machine (SVM), Decision Tree, and also Neural Network (multilayer perceptron) for our purpose. The ensemble data mining approaches have been improved by classification algorithm. The experimental result shows that Naive Bayes algorithm shows better accuracy of 83.5% by splitting techniques (ST), when the data sets is reduced by 70–30 ratio percentage. By cross-validation (CV) decision tree shows better result 78.3% when compared with other classifiers. The experiment is performed on diabetes dataset at UCI repository in Weka tool. The study shows the potential of ensemble predictive model for predicting instance of diabetes using UCI repository diabetes data. The results are compared among various classifiers and accuracy of test results is measured.

Keywords Data mining · Diabetes · Classification algorithm · Naive bayes Support vector machine · Decision tree · Neural network

D. Tamil Priya (✉) · J. Divya Udayan
School of Information Technology and Engineering, VIT University,
Vellore, Tamil Nadu, India
e-mail: tamilpriya.d@vit.ac.in

J. Divya Udayan
e-mail: divya.udayan@vit.ac.in

© Springer Nature Singapore Pte Ltd. 2019
J. D. Peter et al. (eds.), *Advances in Big Data and Cloud Computing*,
Advances in Intelligent Systems and Computing 750,
https://doi.org/10.1007/978-981-13-1882-5_29

1 Introduction

In this era, the major health issue is diabetes and its consequences. Continuous monitoring and diagnosis of the level of glucose in the blood of the patient is essential. By using different learning approaches, we can predict diabetes of a patient and likewise improve the application of algorithms in changing the health care. Now days, diabetes is not only one of the most catastrophic diseases, but also originator of different types of diseases like heart attack, blindness, kidney diseases, etc. It is anticipated that more than half a million of children under the age of 14 suffer from diabetes of type 1 (Juvenile diabetes). In future, it is also anticipated that 318 million adults will have diminished glucose tolerance that situates them in developing more dangerous health conditions. By 2040, the number of the people living with diabetes disease will exceed to 642 million, if the risk of the disease is not under control. The normal identifying process of these diseases is that patients need to consult their doctor and necessity to investigate in the diagnostic center and they are spending more time in diagnosing the disease and getting their reports from hospitals/centers. They also need to have periodical communication with health specialists, and should have support for self-management of their vigorous condition [1, 7].

From previous literature [1], the machine learning approach has been proven to provide better proficiency in predicting the disease. Moreover, it is crucial to predict the disease as early as possible, before it becomes dangerous and leads to provide treatment to the patients. Additionally, it has the ability to mine large amount of data related to the diabetes dataset and extract the hidden knowledge of the data. Perpetually, it has a significant role in diabetes research. The aim of this work is to improve the model which can predict the risk of diabetic level of a patient with a higher accuracy. We also focus to develop an effective predictive model for detecting the diabetes at its primary phase.

This paper works towards various prediction measures to calculate the accuracy of diabetes data. Precision accuracy is computed by the correctly classified instance of data used. To discriminate the relationship between the data collection and their validity, kappa statistic shows the best classification among them.

The remainder of the paper is structured as follows. Section 2 describes related work of predictive models for classification. Section 3 presents the proposed methodology. Section 4 discusses the results and finally, concluding remarks and future work is given in Section 5.

2 Related Works

Data mining classification technique is used [1] to classify various kinds of data. Accordingly, the classification is performed based on the features of predefined set of classes. The detailed study of information about data mining techniques [1] is presented to emphasis on classification of supervised learning techniques. This paper also discusses performance analysis of classification techniques on different kinds of data by using Weka tool. A similar study and experiment is performed on different classification strategies [2] using three different data mining techniques, implemented to compare the performance accuracy of data. There has also been an investigation to measure the execution time of different classification techniques in order to calculate the performance measures for the large volume of data set. The diabetes data is utilized to check and validate the variations among different classification algorithms. Consequently, the technique applied to classification methods has the potential to progress the conventional methods on huge measures of data in a dramatic way, in bioinformatics or other applications of medical science. In [3] different classification techniques are used like Bayesian, NB (Naive Bayes), J48-pruned tree, REP function, Random-Forest, Random-tree, CART (classification and regression tree for machine learning) decision tree, KNN (K-Nearest neighbor) algorithm, and conjunctive rules learning method. Classification process is implemented on different algorithms [3], to measure, analysis the performance and also evaluate the diabetes dataset. A comprehensive study of assessment and performance analysis is done in this paper, which can be used for future research purpose. The analysis would be directed on the diabetes dataset starting with UCI repository. Similarly, it assists the people in recognizing the state of the diseases (illness), and also helps them to identify if they are having diabetics or not, based on the attributes value of data sets. Mining tools [4] resolve a lot of issues related with classification, clustering, association rule, and neural networks. This paper elaborates the classification techniques used by mining tool for large data set. This work also extends to achieve better attainment level of accuracy. Besides, it classifies the data to evaluate the performance of computational accuracy based on correctly and incorrectly classified instances. Moreover, it might use this kind of decision tree algorithms in medical, banking, stock market and various areas. The advantage of the Bayesian model [5] is that it can take into account the uncertainty and can get the scenarios of the system change for the evaluation of diagnosis procedures. In this study, various prediction models are compared and the results indicated that the Bayesian model is much more accurate in diabetes diagnosis.

3 Methodology

Classification algorithms are used to mine representative data and to predict the future data to estimate the accuracy of different predictive model. The benefit of this method is that it can take into account the ambiguity and can get the scenarios of the system change for the evaluation of diagnostic procedures. The detailed steps for classification techniques and their performance analysis for diabetes prediction are discussed.

A. *Classification Techniques*

Now, by using machine learning techniques, we can predict the diabetes in the Patients. We can learn more about how algorithm is used in health care to change dramatic way.

Preprocessing

In preprocessing, discretization, normalization, resampling, quality selection, conversion, also unification of attributes is done by utilizing product. A classification technique, such that association rule mining might additionally make performed with respect to unmitigated data. This assists performing discretization around numeric alternately constant qualities. We do not need to convert the information for classification of data. There need aid some few attributes that have to be changed over to those attributes to 0s and 1s, it sets, similarly as numeric. On other words, we will keep all of the values for the attributes in the data. This implies you might discretize towards eliminating the keyword "numeric" as and substitute it with the set of "binary" values by filtering mechanism [6].

Classification

Machine learning techniques like Decision trees and Lists, additionally instance based classifiers, support vector machines (SVM), Multilayer perceptions, Logistic Regression (LR), and Bayes nets can be accomplished by classification techniques. During classification, following testing process can be taken place: First, evaluates the classifier with respect to that number of the instances for predictions and then evaluates the classifier to predict these set of class about instances with set of a record. Then, evaluates these set of classifier by cross-validation by entering the "Folds" in field. Finally, evaluates the classifier to predict a specific rate of the individual data.

Clustering

By this model, learning similar group of class instances for clustering is accomplished by the given dataset. Moreover, approximation is done by clustering scheme of probability distribution.

Association

This is the learning association rule scheme to implement Apriori algorithm. It meets expectations is best for discrete information; furthermore it will recognize algebraic dependencies among aggregations about attributes assets. It works only with discrete data and will identify algebraic dependencies among groups of attributes.

B. *Performance Analysis of classification algorithms*

Naive Bayes Algorithm

Bayes principle for restrictive likelihood is used in this calculation. It uses every one of attributes encased in the information (data) and inspects them separately. The methodology to build Naive Bayes algorithm will be parallelized. This method defeats various constraints such as omission of composite iterative valuations of factor by means of it can make useful to a massive data set for real-time applications [1, 8].

Naive Bayesian classifier is dependent upon Bayes' hypothesis and the hypothesis about aggregate likelihood. The likelihood of data d with vector $V = \{s_1, s_2...s_n\}$ belongs to theory t is given in Eq. (1),

$$L(t_1|s_1) = \frac{L(s_1|t_1)L(t_1)}{L(s_1|t_1)L(t_1) + L(s_2|t_2)L(t_2)} \tag{1}$$

Here, $L(t_1|s_i)$ is posterior likelihood, whereas $L(t_1)$ is the prior likelihood accompanying the hypothesis t_1. For each different hypotheses t, the formula is given in Eq. (2),

$$L(S_i)^n = \sum_{i=1}^{n} L(s_i|t_j)L(t_j) \tag{2}$$

Thus we have Eq. (3),

$$L(t_1|s_1) = \frac{L(s_1|t_1)L(t_1)}{L(s_i)} \tag{3}$$

Support Vector Machine (SMO-SVM)

This algorithm may be regularly utilized to probing those quadratic programming issues that emerge throughout SVM learning process. SMO usage heuristics to segment those preparing issue under more diminutive issues that have an opportunity should be handled analytically. It replaces persistently all absent values furthermore transform insignificant features into double ones. Similarly, normalizes every one features by avoidance which aids in prompting the training process [6]. It is trained by resolving a quadratic programming problem, which is expressed in Eq. (4) as follows:

$$\max_{a} \sum_{i=1}^{n} a_i - \frac{1}{2} \sum_{i=1}^{n} \sum_{j=1}^{n} v_j f(u_i, u_j) a_i a_j \tag{4}$$

Eq. (5) is subjected to, $0 \le a_i \le S$, where $i = 1, 2... n$,

$$\sum_{i-1}^{n} v_i a_i = 0, \tag{5}$$

where S is a SVM parameter and $f(u_i, u_j)$ is the Kernel function and variable a_i are Lagrange multipliers.

J-48 Decision tree

Decision tree remains a tree structure which needs that manifestation of a flow chart. It could be used for classification techniques and method to build predictive model by using an illustration of nodes and internodes. Root and internal nodes (hub) would be the test cases for model and leaf nodes deliberated as outcomes (class variables). Dependent on those attribute values of the prevailing training data set, decision tree is created in order to classify new item. Every node of the tree is produced towards the calculation of the maximum information gain of all attributes [7].

Algorithm 1:
J48 - Decision Tree
Input:
Training set C;
Output:
A decision Tree T
Method:
Create a NodeDT (*N)
{
T = ϕ;
T = R; //label with splitting attribute
T = R <= R + a; //for each respective label and predicate split
Do
C = 'D 'data is created by predicate splitting to C;
When ending point is attained by this path, then
T' = L; //apt class is labeled
then
S' = BuildTD(C);
T = S' is added to a;
}

When it build a tree, J48-decision tree ignore the absent values. The ultimate idea overdue is the separations of records (data) into number of series established by attribute values designed for that item that present in the learning (training) model. J48 permits classification order by means of decision trees or rules generated by them [6, 8]. J48 decision tree structure is shown in the Fig. 1. Overhead, in the structure of tree, with a number of leaves and the number of nodes in the tree size of the tree. This yield provides for estimates of the tree's predictive performance, produced toward calculation unit. It outputs on anticipate the unpretentious population of the instances under the chosen test unit.

Algorithm 2:
Classification algorithm for predicting diabetes
Load t; //training data set
Preprocessing t;
Operation on dataset D:
if X ≤ Y is true: $\lambda(X) \le \lambda(Y) \in$ (D,A) is true; //Replacing Missing values.
$i = X_i - min(X) \ max(X) - min(X)$; //Normalization of values.

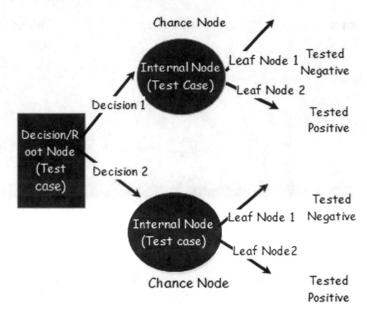

Fig. 1 J48 decision tree

Input:
D to the system; //diagnose the disease
Build model 'M';

   Train D;
   Test D by 'M';

Output:
Get R; //Evaluation result
Diagnoses result of predictive model is obtained.

Neural Networks (Multilayer Perceptron)

A neural system might a chance to be used to mean a nonlinear mapping concerning the input and output trajectories. Neural networks need aid those common signal-processing advances around every one. Similarly in engineering, neural networks help two critical functions: as pattern classifiers and as nonlinear adaptive filters. A common network consists of three layers; one is an input layer, another is more hidden layers and an output layer.

The multilayer perceptron (MLP) is an illustration of an artificial neural network that is used comprehensively should take care of amount of different problems, plus pattern recognition and interpolation. Each layer is made for neurons, which are intersected for one another (by weights), respectively for each neuron, a specific mathematical function so-called the activation function receives input from former

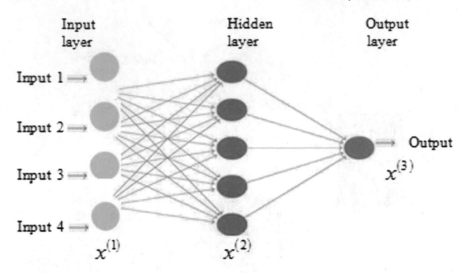

Fig. 2 A multilayer perceptron (neural network) diagram with input and output unit

layers and breeds output for the following layer. In the research, the actuation capacity utilized may be that hyperbolic tangent sigmoid transfer function [13].

Figure 2 signifies multilayer perceptron with three levels. The first (green) level, represented as $x^{(1)} = \left(x_1^{(1)}, x_2^{(2)}, x_3^{(3)} \right)^T$, termed as input layer since its nodes are designed by means covariate(features) $y = (y_1, y_2, y_3)$, so that $a^{(1)} = x$.

The second (blue) level, represented as $x^{(2)} = \left(x_1^{(2)}, x_2^{(2)}, x_3^{(2)} \right)^T$, termed as hidden layer, used to calculate the value of its nodes, but it cannot be viewed. $x^{(2)}$ is the component specified by a nonlinear function useful to a linear combination of the nodes of the former layer, therefore,

$$x_1^{(2)} = f\left(\Theta_{11}^{(1)} x_1^{(1)} + \Theta_{12}^{(1)} x_2^{(1)} + \Theta_{13}^{(1)} x_3^{(1)} \right)$$

$$x_2^{(2)} = f\left(\Theta_{21}^{(1)} x_1^{(1)} + \Theta_{22}^{(1)} x_2^{(1)} + \Theta_{23}^{(1)} x_3^{(1)} \right)$$

$$x_3^{(2)} = f\left(\Theta_{31}^{(1)} x_1^{(1)} + \Theta_{32}^{(1)} x_2^{(1)} + \Theta_{33}^{(1)} x_3^{(1)} \right),$$

where
$f(x) = 1/\left(1 + e^{-z} \right)$ is the sigmoid function.

The third (green) level, represented as $x^{(3)} = \left(x_1^{(3)} \right)$, termed as output layer since it yields the hypothesis function $g_\Theta(x)$, over which a nonlinear function is applied to a linear combination of the nodes of the former layer is calculated by Eq. (6),

$$g_\Theta(x) = x_1^{(3)} = f\left(\Theta_{11}^{(2)} x_1^{(2)} + \Theta_{12}^{(2)} x_2^{(2)} + \Theta_{13}^{(2)} x_3^{(2)}\right) \tag{6}$$

So, $x^{(j)}$ represents number of elements in level j, $x_n^{(j)}$ denotes the nth unit in level j and $\Theta^{(j)}$ represents the parameters matrix monitoring the mapping function from level j to level $j+1$.

Figure 3 shows the system architecture of the classification techniques for diabetes prediction, the classification algorithm applied in machine learning approach in diagnosing the disease to build the predictive model. The classification process of our proposed system is shown in Fig. 4 that indicates how execution phases of classification techniques applied to evaluate the performance of the algorithm.

Accuracy is calculated by; Tp is used to compute True positive rate, Fp for False positive rate, P for precision, F for F-measures and R for recall. Following Eqs. (7)–(13) are used as the evaluation measures for classification techniques:

$$\text{Accuracy} = \frac{\text{Tp} + \text{Fn}}{\text{Tp} + \text{Fp} + \text{Tn} + \text{Fn}} \tag{7}$$

$$\text{Positive} - \text{Precision} = \frac{\text{Fp}}{\text{Tp} + \text{Fp}} \tag{8}$$

$$\text{Negative} - \text{Precision} = \frac{\text{Fp}}{\text{Tn} + \text{Fn}} \tag{9}$$

$$\text{Error} - \text{rate} = \frac{\text{Fp} + \text{Fn}}{\text{Tp} + \text{Fp} + \text{Tn} + \text{Fn}} \tag{10}$$

$$F = \frac{2 * P * R}{(P + R)} \tag{11}$$

$$R = \frac{\text{Tp}}{\text{Tp} + \text{Fn}} \tag{12}$$

$$P = \frac{\text{Tn}}{\text{Fp} + \text{Tn}} \tag{13}$$

The equation to calculate mean absolute error and relative absolute error is given below.

The mean absolute error M_i of a specific program i is estimated by the Eq. (14):

$$M_i = \frac{1}{n} \sum_{j=1}^{n} |\alpha_{(ij)} - \beta_j| \tag{14}$$

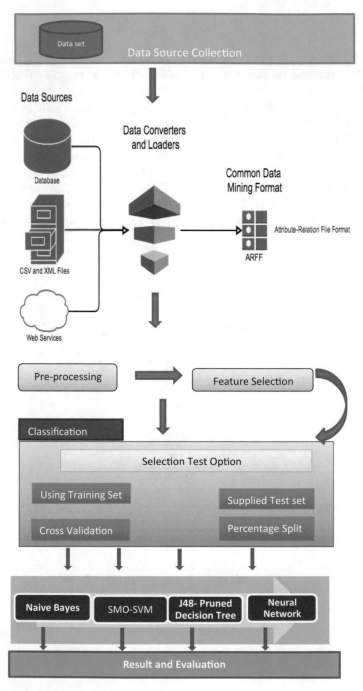

Fig. 3 System architecture of predictive models

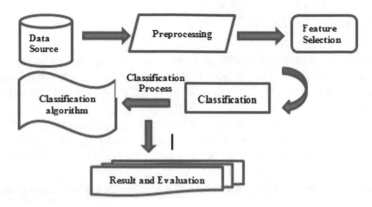

Fig. 4 Pipeline of the classification process executed in our proposed approach

Out of n sample cases, $\alpha_{(ij)}$ is the specific estimated value define by program i for sample case j, where β_j is the target value for sample case j. But $\alpha_{(ij)} = \beta_j$ and $M_i = 0$ is used for perfect fit. Therefore, M_i index which ranges from 0 to infinity, with value 0 resultant to be ideal.

The relative absolute error is fairly the average of real values which is related simple predictor. For instance, the error is the total absolute error in its place of the total squared error. As a result, the relative absolute error earns the total absolute error and normalize it by dividing the total absolute error of the simple predictor.

Statistically, the relative absolute error M_i of an individual program i is evaluated by the following Eq. (15),

$$M_i = \frac{\sum_{j=1}^{n} |\alpha_{(ij)} - \beta_j|}{\sum_{j=1}^{n} |\beta_j - \bar{\beta}|} \tag{15}$$

Out of n sample cases, where $\alpha_{(ij)}$ the specific value is estimated value define by specific program i for sample case j. β_j the target value for sample case j and $\bar{\beta}$ is given by Eq. (16):

$$\bar{\beta} = 1/n \sum_{i=1}^{n} \beta_j \tag{16}$$

The root mean squared error is also calculated which is the square root of Eq. (14) and Root Relative Squared Error is square root of Eq. (16).

4 Experimental Results

The data has been analyzed by different techniques to evaluate the performance of classification algorithm. Data set used in our research has been obtained from PIDD (Pima Indians Diabetes Database) and the NIDDKD (National Institute of Diabetes and Digestive and Kidney Diseases) at UCI.

The data set has 768 instances and 9 attribute; pregnant times, BP (blood pressure in mm Hg), thickness of skin fold (in mm), serum insulin (in mu U/ml), BMI (Body mass index in weight in kg/(height in m)^2), pedigree function of diabetes, Age (in years), outcome (Class variable (0 or 1)). We have done a case study of two methods for predicting diabetics. First, using reduction techniques and next using validation techniques. Details are as follows:

C. *Predicting Diabetes Using Reduction Techniques*

Classification is the technique used to extract semantic meaning from data or to predict new data. Classification involves two processes. First, the given data is trained by classification algorithms. Second, testing the input data to estimate its accuracy.

Experiments with Naive Bayes Algorithm

The technique of Naive Bayes classification algorithm is predominantly appropriate to high dimensionality of information source. Naive Bayes model identifies the characteristics diabetes disease of patients. The reason behind choosing the Naive Bayes or Bayes' Rule is the establishment of several machine learning and data mining approaches. The use of this algorithm is to create models that have an ability to predict level of the diseases. This model gives better results according to our findings [13]. Table 1 shows the evaluation summary of the test results using Naive Bayes algorithm, in which data are tested by reduction techniques, splits test like 40:60, 50:50, 60:40, 70:30 (in percentage) first part splits 40, 50, 60, 70% as trained set and second part splits are testing data used in the calculation of accuracy and bold numeric indicates, 70–30% splitting shows high accuracy when compared with other splitting process.

Experiments with SMO (sequential minimal optimization algorithm)—SVM (Support Vector Machines)

During the training of SMO (sequential minimal optimization algorithm)—SVM (support vector machine) the training data set are partitioned into smaller ones that can be computed mathematically and used to solve mathematical problem. We have used 40:60, 50:50, 60:40, 70:30% of split techniques to build the model using this algorithm. 40, 50, 60, 70% of the data set have been used to train the data and other 60, 50, 70, and 30% of the data set have been used to test the model.

The results of the various split percentage of test model of SMO-SVM is shown in the Table 2, that shows when we apply the reduced percent of data set in testing model shows good performance in test results and bold numeric indicates, 60–40% splitting shows high accuracy for SMO-SVM algorithm when compared with other algorithms.

Table 1 Evaluation summary of the test results using naive Bayes approach

Evaluation results	40:60	50:50	60:40	70:30
Correctly classified instances	82.6464%	83.0729%	83.0619%	**83.4783%**
Incorrectly classified instance	17.3536%	16.9271%	16.9381%	16.5217%
Kappa statistic	0.6125	0.6184	0.6167	0.6129
Mean absolute error	0.2662	0.26	0.2494	0.2435
Root mean squared error	0.3603	0.3583	0.3539	0.3461
Relative absolute error	58.24%	57.15%	54.94%	54.17%
Root relative squared error	76.09%	75.75%	74.57%	74.18%
Total number of instances	461 (381 + 80)	384 (319 + 65)	307 (255 + 52)	230 (192 + 38)

Table 2 Evaluation summary of the test results using support vector machine

Evaluation results	40:60	50:50	60:40	70:30
Correctly classified instances	79.6095%	79.6875%	**80.1303%**	80%
Incorrectly classified instance	20.3905%	0.3125%	19.8697%	20%
Kappa statistic	0.5312	0.5412	0.534	0.5051
Mean absolute error	0.2039	0.2031	0.1987	0.2
Root mean squared error	0.4516	0.4507	0.4458	0.4472
Relative absolute error	44.6113%	44.6481%	43.772%	44.4986%
Root relative squared error	95.3563%	95.2691%	93.9335%	95.8425%
Total number of instances	461 (367 + 94)	384 (306 + 78)	307 (246 + 61)	230 (184 + 46)

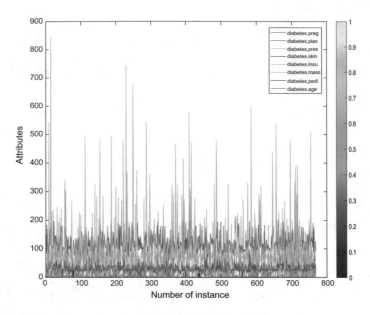

Fig. 5 Graphical representation of statistical result analysis of diabetes dataset

Performance measure of the naive Bayes algorithm by reduction techniques is depicted in the Table 3, measures like TP rate, FP rate, Precision, Recall, F-measure, MCC, ROC area, PRC area is computed based class outcome and bold numeric indicates which has high performance measures when compared with other splitting techniques and algorithms. The performance measure of SMO-SVM algorithm is shown Table 4, is computed based on class outcomes and high performance measures are indicated as bold numeric compared with others.

Figure 5 shows graphical representation of statistical result graph of training model of test results created by attributes of diabetes dataset. Total number of instances for various classification algorithms is calculated by confusion matrix, which analysis how fine the classifier can recognize the attributes of different classes.

Experiments with J48-Decision Tree

J48 classifier is a simple decision tree for classification. It builds a binary tree. Furthermost, convenient method in classification problem is the decision tree. With this method, a tree is created to model the classification process. Once the tree is built, it is applied to each tuple in the database and a result in classification for that tuple. J48 is an algorithm in machine learning technique [3]. Evaluation summary test results of J48-decision tree algorithm is shown in the Table 5 has high accuracy for 70–30% splitting techniques, indicated by bold numeric. Table 6, shows the evaluation summary test results of neural network (multilayer perceptron) of predictive model which achieve high accuracy rate for 50–50% spitting when compared with others.

Table 3 Performance measures of Naive Bayes algorithm by reduction technique (in %)

Reduction techniques	TP rate	FP rate	Precision	Recall	F-Measure	MCC	ROC area	PRC area	Class
40:60	0.866	0.252	0.872	0.866	0.869	0.612	0.883	0.932	tested_negative
	0.748	0.134	0.739	0.748	0.744	0.612	0.883	0.786	tested_positive
50:50	0.878	0.264	0.868	0.878	0.873	0.619	0.883	0.932	tested_negative
	0.736	0.122	0.754	0.736	0.745	0.619	0.883	0.782	tested_positive
60:40	0.891	0.286	0.857	0.891	0.874	0.618	0.886	0.933	tested_negative
	0.714	0.109	0.773	0.714	0.743	0.618	0.886	0.790	tested_positive
70:30	0.886	0.278	**0.875**	0.886	**0.881**	0.613	**0.890**	0.942	tested_negative
	0.722	0.114	0.743	0.722	0.732	0.613	0.890	0.794	tested_positive

Table 4 Performance measures of smo-support vector machine algorithm by reduction technique (in %)

Reduction techniques	TP rate	FP rate	Precision	Recall	F-Measure	MCC	ROC area	PRC area	Class
40:60	0.873	0.355	0.829	0.873	0.850	0.533	0.759	0.808	tested_negative
	0.645	0.127	0.719	0.645	0.680	0.533	0.759	0.583	tested_positive
50:50	0.855	0.318	**0.842**	0.855	0.848	0.541	**0.769**	0.816	tested_negative
	0.682	0.145	0.704	0.682	0.693	0.541	0.769	0.587	tested_positive
60:40	0.906	0.400	0.813	0.906	0.857	0.542	0.753	0.799	tested_negative
	0.600	0.094	0.768	0.600	0.674	0.542	0.753	0.598	tested_positive
70:30	0.905	0.431	0.822	0.905	**0.861**	0.513	0.737	0.809	tested_negative
	0.569	0.095	0.732	0.569	0.641	0.513	0.737	0.552	tested_positive

Table 5 Evaluation summary of the test results using of J48-decision tree algorithm

Evaluation results	40:60	50:50	60:40	70:30
Correctly classified instances	80.2603%	78.3854%	81.1075%	**81.3043%**
Incorrectly classified instance	19.7397%	21.6146%	18.8925%	18.6957%
Kappa statistic	0.5269	0.4992	0.5559	0.5355
Mean absolute error	0.3112	0.3193	0.2818	0.2929
Root mean squared error	0.384	0.4006	0.371	0.3719
Relative absolute error	68.0959%	70.1759%	62.0733%	65.1711%
Root relative squared error	81.0962%	84.6854%	78.1825%	79.6982%
Total number of instances	461 (370+91)	384 (301+83)	307 (249+58)	230 (187+43)

Table 6 Evaluation summary of the test results using of neural network (multilayer perceptron)

Evaluation results	40:60	50:50	60:40	70:30
Correctly classified instances	77.6573%	**78.9063%**	78.1759%	78.6957%
Incorrectly classified instance	22.3427%	21.0938%	21.8241%	21.3043%
Kappa statistic	0.4838	0.476	0.4755	0.4951
Mean absolute error	0.2928	0.2689	0.2669	0.2688
Root mean squared error	0.4199	0.3955	0.3832	0.3651
Relative absolute error	64.0614%	59.0965%	58.7985%	59.817%
Root relative squared error	88.6711%	83.6043%	80.7571%	78.24%
Total number of instances	461 (358+103)	384 (303+81)	307 (240+67)	230 (181+49)

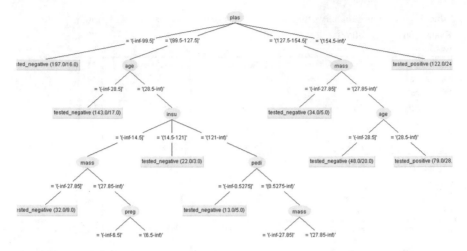

Fig. 6 Ontology of J48 decision tree algorithm for predictive model

Experiments with Neural Networks

Neural networks are one of the learning algorithms used in machine learning. It consists of different layers for evaluating and knowledge data. Every hidden layer strains to identify patterns of data. Every hidden layer strains to identify patterns of data. When a pattern is detected the next hidden layer is activated and so on. The more layers in a neural network, the more is learned and the more accurate the pattern is detected. Neural Networks learn and attribute weights to the connections between the different neurons each time the network processes data. Ontology of J48-decision tree algorithm is shown in Fig. 6.

Table 7 shows the performance measures of j48-decision tree algorithm, in which data are tested by reduction techniques, splits test like 40:60, 50:50, 60:40, 70:30 (in percentage) first part splits 40, 50, 60, 70% as trained set and second part splits are testing data used in the calculation of accuracy and high performance measures are indicated in bold numeric.

Table 8 shows the performance measures of neural network in which data are tested by reduction techniques, splits test like 40:60, 50:50, 60:40, 70:30 (in percentage) first part splits 40, 50, 60,70% as trained set and second part splits are testing data used in the calculation of accuracy by the class attributes and high performance measures are indicated in bold numeric.

D. Predicting Diabetes Using Validation techniques by machine learning approach

To evaluate the performance of diabetes dataset at UCI repository, the two modes of analysis techniques has been followed: (i) Cross-Validation of K-fold mode and (ii) Split percentage mode (reduction techniques). In K-fold Cross-Validation mode, the datasets are divided into K-disjoint unit of objects in random manner, then the machine learning algorithm is trained by K-1 blocks unit and the remaining blocks units are used to test the performance of algorithm.

Table 7 Performance measures of J48-decision tree algorithm by reduction technique (in %)

Reduction techniques	TP rate	FP rate	Precision	Recall	F-Measure	MCC	ROC area	PRC area	Class
40:60	0.918	0.426	0.810	0.918	0.861	0.539	0.828	0.863	tested_negative
	0.574	0.082	0.781	0.574	0.662	0.539	0.828	0.713	tested_positive
50:50	0.871	0.388	0.816	0.871	0.843	0.502	0.801	0.854	tested_negative
	0.612	0.129	0.705	0.612	0.656	0.502	0.801	0.653	tested_positive
60:40	0.916	0.390	0.819	0.916	0.864	0.566	0.859	0.894	tested_negative
	0.610	0.084	0.790	0.610	0.688	0.566	0.859	0.740	tested_positive
70:30	0.918	0.417	**0.829**	0.918	**0.871**	0.545	**0.831**	0.878	tested_negative
	0.583	0.082	0.764	0.583	0.661	0.545	0.831	0.699	tested_positive

Table 8 Performance measures of neural network by reduction technique (in %)

Reduction techniques	TP rate	FP rate	Precision	Recall	F-Measure	MCC	ROC area	PRC area	Class
40:60	0.863	0.394	0.812	0.863	0.837	0.486	0.797	0.882	tested_negative
	0.606	0.137	0.691	0.606	0.646	0.486	0.797	0.642	tested_positive
50:50	0.941	0.512	0.784	0.941	0.856	0.504	0.835	0.912	tested_negative
	0.488	0.059	0.808	0.488	0.609	0.504	0.835	0.685	tested_positive
60:40	0.916	0.476	0.787	0.916	0.847	0.492	0.856	0.917	tested_negative
	0.524	0.084	0.764	0.524	0.621	0.492	0.856	0.754	tested_positive
70:30	0.861	0.375	**0.834**	0.861	**0.847**	0.496	**0.875**	0.939	tested_negative
	0.625	0.139	0.672	0.625	0.647	0.496	0.875	0.704	tested_positive

The classification process is repeated k-times and at last, the recorded measures are averaged. In general, choose the K-fold as $K = 10$, otherwise, size of fold is depending on the size of original data sets used. In split percentage mode (reduction techniques), the datasets are divided into two disjoint units of datasets. The first set of units, in which machine learning algorithm tried to extract the data knowledge from so-called training set. The extracted data knowledge might be tested against the second set of data in which it is so-called test data.

In general, it has 66% units of the objects of original dataset as training set and remaining set of data as test set of objects. Formerly, when the tests is passed out by using the particular datasets and then by means of possible classification algorithm and test modes, the results are calculated and overall comparison is made over it. Evaluation summary test results of K-fold cross-validation mode ($K = 10$) of different classifiers is shown in Table 9, shows that J-48 decision tree classifier shows good accuracy results that are given by correctly classified instances and high test results are indicated by bold numeric.

The Comparison weighted average measures shown in Table 10 indicated that, for SMO-SVM the accuracy measures such as TP rate, FP rate, Precision, Recall, F-measure is good compared to other algorithms and MCC, ROC area, PRC area is good for Naive Bayes for predictive analysis model and it is denoted by bold numeric.

Figure 7. ROC area of diabetes in predictive model analysis. The time taken to build the model for classifier; Naive Bayes, SMO-SVM, J48-decision tree, and Neural network (Multilayer perceptron) is 0 s, 0.08, 0.02, and 1.11 s as execution time by validation techniques (cross-validation of fold $K = 10$). Comparison results of classification algorithm by reduction techniques is shown in the Fig. 8. Figure 9 depicts the performance measures of different classification algorithm by validation techniques ($k = 10$ folds).

In our work to evaluate the performance of classifiers by splitting (reduction) techniques, the split test of units is like 40:60, 50:50, 60:40, and, 70:30% are used. The time taken to build the model by execution time for Naive Bayes algorithm is 0.02 s and for remaining split it is 0 s. The time taken for Support vector machine is 0.06 s for 40:60 and 50:50% and for remaining test split it is 0.05 s. For decision tree to build the model, the execution time is 0 s for all split tests. This shows that, it is the good predictive techniques to build the model to analysis the performance data set. And for the neural network (multilayer perceptron), the execution time to

Table 9 Performance measures of classification algorithm by validation ($k = 10$ fold) techniques

Classification algorithm	TP rate	FP rate	Precision	Recall	F-Measure	MCC	ROC area	PRC area	Class
Naïve Bayes	0.830	0.317	0.830	0.830	0.830	**0.513**	**0.846**	**0.908**	tested_negative
	0.683	0.170	0.683	0.683	0.683	0.513	0.846	0.741	tested_positive
SMO–SVM	**0.872**	**0.407**	**0.800**	**0.872**	**0.834**	0.489	0.733	0.781	tested_negative
	0.593	0.128	0.713	0.593	0.648	0.489	0.733	0.565	tested_positive
J48-decision tree	0.864	0.369	0.814	0.864	0.838	0.510	0.801	0.845	tested_negative
	0.631	0.136	0.713	0.631	0.669	0.510	0.801	0.666	tested_positive
Neural networks	0.846	0.403	0.797	0.846	0.821	0.457	0.824	0.899	tested_negative
	0.597	0.154	0.675	0.597	0.634	0.457	0.824	0.694	tested_positive

Table 10 Evaluation summary of test results by validation techniques ($k = 10$) for classification algorithm

Evaluation results	Naive Bayes	SMO-SVM	J48-Decision Tree	Neural Network
Correctly classified instances	77.8646%	77.474%	**78.2552%**	75.9115%
Incorrectly classified instance	22.1354%	22.526%	21.7448%	24.0885
Kappa statistic	**0.5128**	0.4841	0.5082	0.4552
Mean absolute error	0.2737	0.2253	**0.3057**	0.2830
Root mean squared error	0.3905	**0.4746**	0.4014	0.4086
Relative absolute error	60.2252%	49.5616%	**67.2489%**	62.2687%
Root relative squared error	81.9274%	99.5746%	**84.2232%**	85.7304%
Total number of instances	768 (598 + 170)	768 (595 + 173)	768 (601 + 167)	768 (583 + 185)

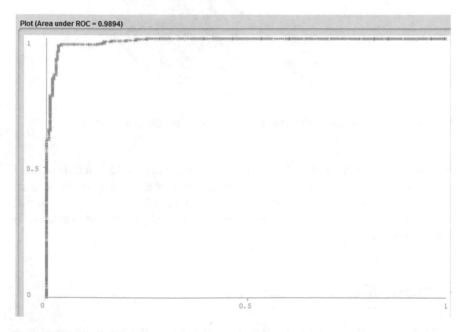

Plot (Area under ROC = 0.9894)

Fig. 7 ROC area of diabetes dataset in predictive model analysis

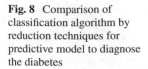

Fig. 8 Comparison of classification algorithm by reduction techniques for predictive model to diagnose the diabetes

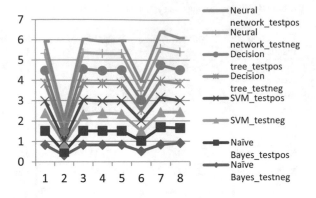

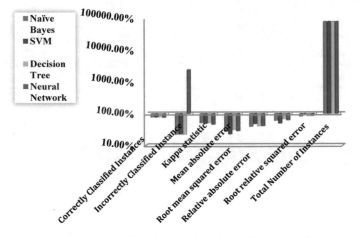

Fig. 9 Performance measures of classification algorithm by validation techniques

build the model for 40:60% is 1.28 s and for the 50:50 and 60:40 split tests is 1.02 s and for 70:30% is 0.02 s. In our work, we concentrate on calculating the F-measure, Precision and ROC area of different classification algorithms. Naive Bayes algorithm shows better result on Precision, F-measure and ROC of performance measures by predictive model.

5 Conclusion

In our research, the data classification for knowledge data discovery of data mining approach is defined accurately and efficiently and classification techniques is adopted by algorithms namely Naive Bayes, SMO-SVM, J48-decision tree, and, neural networks. We investigated various techniques and conducted experiments on the diabetes dataset from UCI learning repository to find the best classifier for diabetes diagnosis. Precision, F-measure, ROC area is very high for Naive Bayes algorithm by splitting (reduction techniques) mode, compared with other algorithms used in our study. By validation mode, SMO-SVM algorithm shows better results when compared with others. It helps in identifying the state of the disease and also helps people to identify if they are diabetic, based on the attributes. These classifiers are largely used in the fields of data mining, biomedical engineering, and diagnosing the patients in medicine.

Acknowledgements The authors thank VIT University for providing "VIT SEED GRANT" for carrying out this research work.

References

1. Patil, T.R., Sherekar, S.S.: Performance analysis of Naive Bayes and J48 classification algorithm for data classification. Int. J. Comput. Sci. Appl. **6**(2), 256–261 (2013)
2. David, S.K., Saeb, A.T.M., Al Rubeaan, K.: Comparative analysis of data mining tools and classification techniques using weka in medical bioinformatics. Comput. Eng. Intell. Syst. **4**(13), 28–38 (2013)
3. Rahman, R.M., Afroz, F.: Comparison of various classification techniques using different data mining tools for diabetes diagnosis. J. Softw. Eng. Appl. **6**(03), 85 (2013)
4. Sujata, Priyanka Shetty, S.R.: Performance Analysis of Different Classification Methods in Data Mining for Diabetes Dataset Using WEKA Tool. Int. J. Recent Innov. Trends Comput. Commun. **3**(3), 1168–1173 (2015)
5. Sharma, T.C., Jain, M.: WEKA approach for comparative study of classification algorithm. Int. J. Adv. Res. Comput. Commun. Eng. **2**(4), 1925–1931 (2013)
6. Kumari, M., Vohra, R., Arora, A.: Prediction of Diabetes Using Bayesian Network (2014)
7. Salas-Zárate, M.d.P., et al.: Sentiment analysis on tweets about diabetes: an aspect-level approach. Comput. Math. Methods Med. (2017)
8. Mohammadi, M., Hosseini, M., Tabatabaee, H.: Using Bayesian network for the prediction and diagnosis of diabetes. Bull. Env. Pharmacol. Life Sci. **4**, 109–114 (2015)

Challenges and Applications of Wireless Sensor Networks in Smart Farming—A Survey

T. Rajasekaran and S. Anandamurugan

Abstract Human survival is a huge task of their lives in the society. Agriculture is the most important role played by the survival of human civilization. The technological advancement in wireless communication and reduction in size of sensor is innovatively projects in the various fields such as environmental monitoring, precision farming, health care, military, smart home, etc. This paper provides an insight into various needs of wireless sensor technologies, wireless sensor motes used in agriculture and challenges involved in deployment of Wireless Sensor Network (WSN). Smart Farming (SF) has been played a major role to enhance more production in the field of agriculture. This article not only focuses on smart farming but also compared with traditional methods in agriculture.

Keywords WSN · Motes · Irrigation · Agriculture · Smart farming
Traditional farming

1 Introduction

Agriculture is playing a vital role in the development of human civilization. Increasing population, resource shortage, and degradation of the ecological environment have caused shortage and sustainability problems with the food supplies. Most of the farmers in developed and developing nations are still following traditional farming methods. In the beginning of this decade, there is a transformation from traditional farming to modern agriculture practice, also called smart farming, which helps to

T. Rajasekaran (✉)
Department of Computer Science Engineering, KPR Institute
of Engineering and Technology, Coimbatore, India
e-mail: rajasekaran30@gmail.com

S. Anandamurugan
Department of Information Technology,
Kongu Engineering College, Erode, India
e-mail: dranandamurugan@gmail.com

© Springer Nature Singapore Pte Ltd. 2019
J. D. Peter et al. (eds.), *Advances in Big Data and Cloud Computing*,
Advances in Intelligent Systems and Computing 750,
https://doi.org/10.1007/978-981-13-1882-5_30

improve the crop management and sustainable agricultural development [1]. SF assist the farmers in incorporating the information and communication technologies to agricultural equipments, increase the productivity of agriculture and reduce the human assistance for cultivating, harvesting of crops. SF enables to bring enormous amount of data and information from farm which helps the farmer to take optimal decision to perform farming activities [2]. The goal of this paper is to examine the importance of the smart farming in wireless sensors for automated irrigation. The rest of this paper is organized as follows: comparison of Traditional and Smart Farming, Smart Farming Technologies, wireless sensor network and key issues in WSN, wireless sensor motes used for agriculture domain, application of WSN in autonomous irrigation and conclusion.

2 Comparison of Traditional Farming and Smart Farming

Traditional Farming (TF) is the common method followed by the farmers for cultivating the crop, which lacks in an effective utilization of resources like man power and water. Traditional way of flowing water to the cultivating land is one of the common irrigation methods and there are various other methods farmers following today. Agricultural monitoring followed in traditional method results in high investment cost and man-made destruction. SF is an alternative method for effective utilization of water to cultivate a crop. SF, addresses the limitations of traditional farming through real-time monitoring, minimization of man power, time, accurate estimation of required water for irrigation and protects the crop from disaster like disease, flood, etc., through early detection [3]. Table 1 summarizes the key differences of TF and SF.

Table 1 Traditional and smart farming comparison

Agricultural requirements	Traditional farming	Smart farming
Need of staffing	Yes	No
Water utilization	High	Low
Field monitoring	High	Low
Data acquisition	Low	High
Cost	High	Low
Yield	Low	High
Time	More	Less

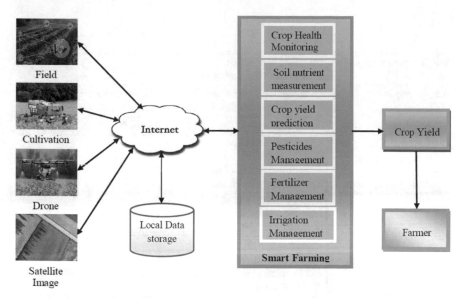

Fig. 1 Smart farming architecture

3 Technologies Applied in Smart Farming

Smart farming involves many tools and technologies (see Fig. 1) which are used for automation and real time monitoring of farming activities. These technologies make agriculture a more beneficial for farmers by curtailing the crop production and maintenance cost and increases the farmer's income through high crop yield. Table 2 shows the SF technologies and its utilities.

4 Deploying Challenges of WSN

WSN gained worldwide attention in recent years due to advancement in Micro-Electro-Mechanical Systems (MEMS), which ease the development of smart sensors [4]. WSN poses many challenges that need to be handled before deploying in an agricultural land; this survey highlights the critical issues of WSN in agriculture domain some of them are listed below.

Table 2 SF technologies

Technologies	Utility
Wireless sensor networks	Irrigation automation Environmental monitoring
Remote sensing	Evaluates the various levels of soil moisture and nutrients, crop health and disease through collected images
Variable rate technology	Apply seed or fertilizer based on soil nutrient
Artificial intelligence	Monitoring condition of crop Pest detection Crop disease identification Plant species classification
Mobile technology	Remote farm monitoring Farm equipment monitoring
Drone	Crop data generation and surveillance of cultivation lands Capturing site images Observes failure in crop plantation Autonomous pest identification Autonomous pesticides spraying

4.1 Energy Consumption

WSN is composed of sensor nodes and each node has the responsibility of event detection such as sensing, data processing, transmission, and data routing. To accomplish all the above actions each node consumes energy. Efficient energy management strategy is applied to enhance the life time of the battery [5–7]. Alternatively, renewable energy sources such as solar power or kinetic energy are adopted for longer life of sensor nodes. Moreover, in many agricultural applications, sensors are repowered frequently by changing the battery based on their placement [8].

4.2 Data Collection

In traditional approaches, WSN requires data to be collected and transferred from sensor node to centralized base station, which consumes huge amount of the battery life. Techniques like aggregation and compression are needed for real-time data transfer and event-based application to prune the energy loss by reducing the amount of data transferred.

4.3 Transmission Range

Data transmission failure in WSN occurs due to ecological effects. In the event of data transmission, most of the wireless sensor communication technologies covers only short range which inference of additional sensors and routers in a WSN [9]. Communication range in the network can be prolonged by adopting multi-tire, ad hoc and mesh network topologies.

4.4 Data Security

Different types of threats that spoil the integrity of data are spoofing, sinkhole attacks, Sybil attacks, Denial of service attack and jamming. WSN afford various access control mechanisms [10] and anomaly detection techniques [11] to secure sensed data.

4.5 Fault Tolerance

Fault tolerance is one of the important issues in WSN. Since, the sensor nodes are placed in open environment, it experience failure of problems due to various factors like physical damage, power depletion, radio inference, asymmetric communication link, blockage and collision. Various fault tolerance mechanisms such as redundancy-based mechanisms, clustering-based mechanisms, and deployment-based mechanisms are employed to increase the reliability of WSN [12].

4.6 Sensor Node Size and Placement

Sensor node size and placement is one of the critical issues in WSN. The size of the sensor node must be small and appropriate for deployment. Performances and lifetime of WSN primarily depends on deployment of sensor node. WSN deployment method falls under two categories such as random and deterministic deployments. Random deployment strategy used for deployment in large scale open environments regions and deterministic strategy (point to point) used for small-scale deployments [13, 14]. Several issues and proposed measures to overcome the issues are presented in Table 3.

Table 3 Issues, causes and proposed measures in WSN

Issues	Causes	Proposed measures
Energy consumption	Sensing Data processing Data transmission	Renewable energy Solar power Kinetic energy
Data collection	Sensing of data	Aggregation Compression
Transmission range	Ecological effect	Multi-tire Ad hoc network Mesh network
Data security	Spoofing Sinkhole Sybil Denial of service Jamming	Access control methods Anomaly detection techniques
Fault tolerance	Physical damage Power depletion Blockage Radio Inference Collision	Redundancy mechanism Clustering techniques
Sensor node placement	Size of sensor node	Deterministic deployment Random deployment

5 Wireless Sensor Motes Used for Smart Farming

Wireless Sensor Mote (WSM) is a basic building block of WSN, which can be deployed in various application domains to sense and transmit the data to the network controller for further processing and selection of motes should be application specific, problem, and distribution pattern [15, 16]. WSN in the form of small sensor motes [16] are used for environmental data collection. To deploy the sensor motes in the agricultural field, the sensor motes should be compact in size, cost, energy efficient, and easy to access. Utilized WSN mote are capable of gathering information from sensor and communicates between other nodes in the networks [5, 17]. Table 4 shows some of the available sensor motes in the market which has been used for agriculture application.

6 Applications of WSN in Agricultural Irrigation

Nowadays, WSN is widely used to overcome many real-world problems. One of the interesting fields is agriculture. This section reviews the use of WSN in automated irrigation.

Krishna et al. developed the automated irrigation systems to reduce the wastage of water by supplying the required amount of water needed for the growth of normal

Table 4 Sensor motes used for agriculture

Mote platform	Onboard sensors	Applications in agriculture
TelosB	Temperature Humidity Light	Sensing light Sensing humidity Sensing temperature GPS positioning
MicaZ/Mica2	Temperature Humidity Light	Sensing temperature Sensing light Measuring pressure Sensing humidity
IRIS	Temperature Humidity Soil moisture	Sensing temperature Sensing light Sensing humidity Measuring soil moisture Controlling water flow
Sun spot	Temperature Light	Sensing temperature Sensing light
EZ430-RF2480/EZ430-RF2500	Temperature Humidity Light	Sensing light Sensing humidity Sensing temperature

plant. The irrigation controller, irrigates the substrate of the plotted plants to a set point (volumetric water content, u) and maintains the water content to that set point [18].

Dursun et al. [19] developed wireless irrigation control system for monitoring the water content of the soil. It is specifically designed for site-specific irrigation that has the three units namely Base Station Unit (BSU), Valve Unit (VU), and Sensor Unit (SU) and applied to irrigate dwarf cherry trees located in central Anatolia and it has advantages such as preventing moisture stress of trees, diminishing of excessive water usage, ensuring of rapid growing weeds, and derogating salification.

Liang et al. [20] designed a real-time soil moisture prediction system based on GPRS and WSN. Soil moisture sensors, collects the moisture data and transmitted over GPRS network. The data is processed and analyzed by genetic BP neural network and gives real-time prediction by simulated annealing algorithm.

Goumopoulos et al. [21] designed a decision support system for autonomous closed-loop zone-specific irrigation control in a green house integrating with a Wireless Sensor/Actuator Network (WSAN). The system uses an ontology to focus attention on flexibility and adaptability of system by incorporating machine learning algorithms which is used for inducing new rules by analyzing logged datasets to accurately determine significant thresholds of plant-based parameters and for extracting new knowledge and extending the system ontology.

Xiang et al. [22] introduced automatic drip irrigation control system based on ZigBee WSN and fuzzy control. The system collects soil moisture, temperature; light intensity information and sends over low power ZigBee based wireless communication technology. The system inputs this information to fuzzy controller and creates a fuzzy rule to control the irrigation automatically.

Joaquin et al. developed automated irrigation system using a WSN and GPRS. This system has distributed wireless network of soil moisture and temperature sensors placed in the root zone of the plants. The sensor information are collected through gateway and transmitted to the web page. This algorithm sets the threshold value for temperature and soil moisture which is programmed into a micro-controller based gateway to control the quantity of water for irrigation. This automated system saves the 90% of water compared with traditional irrigation practices [23].

Giusti et al. demonstrate a Fuzzy Decision Support System (FDSS) which needs to enhance the performance of an existing irrigation method, based on the IRRINET model, by describing a protocol for the field implementation of a fully automated irrigation system and incorporating a fuzzy soil moisture model. This model estimates the soil moisture based on Growing Degree Days (GDD), total water applied to the crop, and crop evapotranspiration (ETc). The model FDSS, decides whether an irrigation is needed and determines its amount by a set of rules involving the variation of growing degree days (D T sum), the two-day ahead rain forecast (RF), and the crop evapotranspiration (ETc) [24].

7 Conclusion and Future Perspectives

This paper highlights the benefits of smart farming over traditional framing and various technologies and applications of SF. By deploying SF, farmers can gain more profit, high yield, ease agricultural land monitoring, and effective utilization of water. A few deployment issues of WSN in agriculture are highlighted and still there are many issues to be fixed in WSN applications, different WSM. Finally, different WSM and various approach for autonomous irrigation has been presented towards agriculture domain. In future, whole agriculture system may automated to build sustainable agriculture using technologies such as Internet of things, fog computing and cloud computing which reduces the time consumption and resource utilization. Furthermore, these technologies provide remote control farm management, warehouse management, and intelligent decision-making to enhance the farmers profitability. Farmers can also measure the yield of crop and calculate the profit using machine learning and deep learning techniques which make the agriculture profitable.

References

1. Maohua, W.: Possible adoption of precision agriculture for developing countries at the threshold of the new millennium. Comput. Electron. Agric. **30**, 45–50 (2001)
2. O'Grady, M.J., O'Hare, G.M.P.: Modeling the smart farm. Inf. Process. Agric. **4**, 179–187 (2017)
3. Nikolidakis, S.A., Kandris, D., Vergados, D.D., Douligeris, C.: Energy efficient automated control of irrigation in agriculture by using wireless sensor networks. Comput. Electron. Agric. **113**, 154–163 (2015)

4. Yick, J., Mukherjee, B., Ghosal, D.: Wireless sensor network survey. Comput. Netw. **52**, 2292–2330 (2008)
5. Ruiz-Garcia, L., Lunadei, L., Barreiro, P., Robla, J.I.: A review of wireless sensor technologies and applications in agriculture and food industry: state of the art and current trends. Sensors **9**(6), 4728–4750 (2009)
6. Ojha, T., Misra, S., Raghuwanshi, N.S.: Wireless sensor networks for agriculture: the state-of-the-art in practice and future challenges. Comput. Electron. Agric. **118**, 66–84 (2015)
7. Aqeel-ur-Rehman, Abbasi, A.Z., Islam, N., Shaikh, Z.A.: A review of wireless sensors and networks applications in agriculture. Comput. Stand. Interfaces **36**, 263–270 (2014)
8. Jawad, H.M., Nordin, R., Gharghan, S.K., Jawad, A.M., Ismail, M.: Energy-efficient wireless sensor networks for precision agriculture: a review. Sensors **17**(8), 1781 (2017)
9. Misra, S., Kumar, M.P., Obaidat, M.S.: Connectivity preserving localized coverage algorithm for area monitoring using wireless sensor networks. Comput. Commun. **34**(12), 1484–1496 (2011)
10. Misra, S., Vaish, A.: Reputation-based role assignment for role-based access control in wireless sensor networks. Comput. Commun. **34**(3), 281–294 (2011)
11. Karapistoli, E., Sarigiannidis, P., Economides, A.A.: SRNET: a real-time, cross-based anomaly detection and visualization system for wireless sensor networks. In: Proceedings of the Tenth Workshop on Visualization for Cyber Security, pp. 49–56 (2013)
12. Chouikhi, S., Elkorbi, I., Ghamri-Doudane, Y., Saidane, L.A.: A survey on fault tolerance in small and large scale wireless sensor networks. Comput. Commun. **69**, 22–37 (2015)
13. Corke, P., Hrabar, S., Peterson, R., Saripalli, D., Rus, S., Sukhatme, G.: Autonomous deployment and repair of a sensor network using an unmanned aerial vehicle. In: IEEE international conference on robotics and automation, pp. 3602–3608 (2004)
14. Chang, C.-Y., Chen, Y.-C., Chang, H.-R.: Obstacle-resistant deployment algorithms for wireless sensor networks. IEEE Trans. Veh. Technol. **58**(6), 2925–2941 (2009)
15. Baggio, A.: Wireless sensor networks in precision agriculture. In: ACM Workshop on Real-World Wireless Sensor Networks (REALWSN2005), Stockholm, Sweden, (2005)
16. Nanda, K., Babu, H., Selvakumar, D.: Smartmote—an innovative autonomous wireless sensor node architecture. In: 2014 IEEE International Conference on Electronics, Computing and Communication Technologies (IEEE CONECCT), pp. 1–6 (2014)
17. Li, Z., Wang, N., Franzen, A., Godsey, C., Zhang, H., Li, X.: Practical deployment of an in-field soil property wireless sensor network. Comput. Stand. Interfaces **36**(2), 278–287 (2014)
18. Xiang, X.: Design of fuzzy drip irrigation control system based on ZigBee wireless sensor network. In: International (CCTA 2010), pp. 495–501
19. Dursun, M., Ozden, S.: A wireless application of drip irrigation automation supported by soil moisture sensors. Sci. Res. Essays **6**(7), pp. 1573–1582 (2011)
20. Liang, R., Ding, Y., Zhang, X.: A real-time prediction system of soil moisture content using genetic neural network based on annealing algorithm. In: IEEE International Conference on Automation and Logistics (ICAL), pp. 2781–2785 (2008)
21. Goumopoulos, C., O'Flynn, B., Kameas, A.: Automated zone-specific irrigation with wireless sensor/actuator network and adaptable decision support. Comput. Electron. Agric. **105**, 20–33 (2014)
22. Xiang, X.: Design of fuzzy drip irrigation control system based on ZigBee wireless sensor network. In: International (CCTA 2010), pp. 495–501 (2010)
23. Gutierrez, J., Villa-Medina, J.F., Nieto-Garibay, A., Porta-Gandara, M.A.: Automated Irrigation System Using a Wireless Sensor Network and GPRS Module. IEEE Trans. Instrum. Meas. **63**, 166–176 (2013)
24. Giusti, E., Marsili-Libelli, S.: A fuzzy decision support system for irrigation and water conservation in agriculture. Environ. Model. Softw. **63**, 73–86 (2015)

A Provable and Secure Key Exchange Protocol Based on the Elliptical Curve Diffe–Hellman for WSN

Ummer Iqbal and Saima Shafi

Abstract Key Exchange serves as bedrock of all cryptographic primitives. As WSN are resource constraint by virtue of limited resources, traditional cryptographic protocols are not optimal. Elliptical Curve Cryptosystems have shown significant computational advantage than other systems. Elliptical Curve Diffe–Hellman is a standard protocol for establishing shared keys. However, it is susceptible to Man-in-the-Middle Attack because there is no authentication between two parties. In this paper, Formal security validation of ECDH has been performed on AVISPA. An enhanced authenticated key exchange protocol based on ECC has been presented to overcome the limitations of ECDH. The proposed protocol provides a prefect resilience to lack of authentication based attacks like Man-in-the-Middle Attack. The protocol has been designed on the principle of ECDH. The Formal security validation of the developed protocol has also been done using AVISPA.

Keywords WSN · Key exchange · ECC · ECDH · AVISPA

1 Introduction

Wireless Sensor Networks has a broad horizon of applications which includes environmental monitoring, Battlefield Surveillance, and Healthcare [1]. As WSN involves the exchange of critical data, security of such systems become paramount. The resource constraint nature of WSN inhibits the usage of traditional security protocols. A typical WSN node is characterized by 4 KB of RAM, 128 Kb of ROM [2, 3]. Such a mote is powered by two lithium cells with 2000 mAH energy, which may just last for few days. Thus the design principle of the security protocols in WSN must limit the duty cycle and conserve energy. Key Exchange serves as a pivotal

U. Iqbal (✉)
National Institute of Electronics and Information Technology, Jammu and Kashmir, India
e-mail: ummer@nielit.gov.in

S. Shafi
Shri Sukhmani Institute of Engineering and Technology, Chandigarh, India

© Springer Nature Singapore Pte Ltd. 2019
J. D. Peter et al. (eds.), *Advances in Big Data and Cloud Computing*,
Advances in Intelligent Systems and Computing 750,
https://doi.org/10.1007/978-981-13-1882-5_31

component of all security services like authentication, integrity, Confidentiality. A Key exchange process typically comprises of Key generation and Key distribution. Primarily a Key exchange process must decrease the computational Cost as well the Storage requirement. Key Management schemes for WSN comprises of Single Network Key, Pairwise Key Establishment, Trusted Third Party, Public Key Schemes, and Pre-Distribution [4].

In a Single Network Key mechanism, a master key is predeployed in all nodes in the network. A lightweight cipher is then used with the key to provide other cryptographic primitives. The major advantage of this scheme is due to its simple design as it includes low complexity overhead and minimal storage. However, this scheme suffers.

From node capture attack. TinySEC [5] is a one such a framework based on this technique. However, a compromise of a single node comprises the complete security of the network. A pairwise key establishment involves the establishment of a pair of keys between every node in the network. This mechanism does help in overcoming node capture issue however it is not feasible due to its high storage demand. For a network of N nodes, each node needs to store in N-1 keys. A trusted base station can be used for overcoming high memory requirements of a pairwise key establishment. In this case, each node has a master key with Trusted Third Party (TTP) which can be used for generating a session key between the nodes. However such an approach is subjected to high computational cost and single point of failure. In a Key pre-distribution scheme, set of odd keys are predeployed in a node. A node uses a discovery process to set up a key. This approach is based on the probabilistic method and does not always guarantee an establishment of an a pairwise key. Key exchange mechanism can also be based on public key cryptography. Traditional public key cryptosystems like RSA have not been found computationally efficient for WSN. However, ECC based cryptosystems have shown more promise [6, 7]. A 160-bit key in ECC provides a same level of security as that in 1024 bits in RSA [8]. The elliptical curve optimization techniques like Barret reduction, Sliding window, Shamir's trick has resulted in 10 performance enhancement in terms of computational effect. Diffie Hellman Key Exchange based on ECC (ECDH) has shown a lot of promise for the Key establishment in WSN. However, a big drawback of ECDH is that it suffers from a Man-in-the-Middle attack due to lack of authentication between two parties [9].

In this paper, Formal verification of ECDH protocol on AVISPA has been carried out. A protocol has also been proposed to overcome the limitations of ECDH. Section 2 provides the preliminaries about Elliptical Curve Cryptography and Avispa. In Sect. 3, ECDH design and its formal verification has been presented. Section 4 presents the proposed protocol and its formal verification.

2 Preliminaries

2.1 Elliptical Curve Cryptography

Elliptical Curve Cryptography was proposed by Victor Miller and Niel Kolbitz in 1985 as an alternate form of public key cryptography [10, 11]. Some of the important definitions on ECC are listed as below:

Definition 1 Elliptical Curve are cubic curves defined over a prime field F_p which satisfy the following equation:

$$y^2 = x^3 + ax + b \tag{1}$$

where $x, y, a, b \in F_p$ and $\Delta = 4a3 + 27b_2 \neq 0$. The set of points also includes a point at infinity (O).

Definition 2: Given an integer K and a point $P(x, y)$ on the elliptical Curve $E(F_p)$, the scalar multiplication is defined as $K \cdot P = K + K + K + \cdots + K$ (K times), where + is a point addition.

Definition 3: Given the point $P(x, y)$ and $Q(x, y)$ such that $Q = K \cdot P$ where $P, Q \in E(F_p)$ and K is a scalar, it is computationally infeasible to find K.

2.2 Avispa

Avispa is an automated validation tool for security protocols. It is a push button tool based on Dolev and Yao [12] model. Dolev and Yoa model gives a complete control of the communication channel to the intruder. The intruder in this case has a capacity to forward, modify, and change messages but cannot overdue the computational strength of an algorithm. In Avispa, the protocols are modeled using HLPSL, which is role-based formal language which comprising of roles compositions, security models, etc. The protocol is modeled in HLPSL which is then converted into IF format using a HLPSLtoOF translator. The IF format is then passed into various AVISPA backend's which include OFMC, Claste, SATMC, and T$SP. These backend's check a protocol against various active and passive attacks.

3 Formal Verification of Elliptic Curve Diffe–Hellman

3.1 Elliptic Curve Diffe–Hellman

Let N_i and N_j be the two nodes in a WSN, who want to establish a shared key using ECDH. Let $E(F_p)$ be the elliptical Curve with domain parameters (n, a, b, G, h).

1. N_i chooses a random number N_a, computes $N_a \cdot G$ and sends it to N_j

$$N_i \rightarrow N_j : N_a \cdot G \qquad\qquad (2)$$

2. N_j chooses a random number Y, computes $Y \cdot G$ and Sends it to N_i

$$N_j \rightarrow N_i : N_b \cdot G \qquad\qquad (3)$$

3. N_i generates the key $K_{ij} = N_a \cdot (N_b \cdot G)$ and N_j generates $K_{ji} = N_b \cdot (N_a \cdot G)$. The shared key generated between N_i and N_j is $K_{ij} = K_{ji} = N_a \cdot N_b \cdot G$.

3.2 AVISPA Verification

The HLPSL implementation of ECDH is shown in Fig. 1, N_i sends Mul($G \cdot N_a$) to N_j and on receiving it, N_j sends Mul($G \cdot N_b$) to N_i. The Key generated by N_i is Mul(Mul($N_b \cdot G$) $\cdot N_a$). N_i sends a nonce Nsecret using the key Mul(Mul($N_b \cdot G$) $\cdot N_a$) with a secrecy goal identified by sec_dhvalue. sec_dhvalue specifies that the Nsecret must remain a secret between N_i and N_j. The environment section comprises of composition of one or more sessions and the intruder Knowledge. The simulation output with OFMC backend is shown in Fig. 2. Ofmc backends detects the protocol as UNSAFE as given in the attack trace. The attack trace has been animated in Fig. 3. We can figure out that that the intruder is the able to establish a half key Mul(Mul(g. nounce-1). var-x-1) with Alice. Similarly, it can also establish the half key with the BOB resulting in Man in the Middle Attack. This Attack Primarily occurs due to lack of authentication Between two parties.

4 Proposed Protocol

The Notations used in the proposed protocol are tabulated in Table 1.

The sequence of steps in the proposed protocol is given below:

1. N_i computes $(X + N_a) \cdot PN_j$ and sends it to N_j

$$N_{i-} \rightarrow N_j : (X + N_a) \cdot PN_j. \qquad\qquad (3)$$

```
role
role_A(A:agent,B:agent,G:text,H:function,SND,RCV:channel(dy))
played_by A
def=
local
  State:nat,Nb:text,Na:text,NSecret:text
  init
  State := 0
  transition
1. State=0 /\ RCV(start) =|> State':=1 /\ Na':=new() /\ SND(H(G.Na'))
2. State=1 /\ RCV(H(G.Nb')) =|> State':=2 /\ NSecret':=new() /\
SND({NSecret'}_H(H(G.Na).Nb')) /\  secret (NSecret' , sec_dhvalue , {A,B})
end role
role role_B(A:agent,B:agent,G:text,H:function,SND,RCV:channel(dy))
played_by B
def=
  local
    State:nat,Nb:text,Na:text,NSecret:text
  init
    State := 0
  transition
1. State=0 /\ RCV(H(G.Na')) =|> State':=1 /\ Nb':=new() /\  SND(H(G.Nb'))
3. State=1 /\ RCV({NSecret'}_H(H(G.Na).Nb)) =|> State':=2
end role
role session(A:agent,B:agent,G:text,H:function)
def=
  local
    SND2,RCV2,SND1,RCV1:channel(dy)
  composition
role_B(A,B,G,H,SND2,RCV2) /\ role_A(A,B,G,H,SND1,RCV1)
end role
role environment()
def=
  const
bob:agent,h:function,alice:agent,g:text, ni:text ,sec_dhvalue : protocol_id
intruder_knowledge = {alice,bob,g,h}
  composition
session(alice,bob,g,h)/\ session(i,alice,g,h) /\ session(alice, i,g,h)
end role
goal
secrecy_of sec_dhvalue
end goal
environment()
```

Fig. 1 HLPSL script for ECDH

2. N_j receives $(X + N_a) \cdot PN_j$ and performs its scalar multiplication with Y^{-1}

$$I(x, y) = \left[(X + N_a) \cdot PN_j\right] * Y^{-1}$$
$$I(x, y) = \left[(X + N_a) \cdot Y \cdot G\right] * Y^{-1}$$
$$I(x, y) = \left[(X + N_a) \cdot G\right]$$
$$I(x, y) = PN_i + N_a \cdot G. \tag{4}$$

```
% OFMC
% Version of 2006/02/13
SUMMARY
  UNSAFE
DETAILS
  ATTACK_FOUND
PROTOCOL
  /home/span/span/testsuite/results/diffe-hellman.if
GOAL
  secrecy_of_sec_dhvalue
BACKEND
  OFMC
COMMENTS
STATISTICS
  parseTime: 0.00s
  searchTime: 0.00s
  visitedNodes: 3 nodes
  depth: 1 plies
ATTACK TRACE
i -> (alice,3): start
(alice,3) -> i: h(g.Na(1))
i -> (alice,3): h(g.x240)
(alice,3) -> i: {NSecret(2)}_(h(h(g.Na(1)).x240))
i -> (i,17): NSecret(2)
i -> (i,17): NSecret(2)
```

Fig. 2 OFMC backend result of ECDH

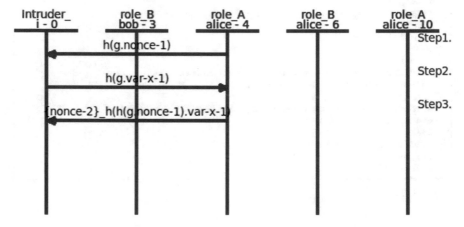

Fig. 3 SPAN animation of the attack

N_j performs a point addition of (4) with $(-PN_i)$ and uses the elliptical curve properties, $P + (-P) = O$ and $P + O = P$ to get $Q(x, y)$ in (5)

$$Q(x, y) = I(x, y) + (-PN_i)$$
$$Q(x, y) = PN_i + N_a \cdot G + (-PN_i)$$
$$Q(x, y) = N_a \cdot G. \tag{5}$$

Table 1 Symbol description

Symbol	Description
N_i	Node i
N_j	Node j
X	Private key of N_i
PN_i	Public key of N_i
Y	Private key of N_j
PN_j	Public key of N_j
N_a	Random number choosen by N_i
N_b	Random number choosen by N_j
G	Generator point

N_j Computes $P(x, y) = N_b \cdot G$, where N_b is a random number and generates a session key K_{ji}

$$K_{ji}(x, y) = P + Q$$
$$K_{ji}(x, y) = N_a \cdot G + N_b \cdot G. \qquad (6)$$

N_i performs a scalar multiplication of Q with a random number N_b and sends it to N_j

$$N_j \rightarrow N_i : Q \cdot N_b \qquad (7)$$

3. N_i receives: $Q \cdot N_b$ and performs its scalar multiplication with N_a^-

$$R(x, y) = [Q \cdot N_b] \cdot N_a^{-1}$$
$$R(x, y) = N_a \cdot G \cdot N_b \cdot N_a^{-1}$$
$$R(x, y) = N_b \cdot G. \qquad (8)$$

N_i computes $S(x, y) = N_a \cdot G$ and generates the session key $K_{ij} = R(x, y) + S(x, y)$

$$K_{ij} = N_a \cdot G + N_b \cdot G, \; K_{ij} = K_{ji}. \qquad (9)$$

The HPSL2 script and OFMC backend output of the proposed protocol is given in Figs. 4 and 5. From Fig. 5, we understand the proposed protocol is SAFE.

```
role role_A(A:agent,B:agent,G:text,H:function,SND,RCV:channel(dy))
played_by A
def=
   local
      State:nat,Nb:text,Na:text,Nid:text,PKa:text, PKb: text , En:text, Em: text,
R:text , S:text , Kab : text , Nsecret :text
   init
      State := 0
   transition
      1. State=0 /\ RCV(start) =|> State':=1 /\ X':=new() /\  PKa':=H(G.X') /\
SND(PKa') /\ secret (X' , private_A, {A})
      2. State=1 /\ RCV(PKb') =|> State':=2 /\ Na':=new() /\  En':= H(xor(Na',X).PKb)
/\ SND (En')
                                        4. State=2 /\ RCV(Em') =|> State':=3 /\ R' :=
H(Em'.inv(Na)) /\  S' := H(Na.G) /\ Kab':= H(Q.R) /\ Nsecret' := new() /\
      SND({Nsecret'}_Kab) /\  secret (Nsecret' ,sec_dhvalue , {A,B})
end role
role role_B(A:agent,B:agent,G:text,H:function,SND,RCV:channel(dy))
played_by B
def=
   local
      State:nat,Nb:text,Na:text,NSecret:text, PKa:text , PKb:text , En: text ,Em:text
, I:text, IP:text , Q:text , Kba : text , S:text
   init
      State := 0
   transition
      1. State=0 /\ RCV(PKa') =|> State':=1 /\ Y':=new() /\ PKb':= H(G.Y') /\ secret (Y'
, private_B , {B})
      3. State=1 /\ RCV(En') =|> State':=2  /\ I' := H(inv(Y).En) /\ IP':=
H(inv(Y).G) /\ â€žQ'= H(I'.IP')/\ Nb' := new() /\ R':= H(Nb.G) /\ Kba':= H(R.Q)
   Em' :=H(Q.Nb') /\ SND (Em')
end role

role session(A:agent,B:agent,G:text,H:function)
def=
   local
      SND2,RCV2,SND1,RCV1:channel(dy)
   composition
      role_B(A,B,G,H,SND2,RCV2) /\ role_A(A,B,G,H,SND1,RCV1)
end role
role environment()
def=
   const
      bob:agent,h:function,alice:agent,g:text, ni:text ,sec_dhvalue : protocol_id
,private_A : protocol_id , private_B : protocol_id , rfd_to_mib: protocol_id
   intruder_knowledge = {alice,bob,g,h}
   composition
      session(alice,bob,g,h)/\ session(i,alice,g,h) /\ session(alice, i,g,h)
end role
goal
secrecy_of sec_dhvalue
secrecy_of private_A
secrecy_of private_B
end goal
enviroment()
```

Fig. 4 HLPSL script for the proposed protocol

5 Conclusion

The paper highlighted various Key exchange mechanisms which can be employed in
WSN. It was pointed that Diffe–Hellman protocol based on ECC (ECDH) is suitable
for Key exchange in WSN. ECDH protocol was modeled using HLPSL script and
analyzed on OFMC backend of AVISPA against various active and passive attacks.

```
% OFMC
% Version of 2006/02/13
SUMMARY
   SAFE
DETAILS
   BOUNDED_NUMBER_OF_SESSIONS
PROTOCOL
   /home/span/span/testsuite/results/proposed-protocol.if
GOAL
secrecy_of sec_dhvalue
secrecy_of private_A
secrecy_of private_B

BACKEND
   OFMC
COMMENTS
STATISTICS
   parseTime: 0.00s
   searchTime: 0.01s
   visitedNodes: 16 nodes
   depth: 4 plies
```

Fig. 5 OFMC backend output of the proposed protocol

ECDH protocol was found UNSAFE on AVISPA as Man in the Middle Attack was carried out on it. A more secure protocol was proposed whose security validation revealed it as SAFE on OFMC Backend.

References

1. Akyildiz, I.F., Su, W., Sankarasubramaniam, Y., Cayirci, E.: A Survey on sensor networks. IEEE Commun. Mag. **40**(8), 102–114 (2002)
2. Sanchez-Rosario, F.: A Low Consumption Real Time Environmental Monitoring System for Smart Cities Based on ZigBee Wireless Sensor Network. IEEE (2015) 978-1-4799-5344-8
3. Perrig, A., Stankovic, J., Wagner, D.: Security in wireless sensor networks. Commun. ACM **47**(6), 53–57
4. Xiao, Y., Ravi, V.K., Sun, B.: A survey of key management schemes in wireless sensor networks. Elsevier J. Comput. Commun. **30**, 2314–2341 (2007)
5. Karlof, C., Sastry, N., Wagner, D.: TinySec: Link Layer Security Architecture for Wireless Sensor Networks. Sensys., Baltimore, MD (2004)
6. Du, W., Wang, R., Ning, P.: An efficient scheme for authenticating public keys in sensor networks. In: 6th. ACM, MobiHoc-05, pp. 58–67 (2005)
7. Mallan, D.J., Welish, M., Smith, D.M: Implementing public key infrastructure for sensor networks. Trans. Sens. Netw. **4** (2008)
8. Gura, N., Patel, A., Wander, A.S, Eberle, H., Chang Shantz, S.: Comparing elliptic curve cryptography and RSA on 8-bit CPUs. Cryptographic Hardware Embed. Syst. **3156**, 119–132. Springer (2004)

9. Huang, X. Shah, P.G., Sharma, D: Protecting from attacking the man-in-middle in wireless sensor networks with elliptic curve cryptography key exchange. In: 4th IEEE International Conference on Network and System Security, pp. 588–593 (2010)
10. Menzes, B.: Network security and cryptography. Cengage Learning (2010)
11. Hankerson, D., et al.: Guide to Elliptic Curve Cryptography. Springer
12. AVISPA Web tool: Automated validation of internet security protocols and applications, www.avispa-project.org, Last Accessed on (2018)

A Novel Background Normalization Technique with Textural Pattern Analysis for Multiple Target Tracking in Video

D. Mohanapriya and K. Mahesh

Abstract Visual tracking plays a central part in various computer vision applications, such as editing in video, surveillance, and computer interaction. It is nothing but detecting the targeted moving object in the video dataset. The basic principle of video tracking is that the frames were retrieved from the video dataset for performing the comparison operation to detect the moving object present in the video. There are lot of techniques were emerged in research in order to tract the object effectively. Since there is a lack in accuracy and other parameters related with processing techniques. The researchers involved in video tracking field facing a problems in tracking the targeted object occurs due to the presence of shadow in the video dataset. So, it is necessary to eliminate the shadows of the targeted object in the video. In this paper novel algorithm are proposed to eliminate the shadow pixels and track the target regions from the video frame and also to get the result with better accuracy. BCP-GPP is a new method used for the extraction of background and retrieval of pattern from moving object. The extracted target is used for classifying the regions from the target using the novel Machine Learning Classification (MLC) algorithm. It is used for matching the grid as the tracked region and provides the binary label for separating the background and the foreground. Then the target is tracked using the blob-based extraction technique. Then the performance of the video tracking system is analyzed using several parameters such as sensitivity, specificity, and accuracy.

Keywords Video editing · Video tracking · Shadows · Accuracy · Object tracking

D. Mohanapriya (✉) · K. Mahesh
Dept. of Computer Applications, Alagappa University, Karaikudi, Tamil Nadu, India
e-mail: mohanapriya.researchscholar@gmail.com

K. Mahesh
e-mail: mahesh.alagappa@gmail.com

© Springer Nature Singapore Pte Ltd. 2019
J. D. Peter et al. (eds.), *Advances in Big Data and Cloud Computing*,
Advances in Intelligent Systems and Computing 750,
https://doi.org/10.1007/978-981-13-1882-5_32

1 Introduction

Video Tracking is tracing of moving object or multiple objects over time in a given video sequence. Applications were Human computer interaction, Communication in Video and Compression of Video, augmented reality, medical imaging, traffic control, video editing, and security surveillance. It is considered as a time consuming process due to the amount of data that is contained in video. There are lot of techniques were emerged in research in order to tract the object effectively. Since there is a lack in accuracy and other parameters related with processing techniques. In the video tracking process, issues arise with segregation of background from the foreground in a frequently altering environment. Utilization of bounding box alone is not sufficient for the video tracking process for achieving the better performance. Some techniques related to preprocessing, feature extraction and feature analysis are required in order to achieve the better performance results. The overall steps involved in the general video tracking system are represented in Fig. 1. From the figure it is clear that the video traffic is captured using the camera [1]. The captured video frames are converted and saved in the digital format. After the estimation of suitable target area, the segmentation process segments the target area from the original video frame. The features from the extracted target area are obtained using the feature extraction algorithm. By exploiting the extracted features, the classification algorithm classifies the video frames and identifies the object for enabling the tracking operation.

1.1 Motivation

In traditional tracking, they utilize a general pixel clustering process based on thresholding effect of image pixel intensity only by enhancing the contrast level for segmentation. In this type of segmentation algorithm, they extract the cluster separation by using mean feature of image pixel. Since these have some limitations in the image clustering due to Multi-label segmentation problem in video frames. Since it is not in clear view of object which may include noise and distortion to that image; for that image analysis, we have to restore the resolution of image. Then for image segmentation and classification, they applied traditional clustering algorithm like FCM, K-Means clustering, etc., which perform clustering of image pixel for selected pixel range. This may include error rate due to verify the intensity of image pixel. To rec-

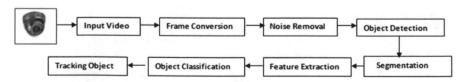

Fig. 1 Phases of video tracking

tify this problem, we propose a novel background normalization technique which is based on textural pattern analysis. In texture based system, there are several methods perform texture pattern extraction to verify the feature matching for target region analysis.

1.2 Objectives

- To rectify the inadequacies in object tracking such as occlusion, illumination changes, shadows, etc., using novel background normalization technique [2].
- To discriminate disparate background environment, a background clustering approach-based Background-Based Chain Pattern (BCP) technique is used.
- To recognize the similarity between the patterns, a Gradient Prediction Pattern Extraction (GPP) is used.
- To accomplish an efficacious multi-object tracking using proposed work.

This paper is structured as follows: Sect. 2 discusses the related works on visual tracking and their issues. Section 3 describes the proposed method BCP-GPP. Section 5 illustrates the performance analysis of proposed algorithm with existing segmentation techniques. Finally, Sect. 6 concluded with conclusion.

2 Related Work

Visual tracking is the main scenario in the computer vision under varying conditions such as different pose, sudden illumination changes, and the low contrast.

3 Proposed Work

The robust proposed texture pattern based background normalization method utilizes the consecutive processes to track the target in the given video frame. Figure 2 shows the overall flow of proposed method.

The author collected the frames from the given video for performing the techniques involved in the proposed work. In preprocessing technique Clang Boundary Filter (CBF) Technique removes the noise present in the frames. The patterns were extracted from the frames using novel Background-Based Chain Pattern technique. Then the obtained patterns were analyzed using Gradient Prediction Pattern Extraction (GPP) technique in order to check whether the movement is present in the video or not. Then the pattern is converted into binary form based on the conditions used in PGP technique. Then, the region properties were extracted and blob in motion was implemented. Finally, a motion in the given video is tracked.

Techniques	Author	Observation
Principal component analysis (PCA)	Wang et al. [3]	Inventive online object tracking methodology by means of incorporating both sparse prototypes and PCA
Block orthogonal matching pursuit (BOMP)	Zhang et al. [4]	Endorsed a stretchy appearance model that is constructed on the basis of sparse representation structure using BOMP
New technique for multi-oriented scene text line detection and tracking in video	Wu et al. [9]	A novel mechanism by means of utilizing both spatial as well as temporal information for identifying and tracking video texts originated from diversified orientation
Markov random field (MRF)	Li et al. [5]	Bayesian approach is incorporated with Markov Random Field (MRF) to recognize those character inferred within text along with its in-built dependencies
Flexible structured sparse representation	Bai and Li [6]	A sparse representation structure framed on the basis of complex nonlinear appearance model by deliberating structured union subspaces library for treating occlusions
Snooper text detecting mechanism	Minetto et al. [7]	Snooper Text detecting mechanism for recognizing the candidate characters embedded within images through employment of toggle-mapping image segmenting approach
Bayesian MTT model and a two-step MHT algorithm	A. Makris and C. Prieur [8]	Bayesian multiple-hypothesis tracking of merging and splitting targets in this work tracking accuracy and computation efficiency was improved

3.1   Clang Boundary Filter (CBF)

It is initial step in video tracking in which input of video is given and it converted into frames in order to acquired for tracking the object using CBF technique. After conversion of frames sequence of imageries is obtained from the given video file, dragon fly optimization methodology is utilized to obtain the better boundaries of the recognized object.

3.2   Background-Based Chain Pattern (BCP)

The enhanced preprocessed image is taken as input for the BCP approach in order to discriminate the foreground portion from the background part. This enhanced image obtained from each frame. From each frame object is figured out and recognized.

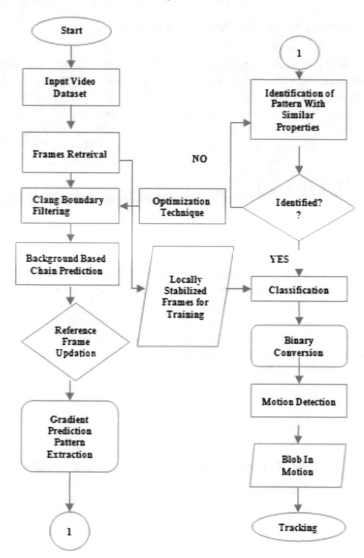

Fig. 2 Overall of proposed work

Finally BCP collects all recognized object from all frames and then background and foreground of object is discriminated after recognition of pointed object is masked. Finally it converted into binary form.

3.3 Gradient Pattern Prediction Extraction (GPP)

The recognized and masked object from BCP tracks the movement of an object in a given video, i.e., the entire image is sectioned and processed through an acquisition of patterns abstracted. GPP approach is used for recognizing the similarity between patterns obtained so far. Then comparison between standardized frame obtained in prior and those processed patterns detects the motion exposed by an object and the object concerned gets identified within a blob. Hence, the system is resistant towards illumination variations interpreted in tracking an object residing within a dynamic background gets tracked in a robust manner.

3.4 Machine Learning Classification (MLC)

MLC is distance between the obtained (extracted) features for the entire image and target region plays a most important role in the classification process. The features are extracted from GPP algorithm are assigned as input values to the classification algorithm. The features of masked region and target with fixed size are reshaped. The matching process between target and input is taken for the alternate pixels. The distance for all grouped region is estimated and least distance is calculated. The above steps are repeated for all frames.

4 Results and Discussion

The proposed method is implemented with windows 8 operating system with Core-i3 and 2 GB.

4.1 Dataset

In this research, there are two videos that are used to evaluate the proposed method. We have taken video from CAVIAR dataset.

4.2 Experimental Results

The video sequence is about 24 frames/s with the dimension of 384 × 288. The normal length of the video sequence is about 100 s and the complete data size is 210 MB in MPEG4-format.

Fig. 3 Comparison graph of precision and recall

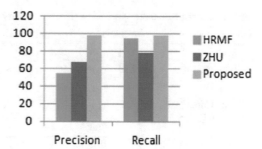

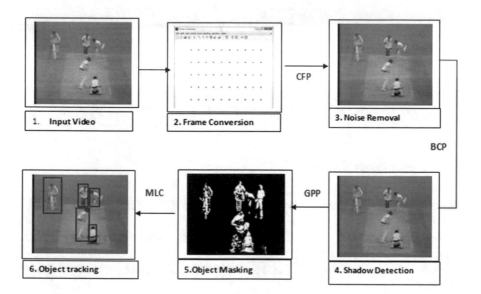

5 Performance Analysis

In this section performance analysis of proposed method is illustrated. The target region detected are calculated with parameters are accuracy, precision, recall, F-measure, success rate, and Failure rate.

Accuracy: Accuracy of Proposed method gives 98.2% compared with existing one (85.3%) it improves by 12.91%.

Precision: The Proposed method gives the precision value of 96.98%. compared with the existing HMRF (64.9%) method offers 38% improvement.

Recall: The Proposed method gives the precision value of 96.7%. compared with the existing HMRF (65.6%) method it improved by 34% (Table 1).

F-measure: The Proposed method provides the 84.53% compared with the existing HMRF (79.5%) method. It improved by 6% (Fig. 3).

Table 1 Comparison table of precision, recall and F-measure

Tracking methods	Precision	Recall	F1 measure
HRMF	64.9	65.6	6.08
BCP-GPP	97.93	97.58	6.19

Table 2 Comparison table of success and failure rate

Methods	Success rate	Failure rate
Frag	3.82	101.3
IVT	4.28	112.1
L1	4.28	160.9
MIL	50.85	64.8
MTT	55.93	23.9
CT	86.92	12.7
Struck	87.32	13.7
PartT	56.93	21.3
MVS	92.14	5
Proposed	**97.93**	4.83

5.1 Complexity Analysis

The complexity of proposed method is texture pattern (P) and the Boundary (B) like $O(P+B)$.

Success Rate: It is the proportion of the number of successful frames to the total frames as Success rate $= \frac{No.of\ successful\ frames}{Total\ no.of\ frames}$.

Failure Rate: It is the fraction of the number of failure frames to the total frames (Table 2).

6 Conclusion

Many computer vision based applications necessitated a perfect object tracking scenario that serves in monitoring the activity of an object. The limitations observed in tracking an object gets alleviated through a deployment of a novel background normalization methodology formulated on the basis of abstracting textural pattern of an image. A Background based chain Prediction approach is devised to suppress the background region from foreground region by means of generating a binary image. The textural pattern of the image is recognized through deployment of Gradient Prediction Pattern (PGP) approach. Comparison between standardized frame obtained in prior and those processed patterns detects the motion exposed by an object and the object concerned gets identified within a blob.

References

1. Mohanapriya, D., Mahesh, K.: Robust video tracking system with shadow suppression based on feature extraction. Aust. J. Basic Appl. Sci. **10**(12), 307–311 (2016)
2. Mohanapriya, D., Mahesh, K.: A novel foreground region analysis using NCP-DBP texture pattern for robust visual tracking. Springer Multimedia Tools Appl. Int. J. **76**(24), pp. 25731–25748 (2017)
3. Wang, D., Lu, H., Yang, M.-H.: Online object tracking with sparse prototypes. IEEE Trans. Image Process. **22**, 314–325 (2013)
4. Zhang, T., Ghanem, B., Liu, S., Ahuja, N.: Robust visual tracking via structured multi-task sparse learning. Int. J. Comput. Vision **101**, 367–383 (2013)
5. Li, Y., Jia, W., Shen, C., van Den Hengel, A: Characterness: an indicator of text in the wild. IEEE Trans. Image Process. **23**, 1666–1667 (2014)
6. Bai, T., Li, Y.: Robust visual tracking using flexible structured sparse representation. IEEE Trans. Ind. Inform.**10**, 538–547 (2014)
7. Minetto, R., Thome, N., Cord, M., Leite, N.J., Stolfi, J.: Snoopertext: a text detection system for automatic indexing of urban scenes. Comput. Vision Image Und. **122**, 92–104 (2014)
8. Makris, A., Prieur, C.: Bayesian multiple-hypothesis tracking of merging and splitting targets. IEEE Trans. Geosci. Remote Sens. **52**, 7684–7694 (2014)
9. Wu, L., Shivakumara, P., Lu, T., Tan, C.L.: A new technique for multi-oriented scene text line detection and tracking in video. IEEE Trans. Multimedia **17**, 1137–1152 (2015)

Latency Aware Reliable Intrusion Detection System for Ensuring Network Security

L. Sheeba and V. S. Meenakshi

Abstract In network communication based applications, Intrusion detection system plays a vital role. In these applications, so as to collapse the system, the malevolent nodes may create enormous amount of traffic. In order to identify and prevent the intrusion activities, there are numerous researches that are presented by diverse researchers. LEoNIDS: architecture solves the energy-latency tradeoff by giving minimum power utilization as well as minimum detection latency all at once. However LEoNIDS architecture focus on the latency but not the precise detection of intrusion. This work resolves these issues by presenting the technique called Attack Feature based Fast and Accurate Intrusion Detection System (AF-FAIDS). This work presents a solution that enables intrusion detection in an effective way with enhanced delay and latency parameter. This also track and eliminate the intrusion attack existing in the system in an effective way with guaranteed energy saving. By using Machine learning, methods like Support Vector Machine, attack detection ratio is enhanced that would learn the attacks features in an effective way. Those features, which are taken in this presented technique, are Average Length of IP Flow, One-Way Connection Density, Incoming and Outgoing Ratio of IP Packets, Entropy of Protocols as well as Length in IP Flow. This technique provides the precise and quicker identification of intrusion attacks that will decrease the latency. The technical work is carried out in the NS2 environment under numerous performance measures and it is confirmed that the presented technique gives improved outcome when compared with the previous research methodologies.

Keywords Intrusion detection system · High throughput · Low latency
Attack feature · Learning method

L. Sheeba (✉)
Bharathiyar University, Coimbatore, India
e-mail: sheebabcom@gmail.com

V. S. Meenakshi
Chikkanna Government Arts College, Tirupur, India
e-mail: meenasri70@yahoo.com

© Springer Nature Singapore Pte Ltd. 2019
J. D. Peter et al. (eds.), *Advances in Big Data and Cloud Computing*,
Advances in Intelligent Systems and Computing 750,
https://doi.org/10.1007/978-981-13-1882-5_33

1 Introduction

Intrusion detection (ID) is one among the ways of security management systems for computers and networks. It collects and analyzes info from numerous fields within a computer or a network for finding probable security holes that comprise intrusions (the assaults that come from outside of the organization) as well as misuse (the assaults that come from the inside of the organization) [1]. It utilizes susceptibility valuation (alias scanning) that is a technology implemented for examining the computer system or network security.

Intrusion Detection System (IDS) may be a hardware system or software system, which systematize the procedure of observing and examining the events, which happen in a network of computers for the purpose of detecting the malevolent activity [2]. ID system have turn out to be an essential inclusion to security infrastructure of numerous organizations, as the criticality of attacks taking place in the network has augmented radically [3]. This system provides the security to the organization to safeguard their systems from the dangers, which arise with increase in network connectivity as well as dependency on information systems. Provided the degree as well as the characteristics of advanced network security attacks, the question posed for security experts must not be whether to utilize intrusion detection however rather which of the features of intrusion detection could be utilized [4].

Intrusions may happen by the following reasons: Assailants accessing the systems, Legal users who abuse the privileges provided to them and Legal users of the systems who try to obtain added rights to which they have not received authorization [5]. In order to identify and deflect the attacks, IDS considers the network- or host-based method. In both the cases, these products try to find attack signatures (particular patterns), which specify malevolent or mistrustful intent. In network based, IDS try to identify these patterns in network traffic. In host-based, IDS tries to identify the attack signatures in log files. For finding diverse kinds of network intrusions, numerous techniques were designed; on the other hand, there is no heuristic for verifying the accurateness of their outcomes [6]. The precise efficacy of a network intrusion detection system's capability for finding out malicious sources could not be stated except a concise measurement of performance is existing.

For distinguishing amid intrusive and non-intrusive network traffic, numerous previous intrusion detection systems depend upon human specialists [7]. Human intrusions are needed to create, test, and deploy the signatures on the examined datasets/databases. Therefore, it might take hours or days to identify/produce a novel signature for an attack that might take in infeasible time to handle speedy attacks. Information security while using Internet is a primary concern to the system users and administrators. As the compromised systems created a severe effect on business as well as personal networks, ID has progressed as a main research issue for researchers, network administrators and cyber-security officers. Numerous systems have been implemented to stop the Internet-based attacks, since there are a lot of risks associated with network attacks underneath the Internet environment. Predominantly, IDSs support the network to counter attack the external attacks. To be exact, the objective of IDSs is to offer a security mechanism to provoke the attacks on computer systems connected to Internet. Intrusion Detection Systems are capable of identifying diverse kinds of malevolent network communications as well as computer systems usage, while the traditional firewall could not carryout these tasks [8].

The research technique presents the new methods that could track and eliminate the intrusion attacks exist on the system in an efficient way with guaranteed energy savings. In the subsequent sub sections, the methods presented in this research are depicted. In this research, by introducing the machine learning methods such as Support Vector Machine, attack detection ratio is enhanced, that will learn the attacks features in an effective way. The attack features taken in this presented system are Average Length of IP Flow, One-Way Connection Density, Incoming and Outgoing Ratio of IP Packets, Entropy of Protocols as well as Length in IP Flow. This technique guarantees the precise as well as speedy identification of intrusion attacks that will decrease the latency.

The complete organization of the presented system is provided along these lines: In this part thorough explanation of the introduction regarding the intrusion detection system is provided. Numerous research works carried out by diverse researchers are explained in the part 2. Thorough discussion of the research technique is provided with appropriate samples and elucidation in part 3. Experimental assessment carried out in MATLAB simulation environment is conversed in part 4. Lastly, complete conclusion of the research is provided.

2 Related Works

Conventional IDSs contain numerous disadvantages for instance, regular updating, time consuming statistical analysis, flexibility, non-adaptive and accurateness. For acquiring data, numerous data obtaining techniques such as sanitized traffic, real traffic, and simulated traffic are utilized. Principally IDS were assessed over a standardized dataset, which is KddCup99. Overheads in training (regular update, time consumption and not capable to identify new attack) and low enhancement in performance (detection rate, false negative, false positive) are considered to be the present problems in IDS utilizing ANN [9].

Numerous approaches are utilized for implementing IDS such as, Train data, Data set generation, and Real-time prediction. A novel system is presented wherein data are utilized as data set of Real Time in this manner. The system is utilized to design the Real-Time Host-based attacks. It is suitable for online environment. Previous methods for ID are typically for offline data and consequently one system is created that could work online [10].

According to [11], it is known that anomaly detection utilizing neural network is as well desirable. In case of anomaly intrusion detection by utilizing a back propagation neural network, "Behavior" is considered as parameters. When a neural network stratifies normal traffic appropriately, and identifies named as well as unnamed attacks deprived of utilizing a big volume of training data. DARPA Intrusion Detection Evaluation datasets are utilized for the purpose of training and testing of the neural network. It is probable to obtain classification rate of 88% for named as well as unnamed attacks by performing the experiments. The following are the steps to build Artificial Neural Network (ANN) based Anomaly Detection System: Pre-process training/testing dataset, Create Training/testing dataset, Identify the structure of the neural network [12], Test neural network, Train neural network. For training as well as learning IDS, Neural networks are utilized as an effective method. The major issue with today's IDS is that they create numerous false alarms, and this consumes numerous system administrator's time as well as resources. Furthermore, during the implementation of IDS or ADS, it is not essential to allot large volume of training information to Neural Network for categorizing the traffic appropriately.

Previous system contains some disadvantages such as, it is dependent upon offline system that was not that much effective and precise and it is utilizing static offline database. A method is utilized for implementing ANN is moreover split into two segments: Training & Testing. When a system is implemented dependent upon the newly introduced technique, it could encompass numerous benefits to IDS such as, it could identify nearly each and every kind of attacks (intrusion), more safe, consistent, exact and effective compared to the existing one, and execution time is minimum [13].

Integrity, Confidentiality, and availability of the system could increase the powerlessness of system to security attacks, threats, and intrusion. For intrusion detection, diverse kinds of neural network are utilized for instance, multilayered feed forward neural network, Self-Organizing feature map [14], Elman back propagation, Cascaded forward back propagation. From the KDD CUP 99 data set, the data is taken for intrusion detection. In this dataset, there are diverse categories of attacks for instance, Prob, Denial of service attack, Remote to user attack (R2U), User to root attack (U2R). Elman Back Propagation (ELBP) IDS provides lesser false positive and false negative rates, superior classification rate when matched up with Cascaded Forward Back. Propagation (CFBP), Multilayer Feed Forward (MLFF), and self-organizing feature maps (SOFM) [15].

Multilayer Feed Forward (MLFF) is presented in [16] that depicts that it is probable to get a novel prevailing neural network by means of a conventional architecture of multilayer feed forward neural network (MLF) as well as the superior operation of the MVN. Its training does not need a derivative of the activation function in addition its function goes beyond the functionality of MLF encompassing the identical amount of layers and neurons. These benefits of MLMVN are definite by assessing by means of parity n, two spirals and "sonar" benchmarks and the Mackey–Glass time series prediction. Relative analysis of Cascaded Forward Back Propagation neural network based IDS and a Hybrid neural network based IDS is carried out in [17], in which cascade connections of two diverse kinds of neural networks, known as Cascaded Forward Back Propagation and Self-Organizing Feature Map, are utilized for ID. In this analysis well-organized KDD CUP 99 dataset is utilized.

The creation of self-organizing feature maps (SOFM) is assessed by means of the direct enhancement of a cost function via a genetic algorithm (GA). The ensuing SOFM is anticipated to create concurrently a topologically right mapping amid input and output spaces and a less quantization error. The research technique takes on a cost (fitness) function that is a weighted mixture of indices, which gauge these two facets of the map quality, specially, the quantization error as well as the Pearson correlation coefficient amid the equivalent distances in input and output spaces. The resultant maps are matched up with those that are produced by the Kohonen's self-organizing map (SOM) technique in regard to the Weighted Topological Error (WTE), Quantization Error (QE) and the Pearson correlation coefficient (PCC) indices. The experimentations prove the research method yields superior values of the quality indices and is more vigorous to outliers.

In [18] authors proposed a new fuzziness-based semi-supervised learning technique by using unlabeled samples combined with supervised learning algorithm for improving the performance of the classifier for the IDSs. A foremost objective of the IDS is to attain accurateness. As a result, certain machine learning methods are required to be formed with the intension that the performance of the system could get enhanced. Certain generally utilized machine learning methods are conversed in [19] that are Single Classifiers, Pattern Classification, Hybrid Classifiers and Ensemble Classifiers.

3 Fast and Accurate Intrusion Detection System

In the communication networking field, IDS is an important field for identifying and preventing the attacks in numerous ways. Identifying the deviation in diverse traffic levels is the most active concern that must be considered on the identification of intrusion process. By means of machine learning algorithms, there is numerous researches concentrates on enhancing the performance of IDS. On the other hand, the previous research is inclined to have more disadvantages in which it does not focus regarding the precise detection of intrusion where it just focus on the latency.

The foremost objective of research method is to present the methods that could be able to carry out intrusion detection in an effective way with enhanced delay as well as latency parameters and could track and eliminate the intrusion attacks exist on the system in the effective way with guaranteed energy savings. In this research, by presenting the machine learning methods like Support Vector Machine, attack detection ratio is enhanced that would learn the attacks features in an effective way. Those features, which are taken in this presented technique, are Average Length of IP Flow, One-Way Connection Density, Incoming and Outgoing Ratio of IP Packets, Entropy of Protocols as well as Length in IP Flow. This technique provides the precise and quicker identification of intrusion attacks that will decrease the latency.

3.1 Attack Detection

With the intention of setting up if the system is under the attack or not, each and every system observes the network traffic. As an alternative, in this technique distributed agents are utilized. They specify resource consumption levels to the master at consistent intervals. These values are matched up in contradiction of threshold values configured by the administrator. The system is taken as under attack when the limits are surpassed. This method is alike to the modus operandi utilized in medicine. A doctor does not take decision on the patient status dependent upon the food the patient has freshly taken. Rather, indications for instance a raised temperature are utilized for identifying an infection. If needed additional action would be taken. No treatment will be given in case no indications are there. As drugs could have serious side effects, this is considered as significant. Supremely treatment starts while the patient indicates the primary indication show ever beforehand he/she gets totally sick.

Transmitting this technique to the region of DDoS attacks denotes that the action is taken only when the system turn out to be burdened because of an attack. We do not consider ineffectual attacks nor do we take any action while the system is regarded to be functioning usually. This decreases the chance that genuine requests are dropped owing to a false positive as may take place with systems considering only the traffic metrics. Once the system is identified to be under attack, countermeasures are instantaneously taken, agreeing that certain benevolent requests may be dropped. On the other hand, since already the system is under attack this is more satisfactory than being not capable to reply any requests in the least.

3.2 Feature Extraction

Choosing suitable features vector is an important thing for identifying DDoS attack with SVM. For checking the features vector, we keep two principles. Primarily, instead of absolute value, select relative value as feature. Taking the reality that diverse nodes in Internet contain numerous flux densities, numerous users, and the greater flux, we select relative value to be the feature, which is free from the network flow. Next, select the parameter, which could distinguish normal flow and attack flow proficiently in the form of feature. By keeping the two principles, we agree eight features to act as a features vector. Each and every features studied in the research are computed at interval $T = 1$ s when there is not note.

3.2.1 One-Way Connection Density (OWCD)

OWCD, replicate the flow exception [9]. In Internet, when a packet is transferred to target, an equivalent returning packet from the target could arrive back to the source.

Two-Way Connection (TWC) is called the connection amid the two packets is. Each and every packet amid the source and the target that is affirmed by a TWC is in the TWC. In contrast, an IP packet deprived of an equivalent returning packet comprises a One-Way Connection (OWC). The proportion of OWC packets to each and every packet is known as One-Way Connection Density (OWCD) in a sampling interval T,

$$OWCD = \frac{\sum OWC\,packets}{\sum IP\,packets} \cdot 100\%$$

OWCD in DDoS attack flow along with spoofing source IP address will rise noticeably. Typically, OWCD is underneath 30 in normal flow, on the other hand it is near to 100 in attack flow.

3.2.2 Average Length of IP Flow

IP Flow, in network analysis, an idea is utilized extensively, denotes that a set of packets contains an identical five-element-group (source IP address, destination IP address, source port, destination port, protocol). A five-element-group could agree an IP flow distinctively. Simply, three categories of IP flow are taken—TCP, UDP and ICMP. The amount of packets is possessed by some IP flow is known as Length of IP flow. In the interval T, the average length of IP flow, L_{ave_flow} is computed as,

$$L_{ave\_flow} = \frac{\sum IP\,packets}{\sum IP\,flows}$$

L_{ave_flow} is typically 5–10 in normal flow, nearing to 1 in attack flow.

3.2.3 Incoming and Outgoing Ratio of IP Packets

The proportion amid incoming and outgoing packets (R_{io}) is constant. Since user of Internet, a general autonomy network contains greater R_{io}, on the other hand it is typically below 15 in normal flow. On the other hand, in DDoS attack flow, as the attack packets are arriving inward, R_{io} will rise to 1000 rapidly.

$$R_{io} = \frac{\sum incoming\,IP\,packets}{\sum outgoing\,IP\,packets}$$

3.2.4 Entropy of Length in IP Flow

Entropy could replicate the randomicity associated with a stochastic variable. Considering ith heft of a discrete random variable x be p_i, the entropy of x is expressed as,

$$E_x = -p_i \sum \log_2 p_i$$

Large E_x denotes x contains big randomicity as well as a lot of information quantity. In normal flow, as randomicity of source IP address is relatively small, L_{ave_flow} is considerably big. Alternatively, comprising numerous packets with several spoofing source IP address, L_{ave_flow} is relatively small in attack flow, typically be inclined to 1. Therefore, L_{ave_flow} and its entropy E_{flow_length} could replicate the randomicity of the source IP address. E_{flow_length} is minor in normal stream than in attack stream. Consider a length set $(l_{f1},\ldots, l_{fn})$ refer to the length of IP flow $(f_1,\ldots,f_n)$ in a interval T. Presuming that $p_i = l_{fi}/(\Sigma \text{ IP packets})$, the entropy of L_{ave_flow} is expressed as,

$$E_{flow\_length} = -p_i \sum \log_2 p_i.$$

In case of normal flow, E_{flow_length} comes to 2–4; in attack flow, E_{flow_length} comes to 8–10.

3.2.5   Entropy of Protocols

Typically, in case of normal stream, the packet ratios of the three protocols, TCP\UDP\ICMP, are static. Hence the randomicities of the three protocols are static as well. The protocol entropy is a constant value, approximately 0.43. On the other hand in attack stream, entire bandwidth gets swamped by attack packets, and the protocol entropy will be inclined to 0. Consider p_t, p_u, p_i are the TCP, UPD and ICMP packets ratio correspondingly. The protocol entropy is defined as,

$$E_p = -p_t \log_2 p_t - p_u \log_2 p_u - p_i \log_2 p_i$$

3.3   Attack Detection Using SVM Learning Approach

In this research, attack detection ratio is increased by presenting the machine learning methods known as Support Vector Machine that would learn the attacks features in an ideal way. Support vector machine (SVM) is typically a supervised algorithm and it is used for prediction purposes in any input provided. This SVM that is best separating hyper plane amidst the two classes of data for the prediction of Intrusion Detection System (IDS) parameters. SVM models are utilized for giving superior prediction outcomes.

Algorithm 1 Support Vector Machine based Attack Learning
   Input: Training Attack feature set (D) which is represented as like in Eq. (1).

$$D = \{x_i, y_i\}_{i=1}^{N}, \quad x \in R^n, y \in \{-1, 1\}, \tag{1}$$

Where,

D → Attack feature set,

x and y –> input variables

The hyper plane separation procedure can be done as like given in Eq. (2).

$$y^i\left[\left|w^T x^i + b\right|\right] \geq 1 \quad i = 1 \text{ to } N \tag{2}$$

w^T and b refer to isolated variables

To reduce the error, we can use given below Eqs. (3) and (4).

$$\Phi(w) = \frac{1}{2}||w||^2 \tag{3}$$

Estimating function is defined as

$$F(x) = \sum_{i=1}^{nsv} (x_i, y_i) k(x_i, y_i) + b \tag{4}$$

SVM Algorithm Procedure

Provided Attack feature set D=$(x_1, y_1),\ldots, (x_n, y_n)$

Initialize vector v=0, b=0; class)// v-vector and b-bias

Train an initial SVM and learn the model

For each $x_i \in X$ do // xi is a vector with features defining example i

Categorize x_i with f (x_i)

If y_i f (x_i) < 1 // prediction class label

Find w′, b′ for known data // w′, b′ for new features

Add x_i to known data

Reduce the error function employing Eq. (3) and estimate utilizing Eq. (4)

If the prediction is incorrect then retrain

Repeat

End

Classify attributes either as normal or abnormal

3.3.1 Training SVM

Training data set comprises four subsets called $T0$, $T1$, $T2$, and $T3$ that specify Normal, Light, Medium, Heavy data correspondingly. Totally there exist 400 normal data in $T0$ and 300 attack data in $T1$, $T2$, and $T3$ respectively. $T1$, $T2$, and $T3$ refer to mixed three types of attack dada, comprising 100 SYN Flood, 100 UDP Flood, and ICMP Flood data correspondingly in keeping with the label. Simply, totally there may be 1300 data present in the training set. Taking out RLT and TRA features from the training set and training SVM correspondingly, we could obtain 2 × 6 SVMs, owing to utilizing 1-v-1 SVM.

Table 1 Composition of testing data in experiment 1

Sorts	Composition of testing data	Sum of data
Normal	Normal data	1200
Light	400 SYN+400 UDP+400 ICMP	1200
Medium	400 SYN+400 UDP+400 ICMP	1200
Heavy	400 SYN+400 UDP+400 ICMP	1200

3.3.2 Testing the Attack Data

With the aim of verifying the training outcomes, carry out two experiments as well as utilize two testing datasets. Primary dataset that contain the category labels is gathered with the identical means of training data. Next dataset is from MIT Lincoln Lab and has not the category labels. Totally there exists four categories of data, Normal, Light, Medium, Heavy, in experiment I, and each sort of dada comprises 1200 data, sum is 4800. The composition associated with checking sets is depicted in Table 1.

4 Experimental Results

For experimental assessment, the NS-2 simulator is utilized for assessing the performance of the presented Attack Feature based Fast and Accurate Intrusion Detection System (AF-FAIDS). This simulations model a network comprising of 100 sensor nodes positioned arbitrarily within a 100 × 100 m area. The two categories of sensor nodes in the simulations are described as: well-behaved nodes and malevolent nodes. The malevolent nodes could launch DDOS attacks in the simulated settings. The BS contains limitless energy. The amount of selected CH is set to 10% for one interval. By matching up the presented method AF-FAIDS with the previous method LEoNIDS, the performance is examined. For assessing the proposed method, the parameters like packet loss, packet delivery ratio, energy consumption, end-to-end delay, and mean packet latency utilized in this work are stated in the Table 2.

4.1 Packet Loss

The overall amount of data packets lost legally or via malevolent action deprived of any notification. The graphical picture of packet loss rate is shown in Fig. 1 that depicts the AF-FAIDS technique contains less packet loss rate while matched up with the previous methods LEoNIDS.

The previous method does not concentrate on dissimilarity amid the legal nodes and the malevolent since it take search sensor node with greater traffic deviation as the

Table 2 Simulation parameters

Simulation parameters	Values
Channel	Wireless channel
Mac	802.11
Antenna type	Omni antenna
Routing protocol	AODV
Initial energy	100 J
Traffic type	CBR
Agent	UDP
Simulation area	100 × 100 m
Number of nodes	100

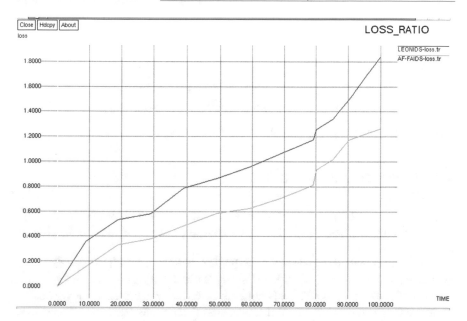

Fig. 1 Comparison of packet loss in different trust model

malevolent. The research method finds the individual malevolent nodes dependent upon the bias and variance value as a consequence the packet drop by the legal nodes could be evaded. The experimentation outcome proves that the presented AF-FAIDS contain less packet loss rate while matched up with the previous LEoNIDS.

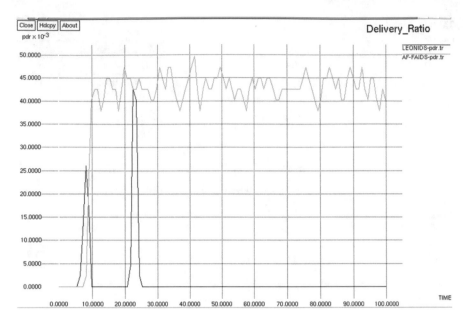

Fig. 2 Comparison of packet delivery ratio for different trust system

4.2 Packet Delivery Ratio (PDR)

PDR is the proportion of the entire amount of data packets obtained to the entire amount of data packets transferred. This exemplifies the delivered data level to the target.

The performance of the presented AF-FAIDS matched up with LEoNIDS regarding the amount of rounds and Packet Delivery Ratio (PDR) is depicted in Fig. 2. The quantity of packets that is efficiently obtained at the target deprived of any packet loss or failure for the presented AF-FAIDS is greater that depicts greater PDR outcomes.

4.3 Energy Consumption

The average energy utilized by every node for the duration of the provided simulation time is stated in Joules (J).

The graphical depiction of energy utilization for diverse trust models used in wireless sensor network of military applications is depicted in Fig. 3. The AF-FAIDStechnique contains less energy utilization while matched up with the previous method LEoNIDS.

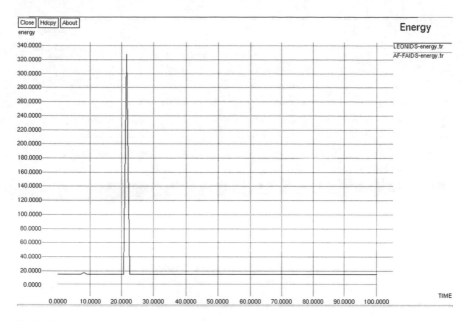

Fig. 3 Comparison made of energy consumption of various trust systems

4.4 End-to-End Delay

It represents the delay undergone by the data packet for the period of transmission from source to BS, comprising processing, queuing, and propagation delay. The pictorial depiction of end-to-end delay for diverse trust models used in wireless sensor network in military applications is depicted in Fig. 4. When hop to hop count distance value is greater, it brings about greater end-to-end delay for the period of path communication. Dependent upon this hop-to-hop count distance data transmission is carried out from source to target in AF-FAIDS system, therefore it outcomes minimum end-to-end delay. In the presented AF-FAIDS system, a greater hop to hop count distance path is not considered for the purpose of data transmission.

The AF-FAIDS technique contains less end-to-end delay while matched up with the previous method LEoNIDS. In AF-FAIDS system, greater hop to hop count distance paths is not considered for the purpose of data transmission and that path is taken as attack path.

4.5 Mean Packet Latency

The mean packet latency for those packets, which attained the target is lesser for AF-FAIDSs it is able to choose the shortest route with the least amount of hops.

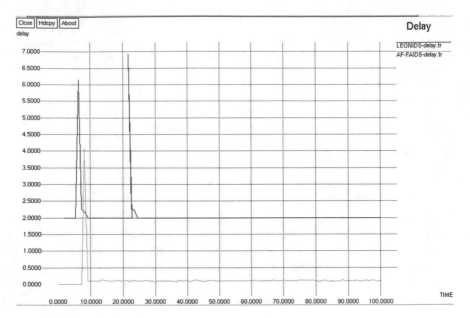

Fig. 4 End-to-end delay comparisons made for various trust systems

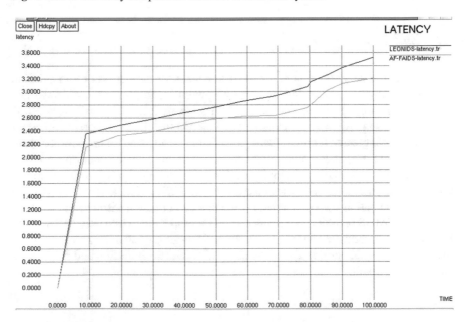

Fig. 5 Mean packet latency

In the research technique, mean packet latency is decreased because of decreased malevolent attacks. Figure 5 depicts the pictorial depiction of the mean packet latency.

Table 3 Simulation parameter metric values

Time	Performance metrics									
	Packet loss ratio		Delivery ratio		Energy		Delay		Latency	
	LEONIDS	AF-FAIDS	LEONIDS	AF-FAIDS	LEONIDS	AF-FAIDS	LEONIDS	AF-FAIDS	LEONIDS	AF-FAIDS
10	0.38	0.18	5	40	18	2	2	0.3	1.39	2.2
20	0.56	0.36	0	46	18	2	2	0.3	2.5	2.38
30	0.6	0.4	0	43.5	18	2	2	0.3	2.6	2.4
40	0.8	0.5	0	45	18	2	2	0.3	2.7	2.5
50	0.88	0.6	0	47	18	2	2	0.3	2.79	2.6
60	0.98	0.63	0	45	18	2	2	0.3	2.85	2.62
70	1.19	0.74	0	43	18	2	2	0.3	2.95	2.63
80	1.2	0.95	0	38	18	2	2	0.3	3.1	2.8
90	1.5	1.18	0	45	18	2	2	0.3	3.39	3.16
100	1.83	1.25	0	40	18	2	2	0.3	3.5	3.2
Average	0.992	0.679	0.5	43.25	18	2	2	0.3	2.777	2.649
% improvement	31.55242		8550		88.88889		85		4.609291	

The AF-FAIDS technique contains less packet latency while matched up with the previous method LEoNIDS. In AF-FAIDS system, greater hop to hop count distance paths is not considered for the purpose of data transmission and that path is taken as attack path.

4.6 Numerical Analysis

In the following Table 3, numerical values of the proposed simulation metrics obtained are given.

From this Table 1, it can be shown that the newly introduced research strategy provides the optimal results in comparison with the available research methodologies. This comparison proves that the newly introduced technique AF-FAIDS is improved in its performance than the existing method LEPNIDS. AF-FAIDS shows 31.55% decreased loss ratio than LEONIDS, 85,550% increased packet delivery ration than LEONIDS, 88.89% reduced energy consumption than LEONIDS, 85% reduced delay than LEONIDS and 4.6% reduced delay than LEONIDS.

5 Conclusion

The foremost objective of research method is to present the methods that could be able to carry out intrusion detection in an effective way with enhanced delay as well as latency parameters and could track and eliminate the intrusion attacks exist on the system in the effective way with guaranteed energy savings. In this research, by presenting the machine learning methods like Support Vector Machine, attack detection ratio is enhanced that would learn the attacks features in an effective way. Those features, which are taken in this presented technique, are Average Length of IP Flow, One-Way Connection Density, Incoming and Outgoing Ratio of IP Packets, Entropy of Protocols as well as Length in IP Flow. This technique provides the precise and quicker identification of intrusion attacks that will decrease the latency.

Acknowledgements I would like to show my warm thanks to Dr. V. S. Meenakshi , Research supervisor who's guidance proved to be a milestone effort toward the success of my research. I also wish to pay my special acknowledgement with high sense of appreciation, for the love and support of my family and my parents Mr. K. Lakshmanan and Mrs. L. Kalaiselvi and also my heartfelt thanks to my husband D. Selvaraj without whom the research is not possible. I would also like to extend my thanks to my friends who assisted me throughout the completion of this research and made me achieve my goal. Finally, wishing to recognize the valuable help of all provided during my research.

References

1. Cárdenas, A.A., Berthier, R., Bobba, R.B., Huh, J.H., Jetcheva, J.G., Grochocki, D., Sanders, W.H.: A framework for evaluating intrusion detection architectures in advanced metering infrastructures. IEEE Trans. Smart Grid **5**(2), 906–915 (2014)
2. Depren, O., Topallar, M., Anarim, E., Ciliz, M.K.: An intelligent intrusion detection system (IDS) for anomaly and misuse detection in computer networks. Expert Syst. Appl. **29**(4), 713–722 (2005)
3. Vasudevan, A., Harshini, E., Selvakumar, S.: SSENet-2011: a network intrusion detection system dataset and its comparison with KDD CUP 99 dataset. In: IEEE International Conference on Internet (AH-ICI), Second Asian Himalayas, pp. 1–5 (2011)
4. Garcia-Teodoro, P., Diaz-Verdejo, J., Maciá-Fernández, G., Vázquez, E.: Anomaly-based network intrusion detection: techniques, systems and challenges. Comput. Secur. **28**(1–2), 18–28 (2009)
5. Duhan, S.: Intrusion Detection System in Wireless Sensor Networks: A Comprehensive Review, pp. 2707–2713 (2016)
6. Srinivasan, T., Vivek V., Chandrasekar, R.: A self-organized agent-based architecture for power-aware intrusion detection in wireless ad-hoc networks. In: IEEE International Conference on Computing & Informatics, pp. 1–6 (2006)
7. Bahrololum, M., Salahi, E., Khaleghi, M.: Machine learning techniques for feature reduction in intrusion detection systems: a comparison. In: IEEE Fourth International Conference on Computer Sciences and Convergence Information Technology, pp. 1091–1095 (2009)
8. Hu, X., Runzi B.: Research on intrusion detection model of wireless sensor network. In: IEEE International Conference on Computer Science and Service System (CSSS), pp. 3471—3474 (2011)
9. Iftikhar, A., Azween, B.A., Abdullah, S.A.: Artificial neural network approaches to intrusion detection: a review. Telecommunications and Informatics Book as ACM guide Included in ISI/SCI Web of Science and Web of Knowledge (2009)
10. Badgujar, T., More, P.: A review for an intrusion detection system combined with neural network. Int. J. **4**(3) (2014)
11. Pradhan, M., Pradhan, S.K., Sahu, S.K.: Anomaly detection using artificial neural network. Int. J. Eng. Sci. Emerg. Technol. **2**(1), 29–36 (2012)
12. Mukkamala, S., Janoski, G., Sung, A.: Intrusion detection using neural networks and support vector machines. IEEE Int. Joint Conf. Neural Netw. **2**, 1702–1707 (2002)
13. Somwanshi, P.D., Chaware, S.M.: A review on: advanced artificial neural networks (ANN) approach for IDS by layered method. Int. J. Comput. Sci. Inf. Technol. (IJCSIT) **5**(4), 5129–5131 (2014)
14. Maia, J.E.B., Barreto, G.A., Coelho, A.L.: On self-organizing feature map (SOFM) formation by direct optimization through a genetic algorithm. In: IEEE Eighth International Conference on Hybrid Intelligent Systems, pp. 661–666 (2008)
15. Afrah, N.: A comparative study of different artificial neural networks based intrusion detection systems. Int. J. Sci. Res. Publ. **3**(7) (2013)
16. Aizenberg, I., Claudio, M.: Multilayer feed forward neural network based on multi-valued neurons (MLMVN) and a back propagation learning algorithm. Soft. Comput. **11**(2), 169–183 (2007)
17. Nazir, A.: A comparative study of Cascaded forward back propagation and hybrid SOFM-CFBP neural networks based intrusion detection systems. Int. J. Sci. Eng. Res. **4**(6), 2447–2452 (2013)
18. Ashfaq, R.A.R., Wang, X.Z., Huang, J.Z., Abbas, H., He, Y.L.: Fuzziness based semi-supervised learning approach for intrusion detection system. Inf. Sci. **378**, 484–497 (2017)
19. Khari, M., Karar, A.: Analysis on intrusion detection by machine learning techniques: a review. Int. J. Adv. Res. Comput. Sci. Softw. Eng. **3**(4) (2013)
20. Tsikoudis, N., Papadogiannakis, A., Markatos, E.P.: LEoNIDS: a low-latency and energy-efficient network-level intrusion detection system. IEEE Trans. Emerg. Topics Comput. **4**(1), 142–155 (2016)

IoT-Based Continuous Bedside Monitoring Systems

G. R. Ashisha, X. Anitha Mary, K. Rajasekaran and R. Jegan

Abstract Oxygen saturation is an important parameter for people with breathing and heart problems to monitor their oxygen levels. The method proposed here uses pulse oximeter sensor to measure SpO_2 (oxygen saturation). In this, two LEDs (Red and IR) with different wavelengths are used to obtain the signal from finger. Noninvasive system permits pain-free continuous monitoring of patient with minimum risk of infection. The oxygen saturation (SpO_2) in blood is obtained and it is transmitted wirelessly through Bluetooth and further, these data can be given to the doctor through cloud. Noninvasively obtained SpO_2 is compared with the standard device values. This measurement can provide early warning to potential problems that your body may potentially be facing.

Keywords SpO_2 · Noninvasive · Pulse oximeter · Photo-plethysmogram
Hemoglobin

1 Introduction

An Advanced Bedside Monitoring system uses sensors with wireless transmission capabilities. Noninvasive methods are chosen over the invasive methods because it does not require any surgical cut into the body. Oxygen saturation is defined as the ratio of oxygenated hemoglobin (HbO_2) to total hemoglobin $(Hb + HbO_2)$. Human

G. R. Ashisha (✉) · X. Anitha Mary · K. Rajasekaran · R. Jegan
Electronics and Instrumentation Engineering, Karunya Institute of Technology and Sciences,
Coimbatore, Tamil Nadu, India
e-mail: grashisha27@gmail.com

X. Anitha Mary
e-mail: anithamary@karunya.edu

K. Rajasekaran
e-mail: k_rajasekaran@karunya.edu

R. Jegan
e-mail: jegan@karunya.edu

© Springer Nature Singapore Pte Ltd. 2019   401
J. D. Peter et al. (eds.), *Advances in Big Data and Cloud Computing*,
Advances in Intelligent Systems and Computing 750,
https://doi.org/10.1007/978-981-13-1882-5_34

Fig. 1 Blood oxygenation
[1]

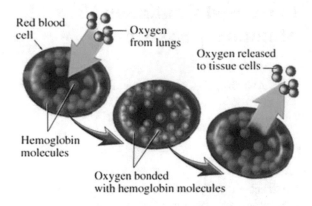

body needs specific balance of oxygen in the blood. SpO$_2$ measurement shows whether the adequate quantity of oxygen is being supplied to the blood. Pulse oximeter is an important method for monitoring blood oxygen saturation (SpO$_2$). Functioning of oximeter depends on the light absorption of deoxygenated hemoglobin and oxygenated hemoglobin. Oxygenated hemoglobin passes only the red light and it takes more infrared light. Deoxygenated hemoglobin allows the infrared light to pass and it absorbs the red light. Wavelength range of the red light is 600–750 nm and the wavelength range of IR (infrared) light is 850–1000 nm. Hence, it is necessary to use different wavelengths for Red and IR LEDs. In invasive method, blood is drawn from the patient to determine the measurement and an extra care should be taken to maintain the samples. Noninvasive method allows infection-free operation with endless online patient monitoring. Figure 1 shows the blood oxygenation in our body [1].

2 Literature Review

A PPG signal can be obtained by using a pulse oximeter which lights up the skin and estimates the changes in light absorption. It is very important to get an artifact-free signal for accurate measurement [2]. Reflection-type pulse oximeter contains transmitter channel and the receiver channel, which receives and processes the signal from the sensor [3]. By choosing the two wavelengths closely, one can minimize the wavelength dependence variation in the measurement [4]. Multi-wavelength approach needs a laser, which is very expensive. Higher accuracy can be achieved by utilizing the Dual-wavelength method than the single wavelength approach [5]. Selections of light sources are very critical for accurate measurement of oxygen saturation [6]. PPG (Photo-plethysmography) signal arises due to the blood volume pulse from the intensity variation of the skin [7]. Power demand of the pulse oximeter system depends on the switching circuit and the power dissipation of the oscillator/LED, which is being used to illuminate both the LEDs (RED and IR) in the system [8]. Oxygen saturation in the blood can be determined by measuring the transmitted light through the finger with known intensity, path length, and the extinction coefficient of a substance (oxyhemoglobin) [9]. Patients are monitored continuously using wearable reflectance pulse oximeter to obtain oxygen saturation [10].

Color of the skin has high influence in the SpO_2 measurement [11]. Ordinary pulse oximeter is so insensate to blood and they can be used safely in human [12]. Pulse oximeter measures the blood oxygen saturation (SpO_2) using Red and IR photoelectric-Plethysmography (PPG) signal [13]. Beer–Lambert law can be applied to the tissues to estimate the amount of oxygen in blood [14]. An optical absorption characteristic of blood yields the information about hemoglobin (Hb) and the arterial blood oxygen saturation (SpO_2) [15].

3   Proposed Method for SpO_2 Monitoring

SpO_2 (arterial oxygen saturation), which is an amount of oxygenated hemoglobin in blood. A normal person must be able to have oxygen saturation of 95–99%. SpO_2 is expressed using the Eq. (1)

$$SpO_2 = \frac{HbO_2}{HbO_2 + Hb} \times 100\% \tag{1}$$

The hemoglobin (Hb) is the nutrient molecule in the red blood cells, which is accountable for carrying the oxygen from the lungs to all the parts of the body. Hemoglobin can be oxygenated (HbO_2) or deoxygenated (Hb).

3.1   Principle of Pulse Oximeter

Pulse oximeter is a device used to measure the oxygen saturation level and the pulse rate. The light source and the photodetector were accommodated inside the pulse oximeter. Two modes of Pulse oximeter are transmission and the reflectance mode. The mode which is used in this proposed method is the reflectance mode (Fig. 2).

In the reflectance mode pulse oximeter, the light source and the diode are on the same side and the transmitted light is reflected back to the photodiode athwart the measurement area. In this proposed method, arterial oxygen saturation measurement is taken at the fingertips of the patient. PPG signal can be obtained very easily from the fingertips and also this location is very convenient to the patients.

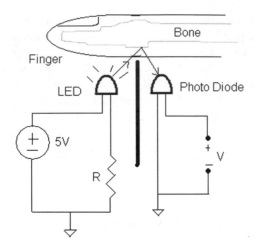

Fig. 2 Reflective-type pulse oximeter

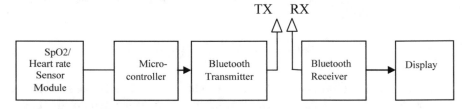

Fig. 3 Block diagram

3.2 Block Diagram

The block diagram of the Continuous Bedside Monitoring System for monitoring Oxygen saturation of the patient is shown in the Fig. 3.

The finger is placed on the sensor board (MAX30102), both the LEDs emits light into the finger. The lights which are reflected back to the sensor are detected by the detector. Red and the IR (infrared) signals which are obtained are processed by an Atmega328 processor and the SpO_2 is calculated. Then, the data is transmitted through the Bluetooth Module.

3.3 Experimental Setup

The MAXREFDES117 is a sensor module which measures both the pulse rate and the pulse oximetry [16]. Method of measuring a person's oxygen saturation is called as pulse oximetry. This module uses the MAX30102 sensor which operates on a

Fig. 4 MAXREFDES117
sensor module

Fig. 5 Experimental setup
to detect SpO$_2$ value

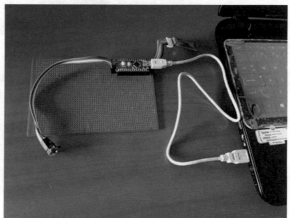

single 1.8 V power supply. 5 V is the input voltage for the internal LEDs of the sensor (Fig. 4).

The sensor communicates with the Microcontroller through I2C Interface. In I2C, clock signal is generated by the master device (microcontroller) and the data has been sent through the Serial data (SDA) for each clock signal.

The sensor is connected to the Atmega328 processor in order to process the signal as shown in Fig. 5. HC-05 is the Bluetooth module which is used for transmitting the value at the frequency of 2.4 GHz. Red, IR, and the SpO$_2$ values are obtained in Arduino IDE and the values are transmitted wirelessly to the receiver through Bluetooth.

Measurement is taken by placing the finger of the patient on the sensor board. Stable placement of the finger gives the better accuracy. Interrupt is given by the

Fig. 6 SpO$_2$ measurement

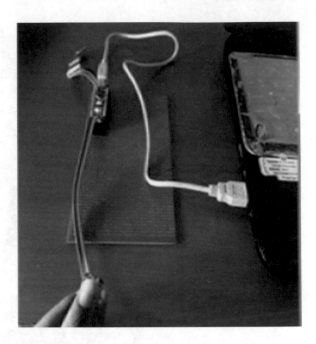

sensor to the processor which reads the data from the sensor. After receiving the interrupt from the sensor, the processor starts to read the data from the sensor. The arterial Oxygen saturation value can determined from the PPG signal. Figure 6 shows the method to measure the oxygen saturation of a person. One of the main advantages of this method is its fast response and it does not require any expert to operate.

Red, IR, and the oxygen saturation values are estimated and displayed in Arduino IDE. To get the better results, averaging of SpO$_2$ is done and then the value is displayed in the Arduino IDE. SpO$_2$ averaging is done by taking the average of 50 values of SpO$_2$ (Fig. 7).

Averaging of SpO$_2$ increases the accuracy of the measurement. Then, the oxygen saturation value is transmitted wirelessly to the receiver through the Bluetooth.

Transparent Serial Wireless Connection is possible with Bluetooth. Default baud rate of HC-O5 is 38,400. 3.3 V is the input voltage of this module and it can be configured as master or slave using AT Commands. Figure 8 shows the transmitted value received by the receiver Bluetooth module. Hence, the values can be monitored from remote places. Bluetooth is considered as very cheapest mode of communication, so it is preferred for the wireless transmission but the transmission distance of HC-05 Bluetooth is 10 m. Hence, it can only allow transmission within a range of 10 m.

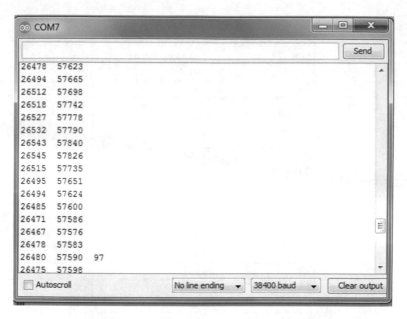

Fig. 7 SpO₂ after averaging

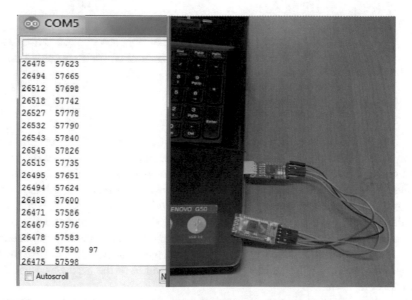

Fig. 8 Bluetooth transmission using HC-05

Table 1 Comparison of sensor value with the standard value

Patient No.	SpO$_2$ value from sensor	SpO2 value from standard device	Error percentage
1	99	97	2.0
2	99	98	1.0
3	98	97	1.0
4	97	97	0
5	98	98	0
6	96	96	0
7	99	98	1.0
8	98	97	1.0
9	95	95	0
10	94	95	1.0
11	97	97	0
12	99	97	2.0
13	98	98	0
14	98	97	1.0
15	97	96	1.0
16	99	99	0
17	95	94	1.0
18	98	96	2.0
19	97	96	1.0
20	98	99	1.0

4   Result and Discussion

The oxygen saturation measurement helps the patient to know their oxygen levels in the blood. If the oxygen saturation is below 95%, then it indicates that the patient has breathing problem. By measuring the Oxygen saturation (SpO$_2$), one can able to identify whether the sufficient oxygen is supplied to the blood or not. Low-level oxygen (hypoxemia) causes respiratory disorder and the tissue hypoxia. Hypoxemia can damage the liver, lungs, and other organs. Early detection and the prevention of hypoxemia are very challenging since the process has a chance to get inaccurate result due to doctor's subjectivity. But the early detection of hypoxia can save the life of a person before it makes harm to them. Maintenance of the hardware is difficult because of the cost.

Measurement taken using pulse oximeter sensor is compared with the standard device. Table 1 shows the sensor value, standard device value, and the error percentage between the sensor value and the standard value. And further, these data can be given to the doctor through cloud.

5 Conclusion

The proposed method provides noninvasive and accurate measurement in biomedical applications. Accurate result can be obtained by the stable placement of the finger. Accuracy of the method depends on the body contact, so the body movement will arise error in the measurement. Accuracy of the result will be affected by the nail polish so it is better to remove the nail polish while taking the measurement. Henna Pigmentation on the hands of the patient is also the main factor which affects the measurement of the proposed system. Future work is to design a wearable device for bedside monitoring system. As future work, the value of oxygen saturation and hemoglobin measurement are done and transmitted to doctor and medical officers through cloud computing.

Acknowledgements The authors are grateful to Karunya Institute of Technology and Sciences for providing needful facilities to do the experiment.

References

1. http://janeshealthykitchen.com/24-ways-increase-blood-oxygenation/
2. Mohan, P.M., Nisha, A.A., Nagarajan, V., Jothi, E.S.J.: Measurement of arterial oxygen saturation (SpO$_2$) using PPG optical sensor. In: International Conference on Communication and Signal Processing (ICCSP), pp. 1136–1140. IEEE (2016)
3. Jarosz, A., Kolacinski, C., Wasowski, J., Szymanski, A., Kurjata-Pfitzner, E., Obrebski, D., Pleskacz, W.A.: Design and measurements of the specialized VLSI circuit for blood oxygen saturation monitoring. In: 14th International Conference the Experience of Designing and Application of CAD Systems in Microelectronics (CADSM), pp. 448–451. IEEE (2017)
4. Gao, F., Peng, Q., Feng, X., Gao, B., Zheng, Y.: Single-wavelength blood oxygen saturation sensing with combined optical absorption and scattering. IEEE Sens. J. **16**(7), 1943–1948 (2016)
5. Peng, Q., Gao, F., Feng, X., Zheng, Y.: Continuous Blood Oxygen Saturation Detection with Single-Wavelength Photoacoustics (2015)
6. Shao, D., Liu, C., Tsow, F., Yang, Y., Du, Z., Iriya, R., Tao, N.: Noncontact monitoring of blood oxygen saturation using camera and dual-wavelength imaging system. IEEE Trans. Biomed. Eng. **63**(6) (2016)
7. Feng, L., Po, L.M., Xu, X.L.Y., Ma, R.: Motion-resistant remote imaging photo plethysmography based on the optical properties of skin. IEEE Trans. Circuits Syst. Video Technol. **25**(5), 879–891 (2015)
8. Maziar, T., Turicchia, L., Sarpeshkar, R.: An ultra-low-power pulse oximeter implemented with an energy-efficient Trans impedance amplifier. IEEE Trans. Biomed. Circuits Syst. **4**(1), 27–38 (2010)
9. Ogunduyile, O.O., Oludayo, O.O., Lall, M.: Development of wearable systems for ubiquitous healthcare service provisioning. APCBEE Procedia **7**, 163–168 (2013)
10. Fu, Yu., Jian, Liu: System design for wearable blood oxygen saturation and pulse measurement device. Procedia Manuf. **3**, 1187–1194 (2015)
11. Mishra, D., Priyadarshini, N., Chakraborty, S., Sarkar, M.: Blood oxygen saturation measurement using polarization-dependent optical sectioning. IEEE Sens. J. **17**(12), 3900–3908 (2017)
12. Zijlstra WG, A Buursma., WP Meeuwsen-Van der Roest (1991) Absorption spectra of human fetal and adult oxyhemoglobin, de-oxyhemoglobin, carboxyhemoglobin, and methemoglobin. Clinical chemistry 37.9: 1633–1638

13. Reddy, K.A., George, B., Mohan, N.M., Kumar, V.J.: A novel method for the measurement of oxygen saturation in arterial blood. In: Instrumentation and Measurement Technology Conference (I2MTC), pp. 1–5. IEEE (2011)
14. Martinez, L.: A non-invasive spectral reflectance method for mapping blood oxygen saturation in wounds. In: Proceedings of 31st Applied Imagery Pattern Recognition Workshop. IEEE (2002)
15. Kraitl, J., Klinger, D., Fricke, D., Timm, U., Ewald, H.: Non-invasive measurement of blood components. In: Advancement in Sensing Technology, pp. 237–262. Springer, Berlin, Heidelberg (2013)
16. MAXREFDES117:http://datasheet.octopart.com/MAXREFDES117%23-Maxim-Integrated-datasheet-66615138.pdf

IoT-Based Vibration Measurement System for Industrial Contactors

J. Jebisha, X. Anitha Mary and K. Rajasekaran

Abstract Vibration analysis is made to study the behavior of the contactors during abnormal conditions. Contactors are electrical devices which are used in many applications such as lighting control, conveyor systems, and so on. Contactors produce noise and vibration during abnormal conditions, which shows that it is a bad contactor. At present, the testing of contactors, whether good or bad, is done manually which is prone to human error. This paper presents the set up which uses vibration sensor to measure the vibration. The sensors used here are accelerometer and vibration sensor. The obtained vibration is analyzed using LabVIEW software. From the analysis made, a threshold is set to differentiate the good contactors from the bad ones. This method is very cheap to implement. It provides better accuracy than many other existing methods. Further, the number of good contactors detected per day will be intimated to the higher officials through cloud computing.

Keywords Vibration · Contactor · LabVIEW · Accelerometer sensor
Contact time

1 Introduction

Contactors are large relay which is used to switch large amount of electrical power. These contactors have multiple contacts in it. They are used in lighting control, pump stations, conveyor systems, air conditioning, and so on. The behavior of these contactors during abnormal condition is studied in this paper. There are various reasons

J. Jebisha (✉) · X. Anitha Mary · K. Rajasekaran
Electronics and Instrumentation Engineering, Karunya Institute of Technology
and Sciences, Coimbatore, Tamil Nadu, India
e-mail: jebisha@karunya.edu.in

X. Anitha Mary
e-mail: anithamary@karunya.edu

K. Rajasekaran
e-mail: k_rajasekaran@karunya.edu

© Springer Nature Singapore Pte Ltd. 2019
J. D. Peter et al. (eds.), *Advances in Big Data and Cloud Computing*,
Advances in Intelligent Systems and Computing 750,
https://doi.org/10.1007/978-981-13-1882-5_35

411

for contactor failure. Some of the reasons are high current, electrodynamic forces during short circuit, overvoltage, continual chattering, aging temperature, power quality, Corrosive environment, voltage and frequency fluctuation, and mechanical shock. Contacts melt, if high current is passed through it. Continual chattering also melts the contacts. Aging causes the coils inside the contactor to crack which causes the insulation to break. Chemicals and vapors in the surrounding environment cause damage in the contactor coil. At present, testing of these contactors is done manually. Hence, they are prone to human error. This paper explains a setup to automatically detect whether the contactor is good or bad. The proposed method acquires vibration from the contactor, analyze it using Lab VIEW software, and detect whether the contactor is good or bad at the production level. Vibration sensors are used here to acquire the vibrations. This method provides an accurate result when compared to many other existing systems. Error percentage is also highly reduced.

2 Literature Survey

Numerous methods have been proposed by various researchers regarding vibrations. A brief review of some contributions is presented in this section.

K. Bouayed has modeled a procedure which is based on the chaining of three analysis methods. First one is the generation of Maxwell forces on magnetic core due to electrical excitation. The two models of ferromagnetic considered are bloc model and laminated model. The model shows good agreement with measurements when the mechanical and acoustic models are compared. If the air gap region of the core is incompressible, then vibration becomes negligible [1]. Hui Zhang, in this paper, investigates the fundamental effects of electrical loads containing non-resistive components such as rectifiers and capacitors. A resistor with a rectifier, a resistor with a rectifier and a capacitor, and a simple charging circuit containing a rectifier and a capacitor are the three types of electrical loads which are being considered. From the results obtained, it is verified that the device performance cannot be generalized to applications involving rectifier and capacitor loads. The results are not directly applicable to piezoelectric devices [2]. In this paper, P. C. Latane detects the catastrophic faults of electric motor at the early stage. Vibration is sensed using vibration sensor and the vibration analysis is done by DSP-based measurement system. Frequency analysis is also done. For large size motors, current-based fault detection is required [3]. The study about the reliability of electronic components during random vibration is studied in this paper. ANSYS software is used to build Finite Element model test bench. This helps in the accurate prediction of induced fatigue life for the analyst. Comparison between simulation and experimental data are done. The mass effect of accelerometer, eigenvectors, and effect of frequency band on fatigue life are also studied [4]. This paper proposes the Generic sensitivity curves of ac coil contactors. The sensitivity of these ac coil contactors depends on voltage sags, under voltage transients, and short interruptions. The various tests which are performed are sensitivity of the contactor to rectangular voltage sags, testing with

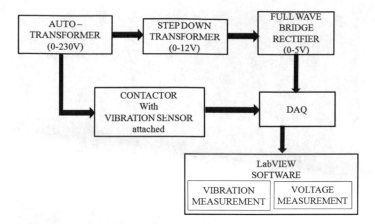

Fig. 1 Block diagram for proposed methodology

two-stage voltage sags, and so on. In spite of performing all these tests, there were fundamental behavior differences between the contactors [5]. Viral K. Patel in this paper uses an embedded accelerometer and an Arduino microcontroller for vibration measurement. The sensor collects 3D vibrations. Bearing fault detection technique is the technique which is used for online bearing condition monitoring. By this module, the fidelity of vibration measured can be maintained [6]. Vibration analysis is done for rotating machines using a DSP-based system. The vibrations from the rotating machines are done online. The system is highly reliable as it is able to isolate the faults. The system is very efficient even in conditions with faults [7]. Impact detection sensor and PVDF sensor are used to study the dynamic characteristics of the AC contactors in this paper. Selection phase angle can affect the closing process of contactors. Hence, a closing strategy is developed to overcome this problem. By using this control strategy, there is a decrease in the impact force. Consistency in the closing process is also guaranteed [8].

3 Proposed Methodology

The setup for the analysis of vibration is shown below. For the acquisition of data, a DAQ and vibration sensor is used. The vibration sensor used here are accelerometer and vibration sensor.

Figure 1, shows the block diagram of the proposed methodology. The setup has an auto-transformer which can be varied from 0 to 230 V. It is stepped down to 12 V using a step-down transformer. A full wave rectifier is used to convert the incoming alternate current to direct current.

It is then given to the Data Acquisition system. The auto-transformer is connected to the contactor with the vibration sensor placed behind it. Again, it is given to the

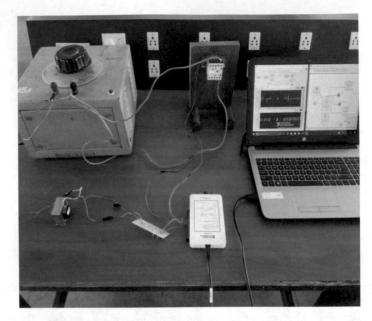

Fig. 2 Experimental setup for acquiring data

Data Acquisition system. Through the LabVIEW software, the acquired signals are viewed, analyzed and the vibration can be viewed. The corresponding voltage can also be viewed.

Figure 2 shows the experimental setup made to acquire the vibration. Figure 3 shows the image of the contactor, which is to be differentiated as good or bad. The experimental setup in the Fig. 2 is explained as follows. The contactor is mounted on a vertical wooden stand on a clamp. The vibration sensor is placed on the wooden stand and behind the contactor. Voltage is passed to the contactors from the auto-transformer. Voltage is increased from 0 to 110 V. The vibration from the contactor is acquired using the sensor. When the voltage is increased, at certain voltage ranges, the vibrations are high and at certain voltage ranges, the vibrations are low. The vibration reading from the sensor is given to the DAQ. The output from DAQ is given to the LabVIEW software from which the vibration analysis is made. The acquired raw signal and the corresponding voltage can be displayed and analyzed in the LabVIEW software. Analysis is made from ten contactors which were provided by the company. The two vibration sensors used are accelerometer and vibration sensor.

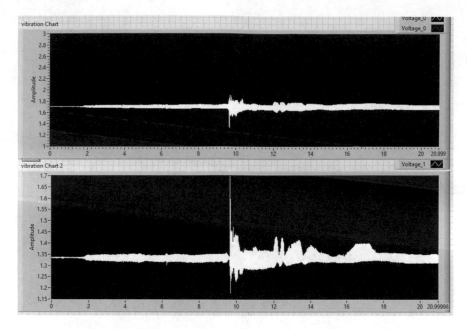

Fig. 3 Signal from bad contactor (at rated hum) using Accelerometer sensor

3.1 Using Accelerometer

Proper acceleration of a body can be measured using an Accelerometer. Accelerometers are used in inertial navigation system, table computers, digital cameras, and so on. The accelerometer used here has three axes, namely x, y, and z. Readings are obtained from the x and y coordinates of the accelerometer. The output from the accelerometer is given to the DAQ as input. The raw signal obtained from the accelerometer for both good and bad contactors at rated and pickup voltage are shown below.

Figure 3 shows the raw signal obtained from bad contactor at rated hum. Rated hum is the voltage range from 45 to 65 V, where the vibration is expected to occur, as specified by the company. The high peak in the graph shows the contact of the two coils inside the contactor. The vibration after the contact is very high which shows the presence of abnormal vibration in the contactor.

Figure 4 shows the raw signal obtained from another bad contactor at pickup hum. Pickup is the voltage range at the ranges of 70 V. After the high peak, the vibrations are high which again shows abnormal vibration in the contactor.

Figure 5 shows the signal obtained for two different good contactors. The ranges of vibration are low after the contact point, showing that there was no or minimal vibration after the contact. The vibration before the contact time is present for both good and bad contactors. This vibration cannot be eliminated as it occurs during

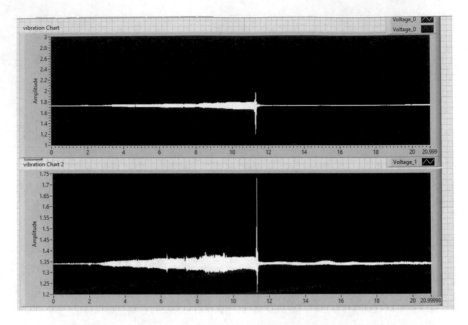

Fig. 4 Signal from bad contactor (at pickup hum) using accelerometer sensor

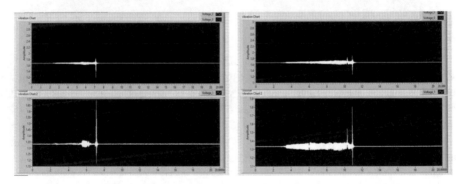

Fig. 5 Signal from good contactor using accelerometer sensor

the contact of the coils inside the contactors. The experiment was conducted for ten contactors which included good and bad contactors in it.

3.2 Using Vibration Sensor

With the same experimental setup, the accelerometer is replaced with vibration sensor. By using piezoelectric effect, the vibration sensor measures the changes in pressure, acceleration, strain, and force. They are used in industries for process control

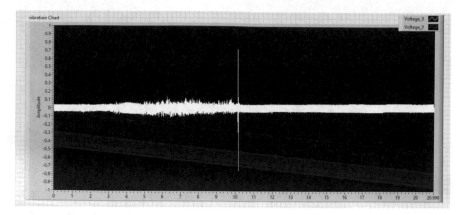

Fig. 6 Signal acquired from contactor at rated voltage using vibration sensor

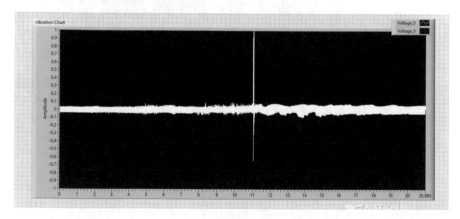

Fig. 7 Signal acquired from contactor at pick-up voltage using vibration sensor

and quality assurance. The vibration is accrued using DAQ and is viewed in the LabVIEW software. The vibration and voltage graph are obtained. The vibration for both good and bad contactor is shown below.

Figure 6 shows the raw vibration signal obtained from the bad contactor at rated hum using vibration sensor. As mentioned earlier, the range of voltage from 45 to 65 V is called rated hum. Contact time of the coils can be seen as a high peak. The vibrations are high after the contact indicating abnormal working condition of the contactor. The vibration before the contact time is high. As mentioned earlier, it occurs during the contact of the coils inside the contactors when energized. This can be clearly viewed in the raw signal that is being captured in the LabVIEW software.

Figure 7 shows the raw signal obtained from a bad contactor using vibration sensor at pickup hum. Pickup hum is the voltage range at the ranges of 70 V. The vibration after the contact time is very high.

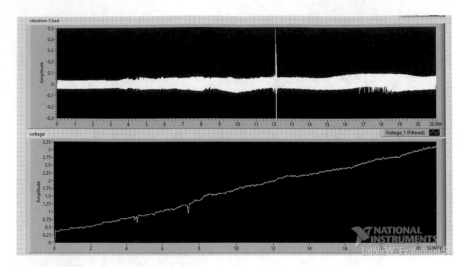

Fig. 8 Signal from good contactor

Figure 8 shows the raw vibration signal for a good contactors using vibration sensor. Vibrations are less after the contact time. There are no significant changes in the vibration, showing that it is a good contactor.

4   Result and Discussion

Now, by using the experimental setup and the raw graph obtained from it, it is possible to differentiate the good and bad contactors by viewing at the graphs. To automate it, the maximum and minimum amplitude for all the ten contactors were analyzed and tabulated for both accelerometer and vibration sensors.

Table 1 gives the maximum and minimum ranges obtained from the x and y coordinates of the accelerometer, respectively. Analysis was made from the observed values. It is observed that the ranges are overlapping. No significant variations could be detected using these values.

Table 2 shows the maximum and minimum amplitude ranges obtained from the vibration sensor. The values are analyzed and found that there is significant variation between good and bad contactors. The contactor with maximum amplitude below 0.075 is considered as good contactor whereas the contactor with maximum amplitude above 0.075 is considered bad. Hence, a LabVIEW program was designed by setting the threshold as 0.075. A LED is setup in the program, which glows when the contactor is a good one. If not it does not glow.

Figure 9 shows the output for good contactor. The raw signal is given as input to the program. The program checks for the threshold value after the contact time.

Here, the maximum amplitude after the contact time is below 0.075. Hence, the LED glows showing that the corresponding contactor is a good contactor.

Figure 10 shows the output for bad contactor. The program checks for the threshold after the contact time in the input raw signal. The maximum amplitude is above the threshold value. So, the LED does not glow indicating that the corresponding contactor is bad.

5 Conclusions

The aim of this project is to differentiate good and bad contactors at the production level. This is done by setting up a threshold to distinguish good contactors from the bad. The process is done in real time. The results are accurate in most of the cases. The setup is very cheap and the results are obtained really fast.

Table 1 Measurement from accelerometer

Contactor No.	X axis maximum amplitude	X axis minimum amplitude	Y axis maximum amplitude	Y axis minimum amplitude
1	1.68	1.67	1.33	1.33
2	1.65	1.64	1.34	1.33
3	1.69	1.69	1.34	1.32
4	1.67	1.65	1.34	1.32
5	1.70	1.68	1.34	1.32
6	1.66	1.65	1.33	1.33
7	1.80	1.60	1.43	1.29
8	1.74	1.72	1.35	1.33
9	1.69	1.67	1.34	1.33
10	1.69	1.67	1.34	1.32

Table 2 Measurement from vibration sensor

Contactor No.	Maximum amplitude	Minimum amplitude
1	0.05	−0.05
2	0.075	−0.085
3	0.095	−0.07
4	0.0625	−0.04
5	0.08	−0.075
6	0.085	−0.01
7	0.075	−0.075
8	0.05	−0.05
9	0.125	−0.025
10	0.05	−0.025

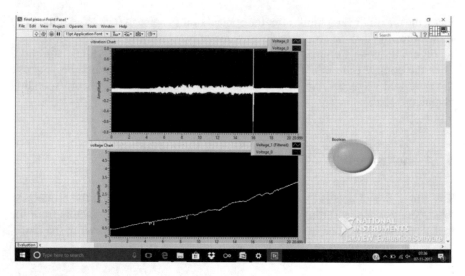

Fig. 9 Output for good contactor

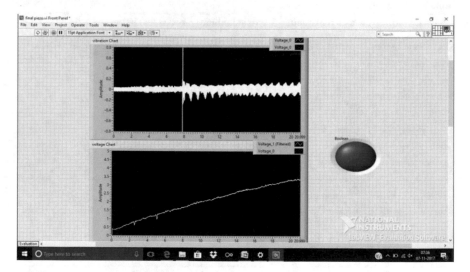

Fig. 10 Output for bad contactor

Acknowledgements The authors would like to thank Karunya Institute of Technology and Sciences (deemed to be University) for providing all the necessary facilities to carry out the experiment successfully.

References

1. Bouayed, K., Mebarek, L., Lanfranchi, V., Chazot, J.D., Marechal, R., Hamdi, M.A.: Noise and vibration of a power transformer under an electrical excitation. Applied Acoustic (2017)
2. Zhang, H., Corr, L.R., Ma, T.: Effects of electrical loads containing non-resistive components on electromagnetic vibration energy harvester performance. Mechanical Systems and Signal Processing, pp 55–66 (2017)
3. Latane, P.C., Punit Urolgin, C., Ruskone, R.: Fault Detection in Electric Motors Using Vibration Analysis and DSP Processor. Int. J. Innovative Res. Sci. Eng. Technol. **5** (2016)
4. Al-Yafawi, A., Patil, S., Yu, D., Park, S., Pitarresi, J., Goo, N.: Random Vibration Test for Electronic Assemblies Fatigue Life Estimation, pp 1–7 (2010)
5. Djokic, S.Z., Milanovic, J.V., Kirschen, D.S.: Sensitivity of AC coil contactors to voltage sags, short interruptions, and undervoltage transients. IEEE Trans. Power Delivery **19**, 1299–1307 (2004)
6. Patel, V.K., Patel, M.N.: Development of smart sensing unit for vibration measurement by embedding accelerometer with the Arduino microcontroller. Int. J. Instrum. Sci. **7**, 1–7 (2017)
7. Betta, G., Liguori, C., Paolillo, A., Pietrosanto, A.: A DSP-based FFT-Analyzer for the Fault Diagnosis of Rotating Machine Based on Vibration Analysis, pp. 572–577 (2001)
8. Shu, L., Xu, C., Wu, G.: Impact Detection of Intelligent AC Contactors and Optimization of the Dynamic Characteristics, pp. 1065–1069 (2015)

Formally Validated Authentication Protocols for WSN

Ummer Iqbal and Saima Shafi

Abstract Authentication is one of the most important cryptographic primitive for safe guarding network entities and data. Data communication within Wireless Sensor Network is primarily limited to Base to node and Node to Base Communication. Methods have been employed to authenticate data transmitted between Node to Base and Base to Node within a WSN. However, such existing methods are primarily based on symmetric cryptography thus providing less flexibility and practicality. Public key cryptography provides flexible and prudent methods for authentication but they are computationally intense making them impractical for WSN. However, ECC and its light weight implementation TinyECC offers high potential for developing computationally efficient security protocols for WSN. In this paper, a brief review of the existing authentication methods in WSN has been presented. Two improved protocols concerning Sink to Reduced Function Device (RFD) and RFD to Sink communication has been presented. The proposed protocol has been formally validated on Avispa. An experimental set up using MicaZ and TinyOS has been created to carry out the performance benchmarking of the proposed protocols.

Keywords WSN · Authentication · ECC · TinyECC · AVISPA · TinyOS

1 Introduction

WSN networks are primarily involved in sensing of various parameters and forwarding it to the Base Station for decision-making. Based on the received sensed data from various nodes in the network, necessary decisions or alarms are raised. Thus source authentication of data becomes an utmost priority for such networks

---

U. Iqbal (✉)
National Institute of Electronics and Information Technology, Srinagar, Jammu and Kashmir, India
e-mail: ummer@nielit.gov.in

S. Shafi
Shri Sukhmani Institute of Engineering and Technology, Chandigarh, India

© Springer Nature Singapore Pte Ltd. 2019
J. D. Peter et al. (eds.), *Advances in Big Data and Cloud Computing*,
Advances in Intelligent Systems and Computing 750,
https://doi.org/10.1007/978-981-13-1882-5_36

[1, 2]. Authentication can be primarily achieved using either symmetric or asymmetric cryptographic methods. As WSN is characterized with low power and processing abilities, symmetric techniques have been found more suitable for WSN. However, the inherent limitation of symmetric cryptography makes these techniques impractical [3, 4]. Authentication protocols in traditional networks are primarily based on public key cryptography. Conventional authentication protocols as applicable to wired and wireless networks cannot be applied directly to a wireless sensor network because of the constraints associated with it. Thus design of new protocols and security architecture to implement authentication within a resource constraint network like WSN poses a great challenge to the researchers [5]. Design of new authentication protocols based on elliptical curve cryptography has been found very efficient for WSN [6].

1.1 Existing Authentication Protocols

Security Protocol for Sensor Networks (SPINS) was presented by Perrig et al. in 2002 [7]. It comprises of Sensor Network Encryption Protocol (SNEP) and μTESLA. SNEP provides integrity as well as authentication but it is based on symmetric technique for key sharing. μTESLA is used for broadcast authentication using delayed disclosure of keys and one-way hash. The major drawback in μTESLA is that it is based on delayed disclosure of keys and requires time synchronization. TinySec [8] is an authentication architecture proposed by Karlof et al. It uses cipher block chaining mode of operation with SkipJack. It has two modes: TinySec-Auth: which is used for authentication and TinySec-AE: provides authentication as well as encryption. It uses single network wide key which is susceptible to node capture attack. MiniSec [9] is another symmetric technique which provides two modes of operation: MiniSec-U and MiniSec-B. MiniSec-U is used for uncast authentication and MiniSec-B is used for Broadcast authentication. C-Sec [10] protocol is another technique with low energy and time consumption. However, it doesn't consider end to end delay and network Scalability. The Comparisons of the existing schemes is given in Table 1.

1.2 Contribution and Outline

In this Paper, a set of lightweight authentication protocols for achieving Sink to RFD and RFD to Sink have been presented. These protocols are based on Elliptical Curve Cryptography. The proposed protocols have validated on AVISPA [11]. The protocols have also been developed on TinyOS [12]. Performance benchmarking of the proposed protocol has also been carried out on MicaZ [13] motes.

Table 1 Comparison of the existing authentication schemes

Protocols	TinySec	MiniSec	SPINS	C-Sec
Authentication	CBC-MAC	CBC-MAC	CBC-MAC	HMAC
Advantages	Flexible, low energy and memory usage	Minimize energy consumption	Provides broadcast authentication	Low energy and time consumptions
Disadvantages	It does not provide any protection against reply attacks. It is susceptible to node capture attack	Encryption modes are limited	It is susceptible to DOS attack. Its performance is not good	It doesn't provide scalability

Table 2 Symbol Description

Symbol	Description
K	Private key of node
B	Private key of base
P_{rfd}	Public key of RFD
P_{sink}	Public key of MIB600
G	Generator point of elliptical curve
H()	Hash function
ID_i	Identity of node I
+	Xor function

2 Proposed Authentication Protocol

The symbols used in the protocols are shown in Table 2. During the initialization, RFD Selects a private key K and computes the public key as $P_{rfd} = K.G$ and Base selects a private key B and computes the public key as $P_{sink} = B.G$.

2.1 Sink to RFD Authentication

In this authentication algorithm, a RFD authenticates Sink.

1 RFD Computes $ID_i * P_{rfd}$ and sends it to the Sink
2 Sink Computes $P = B * ID_i * P_{rfd}$ and Sends it to the RFD
3 RFD Computes $Q = ID_i * K * P_{sink}$.
4 RFD authenticates Sink if P==Q

Protocol	No of scalar multiplication
Base to node	3
Node to base	2

Table 3 No of Scalar Multiplication

2.2 RFD to Sink Authentication

In this authentication algorithm, Sink authenticates RFD.

1: RFD computes $H(ID_i)$ and Computes $[H(ID_i)+K]$ *G
2 RFD Sends $[H(ID_i)+K]$ *G, ID_i to the Sink
3 Sink Stores $[R=[H(ID_i)+K]$ *G and Computes $S=[H(ID_i)*G+P_{rfd}]$
4 Sink Checks if (R==S), if true then RFD is authenticated

3 Formal Verification of Proposed Protocols

Avispa is an automated validation tool for security protocols. It is a push button tool based on love and Yao 7 model. Dolev and Yoa model gives a complete control of the communication channel to the intruder. The intruder in this case has a capacity to forward, modify and change messages but cannot overdue the computational strength of an algorithm The protocols were formally validated on AVISPA platform. The HLPSL Script for Sink to RFD authentication and RFD to Sink Authentication in given in the Figs. 1 and 3. The HLPSL scripts were validated on OFMC backend as shown in the Figs. 2 and 4. Both the protocols are safe against various active and passive attacks.

4 Energy Analysis

Scalar multiplication is the computational intensive operation which primarily decides the energy profile of program. The no of scalar multiplication in the proposed protocols is shown in Table 3. Energy consumed by a scalar multiplication is calculated using Energy = Voltage * Current * Time. For MicaZ mote the values of voltage and current are 3 V and 19 mA. To calculate time, a setup was created as shown in Fig. 5.

```
Rolerole_A(A:agent,B:agent,G:text,H:function,SND,RCV:channel(dy))
played_by A
def=
        local
State:nat,Nb:text,Na:text,Nid:text,PKa:text, PKb: text , En:text, Em: text
        init
State := 0
        transition
1. State=0 /\ RCV(start) =|> State':=1 /\ Na':=new() /\  PKa':=H(G.Na') /\  SND(PKa')
/\ secret (Na' , private_A, {A})
2. State=1 /\ RCV(PKb') =|> State':=2 /\ Nid':=new() /\  En':= H(PKb'.Nid')  /\ SND
(En') /\ request(A,B,mib_to_rfd, En')
4. State=2 /\ RCV(Em') =|> State':=3
end role
role role_B(A:agent,B:agent,G:text,H:function,SND,RCV:channel(dy))
played_by B
def=
        local
State:nat,Nb:text,Na:text,NSecret:text, PKa:text , PKb:text , En: text ,Em:text
        init
                State := 0
        transition
1. State=0 /\ RCV(PKa') =|> State':=1 /\ Nb':=new() /\ PKb':= H(G.Nb') /\ secret (Nb' ,
private_B , {B})
3. State=1 /\ RCV(En') =|> State':=2 /\ Em':= H(En'.Nb) /\ SND (Em') /\
witness(B,A,mib_to_rfd,En')
end role
role session(A:agent,B:agent,G:text,H:function)
def=
        local
SND2,RCV2,SND1,RCV1:channel(dy)
        composition
role_B(A,B,G,H,SND2,RCV2) /\ role_A(A,B,G,H,SND1,RCV1)
end role
role environment()
def=
        const
bob:agent,h:function,alice:agent,g:text, ni:text ,sec_dhvalue : protocol_id
,private_A : protocol_id, private_B : protocol_id , rfd_to_mib: protocol_id
        intruder_knowledge = {alice,bob,g,h}
        composition
session(alice,bob,g,h)/\ session(i,alice,g,h) /\ session(alice, i,g,h)
end role
goal
secrecy_ofsec_dhvalue
secrecy_ofprivate_A
secrecy_ofprivate_B
authentication_onmib_to_rfd
end goal
environment()
```

Fig. 1 HLPSL script for sink to RFD

The NesC Program has been written to compute time taken for various critical operations by integrating TinyECC [14]. The component graph of the developed

```
% OFMC
% Version of 2006/02/13
SUMMARY
  SAFE
DETAILS
  BOUNDED_NUMBER_OF_SESSIONS
PROTOCOL
  /home/span/span/testsuite/results/rfd_to_Mib.if
GOAL
as_specified
BACKEND
  OFMC
COMMENTS
STATISTICS
parseTime: 0.00s
searchTime: 0.01s
visitedNodes: 16 nodes
depth: 4 plies
```

Fig. 2 OFMC output for sink to RFD

Table 4 Time taken for scalar multiplication on various curves

Operation	Time taken (s)		
	Secp192r1	Secp160r1	Secp128r1
Scalar multiplication	6.07	3.90	2.23

program is shown in Fig. 6. TimerC module is used connected using Timer interface. The encapsulated value of Number of ticks is captured by sending it to the serial forwarder. A java program has been written to connect to serial forwarder on port Java application. The time taken for scalar multiplication on various ECC Curves which includes SECP 128r1, Secp160r1, and Secp192r1 is shown in the Table 4. The Energy consumption on MicaZ mote is shown in Figs. 7 and 8.

5 Conclusion

The paper reviewed various authentication mechanism which can be employed in WSN. It was pointed out that most the existing methods are based on symmetric methods thus making them less flexible and impractical. The usage of elliptical curve cryptography for developing authentication protocols for WSN was highlighted. Two authentication protocols for Base to Node and Node to Base communication were presented. The developed protocols were formally validated on Avispa tool. Performance benchmarking of the developed protocols was also carried out by creating an experimental setup based on TinyOS and MicaZ.

```
role role_A(A:agent,B:agent,G:text,H:function,SND,RCV:channel(dy))
played_by A
def=
        local
State:nat,Nb:text,Na:text,Nid:text,PKa:text, PKb: text , En:text, Em: text
        init
State := 0
        transition
1. State=0 /\ RCV(start) =|> State':=1 /\ Na':=new() /\ PKa':=H(G.Na') /\
SND(PKa') /\ secret (Na' , private_A, {A})
2. State=1 /\ RCV(PKb') =|> State':=2 /\ Nid':=new() /\ En':= H(xor(H(Nid'),Na).G)
/\ SND (En') /\ witness(A,B,rfd_to_mib, En')
 4. State=2 /\ RCV(Em') =|> State':=3
end role
role role_B(A:agent,B:agent,G:text,H:function,SND,RCV:channel(dy))
played_by B
def=
        local
State:nat,Nb:text,Na:text,NSecret:text, PKa:text , PKb:text , En: text ,Em:text
        init
State := 0
        transition
1. State=0 /\ RCV(PKa') =|> State':=1 /\ Nb':=new() /\ PKb':= H(G.Nb') /\ secret
(Nb' , private_B , {B})
3. State=1 /\ RCV(En') =|> State':=2  /\ request(B,A,rfd_to_mib,En')
end role
role session(A:agent,B:agent,G:text,H:function)
def=
        local
SND2,RCV2,SND1,RCV1:channel(dy)
        composition
role_B(A,B,G,H,SND2,RCV2) /\ role_A(A,B,G,H,SND1,RCV1)
end role
role environment()
def=
        const
bob:agent,h:function,alice:agent,g:text, ni:text ,sec_dhvalue : protocol_id
,private_A : protocol_id, private_B : protocol_id , rfd_to_mib: protocol_id
        intruder_knowledge = {alice,bob,g,h}
        composition
session(alice,bob,g,h)/\ session(i,alice,g,h) /\ session(alice, i,g,h)
end role

goal
secrecy_ofsec_dhvalue
secrecy_ofprivate_A
secrecy_ofprivate_B
authentication_onrfd_to_mib
end goal
environment()
```

Fig. 3 HLPSL script for RFD to sink

```
% OFMC
% Version of 2006/02/13
SUMMARY
  SAFE
DETAILS
  BOUNDED_NUMBER_OF_SESSIONS
PROTOCOL
  /home/span/span/testsuite/results/rfd_to_Mib.if
GOAL
as_specified
BACKEND
  OFMC
COMMENTS
STATISTICS
parseTime: 0.00s
searchTime: 0.01s
visitedNodes: 16 nodes
depth: 4 plies
```

Fig. 4 OFMC output for RFD to sink

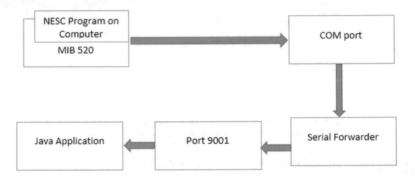

Fig. 5 Set-up for calculating computational time

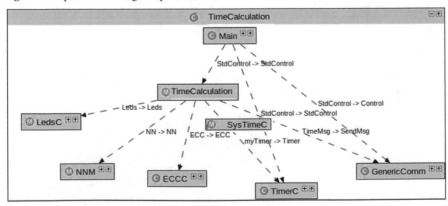

Fig. 6 Component diagram of NesC program for measuring time

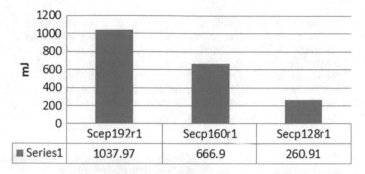

Fig. 7 Energy consumed for base to node protocol

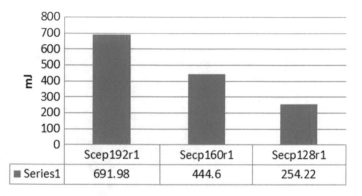

Fig. 8 Energy consumed for node to base protocol

References

1. Sanchez-Rosario, F.: A low consumption real time environmental monitoring system for smart cities based on ZigBee wireless sensor network. IEEE 978-1-4799-5344-8 (2015)
2. Perrig, A., Stankovic, J., Wagner, D.: Security in wireless sensor networks. Commun. ACM **47**(6), pp. 53–57
3. Menezes, A.J., van Oorschot, P.C., Vanstone, S.A.: Handbook of Applied Cryptography. CRC Press, Boca Raton (1997)
4. Menzes, B.: Network Security and Cryptography. Cengage Learning
5. Mallan, D.J., Welish, M., Smith, D.M.: Implementing public key infrastructure for sensor networks. Trans. Sens. Netw. **4**
6. Gura, N., Patel, A., Wander, A.S., Eberle, H., Chang Shantz, S.: Comparing elliptic curve cryptography and RSA on 8-bit CPUs. In: Cryptographic Hardware and Embedded Systems, vol. 3156, pp. 119–132. Springer
7. Perrig, A., Szewczyk, R., Tygar, J.D., Wen, V., Culler, D.E.: SPINS: security protocol for sensor networks. In: Proceedings of 7th International Conference on Mobile Networking And Computing, vol. 8(5), pp 189–199 (2001)
8. Karlof, C., Sastry, N., Wagner, D: TinySec: a link layer security architecture for wireless sensor networks. In: Proceedings of the 2nd International Conference on Embedded Networked Sensor Systems, SenSys, Baltimore, MD, USA, 3–5 Nov 2004, pp. 162–175. ACM

9. Luk, M., Mezzour, G., Perrig, A., Gligor, V.: MiniSec: a secure sensor network communication architecture. In: Proceedings of the 6th International Conference on Information Processing in Sensor Networks (IPSN '07) April 200747948810.1145/1236360.12364212-s2.0-3534889734

10. Mohd, A., Aslam, N., Robertson, W., Phillips, W.: C-sec: energy efficient link layer encryption protocol for wireless sensor networks. In: Proceedings of the 9th Consumer Communications and Networking Conference (CCNC '12). IEEE, Las Vegas, Nevada USA, pp. 219–223 (2012)

11. AVISPA Web tool.: Automated Validation of Internet Security Protocols and Applications, www.avispa-project.org, Last Accessed Jan 2018

12. Levis, P., Gay, D.: TinyOS Programming. Cambridge University Press (2009)

13. Memsic: Xserve User Manual May (2007)

14. Liu, A., Ning et al., P.: Tiny ECC: a configurable library for elliptical curve cryptography in wireless sensor networks. In: 7th International Conference on Information Processing in Sensor Networks SPOTS Track, April 2008

Cryptographically Secure Diffusion Sequences—An Attempt to Prove Sequences Are Random

M. Y. Mohamed Parvees, J. Abdul Samath and B. Parameswaran Bose

Abstract The use of random numbers in day-to-day digital life is increasing drastically to make the digital data more secure in various disciplines, particularly in cryptography, cloud data storage, and big data applications. Generally, all the random numbers or sequences are not truly random enough to be used in various applications of randomness, predominantly in cryptographic applications. Therefore, the sequences generated by pseudorandom number generator (PRNGs) are not cryptographically secure. Hence, this study proposes a concept that the diffusion sequences which are used during cryptographic operations need to be validated for randomness, though the random number generator produces the random sequences. This study discusses the NIST, Diehard and ENT test suite results of random diffusion sequences generated by two improved random number generators namely, Enhanced Chaotic Economic Map (ECEM), and Improved Linear Congruential Generator (ILCG).

Keywords Random number generator · Chaotic map · Diffusion · Confusion
Sequences · Encryption · Security

M. Y. Mohamed Parvees (✉)
Division of Computer & Information Science, Annamalai University, Annamalainagar,
Chidambaram 608002, India
e-mail: yparvees@gmail.com

M. Y. Mohamed Parvees
Research & Development Centre, Bharathiar University, Coimbatore 641046, India

J. Abdul Samath
Department of Computer Science, Government Arts College, Udumalpet 642126, India

B. Parameswaran Bose
Fat Pipe Network Pvt. Ltd., Mettukuppam, Chennai 600009, India

B. Parameswaran Bose
#35, I Main, Indiragandhi Street Udayanagar, Bengaluru, India

© Springer Nature Singapore Pte Ltd. 2019
J. D. Peter et al. (eds.), *Advances in Big Data and Cloud Computing*,
Advances in Intelligent Systems and Computing 750,
https://doi.org/10.1007/978-981-13-1882-5_37

433

1 Introduction

The pseudorandom number generators play immense roles in providing security and privacy-preservation of cloud-based distributed environment to store and process the data, even big data. The cloud services and big data applications use cryptographic algorithms to secure the data [1]. The roles of PRNGs have lots of importance in developing cryptographic algorithms by means of their unpredictability towards adversary attacks. The unpredictability is merely based on randomness of the random numbers or sequences which influences the security of the cryptographic algorithms while trying to break them. The several researchers are presently working on different random number generators to achieve the higher level of security and also to generate the random numbers in a faster manner at a cost low [2–6]. The random sequences generated by PRNGs could not be used in cryptographic algorithms because of their weakness towards randomness. The sequences generated by PRNGs should be cryptographically secure sequences. Some of the researchers work on filters which could be applied on the PRNGs, thereby get cryptographically secure random sequences [4, 7, 8]. The researchers test the randomness of sequences produced by different PRNGs using common three test batteries namely, NIST, Diehard and Entropy test batteries [9–13]. In cryptographic applications, the two basic operations namely, confusion and diffusion are importantly employed to shuffle and diffuse the data. The diffusion sequences could be generated from the PRNGs, but, so far, not yet tested by researchers for their randomness to prove their adequacy of usage in cryptographic algorithms. Hence, this study proposes two different PRNGs, generate diffusion sequences from their random sequences, and test their diffusion sequences towards randomness to prove they are secured cryptographically.

2 Preliminaries

Parvees et al. [14] derived a nonlinear equation called enhanced chaotic economic map which was used for diffusing the image pixels. The equation is

$$x_{n+1} = \cos x_n + k \times \left[a - c - b \times (1 + \gamma) \times (\cos x_n)^\gamma \right], \qquad (1)$$

where a is market demand size >0, b is market price slope >0, c is fixed marginal cost ≥ 0, $\gamma = 4$ (constant), k is adjustment parameter >0. X_0 is the initial parameter. Figure 1a shows the chaotic behavior of the ECEM ranges from 0.63 to 1.00. The input keys and bifurcation range proves the chaotic map has larger keyspace to resist brute-force attacks. Figure 1b shows the positive Lyapunov exponent values which proves that the enhanced map is suitable for cryptic solutions.

Parvees et al. [15] proposed three improved LCGs to generate random sequences. The three different improved LCGs are shown in Eqs. (2), (3) and (4).

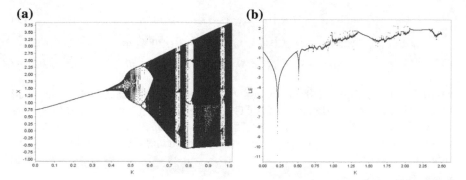

Fig. 1 The bifurcate and Lyapunov graph of ECEM with respect to the control parameter (k). **a** Bifurcation and **b** Lyapunov graph

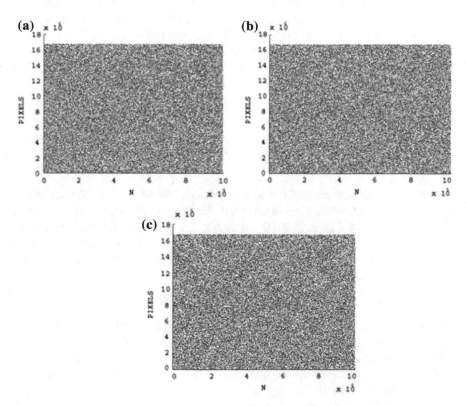

Fig. 2 **a–c** The corresponding three random pixel arrays C_1, C_2 and C_3 using three types of LCGs as mentioned in Eqs. (2–4)

$$x_{n+1} = \{i \times [(a \times x_n) + (i + c)]\} \bmod m \tag{2}$$

$$x_{n+1} = \{i \times [(a \times x_n) + (i \times c)]\} \bmod m \tag{3}$$

$$x_{n+1} = \{i \times [(a \times x_n) + (i \wedge c)]\} \bmod m, \tag{4}$$

where $i \in (0, n)$. Figure 2 shows the graph of random numbers created by improved LCG which depicts randomness of ILCG. The seed values for improved LCGs are $n = 99,999$, $a = 16,777,213$, $x_0 = 996,353,424$, $c = 12,335$, $m = 16,777,214$.

The above Eqs. (2–4) yield different random sequences which could be used to create diffusion sequences.

3 Methodology

In cryptographic image encryption algorithms, the confusion-diffusion is two imperative operations to change the pixels location and alter pixels values. For accomplishing complete encryption, the random sequences for confusion and diffusion are generated from an enhanced chaotic map and improved pseudorandom generator namely, ECEM and ILCG. The enhanced map and PRNG are iterated separately to generate the random sequences and those sequences are converted into cryptographically secure diffusion sequences according to the nature of data. The keys are supplied to nonlinear equations to produce random sequences. In this study, the diffusion sequences are generated to encrypt 16-bit and 24-bit images by diffusion operation.

3.1 Generation of 16-Bit Diffusion Sequence

Actually the ECEM generates the chaotic sequences with 64-bit double value. In this study, the idea is to generate 16-bit cryptographically secure random sequences and also the values should lie between 0 and 65,535 for diffusing 16-bit values present in a DICOM image [14]. Therefore, the diffusion sequences will produce equivalent ciphertext from the plaintext. The algorithm 1 depicts the steps for generating diffusion sequences.

Algorithm 1: Generation of chaotic diffusion sequence

Step 01: Generate a chaotic sequence of length n using Eq. 1 where $n = n + 1111$.
Step 02: To avoid the transient effect, discard first 1111 chaotic elements in the chaotic sequence $C = \{c_1, c_2, c_3, \ldots, c_n\}$ by supplying input parameters a, b, c, γ, k and x_n.
Step 03: Choose the chaotic sequence $X = \{x_{1112}, x_{1113}, x_{1114}, \ldots x_n\}$ from the 1112nd chaotic element.
Step 04: Calculate diffusion sequences $D_i = \text{int}\{[abs(x_i) - \lfloor abs(x_i)\rfloor] \times 10^{16}]$ $\bmod 65535\}$where $D_i \in (0, 65535)$.
Step 05: The integer sequences $D = \{d_1, d_2, d_3, \ldots, d_n\}$ are generated.

3.2 Generation of 24-Bit Diffusion Sequence

The improved LCG generates 64-bit double valued random sequences. To encrypt 24-bit color DICOM pixels' bytes, these double valued sequences have to be converted into three 8-bit byte values which could be used to create final diffusion pixel set P_r. The pixel set P_r can be diffused with plain pixel values. The values of individual diffusion sequences lie between 0 and 255 [15]. Hence, it is essentially required to check whether these diffusion sequences P_r are cryptographically secure or not. Algorithm 2 explains the steps involved in generating 24-bit diffusion sequences.

Algorithm 2: Generation of ILCG diffusion sequence

Step 01: Generate three random 64-bit double valued sequences C_1, C_2 and C_3 using three different improved LCGs as shown in Eq. (2–4).
Step 02: Discard first 1111 elements to have transient effect.
Step 03: Calculate different bytes R_1, G_1, B_1, R_2, G_2, B_2, R_3, G_3 and B_3 values by assimilating three different C_1, C_2, and C_3.
Step 04: To make algorithm more secure, calculate three different byte values P_1, P_2 and P_3 by

$$P_1 = (R_1 << 16) + (G_2 << 08) + B_3,$$
$$P_2 = (R_2 << 16) + (G_3 << 08) + B_1,$$
$$P_3 = (R_3 << 16) + (G_1 << 08) + B_2.$$

Step 05: Finally create the diffusion sequence in the form image pixels P_r by $P_r = P_1 \wedge P_2 \wedge P_3$.

The diffusion sequences generated using ECEM and ILCG are converted into bit streams and tested towards the NIST battery, Diehard battery and ENT Statistical Battery.

4 Experimental Results and Discussion

The diffusion bit sequences generated by ECEM with length of 7,700,000 were tested towards the various statistical tests. Similarly, the diffusion sequence length of 6,000,000 obtained using ILCG is also tested for their randomness. The bit streams are tested using NIST test suite version STS-2.1.2. The STS-2.1.2 'C' source files are downloaded and compiled in Eclipse using MinGW compiler. Tables 1 and 2 illustrate the results of various tests available in NIST test suite. The results are shown in Tables 1 and 2 which prove that the tests are optimistic and the *p-value* for each test is accepting the null hypothesis. That is, if *p*-value is less than 0.01, hence the test rejected the null assumption; if *p*-value ≥ 0.01, accepted null hypothesis. Therefore, the diffusion sequences produced by the ECEM and ILCG are cryptographically secure. Similarly, the other two common tests namely, Diehard and ENT test batteries are tested towards the randomness and the corresponding outcomes are revealed in Tables 3, 4, 5 and 6.

Table 1 The outcome of NIST battery tests for diffusion sequence generated by ECEM

Statistical test name	Proportion	P-value	Result
Frequency test	10/10	0.350485	Succeed
Block-frequency test	10/10	0.739918	Succeed
Cumulative-sums-forward test	9/10	0.122325	Succeed
Cumulative-sums-reverse test	10/10	0.911413	Succeed
Runs test	10/10	0.213309	Succeed
Longest-run test	10/10	0.350485	Succeed
Rank test	10/10	0.213309	Succeed
FFT test	10/10	0.534146	Succeed
Non-overlapping-template test	10/10	0.991468	Succeed
Overlapping-template test	10/10	0.213309	Succeed
Universal Maurer's test	10/10	0.350485	Succeed
Approximate-entropy ($m = 10$) test	10/10	0.350485	Succeed
p-value of serial test 1	10/10	0.534146	Succeed
p-value of serial test 2	10/10	0.350485	Succeed
Linear-complexity test 1	10/10	0.350485	Succeed

Table 2 The outcome of NIST battery tests for diffusion sequence generated by ILCG

Statistical test name	Proportion	P-value	Result
Frequency test	10/10	0.066882	Succeed
Block-frequency test	10/10	0.739918	Succeed
Cumulative-sums-forward test	10/10	0.350485	Succeed
Cumulative-sums-reverse test	10/10	0.739918	Succeed
Runs test	9/10	0.350485	Succeed
Longest-run test	10/10	0.213309	Succeed
Rank test	10/10	0.534146	Succeed
FFT test	10/10	0.122325	Succeed
Non-overlapping-template test	10/10	0.911413	Succeed
Overlapping-template test	10/10	0.739918	Succeed
Universal Maurer's test	10/10	0.911413	Succeed
Approximate-entropy ($m = 10$) test	10/10	0.213309	Succeed
p-value of serial test 1	10/10	0.534146	Succeed
p-value of serial test 2	10/10	0.534146	Succeed
Linear-complexity test 1	10/10	0.534146	Succeed

Table 3 Results of diehard battery tests for diffusion sequence generated by ECEM

Statistical test name	Average p-value	Result
Birthday spacing test	0.160283	Succeed
Overlapping 5-permutation test	0.442425	Succeed
Binary rank 31×31 matrices test	0.321420	Succeed
Binary rank 32×32 matrices test	0.512901	Succeed
Binary rank 06×08 matrices test	0.212151	Succeed
Bit stream test	0.399190	Succeed
Overlapping-pairs-sparse-occupancy test	0.744600	Succeed
Overlapping-quadruples-sparse-occupancy test	0.723100	Succeed
DNA test	0.980200	Succeed
Count the ones-01 test	0.347402	Succeed
Count the ones-02 test	0.919996	Succeed
Parking lot test	0.419058	Succeed
Minimum distance test	0.358596	Succeed
3D spheres test	0.032083	Succeed
Squeeze test	0.186114	Succeed
Overlapping sum test	0.743564	Succeed
Runs-up test	0.336334	Succeed
Runs-down test	0.523293	Succeed
Craps test	0.738980	Succeed

Table 4 The outcome of diehard battery tests for the diffusion sequence generated by ILCG

Statistical test name	Average p-value	Result
Birthday spacing test	0.258059	Succeed
Overlapping 5-permutation test	0.320891	Succeed
Binary rank 31×31 matrices test	0.320891	Succeed
Binary rank 32×32 matrices test	0.370736	Succeed
Binary rank 06×08 matrices test	0.630377	Succeed
Bit stream test	0.387510	Succeed
Overlapping-pairs-sparse-occupancy test	0.768300	Succeed
Overlapping-quadruples-sparse-occupancy test	0.707000	Succeed
DNA test	0.773000	Succeed
Count the ones-01 test	0.304902	Succeed
Count the ones-02 test	0.117650	Succeed

(continued)

Table 4 (continued)

Statistical test name	Average p-value	Result
Parking lot test	0.180042	Succeed
Minimum distance test	0.352174	Succeed
3D spheres test	0.821221	Succeed
Squeeze test	0.833616	Succeed
Overlapping sum test	0.491965	Succeed
Runs-up test	0.328656	Succeed
Runs-down test	0.607940	Succeed
Craps test	0.968010	Succeed

Table 5 Results of ENT test suite for the diffusion sequence generated by ECEM

Statistical tests	Value	Result
Entropy value	7.999982	Succeed
Arithmetic mean value	127.4764	Succeed
Monte Carlo value	3.142789257	Succeed
Chi-square value	244.10	Succeed
Serial correlation coefficient value	0.000212	Succeed

Table 6 Results of the ENT test suite for the diffusion sequence generated by ILCG

Statistical tests	Value	Result
Entropy value	7.999990	Succeed
Arithmetic mean value	127.4862	Succeed
Monte Carlo value	3.14256533	Succeed
Chi-square value	254.49	Succeed
Serial correlation coefficient value	−0.000178	Succeed

This research proposes to check the randomness of the diffusion sequences, that is, the importance is given on diffusion sequences. Any cryptographic algorithm uses two different basic operations, namely confusion and diffusion, to secure the data usually by means of encryption process. During confusion, the plaintext is shuffled. But, the values of the plaintext have been altered during diffusion process. The diffusion sequences undergo some basic arithmetic or XOR operations to alter the plaintext values, thereby provides security. Hence, it is essential to hide the diffusion sequences during their masking operations and the guess could be avoided during the specific operation. So, it is important to validate the security or unpredictability of the diffusion sequences from the adversaries. The several researchers check their random sequences generated using random number generators for the randomness and produce similar results [3, 5, 9, 16]. Though several researchers validate their random number generators, this study particularly proposes to validate the diffusion sequences obtained from chaotic or random sequences generated by enhanced PRNGs.

5 Conclusion

The two pseudorandom sequence generators, namely, enhanced chaotic economic map and improved linear congruential generator are nonlinear systems, but they are deterministic. Hence, these two PRNGs are successfully employed to prove relationship between the random sequences and cryptography. Particularly, the sequences that are involved in actual cryptographic operations are validated towards randomness tests. The diffusion sequences, in the form binary sequences are tested towards different test suites and empirical results are satisfactory while comparing with the literature and NIST and other documentations. So, the diffusion sequences are cryptographically secure random sequences. Hence, these enhanced PRNGs and their diffusion sequences can be utilized in different disciplines, not only in cryptography, wherever the random numbers are playing essential roles.

References

1. Chen, J., Miyaji, A., Su, C.: Distributed pseudo-random number generation and its application to cloud database. In: Huang, X., Zhou, J. (eds.) Information Security Practice and Experience. ISPEC 2014. Lecture Notes in Computer Science, vol. 8434. Springer, Cham (2014)
2. Deng, L.Y., Bowman, D.: Developments in pseudo-random number generators: Pseudo-random number generators. Wiley Interdisc. Rev. Comput. Stat. **9**(5), e1404 (2017). https://doi.org/10.1002/wics.1404
3. Stoyanov, B.P., Kordov, K.: A Novel pseudorandom bit generator based on Chirikov standard map filtered with shrinking rule. Math. Prob. Eng. **2014**, (2014) Article ID 986174, p. 4. https://doi.org/10.1155/2014/986174
4. Patidar, V., Sud K.K., Pareek K.: A pseudo random bit generator based on chaotic logistic map and its statistical testing. Informatica **33**, 441–452 (2009)
5. Stoyanov, B.P., Szczypiorski, K., Kordov, K.: Yet another pseudorandom number generator. Int. J. Electron. Telecommun. **63**(2), 195–199 (2017). https://doi.org/10.1515/eletel-2017-0026
6. Kordov, K., Stoyanov B.:, Least significant bit steganography using Hitzl-Zele Chaotic Map. Int. J. Electron. Telecommun. **63**(4) (2017)
7. Rahimov, H., Babaie, M., Hassanabadi, H.: Improving middle square method RNG using chaotic map. Appl. Math. **2**, 482–486 (2011)
8. Sridevi, R., Philominat, P., Padmapriya, P., Rayappan, J.B.B., Amirtharajan, R.: Logistic and standard coupled mapping on pre and post shuffled images: a method of image encryption. Asian J. Sci. Res. **10**(1), 10–23. (2016). https://doi.org/10.3923/ajsr.2017.10.23
9. Patidar, V.R., Sud, K.K.: A novel pseudo random bit generator based on chaotic standard map and its testing. Electron. J. Theoret. Phys. **6**(20), 327–344 (2009)

10. Rukhin, A., Soto, J., Nechvatal, J., et al.: A Statistical Test Suite for Random and Pseudorandom Number Generators for Cryptographic Application. NIST Special Publication 800-22, Revision 1a (Revised: April 2010), Lawrence E. Bassham III, (2010). http://csrc.nist.gov/groups/ST/toolkit/rng/index.html
11. Marsaglia, G. Diehard: a battery of tests of randomness (1996) http://www.fsu.edu/pub diehard
12. Walker, J.: ENT: a pseudorandom number sequence test program (2008) http://www.fourmilab.ch/random/
13. Soto, J.: Randomness testing of the advanced encryption standard candidate algorithms. NIST Internal Reports 6390 (1999), http://csrc.nist.gov/publications/nistir/ir6390.pdf
14. Parvees, M.Y.M., Samath, J.A.: Bose BP secured medical images—a chaotic pixel scrambling approach. J. Med. Syst. **40**, 232 (2016). https://doi.org/10.1007/s10916-016-0611-5
15. Parvees, M.Y.M., Samath, J.A., Bose, B.P.: Medical images are safe—an enhanced chaotic scrambling approach. J. Med. Syst. **41**, 167 (2017). https://doi.org/10.1007/s10916-017-0809-1
16. Stępień, R., Walczak, J.: Statistical analysis of the LFSR generators in the NIST STS test suite. Comput. Appl. Electr. Eng. **11**, 356–362 (2013)

Luminaire Aware Centralized Outdoor Illumination Role Assignment Scheme: A Smart City Perspective

Titus Issac, Salaja Silas and Elijah Blessing Rajsingh

Abstract In the modern era of smart cities, devices are becoming much smarter, smaller, and energy-efficient. Modern devices are able to communicate and process with the adaptation of Internet of Things. Despite breakthrough in technologies, a majority of the power generated is utilized for outdoor illumination. Existing smart outdoor illumination methods are self-actuated by the onboard sensors in the luminaire leading to power savings. Investigation reveals the need of efficient illumination through the smart collaboration among luminaires. A centralized outdoor illuminating role assignment scheme is proposed to address the collaborative illumination between the heterogeneous luminaires. Simulations and comparisons were carried out on the proposed centralized scheme with the legacy approach and the self-actuated ZB-OLC assignment scheme. The paper concludes with the comparative analysis and simulation results of the effect of role assignment on various types of luminaires, overall power consumption and lifetime analysis of the luminaires with perspective to zones, duration, and neighboring luminaires.

Keywords Smart city · Street lighting · IoT · WSN · WSAN

1 Introduction

Smart city initiative intelligently integrates core services of a city with modern technology [1–3]. Pollution control [4], environmental monitoring [4], smart traffic controls [5], smart home [6], and smart parking [5] are few core services of a smart city initiative. Adaptation of energy efficient modern smart devices quickens the

T. Issac (✉) · S. Silas · E. B. Rajsingh
Karunya Institute of Technology and Sciences, Coimbatore, India
e-mail: titusissac@gmail.com

© Springer Nature Singapore Pte Ltd. 2019
J. D. Peter et al. (eds.), *Advances in Big Data and Cloud Computing*,
Advances in Intelligent Systems and Computing 750,
https://doi.org/10.1007/978-981-13-1882-5_38

transformation into a smart city. Integration of core services using internet of things renders precise control and collaboration between devices, resulting to better energy efficiency and throughput [7]. Despite smart city initiatives, a city utilizes the major part of the power generated for illumination [1, 4, 5]. Sufficient illumination plays a vital role in the development and safety of a city [8]. Illumination is primarily performed by a set of luminaire. A complete set of lighting unit, also known as a luminaire, primarily comprises of one or multiple lighting component. Significant advancement in technology has led to the development of a new generation of luminaires. Additionally, a luminaire could have (i) sensor unit [6, 9], (ii) driver unit [10], (iii) dimmer unit [11], and (iv) communication module [3]. Sensor unit might comprise of sensors such as occupancy sensor [6, 11, 12], temperature sensor, luminosity sensor [10] as per the requirement of the application. Driver unit facilitates to turn on and off, individual lighting bulbs in a luminaire. Dimmer unit is used to realize multi level dimming of the luminaire by controlling the level of power supply to the individual lighting unit in a luminaire. Communication module incorporated in luminaire includes wired technology such as power line communication (PLC) [13] and wireless technologies such as wifi [4], ZigBee [12]. Illumination could be primarily classified as indoor and outdoor illumination. Smart home lighting [6, 14], museum lighting [15], supplemental lighting for growing plants [10] are few examples of indoor illumination. Indoor illumination requires luminaires with less luminous intensity. Outdoor illumination includes the illumination of streets [12], airports, etc. In comparison to indoor illuminations, outdoor illumination requires luminaires that provide high illumination.

Legacy luminaires providing high illumination requires huge amount of power. The limitation of transitioning a city using legacy luminaires to smart city is also feasible with retrofitting modern day luminaire accessories like the processor, dimmer unit, sensors, and communication module [16]. The retrofit adaptation would render significant control over the legacy luminaires [2, 9]. Newer generation luminaires are manufactured with embedded modern luminaire accessories. The modern advancements in luminaires, the inherent high power and illumination level requirements instill the need to investigate on outdoor illumination.

Section 2 analyses the major factors influencing the outdoor illumination based on the literature. Various illumination methods and advancements in technology in the realm of illumination from the literature were discussed in Sect. 3. The proposed luminaire aware outdoor illuminating role assignment is discussed in Sect. 4. Simulation by incorporating all the factors and findings are presented in Sect. 5. Section 6 concludes with a summary of the paper with future directions.

2 Major Factors Influencing Outdoor Illumination

A smart city comprises of heterogeneous types of luminaires ranging from legacy luminaire to sophisticated luminaire Most commonly available lighting components include metal halide, fluorescent bulbs, incandescent bulbs and Light Emitting Diode (LED). The major factors influencing outdoor illumination are as follows.

(i) **Illumination intensity** [12]:

Illumination intensity (L) is the total luminance emitted by the "n" lighting components in a luminaire and is calculated by the Eq. (1) where l is the lumen generated by a watt and p is the power supplied.

$$L = \sum_{i=1}^{n} l_i * p_i \tag{1}$$

Illumination intensity is measured in Lumens. Desired levels of illumination are achieved by (i) dimming the lighting component in a luminaire, (ii) illumination of selective lighting component in a luminaire and (iii) combination of dimming with selective lighting component illumination.

(ii) **Total power consumption** [12]:

Total power consumption (P) of a luminaire is calculated by (2) where p_i is the power requirement of ith lighting component, p_s is the power requirement of sth sensors available in the luminaire, p_r is the power requirement for the radio communication, p_d is the power requirement of the dimmer circuit and p_{mc} is the power requirement of micro controller.

$$P = \sum_{i=1}^{n} (p_i) + \sum_{s=0}^{t} (p_s) + p_r + p_d + p_{mc} \tag{2}$$

The major part of power in a luminaire is used by the lighting components. Dimming circuits and selective illumination of lighting component reduces the overall power consumption in a luminaire.

(iii) **Lifetime expectancy** [12]:

The maximum lifetime of a luminaire is the duration until the illumination intensity drops below the critical lumens. The average expected lifetime of lighting components under specific conditions are predetermined. The total lifetime expectancy of luminaire is calculated by the Eq. (3) where t_i is expected lifetime of ith lighting components.

$$T = \sum_{i=1}^{n} t_i \tag{3}$$

Life expectancy is represented in hours. Selective illumination and dimming the lighting component extends significantly.

(iv) **Total Illumination Duration** [12]:

Total illumination duration (D) is the sum of "n" lighting component's active duration in a luminaire and is calculated by Eq. (4) where d_i is the individual illumination duration of ith lighting component in a luminaire

$$D = \sum_{i=1}^{n} d_i \tag{4}$$

Illuminating luminaires inconsideration with total illumination duration would ensure uniform lifetime degradation of luminaire.

(v) **Temperature** [12]:

Sensor mounted on the luminaire could sense the temperature of the luminaire (H) during illumination. High temperature deteriorates the lifetime of the lighting unit. Legacy illumaires on prolonged operation are susceptible to very high temperatures.

(vi) **Location** [3]:

Conventionally, non-discriminative location based uniform outdoor illumination had been followed. Uniform illumination of luminaires may not be required for all the outdoor regions. Differentiating a region into multiple zones say critical and non-critical zones would enable to assign, zone specific illumination, minimizing the overall power utilization [3]. Critical zone includes places not limiting to tunnels, bridges, street junctions, and tourist spots require sufficient illumination at the highest order. The rest of the city could be classified as non-critical zones, requires sufficient illumination.

(vii) **Duration** [11]:

Continuous illumination may not be required for the entire duration of illumination [11]. It is deemed ideal to split the entire duration into multiple slots for effective illumination. In general, the duration could be classified as (i) Peak hours (PH) and (ii) Non-peak hours (NPH). Peak hours are duration during which it is considered to be full of activity and it requires maximum illumination intensity. Duration with null or minimum activity could be classified as non-peak hours and it does not require maximum illumination intensity from a luminaire.

Luminaire's illumination intensity (L), total power consumption (P), lifetime expectancy (T), temperature (H), total illumination duration (D) location and illuminating durations were identified as the key factor that influencing the outdoor illumination for effective illumination. In consideration of all the identified factors, existing outdoor illumination methods have to be investigated.

3 Related Works

Illumination has been widely discussed in the literature, especially after the adaptation of LED type of luminaires. Boissevain [1] envisions a smart city with lighting of the state of art of LED-based luminaires. Zenulla et al. [2] proposed an urban IoT architecture and the required devices for deploying various applications in the smart cities. The lighting application using the architecture can optimize the intensity of street lamp depending on the climatic conditions and detection of people. Neirotti et al. [4] has proposed smart street lighting that offers additional functions such as air pollution control and wifi connectivity through the luminaire as the recent trends in the smart cities. Guillemin et al. [6] introduced an innovative, integrated controller for indoors lightings that controls (i) heating/cooling system, (ii) ventilation, (iii) blinds on windows and (iv) lighting system based on the presence of the user. Viani et al. [15] utilized particle swarm optimization based evolutionary algorithm for smart illumination of artifacts in a museum and have effectively utilized wireless sensor and actuator unit to dynamically illuminate the museum on demand. Wifi based dimmable luminaires were employed for showcasing the artifacts in the museum. The approach was suitable for indoor scenarios, but large-scale outdoor scenarios were not addressed.

Yen et al. [16] proposed smart DC grid powered LED lighting system using sensors for indoor scenario. DC-powered LED luminaires achieves a higher lifetime and efficiency with lower maintenance compared with conventional luminaires. Atici et al. [13] used PLC between luminaires to dynamically adjust luminous intensity in roads. The inherent cost of PLC and limitation of PLC were highlighted. Modern day luminaires controls and safety illumination measures were not included in the proposal. Basu et al. [9] proposed the realization of smart lighting control in grid integrated smart buildings with various sensor devices. De Paz et al. [17] proposed a predictive centralized architecture for smart lighting application using artificial neural network with the ANOVA statistic method. The ANOVA has been used to predict the traffic and count of people to generate a suitable lighting schedule.

Kaleem et al. [12] substantiated the overall advantages of using LED luminaire over conventional luminaire. ZigBee outdoor light monitoring and control (ZB -OLC) system was proposed and employed on homogeneous LED luminaire to achieve better power efficiency. The LED luminaires were embedded with (i) a microcontroller, (ii) dimmer circuit, (iii) sensors including luminous sensor, occupancy sensor and temperature sensor. Based on data sensed, microcontroller actuates the dimmer circuit. The level of luminance in a luminaire was controlled by the dimmer circuit.

In the literature, most illumination methods employ homogeneous type of luminaires and are self-actuated for outdoor illumination. In a smart city scenario, wide range of luminaires would be used for illumination. Comparative analysis on the effect of the heterogeneous kind of luminaires is to be investigated further. In specific areas, multiple luminaires are mounted at a point in a location and as well as multiple luminaire may be located within the close proximity. Persistent illumination of all the luminaires may not be required for the entire duration in the specific areas. The need for collaborative illuminate by luminaire aware role assignment scheme would further minimize power consumption, increase the lifetime of the luminaire and provide sufficient illumination.

4 Centralized Outdoor Illuminating Role Assignment Scheme (COIRAS)

Role assignment is the process of assigning tasks effectively [18]. The effectiveness and precise control provided by the role assignment in wireless sensor network [19] motivated to adapt it for outdoor illumination. The primary objective of outdoor illumination is to (i) provide adequate illumination, (ii) reduce power consumption, (iii) increase the lifetime of the luminaire. The multi-objective problem is converted into single objective problem by adapting weighted average approach. Equation 5 provides the generic formula to calculate overall role fitness value (π) of a given luminaire containing 'q' lighting elements, and "n" criterions (c).

$$\pi = \sum_{p=1}^{q} \frac{\sum_{i=0}^{n} w_i * c_i}{\sum_{i=0}^{n} w_i} \qquad (5)$$

For preliminary investigations, following the criteria are considered for illuminating role assignment. (i) Total illumination duration measured in hours (D), (ii) power requirement for illumination (P) and (iii) total illumination duration in hours (H). Each objective could be emphasized by providing higher weights to the corresponding weight of the particular criteria. Legacy luminaires were limited to the full illumination role or no illumination role. But, modern-day sophisticated luminaires have given rise to number of complex roles. Complex roles include multiple stage dimming roles of luminaires and selective illumination role in a set of multiple lighting components inside a luminaire. Outdoor illuminating role assignment has the following phases. (i) Registration phase, (ii) Initialization phase, (iii) Evaluation phase and (iv) Reorganization phase.

(i) **Registration phase**:

The participating smart luminaires and the retrofitted legacy luminaires are registered to the central controller with a unique id. The following details are registered (i) type of luminaire, (ii) lighting elements in the luminaire, (iii) location of the luminaire, (iv) total expected lifetime, (v) total illumination duration and (vi) possible illuminating roles.

(ii) **Initialization phase**:

The central controller generates and broadcasts a reorganization init message containing timestamp to all the luminaires. The participating luminaire acknowledges the init message by reverting with an update message containing luminaire's current role, total power consumption, and total illuminating duration.

(iii) **Evaluation phase**:

On expiry of the waiting period after dispatching the init message, all the luminaires are re-evaluated by Eq. (5) and roles based on location and duration are designated to the luminaires in the order of highest role fitness value. The role assignment update message is broadcasted to the luminaires.

(iv) **Reorganization phase**:

The luminaire on receiving the message updates its roles and acknowledgement is dispatched.

Figure 1 depicts the overall functioning of the COIRAS. Initially, luminaires L1-6 register wirelessly with the main controller through the local controller.

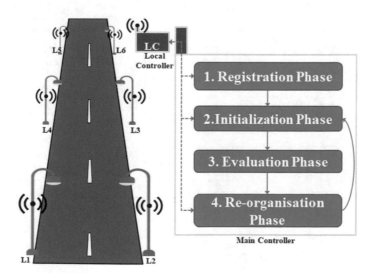

Fig. 1 Overview of the COIRAS

Role assignment is initiated by the central controller and up on receiving the current status from the luminaires, role assignment evaluation is performed. In the reorganization phase, roles are assigned. After the timeout period, phases 2–4 is repeated for reassignment of roles.

5 Simulation Analysis

Simulations were carried out in the Scilab, an open source numerical computing environment. The simulation environment of a smart city was achieved by placing the luminaires randomly onto a two-dimensional plane. Homogenous set of luminaires was not considered for simulation since the newer generation LED type of luminaires out performs the legacy luminaire by having higher lifetime and lower power consumption [1, 20–22]. Also, the possibility of a city assuming to have uniform luminaires across the city is very low.

Hundred randomly chosen luminaires from a set of four retrofitted legacy and two modern luminaires were positioned onto a two dimensional plane. The properties of the luminaires considered are similar to luminaires used for simulation in [12] and are tabulated in Table 1.

Figure 2 is the snapshot of the random dynamic initialization of luminaires in a two-dimensional space. TOL1-4 types of luminaires are legacy-based luminaires

Table 1 Luminaire properties

Label	Type of luminaire	Power rating (P)	Efficiency (lm/W)	Lifetime (h) × 1000
TOL1	Metal Halide	150	80	6
TOL2		250		6
TOL3	LED	70	55	30
TOL4		140		30
TOL5	LED-WD	70	55	30
		49		40
		35		50
TOL6	LED-WD	140	55	30
		105		40
		70		50

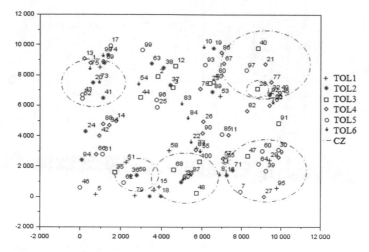

Fig. 2 Snapshot of the random dynamic initialization of luminaires

with two stage roles of lighting. The two roles applicable are (i) maximum luminosity requiring 100% power demand or (iii) no luminosity requiring 0% power demand.

TOL5 and 6 types are smart luminaires with LED dimmer unit. Four stage illumination roles are assumed to be applicable. They are (i) maximum luminosity role requiring 100% power demand, (ii) average luminosity role requiring 75% of power, (iii) minimal power luminosity role requiring 50% of the power and (iv) no luminosity role requiring 0% power demand. We assume all the luminaires could communicate with Zigbee communication modules via its neighbors to the local controllers. Also, the luminaires are assumed to change to its capable roles on demand dynamically. Dotted circular area in the figure depicts the randomly generated critical zone. The rest of the area is taken as non-critical zones.

Tables 2, 3 and 4 tabulate various luminaires type's (LT) power requirement (P) in watts, total required luminaire count (TC), actual luminaire count (C), illumination duration (T) in hours and total power consumed (TP) in hours for the following role assignments (i) legacy method, (ii) ZB-OLC and (iii) COIRAS. Total power consumption by each luminaire (TP) was calculated based on the (i) various zones such as critical (CZ) and non-critical (NCZ), (ii) duration such as peak hour (PH) and non-peak hour (NPH).

Table 2 Illumination analysis- legacy method

LT	P	TC		CZ						NCZ					
		CZ	NCZ	PH			NPH			PH			NPH		
				C	T(H)	TP(W)	C	T(H)	TP(W)	C	T(H)	TP(W)	C	T(H)	TP(W)
TOL1	150	8	17	8	6	7200	8	5	6000	17	6	15,300	17	5	12,750
TOL2	250	1	11	1	6	1500	1	5	1250	11	6	16,500	11	5	13,750
TOL3	70	13	5	13	6	5460	13	5	4550	5	6	2100	5	5	1750
TOL4	140	7	8	7	6	5880	7	5	4900	8	6	6720	8	5	5600
TOL5	70	6	8	6	6	2520	3	5	1050	8	6	3360	8	5	2800
	49			0	6	0	3	5	735	0	6	0	0	5	0
	35			0	6	0	0	5	0	0	6	0	0	5	0
	0			0	6	0	0	5	0	0	6	0	0	5	0
TOL6	140	4	12	4	6	3360	4	5	2800	12	6	10,080	12	5	8400
	98			0	6	0	0	5	0	0	6	0	0	5	0
	50			0	6	0	0	5	0	0	6	0	0	5	0
	0			0	6	0	0	5	0	0	6	0	0	5	0

Table 3 Illumination analysis—ZB-OLC

LT	P	TC		CZ						NCZ					
		CZ	NCZ	PH			NPH			PH			NPH		
				C	T(h)	TP(W)	C	T(h)	TP(W)	C	T(h)	TP(W)	C	T(h)	TP(W)
TOL1	150	8	17	8	6	7200	8	5	6000	17	6	15,300	17	5	12,750
TOL2	250	1	11	1	6	1500	1	5	1250	11	6	16,500	11	5	13,750
TOL3	70	13	5	13	6	5460	13	5	4550	5	6	2100	5	5	1750
TOL4	140	7	8	7	6	5880	7	5	4900	8	6	6720	8	5	5600
TOL5	70	6	8	6	6	2520	3	5	1050	8	6	3360	0	5	0
	49			0	6	0	3	5	735	0	6	0	4	5	980
	35			0	6	0	0	5	0	0	6	0	4	5	700
	0			0	6	0	0	5	0	0	6	0	0	5	0
TOL6	140	4	12	4	6	3360	2	5	1400	12	6	10,080	0	5	0
	98			0	6	0	2	5	980	0	6	0	6	5	2940
	50			0	6	0	0	5	0	0	6	0	6	5	1500
	0			0	6	0	0	5	0	0	6	0	0	5	0

Table 4 Illumination analysis—COIRAS

| LT | P | TC | | CZ | | | | | | NCZ | | | | | |
| | | CZ | NCZ | PH | | | NPH | | | PH | | | NPH | | |
				C	T(h)	TP(W)	C	T(h)	TP(W)	C	T(h)	TP(W)	C	T(h)	TP(W)
TOL1	150	8	17	8	6	7200	8	5	6000	17	6	15,300	15	5	11,250
TOL2	250	1	11	1	6	1500	1	5	1250	11	6	16,500	11	5	13,750
TOL3	70	13	5	13	6	5460	13	5	4550	5	6	2100	5	5	1750
TOL4	140	7	8	7	6	5880	7	5	4900	8	6	6720	8	5	5600
TOL5	70	6	8	6	6	2520	3	5	1050	8	6	3360	0	5	0
	49			0	6	0	3	5	735	0	6	0	1	5	245
	35			0	6	0	0	5	0	0	6	0	3	5	525
	0			0	6	0	0	5	0	0	6	0	4	5	0
TOL6	140	4	12	4	6	3360	2	5	1400	12	6	10,080	0	5	0
	98			0	6	0	2	5	980	0	6	0	6	5	2940
	50			0	6	0	0	5	0	0	6	0	1	5	250
	0			0	6	0	0	5	0	0	6	0	5	5	0

5.1 Findings

Simulation results substantiate the high power consumption of the legacy luminaire in comparison to the new generation LED based luminaires. Legacy method in comparison with ZB-OLC and COIRAS consumes 0.96 and 0.93% more power. COIRAS consumes 0.02% less power over ZB-OLC approach. Uniform lifetime degradation across the luminaires has been reduced due to the adaptation of zones specific and duration specific illumination roles. Adaption of multi stage illumination dimming roles not only reduces the power consumption, but also elongates the lifetime of the luminaire. Collaborative illumination performed in COIRAS has comparatively longer lifetime than ZB-OLC method.

6 Conclusion

Inherent features of legacy and modern luminaires were elaborately discussed in smart city perspective. Significance of outdoor illumination and the factors influencing the outdoor illuminations were analyzed. Outdoor illuminating role assignment scheme for heterogeneous luminaires was proposed and compared with the legacy illumination method and ZB-OLC method. The proposed scheme minimizes power consumption and increases the lifetime of the luminaire with sufficient illumination even in a heterogeneous luminaire scenario. Subsequent decomposition of the outdoor illumination factors such as zones and duration into sub factors would positively influence the overall power consumption and lifetime of the luminaires. Future direction includes identifying and incorporating multiple parameters for role assignment with a perspective on outdoor illumination to reduce the power consumption and increase the lifetime of the luminaires.

References

1. Boissevain, C.: Smart City Lighting. Smart Cities. Springer, Cham, pp. 181–195 (2018)
2. Zanella, A., Bui, N., Castellani, A., Vangelista, L., Zorzi, M.: Internet of things for smart cities. IEEE Internet Things **1**(1), 22–32 (2014)
3. Su, K., Li, J., Fu, H.: Smart city and the applications. In: Electronics, International Conference Communications and Control (ICECC), 9 Sep 2011, pp. 1028–031. IEEE (2011)
4. Neirotti, P., De Marco, A., Cagliano, A.C., Mangano, G., Scorrano, F.: Current trends in smart city initiatives: some stylised facts. Cities **38**, 25–36 (2014)
5. Shahzad, G., Yang, H., Ahmad, A.W., Lee, C.: Energy-efficient intelligent street lighting system using traffic-adaptive control. IEEE Sens. J. **16**(13), 5397–5405 (2016)
6. Guillemin, A., Morel, N.: An innovative lighting controller integrated in a self-adaptive building control system. Energy Build. **33**(5), 477–487 (2001)
7. Kopetz, H.: Internet of things: real time systems. Real Time Syst. Ser. pp. 307–323 (2011)
8. Painter, K.: The influence of street lighting improvements on crime, fear and pedestrian street use, after dark. Landscape Urban Plan. **35**(2), 193–201 (1996)

9. Basu, C., et al.: Sensor-based predictive modeling for smart lighting in grid-integrated buildings. IEEE Sens. J., 4216–4229 (2014)
10. Xu, Y., Chang, Y., Chen, G., Lin, H.: The research on LED supplementary lighting system for plants. Optik-Int. J. Light Electron Opt. **127**(18), 7193–7201 (2016)
11. Chew, I., Kalavally, V., Oo, N.W., Parkkinen, J.: Design of an energy-saving controller for an intelligent LED lighting system. Energy Build. **120**, 1–9 (2016)
12. Kaleem, Z., Yoon, T.M., Lee, C.: Energy efficient outdoor light monitoring and control architecture using embedded system. IEEE Embed. Syst. Lett. **8**(1), 18–21 (2016)
13. Atıcı, C., Ozcelebi, T., Lukkien, J.J.: Exploring user-centered intelligent road lighting design: a road map and future research directions. IEEE Trans. Consum. Electron. **57**(2), 788–793 (2011)
14. Pandharipande, A., Caicedo, D.: Smart indoor lighting systems with luminaire-based sensing: a review of lighting control approaches. Energy Build. **104**, 369–377 (2015)
15. Viani, F., Polo, A., Garofalo, P., Anselmi, N., Salucci, M., Giarola, E.: Evolutionary optimization applied to wireless smart lighting in energy-efficient museums. IEEE Sens. J. **17**(5), 1213–1214 (2017)
16. Tan, Y.K., Huynh, T.P., Wang, Z.: Smart personal sensor network control for energy saving in DC grid powered LED lighting system. IEEE Trans. Smart Grid **4**(2), 669–676 (2013)
17. De Paz, J.F., Bajo, J., Rodríguez, S., Villarrubia, G., Corchado, J.M.: Intelligent system for lighting control in smart cities. Inf. Sci. **372**, 241–255 (2016)
18. Titus, I., Silas, S., Rajsingh, E.B.: Investigations on task and role assignment protocols in wireless sensor network. J. Theoret. Appl. Inf. Technol. **89**(1), 209 (2016)
19. Misra, S., Vaish, A.: Reputation-based role assignment for role-based access control in wireless sensor networks. Comput. Commun. **34**(3), m281–m294 (2011)
20. Campisi, D., Gitto, S., Morea, D.: LED technology in public light system of the Municipality of Rome: an economic and financial analysis. Int. J. Energy Econ. Policy **7**(1) (2017)
21. Komine, T., Nakagawa, M.: Fundamental analysis for visible-light communication system using LED lights. IEEE Trans. Consum. Electron. **50**(1), 100–107 (2004)
22. Cheng, H.H., Huang, D.S., Lin, M.T.: Heat dissipation design and analysis of high power LED array using the finite element method. Microelectron. Reliab. **52**(5), 905–911 (2012)

A Survey on Research Challenges and Applications in Empowering the SDN-Based Internet of Things

Isravel Deva Priya and Salaja Silas

Abstract Many challenges arise due to the increase in the number and variety of Internet-connected devices. Modern data center has immensely improved and gained momentum in providing cloud computing services for handling heterogeneous application and for satisfying various customer needs. Internet of things (IoT) enables numerous number of devices with network connectivity to communicate, to gather, and exchange information for providing smarter services. Management and performance tuning of billions of devices that are geographically distributed has become a challenging and complex task. To overcome the limitations of traditional networking, Software-Defined Networking (SDN) technologies were developed. SDN offers flexibility, agility, programmability, and addresses end-to-end connectivity issues that are prevalent in managing dynamic traffic patterns and real-time traffic flows. This paper surveys and investigates the SDN principles, research challenges, and presents an insight into SDN-based IoT.

Keywords Software-defined networking · Internet of things
Data center networks · Cloud computing

1 Introduction

With the advent of the internet, the modern computer networks have grown tremendously into a complex beast that is extremely difficult to manage and struggles to scale to today's requirements. IoT, cloud computing, and big data introduces new challenges for operating the network. Emerging Internet applications depend on cloud-based services that are geographically spread among different data centers.

I. Deva Priya (✉) · S. Silas
Karunya Institute of Technology and Sciences, Coimbatore, India
e-mail: i.devapriya@gmail.com

© Springer Nature Singapore Pte Ltd. 2019
J. D. Peter et al. (eds.), *Advances in Big Data and Cloud Computing*,
Advances in Intelligent Systems and Computing 750,
https://doi.org/10.1007/978-981-13-1882-5_39

The rapid shift in data center functionality and huge volume of dynamic traffic imposes challenges. The traditional network technologies are not capable of adjusting acceptable strategies to meet the specific requirements of IoT applications in real time [1]. This can be overcome with the SDN model. SDN provides a promising feature that enables automation of network, programmability, and virtualization. SDN is highly scalable and best suited for mega-scale networks.

In traditional IP networks, the control and data planes are joined together and reside in the same networking device which seemed to be the best way for early days Internet. But with the advancement of technology, and dynamic nature of today's Internet, the traditional method of operating the network degrades network performance [2]. Traditional networks implement various sophisticated algorithms and a list of rules on every hardware components to monitor and control the flow of data in the network. The configuration of each device and administering routing paths dynamically are difficult. When the packets are received by the device, it employs a set of rules which are already embedded in its firmware to detect the routing path for that packet as well as destination device address. The current network devices have the limitation on network performance due to high network traffic, which hinders the network performance in terms of speed, scalability, security, and reliability [3]. This limitation can be improved with SDN model. The key feature of SDN is separating the control plane from the data plane that facilitates easier management of all SDN-enabled routers. SDN controller provides an interface for programming the behavior of network devices. The system administrator gains access to control the way packets are forwarded and supports better utilization of network resources. The comparison between traditional IP networks and SDN are tabulated in Table 1. The structure of the traditional IP network is depicted in Fig. 1.

Table 1 Comparison between traditional IP networks and SDN [4]

Characteristics	Traditional IP networks	SDN
Features	Data and control plane are bundled together	Separate data plane and control plane
Configuration	Addition/removal of devices is tedious and error prone	Configuration can be programmed dynamically
Performance	Heterogeneity makes network optimization difficult	With feedback mechanism, optimization is made simpler
Accessibility	Response to load changes, faults and failures are slow	Response to faults and failures are faster
Cost	Management is costly and complex	Simpler and easier manufacturing leads to a low-cost device
Innovation	Difficulties faced while implementing new ideas and design	Encourages the implementation of new ideas via programmability

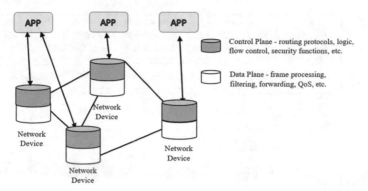

Fig. 1 Traditional IP network

The main objective of this paper is threefold. First, in this paper, the architectural principle of SDN is discussed. Second, the various challenges and research issues faced in the deployment of SDN are presented. Finally, the applications where SDN-based IoT can be implemented for achieving better results is discussed.

2 SDN Architecture

Software-Defined Networking has emerged into an innovative networking standard in the recent few years. SDN addresses the lack of programmability in current network architecture. The factors that are vital for the success of SDN are plane separation, flow-based traffic forwarding, centralized control, and programmability.

Figure 2 shows the SDN architecture which comprises of three layers. The architecture could be easily understood based on a bottom-up approach. The lowest layer is called data plane layer or infrastructure layer [5, 6]. The main responsibility of data plane is data forwarding, filtering, monitoring local information, and gathering statistics. The data plane layer is discussed in Sect. 2.1. One layer above is the control plane layer. The control plane is responsible for taking decision and administer for the data plane devices. The control plane layer is reviewed in Sect. 2.2. The application layer is discussed in 2.3. In the context of Open SDN, the terms northbound API and southbound API distinguishes whether the interface is to the applications or to the devices [7]. Section 2.4 presents the southbound and northbound API which represents the interface from the control plane to data plane and to application layer, respectively.

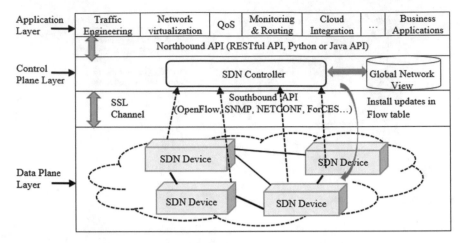

Fig. 2 SDN architecture

2.1 Data Plane Layer

The SDN-enabled devices represent the data plane layer. The network intelligence is removed from the data plane devices and moved to a centralized SDN controller. Data plane SDN devices have no control functionalities and only simple forwarding devices. It handles packet processing, forwarding, and filtering based on QoS function. Forwarding tables of SDN devices are populated by the controller and data plane forward packets as per the instructions given by the control plane through a secure SSL channel.

2.2 Control Plane Layer

The SDN controllers represent the control plane layer. SDN architecture provides a feature that enables a controller to manage multiple data plane devices via a single standard API. Control plane makes decision as to how to handle the packets in routers and switches by analyzing the received packet and routing table. The global view of the entire network topology view is maintained by the controller in order to make decisions for all the forwarding devices. The network behavior can be controlled automatically through the software-based centralized controller. This model provides the system administrator to dynamically program the devices to achieve performance requirements. It ensures that the proper access policies, flow scheduling, error detection, and network security strategies are applied to the appropriate devices. Thus, it helps in reducing, inconsistency or errors during network configuration [8].

2.3 Application Layer

The application layer is also called as management plane. It includes the software services and tools to monitor and configure the functionality remotely. It introduces new features such as security and manageability, network policies, forwarding schemes, and also assists the control layer in network configuration. The application layer receives the network topology and other related information from the controllers and uses that information to provide appropriate guidance to the control layer for making decisions as to how to orchestrate the data plane devices.

2.4 Southbound API and Northbound API

The southbound interface represents the interaction between the SDN controller and the data plane devices. The SDN controller can communicate with the switch via the southbound API. The OpenFlow is commonly adopted southbound interface, which is standardized by the Open Networking Foundation (ONF). OpenFlow is a protocol that defines the interaction of one or more control with OpenFlow-based switches. The OpenFlow protocol provides an interface that allows a control software to program SDN switches and install flow table entries in switches which facilitates the switches to forward traffic. OpenFlow is currently evolving with newer releases that are more powerful with feature sets that support IPv6, Multiprotocol Label Switching (MPLS), metering, and policing [9]. Other southbound API are SNMP, NETCONF, ForCES, ALTO, and I2RS [6] which are also involved in SDN-related activities.

The northbound API represents the interface between the controllers and applications. This enables the programmability of the network. The network programming languages are needed to ease and automate the configuration and management of the network. These northbound interfaces either use a specific programing language (e.g., Java or Python) or a REST-based API. The various network operating system platform software's available are NOX, POX, Ryu, Floodlight, Opendaylight, etc. [10]. Using the northbound API, the topology abstraction of the network can be obtained which is one of the mandatory features in every SDN controller.

3 Research Challenges in SDN

This section highlights research challenges in SDN and developments needed for realizing the vision and accomplishing the promises of SDN. Identifying the limitations of SDN architectures can take a lot of time and effort as it is complex to produce consistent outputs with real deployments and experimentation. Following are some of the major concerns that need to be addressed.

3.1 Resiliency

Resilience is the capacity of the network to sustain a tolerable level of service when confronted with operational challenges [11]. SDN is expected to yield the same levels of availability as legacy and also expected to yield better results than legacy IP network. Proper mitigation strategies are to be activated without human interference to protect the network and to provision support services. SDN are frequently questioned on their actual ability of being resilient to faults such as link failure, node failure, malfunctioned SDN elements, misconfigured devices, and availability of SDN controller [12]. Despite intense investigation by academia, sufficient research is needed in terms of implementations to achieve highly reliable fault tolerant operating SDN network.

3.2 Scalability

From the inception of SDN, scalability has been one of the major concerns. Most of the scalability concerns in SDNs are related to the separation of the control and data planes. When the network grows bigger with the introduction of new SDN devices, the large volume of network request in terms of control traffic in addition to data traffic would overwhelm the controller and as a result, the performance degrades. Reactive network configurations by the controller would introduce extra traffic and bottlenecks. On large-scale networks, controllers have to process millions of flows per second without degrading the QoS requirement. Therefore, these overheads on the control plane and latency on flow table setup are the major scaling concerns in SDN [10]. Still, research efforts are required to address the various challenges that arise due to scalability.

3.3 Controller Placement

In SDN model, the controller is an essential device, and therefore efforts are being dedicated to ensure high availability with distributed, high performance, scalable and flexible programmer friendly software. As the network grows in size, single controller will not be able to accommodate huge traffic from different data plane devices [10]. To address scalability, distributed controller platforms are needed. The issues to address are the number of controller required, placement of controller location, assignment of forwarding devices to controller, etc. Deserving special attention is the latency between forwarding devices and controller instances, fault tolerance, load balancing, reliability, and synchronization [13].

3.4 Security

Cyber-attacks are capable of destroying the entire network infrastructure. Due to the threat of cyber-attacks and the current plethora of digital threats, security and reliability are top significance in SDN [2, 7]. As most of the attacks are launched through local area network, industry experts believe that further investigation is needed to ensure security and dependability. Moreover, from the reliability perspective, the availability of Internet routers is currently a major concern and therefore it is crucial to achieve high levels of availability on SDN control platforms if they are to become the main pillars of networked applications [14]. Therefore, security strategies must focus on protecting the controller and authenticating the application access to the control plane.

3.5 Switch Design

Currently, available switches are very varied and exhibit prominent differences in terms of performance, features, power consumption, protocol specification, security, scalability, and architecture. Various SDN switch designs efficiently work together with TCAMs, such as SRAM, RLDRAM, DRAM, GPU, FPGA, NPs, CPUs, and other specialized network processors [2, 7]. One challenge is to develop switches with large and efficient flow tables to store the rules. A series of ongoing research efforts target a modular and flexible structure of controllers that includes all necessary features.

3.6 Performance

Implementing a high performance, scalable network requires intensive study into the current network performance and limitation of SDN technology with its underlying concepts. Performance can be measured in terms of packet delivery latency, flexible traffic steering, switching speed of traffic, lookup delay, impact of policies, packet sizes, frequency of occurrence of bottlenecks, the impact of configuration on switches, etc. Regardless of how robust, secure, scalable, or interoperable a network is, it is unusable if it lacks performance [15]. Therefore, further investigations and more intelligence are required for improving performance of the network.

4 SDN-Based Internet of Things

SDN and IoT are two major emerging trends that are destined to overlap with one another. The Internet of Things (IoT) has gained a significant momentum in recent years. It presents an image of a future internet and recognized as an ecosystem of connected devices, computing mechanisms and other items to exchange data with higher ease and financial benefits. It has emerged as a most promising technology and is used in a more distributed environment offering a vast amount of applications such as smart wearables, smart home, smart mobility, smart cities, etc. There is a huge demand in terms of performance in delivering the different services to its customers. Experts predict that IoT could connect a colossal 21 billion devices by 2020.

The promise of SDN is a significant improvement in the manageability and flexibility of the network. For SDN to become the standard, growing number of mobile and other connected devices require the infrastructure to be more agile, manageable, and programmable. SDN provides an open environment for application developers to develop innovative tools and software that connect IoT devices more effectively [16]. SDN has the facility to efficiently use the resources and intelligently route traffic. This makes it much easier to handle the tremendous data from IoT. SDNs will eliminate bottlenecks and encourage efficiencies in processing the data generated by IoT without placing a much large workload on the network. SDN provides significant advantages in a number of ways, such as automation, deal with varied and heightened traffic, provide secure end-to-end transmission, dynamic control, routing, etc.

Adequate solutions are required to concurrently deliver traffic from a collection of sensor and several networking resources. SDNs have the ability to optimize the distribution of the workload via powerful control plane [3]. SDN controller can handle specific tasks and allocates paths in an ad hoc manner. Various algorithms are used to optimize the data plane functionality that results in high-speed transmissions and makes more efficient use of the resources.

5 Applications

Rapid evolution of the IoT brings in a lot of applications interest that could benefit from the concepts of SDN. In this section, some of the major use cases to show flexibility, easy management, and deployment of SDN with IoT applications to achieve high-performance requirements are discussed.

5.1 Smart Cities

Smart cities start with a smart public infrastructure which requires expertise that spans many different fields including planning, transport, finance, energy, safety, telecommunications, and many more. IoT shall be able to integrate a large number of different and heterogeneous end systems. IoT enables easy access and interaction with a wide variety of devices such as home appliances, surveillance cameras, monitoring sensors, actuators, displays, vehicles, and so on [17]. IoT can provide access to select portion of data for the progress of providing digital services. IoT associated with SDN can clearly simplify the creation, deployment and ongoing management of the IoT devices. SDN model, when integrated with IoT, can deliver improved performance by dynamic steering of traffic, reducing delay by finding the optimal path for data traffic, fast failure recovery, and contribute for optimizing the energy efficiency and mitigating environmental effects.

5.2 Smart Transportation

Smart transportation provides innovative services and efficient transportation system with the introduction of IoT. Traffic management services analyze traffic behavior and events in order to optimize overall road capacity, reduce travel time and to route vehicles through the shortest path and coordinate traffic lights. Safety services are needed to be smarter to reduce pedestrian and vehicle accidents. Strategic mechanisms are needed to provide real-time delivery of accurate data to take corrective measure. Support for real-time decision-making provided by SDN would make autonomous vehicle a reality in future. Some applications like audio/video streaming which has strict delay requirements and those with high mobility and fluctuations in topology are challenging to handle. The idea of SDN-assisted IoT for smart transportation to assume the existence of controller at road side that can communicate with elements in the data plane and educate vehicles on aspects such as safety, management, comfort, etc. [18]. SDN controller has enough knowledge to make optimal routing decision for achieving enhanced performance.

5.3 Health Care

A modern healthcare system facility introduces IoT technologies for enhancing clinical health care and improves access to medical services. Managing the number of devices connected and the tremendous amount of data collected is a challenging task. Maintaining security and privacy of data, especially when it is being exchanged with other devices is complex. Establishing a smart health care system requires real-time monitoring, analyzing, and prediction of treatment. The system needs to ensure high availability, security to protect sensible data and auto-scaling to ensure that all data is processed and received in the system. SDN with IoT can guarantee QoS requirements [19]. More work is required on models and algorithms to utilize data for the decision-making activities of healthcare diagnosis and medical treatment.

5.4 Agriculture

The IoT-based agricultural system aims to provide stability for supply and demand of agricultural products. It facilitates forecast system and develops environment sensors to improve growth and production of crops by gathering information pertaining to fields, irrigation, fertilizers, etc. The system should be capable of monitoring basic parameters in the fields such as air, soil, plants, pests, and diseases and tools that analyze these data providing alerts and decision support information. IoT sensor nodes are installed throughout the field to gather, process, and transmit measurements of interest to the nearby control center for decision-making. Designing and implementing IoT systems in the domains of analyzing harvest statistics, precision agriculture, and ecological monitoring can be challenging, therefore a systematic approach is needed [20]. To overcome this, the SDN-assisted IoT-based monitoring system with SDN controllers could be deployed to improve the decision-making efficiency.

6 Conclusion

SDN represents an open architecture that facilitates the system to be more flexible with agility, automation, and dynamic provisioning. It has emerged as promising solutions and seen as an evolutionary paradigm shift, but still faces several challenges that hinder its performance and implementation. This paper studies the SDN architecture principles and how it varies from traditional networking. The research challenges that are prevalent in the deployment of SDN are presented. Furthermore, the SDN based IoT with its applications are addressed. Next-generation data networking with SDN requires more new research advances to unleash its real effectiveness and potential.

References

1. Bera, S., Misra, S., Vasilakos, A.V.: Software-defined networking for internet of things: a survey. IEEE Internet Things J. **4**(6), 1994–2008 (2017)
2. Masoudi, R., Ghaffari, A.: Software defined networks: a survey. J. Netw. Comput. Appl. **67**, 1–25 (2016)
3. Hu, F., Hao, Q., Bao, K.: A survey on software defined networking (SDN) and openflow: from concept to implementation. IEEE Commun. Surv. Tutorials **16**(4), 2181–2206 (2014)
4. Singh, S., Jha, R.K.: A survey on software defined networking: architecture for next generation network. J. Netw. Syst. Manage. **25**(2), 321–374 (2017)
5. Braun, W., Menth, M.: Software-defined networking using openflow: protocols, applications and architectural design choices. Future Internet **6**(2), 302–336 (2014)
6. Farhady, H., Lee, H., Nakao, A.: Software-defined networking: a survey. Comput. Netw. **81**, 79–95 (2015)
7. Kreutz, D., Ramos, F.M.V., Verissimo, P., Rothenberg, C.E., Azodolmolky, S., Uhlig, S.: Software-Defined Networking: A Comprehensive Survey, pp. 1–61 (2014)

8. Gong, Y., Huang, W., Wang, W., Lei, Y.: A survey on software defined networking and its applications. Front. Comput. Sci. **9**(6), 827–845 (2015)
9. Jammal, M., Singh, T., Shami, A., Asal, R., Li, Y.: Software defined networking: State of the art and research challenges. Comput. Netw. **72**, 74–98 (2014)
10. Karakus, M., Durresi, A.: A survey: control plane scalability issues and approaches in software-defined networking (SDN). Comput. Netw. **112**, 279–293 (2017)
11. Da Silva, A.S., Smith, P., Mauthe, A., Schaeffer-Filho, A.: Resilience support in software-defined networking: a survey. Comput. Netw. **92**, 189–207 (2015)
12. Sterbenz, J.P.G., et al.: Resilience and survivability in communication networks: strategies, principles, and survey of disciplines. Comput. Netw. **54**(8), 1245–1265 (2010)
13. Heller, B., Sherwood, R., McKeown, N.: The controller placement problem. In: Proceedings of the First Workshop on Hot Topics in Software Defined Networks—HotSDN '12, p. 7 (2012)
14. Chen, M., Qian, Y., Mao, S., Tang, W., Yang, X.: Software-defined mobile networks security. Mobile Netw. Appl. **21**(5), 729–743 (2016)
15. Xie, J., Guo, D., Hu, Z., Qu, T., Lv, P.: Control plane of software defined networks: a survey. Comput. Commun. **67**, 1–10 (2015)
16. Jim, B., Follow, D.: SDN vital to IoT (2014). (Online). Available: https://www.networkworld.com/article/2601926/sdn/sdn-vital-to-iot.html
17. Zanella, A., Bui, N., Castellani, A., Vangelista, L., Zorzi, M.: Internet of things for smart cities. IEEE Internet Things J. **1**(1), 22–32 (2014)
18. Chahal, M., Harit, S., Mishra, K.K., Sangaiah, A.K., Zheng, Z.: A survey on software-defined networking in vehicular ad hoc networks: challenges, applications and use cases. Sustain. Cities Soc. **35**(July), 830–840 (2017)
19. Parsaei, M.R., Mohammadi, R., Javidan, R.: A new adaptive traffic engineering method for telesurgery using ACO algorithm over software defined networks. Eur. Res. Telemed. **6**(3–4), 173–180 (2017)
20. Tomovic, S., Yoshigoe, K., Maljevic, I., Radusinovic, I.: Software-defined fog network architecture for IoT. Wirel. Pers. Commun. **92**(1), 181–196 (2017)

Evaluating the Performance of SQL*Plus with Hive for Business

P. Bhuvaneshwari, A. Nagaraja Rao, T. Aditya Sai Srinivas, D. Jayalakshmi, Ramasubbareddy Somula and K. Govinda

Abstract Implementation of advanced analytics for big data processing in business intelligence is significant towards gaining profits. For processing large-scale data sets efficiently, so many challenges are faced by traditional database system. To overcome the disadvantages present in an existing system, various kinds of a new database have been evolved along with application programs (e.g., MySQL, PostgreSQL, Hive, etc.). Such type of systems store the data in the database, retrieves, and displays the information once it is queried. The time duration varies in the different database for doing the process. This paper evaluates the performance by using SQL*Plus and Hive. In the enterprise business data model, a comparison of both SQL*Plus and Hive for some CRUD operations (Insert, Join, and Retrieve) are estimated. By the work presented in this paper, we conclude if performance is a key, then SQL*Plus is a right choice to use and for processing large datasets hive works better.

Keywords Big data · Business intelligence · CRUD operations · Hive · SQL*plus

P. Bhuvaneshwari (✉) · A. Nagaraja Rao · T. Aditya Sai Srinivas · D. Jayalakshmi · R. Somula
K. Govinda
SCOPE, VIT, Vellore, India
e-mail: thenameisbhuvanapatt@gmail.com

A. Nagaraja Rao
e-mail: nagarajaraoa@vit.ac.in

T. Aditya Sai Srinivas
e-mail: saircew@gmail.com

D. Jayalakshmi
e-mail: jayalakshminandakumar2@gmail.com

R. Somula
e-mail: svramasubbareddy1219@gmail.com

K. Govinda
e-mail: kgovinda@vit.ac.in

© Springer Nature Singapore Pte Ltd. 2019
J. D. Peter et al. (eds.), *Advances in Big Data and Cloud Computing*,
Advances in Intelligent Systems and Computing 750,
https://doi.org/10.1007/978-981-13-1882-5_40

469

1 Introduction

The advancement in big data analytics and business intelligence leads to store and process data for retrieving the valuable information. In business organizations, data plays a major role in increasing profits and value to the enterprise. The data can be generated both internally and externally and act as a primary economic resource for business decision making. The enterprise collects data daily from the various available sources and accumulates it persistently in storage for a long time. Thus, big data emerges and it became inconvenient and cost intensive to store and process the data by using traditional databases. And a lot of researchers have contributed various tools to tackle the challenges in processing the big data. Thus, we analytically experiment the processing time required for accessing the data stored in SQL*plus and Hadoop database. Structured Query Language (SQL) operations are widely used for the construction as well as the processing of data warehouse System. Several SQL-based data warehouse are investigated by worldwide researchers. In this paper, we are going to analyze the performance based on the SQL-Hadoop Framework known as Hive and Oracle-Based SQL called SQL*Plus. In a MapReduce framework, Hadoop splits the files into larger blocks and distribute the task to both phases. It results in high-speed processing of the data. But it is complex for programmers to write MapReduce programs in Java and other languages. Many studies have been investigated and it leads to the innovation of Hadoop ecosystem. Hive is one of the familiar non-relational tools used for handling huge amount of fine-grained data [1] in Hadoop ecosystem. It uses its own compiler to convert the scripts written in other languages. The converted script functions similar to the MapReduce framework. Hive, a data warehouse of an Apache Hadoop, act as an SQL interface and convert the user query (HiveQL) into a Directed Acyclic Graph (DAG). It helps the user to easily modify the data persist in Hadoop Distributed File System (HDFS) and also works well for both storage and quick access to data [2]. Hive is the front end for parsing HiveQL (Similar to SQL) statements where the files are virtually connected to table like structures and the queries are converted into MapReduce jobs. Similarly, SQL*Plus is a Command-line interface for Oracle database which manages and retrieves data using SQL statements. It is an interactive tool which has its own commands and environment to access the Oracle database. For example: In addition, to supporting basic SQL commands, SQL*plus has its own commands like "SET DEFINE". It acts as an excellent tool for loading and extracting data by using SQL and PL/SQL [3]. Each tool performs the similar task in a different manner. Hive is used for querying data which is stored in HDFS, whereas SQL*Plus is utilized in online operations with more read and writes.

Contribution In this paper, we compare the time required for doing additional CRUD operations in Hive and SQL*Plus. The experiment is conducted by using three various sized datasets such as small-, medium-, and large-scale datasets.

Paper Organization The rest of the paper is organized as follows: In Sect. 2, an overview of the existing database model is discussed. In Sect. 3 the comparative analysis of SQL*Plus and Hive for different size data sets are done. Finally, in Sect. 4 the conclusion based on the performance analysis is presented.

2 Existing Database Overview

Each database system has its own model to store and process its own data. These models are crucial for entrepreneurs to determine how a database will function and handle the information it handles. The information stored can be an Enterprise Resource Data (ERP), Customer data, Accounting data and some other business-related data. Database management system consists of a mechanism to store, process, and extract a large amount of data. The data stored in this database can be transformed into valuable information by analyzing properly. The most popular database model used widely in all applications is Relational Model. This database model is very powerful and also faces several issues that the solutions have never been really offered. Recently, NoSQL database is popular as it overcomes the limitations of the relational model and provides additional functionality.

2.1 The Relational Model

Initially, Relational database systems play a lead role in enterprise applications. It has high valued features and capabilities in transaction management [4]. It allows the business organizations to store the data in a structured manner. It relates the data sets and can be selected horizontally. Keys are used to creating the relationship between data. But recently due to the emergence of big data, it could not handle large datasets effectively. It also lacks the efficiency in joining operations. This operation is used to combine the individual tables into the single logical unit. SQL is a common query language used by all the relational systems. The big advantage of using RDBMS is, it has the capability of holding "ACID" properties. It provides accurate and safe transaction of data. For example, in the online purchase, if more customers try to purchase the same product at the same time, then ACID property assures the transaction not to overlap with each other.

The elaboration of ACID property are as follows:

- **Atomicity:** When an update happens in the database, it should be either commit or abort.
- **Consistency:** The integrity of the database is maintained by the changes made for an instance should remain same all over the database.
- **Isolation:** The changes occur in one transaction should not be visible to other transactions before it gets over.
- **Durability:** The changes made in a transaction should be stored permanently in the database and should be retrieved in case of system failure.

Table 1 Comparison of Oracle SQL*Plus and Apache Hive features

S. No.	Oracle SQLplus	Apache Hive
1	Widely used RDBMS	Distributed data processing
2	Commercial implementation	Open source implementation
3	**Server OS**: Linux, Windows, Solaris, Z/OS, AIX, OS X, HP-UX	**Server OS**: all OS with Ja va VM
4	**Partitioning methods**: horizontal partitioning	**Partitioning methods**: sharding
5	Performs better for smaller datasets	Performs better for larger datasets
6	**Transaction concept**: ACID	**Transaction concept**: nothing specific
7	www.oracle.com/database/index.html	Hive.apache.org

2.2 The Model-Less (NoSQL) Approach

As the data increases in size, the failures along with performance degradation also increased. To overcome this issue, entrepreneurs started looking for the alternative database system. The horizontal scaling and failure tolerance nature of NoSQL attracted the enterprise to make use of it initially. The efficiency of NoSQL is also achieved through high performance, unstructured nature, and massive informative processing. Hadoop is an Open-Source framework which enables the certain type of NoSQL databases to hold data across thousands of servers (HBase). HBase is one of the NoSQL database written in Java. Another important property to be noted in NoSQL is "BASE". As the big data is very large in size, it is very hard to attain "ACID". That is the reason, NoSQL is based on "BASE" principle [5].

- **B**asically **A**vailable: Data is distributed in nature.
- Soft State: No guarantee for consistency.
- Eventually Consistent: Even though data is not consistent, the system guarantees the eventuality.

Table 1 describes the features and differences of both database management tools (SQL* PLUS and HIVE). SQL*Plus is domain specific language for stream processing in RDBMS, whereas Hive is suitable for both real time and batch processing.

3 Proposed Method

The business model was simulated on Windows machine to be able to use both Apache and Oracle server. The enterprise source data was initially inserted manually in SQL*Plus and imported as CSV file in Hadoop which stores data in HDFS. Later, the time in seconds was retrieved for each operation and duly noted.

Table 2 Insertion time

Size of the data	Time in SQL*Plus (s)	Time in HIVE (s)
Small	0.01	2.95
Medium	0.01	2.72
Large	0.02	2.73

The important factors to be considered in the implementation of this paper are:

- Size of the data
- SQL*Plus Versus Hive environment
- Execution time.

3.1 Size of the Data

Three different size datasets are used in both Hive and SQL*plus. The data is categorized based on their tuple size.

(a) Small Dataset: 10 rows and 5 columns
(b) Medium Dataset: 100 rows and 5 columns
(c) Large Dataset: 500 rows and 5 columns

I. SQL*plus Versus Hive environment

The simulation is carried out in windows operating system with installed Apache server and SQL*Plus software. The data is manually imported for SQL and for Hive the CSV format was used. We use Apache Hive 1.2.1 version and Oracle SQL*Plus version 9.2.0.1.0 and followed the tutorials to work on the specific hardware.

II. Execution Time

The time required for insertion, joint, and retrieval operations are estimated.

A. Insertion Time

Table 2 shows the time required to insert different size data sets for SQL*Plus and Hive. SQL*Plus is faster for inserting small- and medium-size data. As the size grows it provides high performance, especially for small size of data sets. Figure 1 Illustrates the time required for SQL*Plus and Hive for insertion operation.

B. Join Time

Table 3 shows the time required for joining operation of data sets in Hive and SQL*Plus. Join is used to combine and retrieve the specific data from multiple tables. Figure 2 depicts the time required for SQL*Plus and Hive for Join operation.

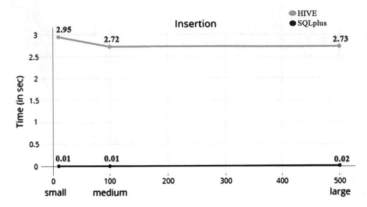

Fig. 1 Chart depicting insertion time

Table 3 Join time

Size of the data	Time in SQLPlus (s)	Time in HIVE (s)
Small	0.02	2.76
Medium	0.08	3.35
Large	0.43	2.98

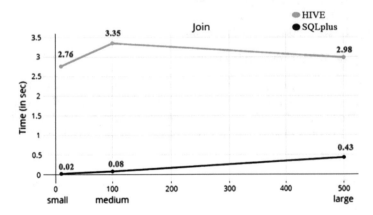

Fig. 2 Chart depicting join time

C. Retrieval Time

Table 4 shows the time required for retrieving data. Hive works faster for retrieving larger data sets than smaller. Figure 3 shows the performance chart for retrieval operation.

Table 4 Retrieval time

Size of the data	Time in SQLPlus (s)	Time in HIVE (s)
Small	0.01	0.17
Medium	0.04	0.148
Large	0.19	0.164

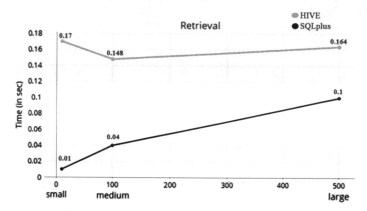

Fig. 3 Chart depicting retrieval time

4 Conclusion Based on the Performed Analysis

In this paper, we have evaluated the optimized performance of both SQL*Plus and Hive for business data. It is concluded that both are suitable for storing and processing data sets for four small business or big enterprise. For a relational database, SQL*Plus performed better. But, for unstructured data and for large datasets Hive works well. For the complex queries with multiple joins, Hive performed notably better than SQL*Plus as its data structure has the capability to hold any type of data. The main intention of this paper is to evaluate the time required for CRUD operation in SQL*Plus and Hive. It is found that Hive is faster in retrieval while comparing to the SQL*Plus. It is also found that SQL*Plus is efficient for updating and modifying the frequent data sets. Further, the investigations for evaluating the performance can be compared and evaluated with other databases.

References

1. Surekha, D., Swamy, G., Venkatramaphanikumar, S.: Real time streaming data storage and processing using storm and analytics with Hive. In: ICACCCT, International Conference on IEEE, pp. 606–610 (2016)
2. Huai, Y., Chauhan, A., Gates, A., Hagleitner, G., Hanson, E.N., O'Malley, O., … Zhang, X.: Major technical advancements in apache hive. In: Proceedings of the ACM SIGMOD International Conference on Management of Data, pp. 1235–1246. ACM (2014)

3. Gennick, J.: Oracle SQL*Plus—The Definitive Guide. O'Reilly Media (1999)
4. Pratt, P.J.: A relational approach to database design. ACM SIGCSE Bull. **17**(1), 184–201 (1985)
5. Abramova, V., Bernardino, J.: NoSQL databases: MongoDB vs Cassandra. In: Proceedings of the International C* Conference on Computer Science and Software Engineering, pp. 14–22. ACM (2013)
6. Guo, Y., Rao, J., Cheng, D., Zhou, X.: ishuffle: Improving Hadoop performance with shuffle-on-write. IEEE Trans. Parallel Distrib. Syst. **28**(6), 1649–1662 (2017)
7. https://hive.apache.org
8. Capriolo, E., Wampler, D., Rutherglen, J: Hive Programming Guide. O'Reilly Media (2012)

Honest Forwarding Node Selection with Less Overhead and Stable Path Formation in MANET

S. Gayathri Devi and A. Marimuthu

Abstract Mobile Ad hoc Network (MANET) has paid considerable attention to wireless communication. MANET is an autonomous collection of self-deployed nodes without any pre-existing infrastructures and nodes are movable that also act as routers. The main characteristics of MANETs are lack of centralized control, lack of association among nodes, rapid mobility of hosts, frequent dynamically varying network topology, shared broadcast radio channel, insecure operating environment, physical vulnerability and limited availability of resources, such as battery power, and bandwidth. In multicast routing, forwarding nodes are chosen by the designed protocol to be present in the path between sender and receiver. Here, sender and receiver can also be a forwarder node. Flooding and ratio of forwarding nodes change must be limited to reduce overhead. To increase security, the estimation of honesty of nodes is significant. The previous work ODMRP-EFNRLP tries to minimize overhead through minimizing redundant data delivery, prediction of link failure and energy-aware node selection. This paper searches On-Demand Multicast Routing Protocol design and the possible fuzzy optimization. In this paper, attacker detection is tested using multi-agent classification to update node's honesty for decision-making. Its results are evaluated. During the evaluation test, the results validate the progress.

Keywords Multicast routing · Fuzzy logic · Honest node selection · Honest value

S. Gayathri Devi (✉) · A. Marimuthu
Department of Computer Science, Government Arts
College (Autonomous), Coimbatore, Tamil Nadu, India
e-mail: gayathridevi212@gmail.com

A. Marimuthu
e-mail: mmuthu2005@gmail.com

A. Marimuthu
Department of Computing, Coimbatore Institute
of Technology, Coimbatore, Tamil Nadu, India

© Springer Nature Singapore Pte Ltd. 2019
J. D. Peter et al. (eds.), *Advances in Big Data and Cloud Computing*,
Advances in Intelligent Systems and Computing 750,
https://doi.org/10.1007/978-981-13-1882-5_41

477

1 Introduction

The best part of MANET applications takes in salvage sites, smart classrooms, video conversation, disaster search, military field, and so forth which insist applicants exchange details using multicast operations. Therefore, multicast routing protocols play a significant role in MANET. One of the significant drawbacks of the multicast protocols is an inherently unstable tree structure. This unstable tree structure assists ad hoc types of networks in updating their link status in response for changing topology continuously. Also, typical multicast trees usually want a link state or distance vector global routing substructure that can end in significant packet loss. Furthermore, continuous topology changes create the frequent exchange of routing vectors or link state tables. This operation causes too much channel and processing overhead that can considerably raise network congestion. As a result, constraints correlated to bandwidth resources, power feeding, and host mobility makes multicast protocol scheme thought-provoking. In response to these complications, several multicast routing protocols have been offered for ad hoc wireless networks.

Multicast transmission refers to the only sender information to be forwarded to multiple receivers [1]. The multicast communication is supported by forming a multicast group id (GID). The multicast designs are destined for group communication and the group is formed dynamically with group members (GMs). Any node can either join the group or leave the group without any notification to the source node. For sending and forwarding the data, the node need not be a GM of that group. The path is recognized through two stages, the path discovery and data re-forward stages. The path discovery is made by broadcasting of control packets like route request (RREQ) and route reply (RREP) packets. The RREQ is broadcast by the origin node. The middle and the destination nodes are to be the receiver of this RREQ packet. The answer of the RREQ is specified only by the mcast receivers. After broadcasting the RREQ, the origin node waits for the RREQ for some preset time. If there is no answer, origin node rebroadcasts the RREQ by incrementing the broadcast id. The forwarding nodes preserve a multicast table. They forward the RREQ by flooding. The RREP is created by the receiver nodes, and hence the forward route is established. In case if a node moves, it does not have alternate routes.

In MANET, nodes do not operate from physically protected places, thus easily fall under attack. Terrible data booster attackers can study the network infrastructure and try to collapse the network. Eavesdroppers, vampires, and traffic reader performance which try to obtain node data can be forbidden using encryption and decryption systems. A public key sharing is weak in the key administration but they secure the network transmission [2]. Intruder detection system secures network against attacks promptly. In MANET, security key sharing and encryption–decryption system increases delay because of computational overhead. The key renewal system attains extra delay and maximizes the processing time on nodes.

The proposed protocol design aims to control the network overhead and minimize the number of forwarding node selection. These forwarding nodes are selected with suitable security measure to find the safety. It has to be good quality to pass up the dishonest nodes as a forwarder. Building honesty which forces to strengthen the network teamwork is most significant one to secure the MANET. Also, honesty measures the security solution that should be active as per the changed honest relationship. Build up honest-based cooperation to eliminate the danger from attackers and improve the network security. The cooperation achieves the node reliability of neighbors to know the honesty. This cooperation develops network outcome because honest nodes reject working with dishonest nodes. Honesty is computed based on the proof updated by the prior communications between nodes, so the initial honest communication level needs to be calculated.

Security reason-based computation of node's honesty is working to compute the honest and dishonest measuring value about the node based on the continuous contact record (CR) which holds out by that node based on straight and Circuitous comments. If the CR is unbeaten, then the node is measured as honest. If not, it is a dishonest node. However, straight contact may be bogus due to link disconnections, channel fault, and so on. Although the honest of a node is computed based on excellent contact certificate provided by surrounding nodes, honest weakens as per time.

This rest of the paper is planned as a tag along. Section 2 gives little notion about multicast network discussion as related works. Implementation and protocol design of fuzzy-based honest node selection (HEFNRLP) is discussed in Sect. 3. Section 4 gives a detailed analysis of HEFNRLP performance. Section 5 speaks about the conclusion of this work.

2 Related Work

Wireless communications are used in the data network to reduce deployment, wiring, and maintenance cost. However, wireless links are susceptible to attacks [3]. The performance issue of trust management protocol design for MANETs addressed in two crucial areas: trust bias minimization and application performance maximization [4]. Wireless Sensor Networks are set up in harsh, neglected, and often adversarial physical environments for specific applications such as military areas and sensing tasks with trustless atmospheres [5]. The network described above can be considered as a distributed system, where broadcast or multicast is a spirited communication primitive [6]. The most important goal of an ad hoc routing protocol is to create a correct and efficient route between any pair of nodes with minimum overhead. The routing overhead is a critical metric [7]. The primary task of the Minimum-Energy Multicasting (MEM) problem in Always Active-Wireless Ad hoc NETworks (AA-WANETs) is to select appropriate forwarding nodes such that a multicast tree with the minimum energy cost can be constructed [8]. A peer-to-peer streaming application allows multicast groups to be designed without any network support and without any other permanent infrastructure dedicated to supporting multicast [9]. Direct application of key distribution schemes of wired networks proves to be energy inefficient for the wireless ad hoc environment [10]. They can be hired in aggressive environments, and the features like use of low power and low maintenance make them the most suited technology for real-time environment [11]. In such single hop networks, security is still a concern, but is more accurately addressed through secure broadcasting and multicasting [10]. An improved dynamic gray-Markov chain prediction measure is proposed to estimate node's trust prediction efficiently. In this trust model, they proposed a trust-enhanced unicast routing protocol and a trust-enhanced multicast routing protocol [12]. The protocol's performance is maximized by relaxing security requirements based on the perceived trust. A composite trust-based public key management (CTPKM) is projected with the goal of maximizing performance while mitigating security vulnerability [13]. A mesh-based multipath routing scheme is suggested to discover all possible secure paths using secure adjacent position trust verification protocol and better link optimal path has found by the Dolphin Echolocation Algorithm for efficient communication in MANET [14]. Previous works related to measure honest of a node do not pay attention much about how to calculate fundamental Honest Value (HV). Therefore, we planned to incorporate the trust-based forwarding node selection in ODMRP-EFNRLP protocol.

3   Design and Implementation

Wireless multicast networks are susceptible to some form of attacks inflicted by malicious nodes; it can be packet dropper, black hole, selective forwarding behavior attack, and so on. An honest, forward node selection is proposed to alleviate the consequence of malicious node due to link disconnection.

3.1   Network Model

Honest is confidence of one node on neighbors group for some transmission. Multicast network is modeled as a graph $|G = \{N, NE, TV\}|$, where TV denotes a node's trust, $N = \{n_1, \ldots n_n\}$ describes the total number of nodes in the network region and $NE = \{ne_1, ne_2, \ldots\ldots ne_n\}, i \neq j$, is the honest connection among nodes between ne_i and ne_i as a straight link that brings TV between 0 and 1, denoted by $TV(ne_{ii}) \in |0, 1|$, and $|ne|$ is the number of neighbors in transmission range.

3.2   Honest Forward Node Selection in the Multicast Group

The steps of node honesty are computed in the network as below.

- Each node in multicast group observes and records the activities of its GM.
- Each one of the multicast group shown in Fig. 1 collects facts about its GM actions from other member nodes and records it for validation.

Fig. 1 Multicast view

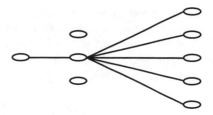

- Each node computes the honesty of all group nodes based on the contacts already recorded.
- Each multicast group node finalizes the best forwarding node to transfer the multicast packets.

3.3 Honest Feasibility Assessment

The Honest Value (HV) is computed for each node to realize its status and credibility for efficient communication. In a multicast network, the source sends data packets to its receivers through selected forwarding node.

- Each node must believe their neighboring nodes.
- Maximum nodes in the network work openly. These nodes provide its participation in the network in the usual manner. Some of the nodes act as malicious behavior.
- Path establishes through forwarding node to multicast receivers the neighbor node honesty and the load which is calculated based on it is Contact Record (CR) among GMs.
- Source sends its data through the honest forwarder. An honest assessment required records about contacts between the neighbors and their related estimated HVs. This implementation discussed to calculate multiple honest estimation methods as Straight Honest Estimation, Circuitous Honest, and Incorporated Honest.

3.3.1 Straight Honest Estimation from One-Hop Group Members

The straight honest behavior is measured by observing the activities of its one-hop group nodes. To calculate straight honest nodes, the HV of all the group nodes is 1 at the opening of the network stage. The record of contacts is observed using continuous tracking of successful and breakdown contacts of the group nodes. A punishment Γ value is merged for status that assists to alleviate the malicious behavior in the system.

Once communication is completed between GM, each node maintains the complete contact (CC) value ($0 \leq CC \leq 1$). Update this value in node CR used to review the node's one-hop honest information. All common nodes broadcast and receive the data packets correctly. Acceptable assessment in multicast GMs speaks about the performance of each one of them. The computation of transmission proportion (TP) is based on the percentage of the data packets sent correctly to the number of nodes who believed to be forwarded. In case data packet corrupted, it does not consider as perfect transmission. Using TP is computed as in (1).

$$TP(i, j) = \frac{\text{Number of packets sent}}{\text{Number of originate packets}} \tag{1}$$

The number of packets sent stands for the increasing count of delivered packets for the node nj. The Originate packets indicate the total count of complete sent packets. Exclusively, the node ni gives its CR about nearby group node nj with good, acceptable assessment score after forwarding data and control packets received from the node, ni. The principle for calculating the straight honest (SH) between the eyewitness i and the observed node, j is given in (2)

$$SH(i, j) = \frac{\text{NOSP}}{\text{NOSP} + \gamma \text{NL}} \tag{2}$$

where NL indicates the packet loss count that says about missing packet forwarding and NOSP is the Total sent packets discussed in the CR. γ is a constant value. Γ is computed the honest of the node, as in (3)

$$\Gamma = \frac{\left(\frac{\text{NOSP}}{\text{NOSP+NL}} + 1\right) \log_{10}}{\log_{10}(2)} + 1 \tag{3}$$

When the number of disobedient raises, then the punishment value Γ also raises which in revolve reduces the HV of the multicast group node.

3.3.2   Circuitous Honest Computation

The Circuitous honest (CH) explained is calculated based on the multicast GMs suggestion, in some conditions where the witness may not watch the performance of the nearby nodes. So some one-hop neighbour node is acting as a witness node it may not be in the transmission range of monitoring node.

The node i has n-one-hop direct GM where $E = \{e_1, \dots e_n\}$. Assume that node i want to set up honest contacts with its second hop GMs. Here, we have considered forwarding count (fc) is 2.

In the Straight Honest computation, node i checks the HV of all its direct multicast GMs and decide to select a forwarder whose HV is greater than limitation (l) which is 0.6. Next, the control packets multicast to all its direct GMs within node i's coverage area. Let j, be the monitoring node; the GM send a reply to i along with the HV of its GMs. The CH of node j estimated as (4).

$$CH(i, j) = SH(i, j) \times SH(n, j) \tag{4}$$

Here, n is the forwarding node. In the Circuitous Honest (CH), nodes of j having HV $< l$ might provide a terrible suggestion about the monitoring node. So, the monitoring node notices and cutoff those attackers.

After certain communications, a set of suggesting nodes updates about j, then the Circuitous Honest (CH) is computed by i based on the suggestion inward from the GMs, and i can build an honest route to node j through forwarding nodes if the Circuitous Honest (CH) value of j is $> l$. If so, node i selects one of its forwarders with maximum HV to send the packet to node j.

The direct GM monitored the GM_j character and sent a suggestion to GM_i. If there are multiple forwarding routes $R, \ldots R$ between node i and j, compute the honest and the influence (IF) of suggesting credential as in (5).

$$CH = \left[\sum_{j=1}^{N} IF_j (SH(N_j, j)) \right] \tag{5}$$

where

$$IF_j = \frac{SH(i, N_j)}{\sum_{i=1}^{N} (i, N_j)}, \quad 0 \leq IF_j \leq 1$$

3.3.3 Incorporated Host

Incorporated honest (IH) is computed based on the suggestions collected from the GMs and SH estimation.

The mean of the straight honest group (SHG) members who are having best status is considered into account as the status by node i in (6),

$$SHG_{avg} = \frac{1}{N-1} \sum_{j=1}^{N} (N_j, j) \tag{6}$$

The Incorporated Honest (IH) of the node j by the node i is in (7)

$$IH(i, j) = \alpha SH(i, j) + (1 - \alpha) SHG_{avg}, \quad 0 \leq \alpha \leq 1 \tag{7}$$

Moreover, the final HV (FHV) is computed as in (8)

$$FHV(i, j) = \beta SHG(i, j) + (1 - \beta) SHG(i, j), \quad 0 \leq \beta \leq 1 \tag{8}$$

Table 1 Honest conclusion of nodes

Stage-of Node honest	Ranges	Fuzzy conclusion of node honest
1	$0 < FHV < 0.5$	Honest attitude very low
2	$0.5 \leq FHV < 0.6$	Honest attitude low
3	$0.6 \leq FHV < 0.7$	Honest attitude medium
4	$0.7 \leq FHV < 0.8$	Honest attitude high
5	$0.8 \leq FHV \leq 1$	Honest attitude very high

3.4   Honest Conclusion Using Fuzzy System

A node's honest estimation still can be optimized using fuzzy conclusion system to get decisions for honest path selection. The inbuilt feature of a wireless network is insecurity. The network contains many paths from sender to multicast receivers. Thus, the honest wireless path selection uses the fuzzy conclusion to select the best path in the multicast networks. The honest node value ranges from 0 to 1 and 0 indicates absolute dishonest, whereas value 1 indicates total honest (Table 1).

After completion of individual transactions, each node computes its multicast GMs about their unbeaten transmission. Initially, each node computes the forwarder to send the data packets to the receivers. At initial the link bandwidth for all nodes initialized equally. No nodes show its misbehaving activity at initial. Many parameters are considered for honest-based routing since some links may have low capacity and HV, such links must not be considered for reliable routing. Honest node link is treated as a selectable link.

The node communicates with its GMs or forwarding node while the incorporated honest >0.5. The node i computes the HV of its members based on watching completed by it. Using fuzzy linguistic expression, fuzzification maps the sharp input values into the equivalent linguistic assessments by the membership function which plans every sharp input onto a linguistic values as (0, 1).

3.5   Routing Process of HEFNRLP

Construct neighbor table using hello message to know each node neighbors. To establish the path and minimizing the control packets through JOIN-QUERY and the JOIN-REPLY using strong forwarding node selection with highest HV whose GM connection is largest. When sender S wishes to send some data to the multicast receivers, but there is no route to it. Hence, S starts broadcasting JOIN-QUERY as a route finding packet every so often along with sequence number. Each node computes the HV of its next node and updates the honest table with monitored value. The HV are optimized using fuzzy membership function. The defuzzification gives the HV to choose honest forwarding nodes. This technique minimizes the routing overhead and increases the security of the network. These JOIN-QUERY messages update the membership list and rebuild the routes. Each node refreshes the routing table periodically. An ultimate receiver receives JOIN-QUERY messages, and instead of sending a reply at once, it waits for a period and caching all queries. Receivers choose highest honest nodes path route for stable routing. Then, it sends JOIN-REPLY packet to the sender, which contains nodes list of the whole route.

4 Performance Analysis

4.1 Simulation Environment

The simulator used for this work is NS2.34 (Network Simulator). The area of the simulation is 1000 m × 1000 m. The time taken for simulation is 200 s. The CBR traffic type and IEEE 802.11 Mac protocol have been used. The Random waypoint mobility model and Two-Ray Ground Radio propagation model have adapted for this work. The queue length is 50 packets, and the packet size is 2000 bytes. The number of nodes is 60. The multicast group size varies from 5 to 25. The multicast sender ranges from 1 to 5. The proposed ODMRP-HEFNRLP is compared with the ODMRP, ODMRP-LM, ODMRP-GM, ODMRP-FNRLP, and ODMRP-EFNRLP protocols. The honest node selection with reduced control packets in multicast network are tested in this implementation. The design of protocol increases the lifetime of a network in HEFNRLP than basic ODMRP. The performance metrics such as packet delivery ratio is considered for the performance analysis of the varying number of multicast senders and multicast group size.

4.2 Results and Discussion

Packet Delivery Ratio (PDR) is calculated using (9). Figure 2 explains the graph of Packet delivery ratio (PDR). The graph shows that the HEFNRLP presents high performance when increasing the group size. The simulation outputs obtained the comparative results of other protocols and HEFNRLP.

$$\text{Packet Delivery Ratio} = \text{received packets/sent packets} * 100 \qquad (9)$$

The network lifetime metric has become the leading quality for estimation of network longevity. In Fig. 3, the network lifetime of ODMRP-HEFNRLP is stable irrespective of number of multicast sender. The stability is achieved by the honest-based path selection.

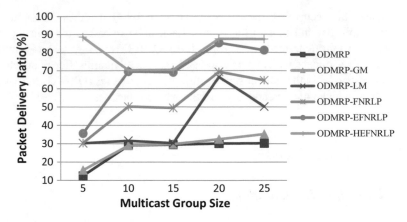

Fig. 2 Packet delivery ratio versus multicast group size

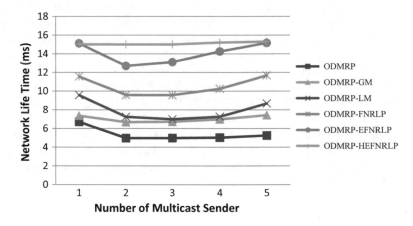

Fig. 3 Network lifetime versus number of multicast sender

5 Conclusion

The path selection in multicasting is demanding due to the vibrant actions of the network. The limitations in these multicast networks are of security, energy, and channel bandwidth constraints [15]. The protocol needs to minimize the security issues, channel usage, and control packet reduction. The mesh-based multicast protocol preserves the connected receivers, so that multiple paths are established from sender to multicast receivers. The ODMRP-HEFNRLP has proposed to control routing overhead and improves security based on fuzzy-based honest node selection. The simulation output shows the high packet delivery ratio with good lifetime. In future, the fuzzy prediction rule will be applied for predicting the node's HV with multiple constraints.

References

1. Cesnavicius, E., Jahn, C.: Broadcasting & Multicasting in IP Networks, 26 July, 2004 (http://www2.cs.uni-paderborn.de/cs/ag-madh/WWW/Teaching/2004SS/AlgInternet/Submissions/13-broadcasting-and-multicasting-in-ip-networks.pdf)
2. Lazos, L., Poovendran, R.: Power proximity based key management for secure multicast in ad hoc networks. Wirel. Netw. 13(1), 127–148 (2007)
3. Zhu, T., Xiao, S., Ping, Y., Towsley, D., Gong, W.: A secure energy routing mechanism for sharing renewable energy in smart microgrid. In: IEEE Conference on Smart Grid Communication (SmartGridComm), Brussels, Belgium, 17–20 Oct 2011
4. Chen, I.-R., Guo, J., Bao, F., Cho, J.-H.: Trust management in mobile ad hoc networks for bias minimization and application performance maximization. Ad Hoc Netw. 19, pp. 59–74 (2014)
5. Kaur, A., Kang, S.S.: Comparative analysis of secure and energy efficient routing protocols in wireless sensor network. Int. J. Eng. Comput. Sci. 3(9), 8389–8393 (2014)
6. Guo, S., Yang, O.: Minimum energy multicast routing for wireless ad-hoc networks with adaptive antennas. In: Proceedings of the 12th IEEE International Conference on Network Protocols (ICNP'04), Berlin, Germany, pp. 151–160, 05–08 Oct 2004
7. Adhvaryu, K.U., Kamboj, P.: Multicast routing protocols with low overhead for MANET. Int. J. Eng. Dev. Res. (IJEDR), 1(1), pp. 1–4 (2013)
8. Han, K., Liu, Y., Luo, J.: Duty-cycle-aware minimum-energy multicasting in wireless sensor networks. IEEE/ACM Trans. Netw. 21(3), 910–923 (2013)
9. Andreev, K., Maggs, B.M., Meyerson, A., Sitaraman, R.K.: Designing overlay multicast networks for streaming. In: Fifteenth Annual ACM Symposium on Parallel Algorithms and Architectures (SPAA)'03, San Diego, California, USA, 7–9 June 2003
10. Fernandes, L.L.: Secure Routing in Wireless Sensor Networks. Report. University of Trento (2007)
11. Simon, Santhosh, Paulose Jacob, K.: Energy optimized secure routing protocol for wireless sensor networks. Int. J. Eng. Innov. Technol. (IJEIT) 3(4), 72–80 (2013)
12. Xia, H., Yu, J., Pan, Z.-K., Cheng, X.-G., Sha, E.H.M.: Applying trust enhancements to reactive routing protocols in mobile ad hoc networks. Wirel. Netw. 22(7), 2239–2257 (2016)
13. Cho, J.-H., Chen, I.-R., Chan, K.S.: Trust threshold based public key management in mobile ad hoc networks. Ad hoc Netw. 44(Issue C), 58–75 (2016)
14. Borkar, G.M., Mahajan, A.R.: A secure and trust based on-demand multipath routing scheme for self-organized mobile ad-hoc networks. Wirel. Netw. 23(8), 2455–2472 (2017)
15. Olagbegi, B.S., Meghanathan, N.: A review of the energy efficient and secure multicast routing protocols for mobile Ad hoc networks. Int. J. Appl. Graph Theory Wirel. Ad Hoc Netw. Sens. Netw. (GRAPH-HOC) 2(2) (2010)

Experimental Study of Gender and Language Variety Identification in Social Media

Vineetha Rebecca Chacko, M. Anand Kumar and K. P. Soman

Abstract Social media has evolved to be a crucial part of life today for everyone. With such a global population communicating with each other, comes the accumulation of large amounts of social media data. This data can be categorized as "Big Data", owing to its large quantity. It contains valuable information in the form of the demographics of authors on online platforms; the analysis of which is required in certain scenarios to maintain decorum in the online community. Here, we have analyzed Twitter data, which is the training data of the PAN@CLEF 2017 shared task contest, to identify the gender, as well as the language variety of the author. It is available in four different languages, namely, English, Spanish, Portuguese, and Arabic. Both Document-Term Matrix (DTM) and Term Frequency-Inverse Document Frequency (TF-IDF) have been used for text representation. The classifiers used are SVM, AdaBoost, Decision Tree, and Random Forest.

Keywords Document Term Matrix · Term Frequency-Inverse Document Frequency · n-grams · Support vector machines · AdaBoost · Random Forest Decision Tree · Author Profiling

1 Introduction

"**Data is the new gold**"—when it comes to something as valuable as gold, and is available in abundance, its handling also gets strenuous. Where traditional data handling techniques failed to tame such large amounts of data, there came up the field

V. Rebecca Chacko (✉) · M. Anand Kumar · K. P. Soman
Centre for Computational Engineering and Networking (CEN),
Amrita School of Engineering, Amrita Vishwa Vidyapeetham, Coimbatore, India
e-mail: vineethachacko7@gmail.com

M. Anand Kumar
e-mail: m_anandkumar@cb.amrita.edu

K. P. Soman
e-mail: kp_soman@amrita.edu

© Springer Nature Singapore Pte Ltd. 2019   489
J. D. Peter et al. (eds.), *Advances in Big Data and Cloud Computing*,
Advances in Intelligent Systems and Computing 750,
https://doi.org/10.1007/978-981-13-1882-5_42

of **Big Data** specialized in handling large volumes of complex data. With its different dimensions like veracity and variety, several challenges like storage, searching through the data, sharing of data, efficient privacy protection, and most importantly, the analysis of data are faced. Analysis is the key ingredient of almost all fields of study that proves, or disproves, all initial hypotheses made by man, about any scenario at hand.

Social media is a concept that came into the limelight in the recent years. Indians got introduced to it all, mostly through platforms like Google Talk, Orkut, etc., which gained popularity at an exponential rate, only to be followed by other social media giants like Facebook, Twitter, and WhatsApp. Along with its advantages, the biggest of which is real-time communication, it has certain negative aspects like cyberstalking, cyberbullying, hacking, and even the spread of fake news. One of the biggest examples of our times is the investigation of the 2016 US election, which was claimed to be rigged by the Russians, and social media played a huge role in it. Summing up all these negative aspects of social media, the aspect that serves as a common helping hand in achieving these heinous goals is **anonymity**. Anonymity is a treacherous enemy to the decorum of any online community.

Since everybody has access to the Internet (except for the North Koreans), people have the freedom to hide behind fake profiles on social media platforms. Hence came up the field of cyber forensics facing the challenge of **Author Profiling**. It is the analysis of the demographics like the nativity, gender, etc., of online authors. PAN@CLEF [1] is a shared task in the field of cyber forensics that started in 2013, which does such analysis. Social media data like Tweets, comments, etc., when analyzed gives information about such fake accounts. Such research has been carried out for data in many languages, but analysis of Indian languages has only made its baby steps till now. One such step in the analysis of Indian languages was the shared task of "Indian Native Language Identification (INLI)", held from 8th to 10th of December 2017, at IISc Bangalore, India [2].

2 Related Work

The most relatable works are the systems submitted [1] for the PAN@CLEF 2017 Author Profiling shared task. Of the 20 teams that submitted their notebooks, the system that secured the first place [3] has used a linear SVM as the classifier. The aspects used for classification are word uni-grams along with character 3–5-g [4]. Other features include POS tags and Twitter handles, along with geographic entities, but these proved to be inefficient in increasing the accuracy. The Twitter 14k and PAN@CLEF 2016 datasets were also used to improve the accuracy but in vain. After trying different tokenization techniques, they settled for the scikit-learn tokenizer, since the former did not improve the average accuracy. Apart from this, they had also analyzed emojis and did POS tagging on the data, only to get lower accuracy than the simple initial model.

The system which secured the second place [5] performed the following data preprocessing—nonsense Tweet removal, where English words' spelling was checked and wrongly spelled words were removed, and reversal of Arabic data. Preprocessing specific to Twitter data includes removal of stop words, punctuation, hashtags, etc. Of the classifiers used, Logistic Regression gave the highest accuracy. Other classifiers used were Random Forest and XGBoost. Linear SVM has also been used but underperformed when compared to Logistic Regression. A combination of classifiers like Logistic regression combined with Voting classifier was also tried but, eventually, the best results were obtained with Logistic Regression.

The system which secured the third place in the PAN@CLEF 2017 shared task [6] uses a MicroTC, which is a generic framework for text classification task, i.e., it works regardless of both domain and language particularities. They have used binary and trivalent parameters for preprocessing. They have used tokenizers and TF-IDF weighting, all to be finally classified using a simple linear SVM. Hence, of the top three papers submitted for PAN@CLEF 2017, classic machine learning algorithms have scored the best. Of the other notebooks submitted, Barathi Ganesh et al. [7] use Vector Space Models for text classification. Bougiatiotis and Krithara [8], submitted for the PAN@CLEF 2016, use stylometric features for the classification.

3 Dataset

The dataset consists of Twitter data in XML format. The corpora are the training dataset of the shared task of Author Profiling, at PAN@CLEF 2017 [1]. The gender and language varieties selected are shown in Table 1. For each variety, the capital of the region where this variety has been used is selected. After selecting the region, Tweets in a radius of 10 km from this region have been collected. PAN, being a team focused on research in this specific area, has overcome challenges like authenticity of authors, because of the previously collected data, where the uniqueness of each author is ensured. The time line of each author is retrieved, which provides information such as the author's official name, her/his language, and also the location.

Table 1 Gender and language variety of PAN@CLEF 2017 corpora

Language	Variety	Gender
English	Australia, Canada, Great Britain, Ireland, New Zealand, United States	Male, Female
Spanish	Argentina, Chile, Colombia, Mexico, Peru, Spain, Venezuela	Male, Female
Portuguese	Brazil, Portugal	Male, Female
Arabic	Egypt, Gulf, Levantine, Maghrebi	Male, Female

For each author, it is ensured that at least 100 tweets are there, with no re-Tweets. It is also ensured that the Tweets are written in the required variety of language. An author is concluded to be using a particular language variety if the Tweets' location is where this language is prevalent. Using a dictionary consisting of proper nouns, automatic assigning of gender is done for an author. Otherwise, it has been done manually. A total of 500 authors' tweets have been collected, for language variety as well as for gender. The training dataset released consists of 60% of the corpus formed, for language variety and gender.

Each author's XML file consists of 100 tweets. For each language, the corresponding truth labels are also provided, for both gender and language variety. The number of XML files in English, Spanish, Portuguese, and Arabic is 3600, 4200, 1200, and 2400, respectively. Hence, there are 3600 × 100 English Tweets, 4200 × 100 Spanish Tweets, 1200 × 100 Portuguese Tweets, and 2400 × 100 Arabic Tweets.

4 Methodology

The basic methodology used for gender identification and language variety identification is given in Fig. 1, implemented using scikit-learn. The corpora given is

Fig. 1 Architecture

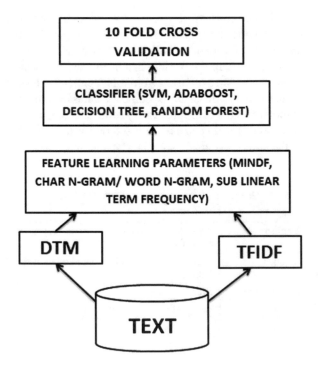

in XML format, which is first converted to text format with the help of Document Object Model of XML data. It is then converted to matrix format either as a DTM or as a TF-IDF matrix. The features of this text are further learned using minimum document frequency, word n-grams, and character n-grams. Using the parameter "sub linear term frequency weighted TFIDF matrix" [3], where instead of normal term frequency, 1 + log(term frequency) is taken, has not improved the accuracy much. Then, the classification is done using Machine Learning algorithms like SVM, AdaBoost, Decision Tree, and Random Forest. Since the test data for PAN@CLEF 2017 [9] Author Profiling has not been released yet, a tenfold cross-validation has been done on the training data and the accuracy is recorded.

4.1 Text Representation and Feature Learning

The first step in NLP is to represent the text in a format on which Machine Learning techniques can be easily applied. The formats in which the text is represented in various ways are Distributed Representation, Distribution Representation [10], and Vector Space Models [7, 11]. Here, the count-based methods of the Distribution Representation are used.

4.1.1 Term Frequency-Inverse Document Frequency (TF-IDF)

TF-IDF matrices are used when frequently used words need not be considered for the final analysis of text [11]. The rows represent documents and the columns represent vocabulary. It gives importance to unique words by transforming the mere count of a word, to the probability of the occurrence of the word in each document, by dividing it with the total number of occurrences in all documents. It can be mathematically expressed as follows:

$$TF - IDF = \log \frac{N}{n_i},\tag{1}$$

where N is the number of documents under consideration and n_i is the number of times the ith word is occurring in the document.

4.1.2 Document-Term Matrix (DTM)

DTMs directly take the count of each word occurring in each document and enters it into the matrix [11]. The rows represent documents and the columns represent the vocabulary. When a document is encountered, the unique words are entered into the vocabulary and the count is entered into the matrix. The same procedure is followed for all documents, hence increasing the vocabulary with each document encountered. It can be expressed mathematically as follows:

$$DTM = \frac{f_{t,d}}{\sum_{t'd} f_{t',d}},\qquad(2)$$

where $f_{t,d}$ is the frequency of the term "t" in document "d".

Author Profiling is a task where the number of words used by authors helps in knowing their gender, as most females use more words to express themselves. Even the information about the usage of stop words can end up being beneficial in Author Profiling. Hence, DTMs are more useful than TF-IDF matrices for this task.

4.1.3 Parameters for Feature Extraction

The parameters used for feature learning are as given below. Their contribution is crucial for the prediction [12].

1. Minimum Document Frequency (*mindf*): It is specified to set a threshold to avoid those words which do not occur in at least "n" documents, where mindf = n.
2. N-gram: It gives the continuous occurrence of "n" characters or words in a text, hence classified as character n-grams and word n-grams, respectively.
3. Analyzer N-grams can be applied for characters as well as for words, which is specified using the "analyzer" parameter. Character n-grams are mostly useful when the language dealt with is unknown.
4. Sublinear Term Frequency: For TF-IDF matrices, a parameter called sublinear term frequency can be specified, where instead of taking the term frequency directly, 1 + log(term frequency) is taken.

Table 2 shows the values taken up by different parameters.

4.2 Classifiers for Author Profiling

Machine Learning is a field in Computer Science which trains computers to learn like humans. It can be applied to different forms of data—speech signals, images, stock market data, weather data and yes, text data also. Even if Deep Learning, which is a part of Machine Learning, has proved to be more efficient than traditional Machine

Table 2 Feature extraction parameters

Parameter	Values
Mindf	1–25
Analyzer	Char, word
Sublinear_tf	True, False
n-gram	1, 2, 3, 4, 5

Learning algorithms, for image data, stock market data, etc., text data analysis still has an affinity for traditional Machine Learning algorithms. The same affinity is observed for the analysis of PAN@CLEF 2017 train data from the top ranking systems [1].

4.2.1 Support Vector Machine

SVM [13] is a supervised learning algorithm, where the machine is trained using a set of training data along with its labels and tested using test data. The accuracy is tested by comparing these true labels and predicted labels. SVMs construct hyperplanes to separate the classes, such that there should be maximum separation between classes. This is done so that when new data is encountered for classification, the chances for making error in the prediction are kept minimal. The hyperplane has two bounding planes which lie on its either sides, and the points falling on these bounding planes are called support vectors.

4.2.2 Decision Tree

Decision Trees [13] are supervised Machine Learning algorithms which are used mainly for classification of data. It has a structure similar to flowcharts, where the top node is the root node and each internal node denotes a test on a feature of the data. The branches represent the outcome of a test and each leaf node holds a class label. It is simple to visualize and understand, and can handle all kinds of data. Nonlinear relations will not affect the persona of the tree. But the algorithm may be biased and may become unstable since a small change in the data will change the structure of the tree.

4.2.3 Random Forest

Random Forest [13] is one of the most popular and efficient Machine Learning algorithms used for regression as well as classification. As the name suggests, it contains a combination of Decision Trees. As the number of Decision Trees increases, the prediction becomes more robust and hence more accurate. Each Decision Tree gives a class to which a data belongs to. The class to which the most number of Decision Trees classify the data to be in is assigned as that particular data's class, and the accuracy is checked. It is capable of handling datasets of higher dimensionality. An important feature is that it handles missing values and maintains the accuracy also. Yet, it is like a black box in the matter of having control over what the model does.

4.2.4   Adaboost

In Machine Learning, bagging algorithms are known to increase the veracity of Machine Learning algorithms [13]. Boosting is necessarily a simple variation of bagging algorithms. It improves the learners as it emphasizes on areas where the classifier is not performing as expected. It involves a repetitive process where the training data is randomly split, to form a bag of data. This bag of data is used for training, and the corresponding model is tested using the entire training data. While forming the next bag of data, the previous wrongly classified data are deliberately included in it, and the same process is repeated as above. These steps when repeated improve the overall accuracy.

Finally, ten fold cross-validation is done on the data, where the data is divided in the ratio of 9:1 for train:test, and the accuracy is recorded and analyzed.

5   Experiments and Results

Here, the results obtained for gender identification as well as for language variety identification have been presented. Ten fold cross validation has been used for all algorithms. Character n-grams have been used, where word n-grams may not be efficient, mostly in unknown languages.

5.1   Gender

The results of the task of gender identification are given in Tables 3 and 4. Table 3 describes the results for Tweets of all the four languages, when the text is represented as DTM. Table 4 gives the results for the same Tweets, when the text is represented as TF-IDF matrix. It can be observed that AdaBoost and Random Forest classifiers stand out in performance.

Table 3   Gender identification: DTM

Language	Classifier			
	SVM	AdaBoost	Decision Tree	Random Forest
English	77.77	76	65.75	74.83
Spanish	72.50	73.16	63.26	69.35
Portuguese	78.25	79.08	71.667	73.916
Arabic	69.708	71.66	66.916	69.79

Table 4   Gender identification: TF-IDF

Language	Classifier			
	SVM	AdaBoost	Decision Tree	Random Forest
English	48.44	76.44	64.65	73.11
Spanish	48.52	73.19	63.45	67.73
Portuguese	46.08	80.75	69.66	71.916
Arabic	51	74	66.04	70.58

Table 5 Language variety identification: DTM

Language	Classifier			
	SVM	AdaBoost	Decision Tree	Random Forest
English	68.61	77.19	73.916	76.08
Spanish	82.38	84.595	84.095	83.904
Portuguese	98.16	98.66	96.75	98.50
Arabic	73.20	73.58	71.708	76.62

Table 6 Language variety identification: TF-IDF

Language	Classifier			
	SVM	AdaBoost	Decision Tree	Random Forest
English	14.25	75.83	72.88	77.83
Spanish	12.619	83.47	81.976	84.09
Portuguese	50.916	98.916	96.58	98.33
Arabic	24.916	72.58	71.54	75.16

5.2 Language Variety

The results of the task of language variety identification are given in Tables 5 and 6. Table 5 describes the results for Tweets of all the four languages, when the text is represented as DTM. Table 6 gives the results for the same Tweets, when the text is represented as TF-IDF matrix. It can be observed that AdaBoost and Random Forest classifiers stand out in performance.

Hence, it can be concluded from these results that although SVM is a good classifier, AdaBoost and Random Forest classifiers outperform SVM in most cases pertaining to this dataset.

6 Conclusion and Future Work

The corpora used here is the PAN@CLEF 2017 corpora for Author Profiling. It consists of Tweets in four languages, namely, English, Spanish, Portuguese, and Arabic in the form of XML files. To identify gender and language variety, the data has been represented as DTM as well as TF-IDF matrix. Feature learning has been done with parameters like *mindf*, word n-gram, and character n-grams. Classic Machine Learning techniques such as SVM, AdaBoost, Decision Tree, and Random Forest have been used for the classification purposes. It has been observed from the results that when using SVM as the classifier, DTM gives more accuracy than TF-IDF for both classifications. Also, the other classifiers— AdaBoost in particular—outperform SVM for this dataset.

Another area of research is to analyze code mixed data obtained from social media platforms like Facebook, Twitter, and WhatsApp. Code mixed data is the combination of two languages, mostly English combined with other native languages, and is the common language of communication in social media. An analysis of the Indian code mixed language, Malayalam–English code mixed data to be specific, is of high prospect as a future work. Deep Learning and Fast Text can be used for text classification [14].

References

1. Rangel, F., Rosso, P., Potthast, M., Stein, B.: Overview of the 5th Author Profiling Task at PAN 2017: Gender and Language Variety Identification in Twitter. CLEF (2017)
2. Anand Kumar, M., Barathi Ganesh, H.B., Singh, S., Soman, K.P., Rosso, P.: Overview of the INLI PAN at FIRE-2017 track on Indian native language identification. CEUR Workshop Proc. **2036**, 99–105 (2017)
3. Basile, A., Dwyer, G., Medvedeva, M., Rawee, J., Haagsma, H., Nissim, M.: N-GrAM: New Groningen Author-profiling Model, Notebook for PAN at CLEF 2017. CLEF (2017)
4. Rangel, F., Franco-Salvador, M., Rosso, P.: A low dimensionality representation for language variety identification. CICLing—Computational Linguistics and Intelligent Text Processing (2016)
5. Martinc, M., Krjanec, I., Zupan, K., Pollak, S.: PAN 2017: Author Profiling—Gender and Language Variety Prediction, Notebook for PAN at CLEF 2017. CLEF (2017)
6. Tellez, E.S., Miranda-Jimnez, S., Grafi, M., Moctezuma, D.: Gender and language-variety identification with MicroTC, Notebook for PAN at CLEF 2017. CLEF (2017)
7. Barathi Ganesh, H.B., Anand Kumar, M., Soman, K.P.: Vector Space Model as Cognitive Space for Text Classification. Notebook for PAN at CLEF 2017. CLEF (2017)
8. Bougiatiotis, K., Krithara, A.: Author profiling using complementary second order attributes and stylometric features. CLEF 2016 Evaluation Labs and Workshop—Working Notes Papers. CLEF (2016)
9. Barathi Ganesh, H.B., Reshma, U., Anand Kumar, M., Soman, K.P. Representation of target classes for text classification—AMRITA-CEN-NLPRusProfiling PAN 2017. In: CEUR Workshop Proceedings, 2036, pp. 25–27 (2017)
10. Barathi Ganesh, H.B., Anand Kumar, M., Soman, K.P. Distributional semantic representation for text classification and information retrieval. In: CEUR Workshop Proceedings, 1737, pp. 126–130 (2016)
11. Turney, P.D., Pantel, P.: From frequency to meaning: vector space models of semantics. J. Artif. Intell. Res. (2010)
12. Medvedeva, M., Kroon, M., Plank, B.: When sparse traditional models outperform dense neural networks: the curious case of discriminating between similar languages. In: Proceedings of the Fourth Workshop on NLP for Similar Languages, Varieties and Dialects, pp. 156–163. Association for Computational Linguistics (2017)
13. https://in.udacity.com/
14. https://www.analyticsvidhya.com/blog/2017/07/word-representations-text-classification-using-fasttext-nlp-facebook/

Paraphrase Identification in Telugu Using Machine Learning

D. Aravinda Reddy, M. Anand Kumar and K. P. Soman

Abstract Paraphrase identification is the task of determining whether two sentences convey similar meaning or not. Here, we have chosen count-based text representation methods, such as term-document matrix and term frequency-inverse document frequency matrix, along with the distributional representation methods of singular value decomposition and non-negative matrix factorization, which is iteratively used with different word share and minimum document frequency values. With the help of the above methods, the system will be able to learn features from the representations. These learned features are then used for measuring phrase-wise similarity between two sentences. The features are given to various machine learning classification algorithms and cross-validation accuracy is obtained. The corpus for this task has been created manually from different news domains. Due to the limitation of unavailability of the parser, only a set of collected data in the corpus has been used for this task. This is a first attempt in the task of paraphrase identification in Telugu language using this approach.

Keywords Paraphrase identification · Count-based methods · Distributional representation methods · Corpus · Classification algorithms

1 Introduction

Paraphrases are two sentences which have similar meaning but are of different syntax. In paraphrases, the semantic equivalence of sentences is given focus, rather

D. Aravinda Reddy (✉) · M. Anand Kumar · K. P. Soman
Centre for Computational Engineering and Networking (CEN), Amrita Vishwa Vidyapeetham, Amrita School of Engineering, Coimbatore, India
e-mail: cb.en.p2cen16004@cb.students.amrita.edu

M. Anand Kumar
e-mail: m_anandkumar@cb.amrita.edu

K. P. Soman
e-mail: kp_soman@amrita.edu

© Springer Nature Singapore Pte Ltd. 2019
J. D. Peter et al. (eds.), *Advances in Big Data and Cloud Computing*,
Advances in Intelligent Systems and Computing 750,
https://doi.org/10.1007/978-981-13-1882-5_43

Professors intel development forum aviskarinchanani shan josh prakatincharu
ప్రాసెసర్లు ఇంటెల్ డెవెలపర్ ఫోరమ్ ఆవిష్కరించాని శాన్ జోస్ ప్రకటించారు .
kotha professor shan josh, californialo intel development forum 2003lo avishkarincharu
కొత్త ప్రాసెసర్ శాన్ జోస్, కాలిఫోర్నియా లో ఇంటెల్ డెవెలపర్ ఫోరమ్ 2003 ఆవిష్కరించారు

Fig. 1 Paraphrase

Jagapathi babu cinemanu chala varaku thana bhujalu mida nadipincharu.
జగపతి బాబు సినిమాను చాలా వరకు తన భుజాల మీదే నడిపించాడు.

Jagapathi babu tama patralanu nayayam chesaru.|
జగపతి బాబు తమ పాత్రలకు న్యాయం చేసారు.

Fig. 2 Non-paraphrase

than its syntax. Paraphrasing is said to be the process of rewriting a sentence with semantic information in different forms. Paraphrasing finds many applications in the field of natural language processing. An example for paraphrases in Telugu and the corresponding roman script is shown below (Fig. 1).

Sentences which are not carrying semantic equivalence are stated as non-paraphrases. Although there exists a small amount of similarity in terms of syntax, between the sentences, they do not convey the same meaning. A simple example of non-paraphrase sentences is shown below (Fig. 2).

One of the main areas which are widely using paraphrase identification is plagiarism detection. Plagiarism, by the definition of Wikipedia, is the unlawful appropriation and stealing of another authors' language, conception, expressions or ideas and representing them as their own work. In the research community, the automatic detection of text reusage plays a prominent role. But major part of the work is focused only on the monolingual comparison (mostly English to English), whereas the largely unexplored part is multilingual domain.

In a similar way, paraphrase identification is also used for information retrieval. The main scope for retrieving information is in the biomedical field and for data security also. In biomedical field, information retrieval involves identifying medicines for a particular disease from the huge data. Paraphrase identification is almost emerging as a crucial aspect in such different fields.

The remaining section of the paper is systematically arranged as follows: Section 2 mentions some of the related works followed by a brief description of the paraphrase corpora for Telugu in Sect. 3. Section 4 explains about the methodology for paraphrase identification. The results of the experiment are given in Sect. 5. Section 6 has drawn the conclusion.

2 Related Works

This section presents previous works in paraphrase identification. Paraphrase identification is a complex task in natural language processing; several kinds of researches have been done on this and still many works are going on. The researchers mainly focus on expanding the existing problem by bringing out all the possibilities of its minute aspects, which lead to the problems' complexity. Some of the famous works related to paraphrase identification in this section present different paraphrase methods based on different techniques. Dolan et al. [1] proposed an 'Unsupervised Construction of Large Paraphrase Corpora' from different news domains. In this approach, an investigation has been done on unsupervised techniques for acquiring monolingual sentence-level paraphrases from a corpus of temporally and topically clustered news articles collected from thousands of web-based news sources. Fernando and Stevenson [2] had taken up a 'Semantic Similarity Approach to Paraphrase Detection'. This work presents a novel approach to the problem of paraphrase identification. Finch et al. [3] proposed machine translation evaluation techniques to determine sentence-level semantic equivalence. The task of machine translation (MT) evaluation is closely related to the task of sentence-level semantic equivalence classification.

A paraphrase detection system based on unfolding recursive autoencoders are explained in. un Choi Lee Yu-N Cheah presented a semantic relatedness measures the shortest path based on synset in WordNet for paraphrase detection. This method used distance per word in the sentence to measure semantic relatedness between two sentences. The proposed method is then evaluated using a paraphrase detection based on the Microsoft Research Paraphrase Corpus, and the reasonable results are shown and compared with other similarity and semantic relatedness. The author suggested that this approach may be used for question answering or content ranking and in text understanding also.

Hoang-Quoac has done paraphrase detection using the identical phrase and similar word matching. The author explains similarity metric (SimMat) which was calculated by matching identical phrases with the maximum length, removing minor words, matching of similar words by Kuhn and Munkre algorithm, calculating related matching metric (RelMat) and finally calculating the SimMat by combining penalty metric and RelMat. The system achieves 77.6% accuracy, which was higher than the previous method.

Praveena et al. [4] proposed an approach of chunking-based Malayalam paraphrase identification using unfolding recursive autoencoders. The main aim of this approach is to examine the semantic similarity of a pair of Malayalam sentences and classify them into paraphrase, non-paraphrases or semi-equivalent paraphrases. This approach involves unsupervised feature learning using recursive autoencoders. In this method, the additional features used are Glove, Word2vec and FastText word embedding with different dimensional vectors.

Mahalaksmi et al. [5] proposed an approach for paraphrase detection for Tamil language using deep learning algorithms. In this approach, the authors performed paraphrase detection for the collected Tamil sentences. Paraphrase detection is the process of detecting the sentences or paragraphs, which is restated using jumbled words in the sentences or different choice of words while preserving the meaning of the same. The real challenge in paraphrase detection is that the semantics of the sentence has to be preserved while restating the sentence. Here, we have used unfolding recursive autoencoders (RAEs), a deep learning algorithm for learning feature vectors for phrases in syntactic trees in an unsupervised way. The experimental results have provided an accuracy of around 65.17% for Tamil sentences.

3 Telugu Paraphrase Corpora

3.1 Paraphrase Corpora for Telugu

Paraphrase corpora for Telugu have been created manually from various Telugu newspapers. The total dataset comprises both paraphrases and non-paraphrases in 4100 pairs. The dataset is then split as 1750 pairs for training from paraphrases and 1750 pairs from non-paraphrases, which are of 3500 sentences in each phase. The validation dataset comprises 300 pairs of each paraphrase and non-paraphrases. Table 1 explains the paraphrase corpora for Telugu dataset and the average number of words present in the corpora.

4 Methodology

In this chapter, we can see the methodology that has been used to complete this task. The methodology which we have followed is given in Fig. 3.

Table 1 Paraphrase corpora for Telugu dataset

	Training dataset (Pairs)	Average no. of words	Test dataset (Pairs)	Average no. of words
Paraphrase	1750	13.58	300	12.34
Non-paraphrase	1750	10.64	300	10.37
Total	3500	12.61	600	11.35

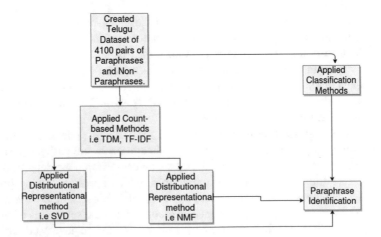

Fig. 3 Methodology

4.1 Feature Extraction for Paraphrase Identification

The feature extraction which is being extracted for this task is from count-based models such as term-document matrix and term frequency-inverse document frequency, distributional methods which are combination of count-based methods such as singular value decomposition and non-negative matrix factorization, and the word share which is taken for different classifiers.

4.1.1 Count-Based Model

Count-based model is the basic start for high-level analysis in natural language processing. In term-document matrix, we will represent frequency of words (or) terms in the documents. The unique words which are present in the documents are used to build the vocabulary. The total number of times a particular word occurs in the document gives the frequency of that particular word. The columns represent words or terms, and the rows represent the documents in a term-document matrix and help in finding document similarity.

$$\textbf{TDM} = \frac{f_{t,d}}{\sum_{t'd} f_{t',d}} \tag{1}$$

where $f_{t,d}$ is the frequency of the term 't' in document 'd'.

Another approach is term frequency-inverse document frequency. With this approach, we are going to find the expected words (stop words) from the document and give low weightage to these words. The main aim is to emphasize on only surprising words. Putting it in another way, this is the strategy to find key words in the corpus or documents. In term frequency, we are going to calculate how many times a word is occurring in the document; it is nothing but term- document matrix.

In inverse document frequency, more weightage is given for surprising words than frequently occurring words in the document.

$$\textbf{TF-IDF} = log N/n_i \tag{2}$$

where

N   —Number of documents.
n_i —Number of times the ith word is occurring in the document.

4.1.2 Distributional Model

In distributional methods, singular value decomposition (SVD) in linear algebra is factorization of a real or complex matrix. SVD is used to reduce the dimension of data with the help of eigenvalues. The data dimensions obtained from the term-document matrix and term frequency-inverse document frequency is very large. The matrix is taken and singular value decomposition is done on it. The main motivation behind the application of singular value decomposition is to capture the essence of a huge matrix into a smaller matrix, for reducing the computational expenses without loosing a crucial part of the matrix efficiency. When we apply SVD, it will decompose into three matrices as shown in the following equation:

$$\textbf{Matrix(M)} = U_{m*r} * \Sigma_{r*r} * V_{r*n}^{T} \tag{3}$$

where

U   Eigenvector matrix of size (m*r).
S   Diagonal matrix which represents eigenvalues.

The decomposition is taken from the top eigenvalues of the diagonal matrix. Non-negative matrix factorization is a similar technique of reducing the high-dimensional matrix to a lower dimensional matrix with help of only two parameters. It will decompose the matrix into weight matrix (W) and coefficient matrix (H). This type of decomposition only takes non-negative values from the two matrices.

$$\textbf{Matrix(V)} = W_{m*r} * H_{r*r} \tag{4}$$

where

W   Weight Matrix of dimension (m*r).
H   Coefficient Matrix of dimension (r*r).

4.1.3   Word Share Features

The word share is one of the features used in the classification algorithm. In this feature, we are able to find out how the words are shared between the pair of sentences in both paraphrases and non-paraphrases. In paraphrase, the amount of shared words in the sentences will be more, since they are related. The probability of words shared in the non-paraphrases will be less, as the sentences are not related. From this feature, we can analyse the word sharing between the sentences.

As the similar way, one more feature is TF-IDF share. In this, how the most surprising words will be having more weightage of the sentences?

4.2   Classifiers Used for Paraphrase Identification

The classification algorithms which have been used in this approach are as follows.

4.2.1 Random Forest Classifier

Random forest classifier is one of the most popular classification algorithms. In machine learning, the random forest classifier is used for regression and classification problems also. Random forest algorithm is a supervised classification learning algorithm. Many classification trees are grown by random forest. The new object which needs to be classified from an input vector is kept down under each tree in the forest. The classification given by each tree is said to be 'votes' for that class. Based on the more number of votes, the forest chooses the classification (over all the trees in the forest).

4.2.2 Decision Tree

The decision tree is a concept that has more affinity towards the rule-based system. From the given training dataset with features and targets, some set of rules will be come up by decision tree that is used to perform the predictions on the test dataset.

4.2.3 Support Vector Machine

Support vector machine (SVM) is a supervised learning algorithm. SVM is one of the powerful techniques for classification and regression outlier detection with an intuitive model. With the help of linear algebra, the hyperplane learning in linear SVM is done by transforming the problem. SVM can be put in other words using the inner product of given any two observations rather than the observations themselves. The inner product is the sum of the multiplication of each pair of input values between two vectors.

4.2.4 K-Nearest Neighbour

The concept of k-nearest neighbour (k-NN) algorithm is the usage of database where all the data points are partitioned into several separate classes to predict the new sample point of a classification.

4.2.5 Naive Bayes

Naive Bayes algorithm is a classification algorithm which depends on the theorem called Bayes, with an assumption of the predictor independence. To put the words in simple, it assumes that a particular feature presence in the class is not related to any other features presence.

5 Results

This section represents the results of the task, and the Sect. 5.1 explains the results which are obtained for term- document matrix. Section. 5.2 explains the results which are obtained for term frequency-inverse document frequency. Section. 5.3 explains the results which are obtained for experiments with different classification algorithms.

5.1 Term-Document Matrix

As mentioned in the methodology about term-document matrix, it is the basic start for NLP and the feature extraction or feature engineering is done using singular value decomposition and non-negative matrix factorization. For obtaining results, we have taken support vector machine as linear kernel. Since singular value decomposition gives the better results, we have chosen the same features for the test. Table 2 represents the results for term-document matrix with tenfold cross-validation accuracy and test accuracy.

The maximum accuracy is obtained for term-document matrix at n-gram = 3 and min-document frequency range = 2 with tenfold cross- validation accuracy = 72.65% for dimensionality reduction using SVD. The feature matrix size for term-document matrix is [3500*16723]. The test accuracy from this method is 67.21%.

5.2 Term Frequency-Inverse Document Frequency

Table 3 explains the results of tenfold cross-validation accuracy and test accuracy for term frequency-inverse document frequency. It is also a basic start for NLP, and the feature extraction or feature engineering is done using singular value decomposition and non-negative matrix factorization. For obtaining results, we have taken the support vector machine as linear kernel. Since singular value decomposition gives the better results, we have chosen the same features for the test. These results which

Table 2 Results for paraphrase identification using TDM

Method	N_Gram	Min_df range	Tenfold cross-validation accuracy(%)	Test accuracy (%)
TDM	3	2	74.91	68.34
TDM-SVD	3	2	72.65	67.21
TDM-NMF	3	2	67.60	62.53

Table 3 Results for paraphrase identification using TF-IDF

Method	N_Gram	Min_df range	Tenfold Cross-validation accuracy(%)	Test accuracy(%)
TF-IDF	3	2	75.37	70.31
TF-IDF-SVD	3	2	74.87	70.10
TF-IDF-NMF	3	2	71.51	66.82

are obtained below are iterated through various numbers of iterations in n-gram and minimum document frequencies, and the top results are shown here.

The maximum accuracy is obtained for term frequency-inverse document frequency at n-gram = 3 and min-document frequency range = 2 with tenfold cross-validation accuracy = 74.87% for dimensionality reduction using SVD. The feature matrix size for term frequency-inverse document frequency is [3500*16743]. The test accuracy from this method is 70.10%.

5.3 Experiments with Different Classification Algorithms

As mentioned in the methodology, the features which are taken for these algorithms are a combination of word share and TF-IDF. As it is a combination of both features, it improves the cross-validation accuracy than the two methods before. The tenfold cross-validation for training dataset of 3500 sentences is divided into training dataset of 2800 sentences and validation dataset of 700 sentences. Table 4 shows the results of Cross-validation accuracy for the various classification algorithms.

where P stands for Paraphrase.

N-P stands for Non-Paraphrase in the above-mentioned table.

From the classification algorithm, the test data is to be predicted using training model. As the test data is used to evaluate the performance of the system. Table 5 shows the results of test accuracy for the various classification algorithms.

Table 4 Cross-validation results of different classifiers

Classifiers	Precision		Recall		F-measure		Overall F-measure	Cross-validation accuracy (%)
	P	N-P	P	N-P	P	N-P		
Random forest classifiers	0.85	0.89	0.87	0.90	0.86	0.89	0.88	0.87
Decision tree	0.79	0.81	0.83	0.86	0.81	0.80	0.81	0.80
Support vector machine	0.91	0.93	0.92	0.93	0.91	0.91	0.91	0.91
K-nearest neighbour	0.83	0.87	0.82	0.84	0.81	0.82	0.82	0.84
Naive Bayes	0.71	0.74	0.73	0.78	0.72	0.76	0.74	0.73

Table 5 Test results of different classifiers

Classifiers	Precision		Recall		F-measure		Overall F-measure	Test accuracy (%)
	P	N-P	P	N-P	P	N-P		
Random forest classifiers	0.64	0.68	0.66	0.69	0.64	0.68	0.66	0.65
Decision tree	0.58	0.60	0.63	0.64	0.60	0.63	0.62	0.61
Support vector machine	0.70	0.72	0.71	0.72	0.70	0.73	0.72	0.71
K-nearest neighbour	0.62	0.66	0.61	0.65	0.60	0.65	0.63	0.62
Naive Bayes	0.59	0.61	0.61	0.63	0.60	0.63	0.62	0.60

6 Conclusion and Future Work

In the proposed work, the main aim is to identify the paraphrases from the dataset. We used a basic count-based and distributional representation methods. From these models, we can conclude that the count-based methods can also be able to acquire good performance with the combination of distributional methods. Machine learning classifiers show better performance than the count-based model. Since, the features that are extracted for this model improves the experiemental results. In this technique, support vector machine acquires better test accuracy of 0.71 because paraphrase identification is a binary classification model.

In this task, the paraphrase identification system determines a binary classification. As future work, there are certain deep learning techniques such as recurrent neural network(RNN) and in some cases, they mention that the convolutional neural network (CNN) will also improve the results further. Word embedding can also be included for the betterment of the accuracy. These can be considered as future work.

References

1. Dolan, B, Quirk, C., Brockett, C: Unsupervised construction of large paraphrase corpora: Exploiting massively parallel news sources. In: Proceedings of the 20th international conference on Computational Linguistics, p. 350. Association for Computational Linguistics (2004)
2. Fernando, S., Stevenson, M.: A semantic similarity approach to paraphrase detection. In: Proceedings of the 11th Annual Research Colloquium of the UK Special Interest Group for Computational Linguistics, pp. 45–52 (2008)
3. Finch, A., Hwang, Y.-S., Sumita, E.: Using machine translation evaluation techniques to determine sentence-level semantic equivalence. In: Proceedings of the Third International Workshop on Paraphrasing (IWP2005), pp. 17–24 (2005)
4. Praveena, R., Anand Kumar, M., Soman, K.P.: Chunking based malayalam paraphrase identification using unfolding recursive autoencoders, 922–928. https://doi.org/10.1109/ICACCI.2017.8125959
5. Mahalaksmi, S., Anand Kumar, M., Soman, K.P.: Paraphrase Detection for Tamil language using Deep learning algorithms. Int. J. Appl. Eng. Res. **10**(17), 13929–13934 (2015)

Q-Genesis: Question Generation System Based on Semantic Relationships

P. Shanthi Bala and G. Aghila

Abstract The prospect of applying the semantic relationships to the question generation system can revolutionize the learning experience. The task of generating questions from the existing information is a tedious task. In this paper, Question generation system based on semantic relationships (Q-Genesis) is proposed to generate more relevant knowledge level questions automatically. It will be useful for the trainer to assess the knowledge level of the learners. This paper also provides the importance of the semantic relationships when generating the questions from the ontology.

Keywords Question generation system · Ontology · Semantic relationships

1 Introduction

Question generation system helps the facilitator to generate sensible questions from structured or unstructured information. Question Generation from unstructured information is the most challenging one that involves Natural Language Understanding and Natural Language Generation. The text is mapped into symbols with natural language understanding and natural language generation. Question generation system map symbols for declarative sentences into interrogative sentences. The symbols are used to represent the semantics of the natural language which can be processed by the system. Existing research towards question generation systems is widely based on template and syntax. Recently, the researchers focused towards semantics based question generation system. The generated questions should help the user to test

P. Shanthi Bala (✉)
Department of Computer Science, Pondicherry University, Puducherry, India
e-mail: shanthibala.cs@gmail.com

G. Aghila
Department of Computer Science and Engineering, National Institute of Technology, Puducherry, India
e-mail: aghilaa@gmail.com

© Springer Nature Singapore Pte Ltd. 2019
J. D. Peter et al. (eds.), *Advances in Big Data and Cloud Computing*,
Advances in Intelligent Systems and Computing 750,
https://doi.org/10.1007/978-981-13-1882-5_44

various cognitive skills such as knowledge, comprehension, application, analysis, synthesis, and evaluation. It will be more useful if the questions are developed on the basis of Bloom's taxonomy [1]. In Semantics based question generation system, Ontologies play a major role. "A specification of a representational vocabulary for a shared domain of discourse which contains set of concepts, properties and inter-relationships of those concepts" is called as ontology [2]. It may be defined as "formal explicit specification of shared conceptualization" [3]. The semantic relationship that exists between concepts provides common understanding of the knowledge. Ontologies provide open world semantics and make domain assumptions explicitly [3]. Semantic relationships that exist in the ontology provide more semantics for an application. It is widely used to accommodate real world knowledge [4]. In this work, the sematic relationships that exist between the concepts are exploited to generate more number of knowledge level questions.

This paper is organized as follows: Sect. 2 discusses about the existing work and Sect. 3 provides information about Q-Genesis. The advantages of the proposed system are discussed in Sect. 4. Section 5 concludes the work.

2 Related Work

Semantic web technologies are widely used for generating, organizing and personalizing web content. The technologies are extended to generate the questions automatically using ontologies. It may be applicable in systems like intelligent tutor system. Apart from this, it helps the trainer to prepare potential questions for learners automatically. The variety of superficial questions can be generated from ontology with semantic relationships. The question generation system is classified into template-based-, syntax-based-, and semantics-based Question Generation (QG) system. Among these, semantics-based question generation system provides better accuracy when generating the questions.

2.1 Template-Based QG System

Self-questioning strategy was proposed by Mastow et al. [5] to generate the questions from narrative sentences. It helps to enhance understanding at the children level. The templates include what, why, how. Chen [6] extended the work to generate the questions such as What-would-happen-when, What-would-happen-if, Why-x and When-would-x-happen questions. In this, templates should be generated manually to generate the questions.

2.2   Syntax-Based QG System

Wyse et al. [7] proposed a system Ceist which is used to generate the questions. In this, suitable sentences are extracted using pattern matching and pre-determined questions are generated using rules. They have used OpenLearn data resource for question generation. Heilman et al. [8] proposed a framework to generate comprehension questions automatically. The declarative sentences are extracted from text and converted using rules and the questions are statistically ranked [9]. The syntax based approaches employ syntactic tree [10] method. Multiple choice questions are generated automatically using NLP methodology [11]. Question generation system is based on syntactic and semantic information which were extracted using Definite Clause Grammar (DCG) [12]. Automatic question generation system is used in educational technologies [13] and dialog systems [14].

2.3   Semantics-Based QG System

Semantic-based method is performed on the semantic representations of the given sentence and helps to map declarative sentences into interrogative sentences. Schwartz et al. [15] introduced content question generator using logical form which can be used to identify the semantic relationships among various segments of a given sentence. The logical form can be used to generate WH questions. Papasalouros et al. [16] proposed an approach to automatically generate multiple choice questions. They have followed the strategies such as class, properties, and terminology. The semantic relationships between concepts are considered to generate the questions. Xu et al. [17] generated rules and used semantic hierarchy of a sentence to generate questions automatically by changing the clause type and converting into interrogative sentence.

Teisma et al. [18] developed Situation Awareness Question Generator (SAQG) for generating questions from ontology. SAQG is used to determine the situation and to generate yes-no type questions. Al-Yahya et al. [19] developed OntoQue engine for generating multiple choice questions from domain ontologies. Analogy questions are generated from ontology [20]. Yao et al. [21] interpreted natural language sentence as Minimal Recursion Semantics (MRS) structure by parsing and through generation, MRS structure can be interpreted as natural language sentence. Yai et al. [22] extended their work and developed MrsQG for transformation of sentences, sentence decomposition and automatic question generation with ranking. An ontology based approach is used for generating analogy questions [20]. In QG system, semantic relationships play a vital role to generate WH questions.

3 Q-Genesis

Question generation system generates questions based on the learning materials to check the learner's knowledge and also helps them to enhance their understanding about a particular field. It helps the trainer to generate various levels of questions which are intended to assess the knowledge level of the learners. The questions generated by the subject experts and making entry by the data entry operator are time consuming and costly process. Hence, there is need of a system which generates the questions automatically. The variety of questions can be generated such as knowledge level, comprehensive level and application level, etc. Knowledge level questions are very popular for assessing the understandability level of the learner. The semantic relationship that exists between two entities plays a vital role in generating knowledge level questions. In this work, an effort has been made to generate knowledge level questions from structured information with the use of semantic relationships.

Q-Genesis helps to generate Bloom's taxonomy level 1 and level 2 questions. Generally, the questions are not only used to assess whether the learners have learned something and also guides the learners to learn sufficient information. The knowledge level questions help the learners to think more creatively. The learners are constantly bombarded with questions; it makes them to think more innovatively and divergently. The generation of questions is a time consuming task for the trainer. Thus, Q-Genesis system helps to generate variety of superficial questions automatically with the use of the semantic relationships.

The architecture Q-Genesis is shown in Fig. 1. In this approach, domain knowledge has been collected from web using crawler and stored in the database. The generation of questions requires robust knowledge about the domain for which the questions to be generated. But given domain knowledge, it is quite impossible to generate questions automatically unless the knowledge can be represented in a proper way. In this approach, the domain knowledge is represented in the form of domain ontologies with the use of OWL parser.

The ontologies represent concepts, semantic relationships, and instances of the particular domain. The ontology is represented in OWL format. It is a standard Web Ontology Language based on the Description Logic representation formalism.

In Description Logic, the following notations have been used.

C(a)   A is an instance of the concept C
R(a, b)   Instance a is related to b with the relationship R

For example,

Leader (Mahatma Gandhi)—Mahatma Gandhi is a leader
Country (India)—India is a country
Situated_in (Gujarat, India)—Gujarat is situated in India

In this, Mahatma Gandhi, India and Gujarat are individuals. Leader and Country are concepts. Situated_in is a binary semantic relationship that exists between the individual Gujarat and India.

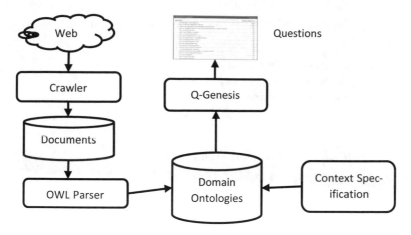

Fig. 1 Architecture of Q-genesis

Table 1 Context
specification

Context	Questions
People	Who
Time	When
Location	Where
Country	Where, what
Things	What
Organization	Where, what
Animal	What
Abstract	What
Transfer	What, who, where
Information	What, who
Emotional	What
Thought	Who, what
Event	What, where

To generate the questions from the ontology, all the concepts, relationships and instances that are defined in ontology will be useful. The number of questions that can be generated is directly proportional to the number of concepts and semantic relationships represented in the ontology. The semantic relationships are not defined properly in the ontology; there is a possibility of generating the definitional questions only. The context specification is incorporated with ontologies to generate the superficial questions. The context specification is shown in Table 1.

The questions to be generated are used to assess the various knowledge levels of the learner. It follows Bloom's Taxonomy. In Q-Genesis, the semantic-based approach is followed for generating Bloom's level 1 and level 2 questions. These questions are generated based on the concepts, semantic relationships and instances.

Three semantic based methods have been formulated to generate the questions. The approaches are:

- Concept-based question generation
- Association-based question generation
- Case-based question generation

3.1 Concept-Based Question Generation

In ontology, the concepts are generally organized in subsumption (class/subclass) hierarchies. The semantic relationships are not explicitly specified, it takes "is-a" relationship between the concepts. In this approach, the key concepts such as sub concepts, super concepts, equivalent concepts and disjoint concepts have been identified from the ontology. The questions such as define, what, compare and contrast type of questions are generated by using the concepts.

3.2 Association-Based Question Generation

In association-based approaches, it identifies the object properties and related concepts. Once the association has been identified from ontology, it generates the questions such as list, what, when, where, list.

3.3 Case-Based Question Generation

Case-based approaches incorporate concepts and semantic relationships with the instances. It can generate the both knowledge level questions and comprehensive level questions. It generates the questions such as who, how, where, when, etc.

A small part of the sample leader's ontology is shown in Fig. 2. In this, the relationships like born_in, located_in and studied_in which specify how the instances are related to other instances.

In this ontology, the following knowledge is represented as follows:

- Born in (Mahatma Gandhi, Gujarat)
- Hypernymy (Gujarat, State)
- Hypernymy(India, country)
- Born in(Gandhi, Porbandar)
- Born in (Gandhi, 1869)
- Studied(Gandhi, law)
- Studied_in(Gandhi, London)

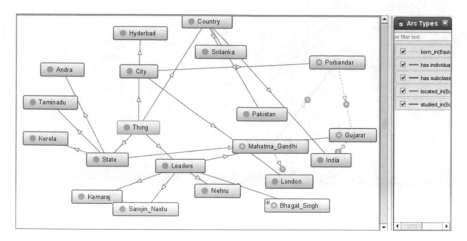

Fig. 2 Leader's antology—a part

Based on the represented knowledge and context specification, the following questions are generated.

- Where was Gandhi born?
- Who was born in Porbandar?
- In which year Gandhi was born?
- What did Gandhi study?
- Where did Gandhi study?

In the similar way, huge numbers of questions are generated from the concepts, relationships and instances which are defined in ontology. The generated questions are verified for the correctness. The Q-Genesis system brings down the human intervention for preparing superficial questions.

4 Discussion

The different ontologies such as GO, OBO, ChEBI, Cheminf, Network, multimedia, design pattern, university, ka and E-Tourism are considered for evaluation. From the results, it is observed that more number of relevant questions is generated with Q-Genesis. The number and relevant question generated is mentioned in Table 2.

The number of questions generated using Q-Genesis is increased. This is due to the inclusion of semantic relationships and context specification with the ontologies. It is also inferred that the number of question generated is directly proportional to the number of concepts and semantic relationships that exists between the given concepts.

From the results, it is observed that the questions which are generated with the use of semantic relationships provide more accuracy.

Table 2 Question generation

S. No	Name of the ontology/dataset	No. of question generated	No. of relevance questions
1	GO	30	21
2	OBO	15	9
3	CHEBI	50	32
4	CHEMINF	67	32
5	NETWORK	34	30
6	MULTIMEDIA	87	78
7	DESIGN PATTERN	79	65
8	UNIVERSITY	45	37
9	KA	37	31
10	E-TOURISM	56	49

5 Conclusion

Question Generation System based on Semantic Relationships (Q-Genesis) system is proposed to generate the knowledge level questions using semantic relationships. The questions have been generated from ontologies using concept, association and case based approaches. From the results, it is observed that the number and complexity of the questions have been increased with the exploitation of the semantic relationships. Q-Genesis provides 86% of the accuracy for computer science and universal ontologies. The accuracy of 77% is achieved with biological and chemical ontologies. In future, effort can be made to generate more complex synthesis and application level questions by employing the semantic relationships.

References

1. Bloom, B.S., David R.K.: Taxonomy of educational objectives: the classification of educational goals. Handbook 1: Cognitive domain (1956)
2. Gruber, Thomas R.: A translation approach to portable ontology specifications. Knowledge Acquisition **5**(2), 199–220 (1993)
3. Borst, P., Akkermans, H., Top, J.: Engineering ontologies. Int. J. Hum Comput Stud. **46**(2), 365–406 (1997)
4. Storey, Vede C.: Understanding semantic relationships. VLDB J. **2**, 455–488 (1993)
5. Mostow, J., Chen, W.: Generating instruction automatically for the reading strategy of self-questioning. Artif. Intell. Educ. AIED, pp. 465–472 (2009)
6. Chen, W., Aist, G., Mostow, J.: Generating questions automatically from informational text. Proceedings of AIED Workshop on Question Generation, pp. 17–24 (2009)
7. Wyse, B., Piwek, P.: Generating questions from openlearn study units, vol 1, pp. 66–73 (2009)
8. Heilman, M., Smith, N.A.: Question generation via over generating transformations and ranking. DTIC Document (2009)

9. Heilman, M., Smith, N.A.: Good question! statistical ranking for question generation. In: Human Language Technologies: The 2010 Annual Conference of the North American Chapter of the Association for Computational Linguistics, pp. 609–617 2010

10. Klein, D., Manning, C.D.: Fast exact inference with a factored model for natural language parsing. In: Advances in Neural Information Processing Systems, pp. 3–10 (2002)

11. Mitkov, R., Ha, L.A.: Computer-aided generation of multiple-choice tests. In: Proceedings of the HLT-NAACL 03 Workshop on Building Educational Applications Using Natural Language Processing, vol. 2, pp. 17–22 (2003)

12. Kunichika, H., Katayama, T., Hirashima, T., Takeuchi, A.: Automated question generation methods for intelligent english learning systems and its evaluation. In: Proceedings of ICCE, pp. 2–5 (2003)

13. Graesser, A.C., Chipman, P., Haynes, B.C., Olney, A.: AutoTutor: an intelligent tutoring system with mixed-initiative dialogue. Educ. IEEE Trans. 48(4), 612–618 (2005)

14. Walker, M.A., Rambow, O., Rogati, M.: SPoT: a trainable sentence planner. In: Proceedings of the Second Meeting of the North American Chapter of the Association for Computational Linguistics on Language Technologies, pp. 1–8 (2001)

15. Schwartz, L., Aikawa, T., Pahud, M.: Dynamic language learning tools. In: InSTIL/ICALL Symposium (2004)

16. Papasalouros, A., Kanaris, K., Kotis, K.: Automatic generation of multiple choice questions from domain ontologies. In: E-Learning, pp. 427–434 (2008)

17. Xu, Yushi, Goldie, A., Seneff, S.: Automatic question generation and answer judging: a Q&A Game for Language Learning, Proceedings SIGSLaTE, pp. 57–60 (2009)

18. Teitsma, M., Sandberg, J., Maris, M., Wielinga, B.: Using an ontology to automatically generate questions for the determination of situations, In: Database and Expert Systems Applications, Springer, pp. 456–463 (2011)

19. Al-Yahya, M.: OntoQue: a question generation engine for educational assessment based on domain ontologies. IEEE 11th International Conference on Advanced Learning Technologies (ICALT), pp. 393–395 (2011)

20. Alsubait, T., Parsia, B., Sattler, U.: Automatic generation of analogy questions for student assessment: an Ontology-based approach. Res. Learning Technol. 20, 95–101 (2012)

21. Yao, X., Bouma, G., Zhang, Y.: Semantics-based question generation and implementation. Dialogue Discourse 3(2), 11–42 (2012)

22. Yao, X. Zhang, Y.: Question generation with minimal recursion semantics. In: Proceedings of QG2010: The Third Workshop on Question Generation, pp. 68 2010

Detection of Duplicates in Quora and Twitter Corpus

Sujith Viswanathan, Nikhil Damodaran, Anson Simon, Anon George, M. Anand Kumar and K. P. Soman

Abstract Detection of duplicate sentences from a corpus containing a pair of sentences deals with identifying whether two sentences in the pair convey the same meaning or not. This detection of duplicates helps in deduplication, a process in which duplicates are removed. Traditional natural language processing techniques are less accurate in identifying similarity between sentences, such similar sentences can also be referred as paraphrases. Using Quora and Twitter paraphrase corpus, we explored various approaches including several machine learning algorithms to obtain a liable approach that can identify the duplicate sentences given a pair of sentences. This paper discusses the performance of six supervised machine learning algorithms in two different paraphrase corpus, and it focuses on analyzing how accurately the algorithms classify sentences present in the corpus as duplicates and non-duplicates.

Keywords Deduplication · Natural language processing · Paraphrase · Quora
Twitter · Machine learning

1 Introduction

In natural language processing, the task of detecting paraphrase texts requires deep semantic level understanding. A paraphrase is nothing but restatement of paragraph, article, or sentence. Paraphrase detection means identification of duplicate phrases, sentences or documents of either arbitrary or same lengths. Some of the applications of paraphrase detection in various languages [1] includes plagiarism detection for fiction, nonfiction, and scientific articles [10].

S. Viswanathan (✉) · N. Damodaran · A. Simon · A. George · M. Anand Kumar · K. P. Soman
Center for Computational Engineering and Networking (CEN),
Amrita Vishwa Vidyapeetham, Amrita School of Engineering, Coimbatore, India
e-mail: cb.en.p2cen16016@cb.students.amrita.edu

M. Anand Kumar
e-mail: m_anandkumar@cb.amrita.edu

K. P. Soman
e-mail: kp_soman@amrita.edu

© Springer Nature Singapore Pte Ltd. 2019
J. D. Peter et al. (eds.), *Advances in Big Data and Cloud Computing*,
Advances in Intelligent Systems and Computing 750,
https://doi.org/10.1007/978-981-13-1882-5_45

The areas where paraphrases and its identification are highly relevant are social media networks like Twitter, Facebook, and question-answering forums like Quora, Stack-exchange. The identification of duplicate sentences will be helpful for question-answering forums like Quora because if a user asks any question and if it is previously asked by any other user, then the new question can be answered directly from the previously obtained responses. Previous works in the field of paraphrase detection had used vector-similarity measures to differentiate between two sentences [2]. Vectors are compared in their entirety to measure the similarity metric which indicates whether two sentences are identical. Analyses of the difference between two sentences were done on the basis of syntactic parsing trees [11], which help in having a better semantic analysis. The syntactic parsing tree cannot be incorporated to social media data like Twitter since the tweets will not be grammatically correct. Some approaches used for accomplishing the duplicate detection tasks include the usage of recursive autoencoders, heuristic similarity, and probabilistic inference. Researchers have also worked on paraphrase detection with the use of supervised machine learning techniques for which various features like word similarity information obtained from WordNet [4] or from word clustering methods are used. Semantic similarity approaches by means of matrix similarity between two sentences with word–word similarities where also considered for duplicate detection task. Socher et al. [11] proposed unsupervised recursive autoencoders in his paper where each sentence is parsed pairwise and their embedding was found out for each node of parse tree to construct a similarity matrix. Dynamic pooling technique was applied to convert matrix of variable size into fixed size. This matrix was used as an input for the soft-max classifier which helped in classifying sentences. There are several approaches that focused on paraphrase detection in an unsupervised manner [7] by using a text to text generation mechanism from a given input sentence [3].

This paper proposes the use of various machine learning techniques such as Random Forests, Logistic Regression, Support Vector Machine (SVM), and Decision Tree for the given classifying pair of sentences into paraphrases or nonparaphrases by using word share and TF-IDF word share as primary features. The TF-IDF is prescribed as an important feature for similarity measurement of sentences [5]. The other features used are word count and character count. The abovementioned machine learning approaches were incorporated by us to identify the best approach that can identify the paraphrases in both the corpus. Two datasets are used for implementing the deduplication task. Twitter paraphrase corpus was presented in [12]. The following table shows some statistics associated with the data. The Twitter corpus includes tweets that are preprocessed with techniques like tokenization that separated the tokens, Part Of Speech (POS) and named entity tagging. The training set comprises 11530 sentence pairs from more than 500 topics on Twitter in which the hash tags were excluded. The test set consists of 972 pairs of tweets from twenty randomly sampled Twitter topics. Expert labeling was used as the labels for train and test set. The second dataset is the Quora question pairs dataset. The dataset released by Quora is also related to the problem of identifying duplicate questions. The dataset consists of 404,300 lines of duplicate question pairs in which each line contains IDs for each question in the pair and for the full text for each question.

Table 1 Quora dataset statistics

Dataset	Duplicate sentence pair	Non-duplicate sentence pair
Train-304300	113113	191187
Test-100000	36150	63850

Table 2 Twitter dataset statistics

Dataset	Duplicate sentence pair	Non-duplicate sentence pair
Train-11530	3996	7534
Test-972	175	797

A binary value present indicates the label to know whether the line truly contains a duplicate pair. For both datasets, the sentence pair has a label which is either 0 or 1. A label 1 indicates that the sentence pair is a duplicate, while a label 0 indicates that it is not a duplicate sentence pair (Tables 1 and 2).

2 Feature Extraction

Feature extraction in Twitter and Quora corpus from both train and test dataset includes the retrieval of features such as character length and the number of words. A word share count which resembles the frequency of occurrence of a word among the tweets or question pairs is an essential feature to be considered. The Term Frequency-Inverse Document Frequency (TF-IDF) word share between the sentence pairs are another important feature considered. Initially, we stripped the string from leading and trailing white space, and then converted them into lower cases. We also removed Not A Number (NAN) values in character count by replacing them with zeros.

There are a total of six main features that were utilized for this work. Number of words, character count, word share ratio, and TF-IDF share ratio of sentence 1 and sentence 2 in a sentence pair are the used features for both corpus. The number of words and character count is extracted using basic string functions. Word share ratio is calculated as the ratio of the number of words appearing in common to both sentences to the total number of words in both sentences in the pair. The result is a normalized value measuring how much word overlap is between sentence pairs which will be either a tweet pair or a question pair.

For TF-IDF factors, first, we combined all questions into one big corpus and then calculated weight for each word. The weight is calculated by taking the inverse of word count added with epsilon which will be randomly assigned as 5000, such that the less common words get higher weights. We ignored words that only appear once in the whole corpus for the reason that it may be typographical error. Then we calculated the TF-IDF word share norm using the ratio between share weight to the total weight. Share weight is the total weight of the words appearing in both sentences in a pair. Total weight is the total weight of every word in sentences. The result we obtain is a normalized TF-IDF weight share. Since TF-IDF discounts on the common words that appear universally in our language, this measure represents more closely to how similar are two tweets or questions.

3 Methodology

To implement a model that classifies the question pairs and tweet pairs as duplicate or non-duplicate, we considered several machine learning approaches to find out the best suited approach that can classify the Quora question pairs and Twitter tweet pairs more accurately. We incorporated algorithms that follow a supervised learning approach in which we have a set of targets which has to be predicted by the model with the help of various variables.

3.1 Machine Learning Algorithms

This is a binary classification problem. We used supervised learning algorithms. Initially, we cleaned the strings by removing leading and trailing white spaces and then converted them to lower case. Feature extraction is carried after this from the data. These features include word count, character count, word share, and TF-IDF share. The algorithms we incorporated are Random Forest, Logistic Regression, Decision Trees, K-Nearest Neighbors, Naive Bayes, and Support Vector Machine. Except for the Naive Bayes model, we have normalized all feature values before performing the fitting of the model.

- **Logistic Regression**: Logistic Regression is a classification methodology. It is used for predicting an outcome from independent variables. It predicts the probability for the outcome to fall into any of the available class. Logistic Regression predicts the probability of occurrence of an event by fitting data to a logistic function.
- **Decision Tree**: A model which uses a tree to do binary classification at each node, and at the leaves of the tree a label prediction will be returned. Decision tree creates a model that can predict class with the help of learned decision rules from previously trained data.
- **Support Vector Machine**: An algorithm which outputs an optimal hyperplane to separate labeled training data. We implemented Grid Search cross-validation to find the best parameters (C, kernel), which are (1000, sigmoid). We set max iteration to 500 to avoid out of memory issue. SVM deals with finding an optimal hyperplane and reduction of the errors in classification based on it.

- **K-Nearest Neighbor**: It is an algorithm which uses the average of k-nearest data points to predict the testing data value. It predicts the label by finding the neighbor class which is nearest with the help of distance measures. The distance here is usually 2D Euclidean distance. We performed a Grid Search cross-validation to find best parameters (n neighbors, weights), which are (8, uniform).
- **Naive Bayes**: Naive Bayes is a classification algorithm which is used for predictive modeling. This algorithm can be applied to binary or multi-class classification. Naive Bayes includes the probabilities of each class in the training dataset called Class Probabilities and the conditional probabilities of each input value given each class value.
- **Random Forest**: Random Forest is performed using an ensemble of decision trees to reduce overfitting of one-tree model. The training algorithm for Random Forest applies the general technique of bagging, which means selecting a random sample of training data and fitting trees to these samples. Finally, we average the result of all trees to make prediction.

4 Feature Visualization

Feature Visualization deals with the representation of the features obtained from the experiments performed on the two available datasets. This helps in creating a better understanding of various unknown statistics about the dataset. The detection of duplicate sentences can be made easy with this proper visualization of the extracted features. In this section, we closely observe and represent the characteristics of the Twitter and Quora corpus. We visualize the features such as word frequency, number of characters in a sentence, number of words in a sentence, word share ratio among sentence pair, and TF-IDF share ratio among the sentences in a sentence pair for both the corpus.

4.1 Twitter Paraphrase Corpus

A Twitter corpus consists of tweet pairs which are posted by various users in Twitter. The tweet pairs will be domain independent. The feature visualization of the Twitter corpus includes visualizing required characteristics associated with the two tweets among a tweet pair. Few earlier approaches on Twitter datasets considered sentence level annotation mechanism for paraphrase detection [13] while some approaches focused on paraphrase detection using unsupervised feature learning methodology [8]. In our approach, the most frequently used words and phrases present in tweet1 and tweet2 among tweet pair are represented in the form of a word cloud graph from training data. The bigger words or phrases in the word cloud graph will indicate the frequently occurring words and phrases in the available corpus. We separated tweet1 and tweet2 into two and used word cloud library to generate plots. When we observed the word cloud graphs, we noticed that there are several words appearing in both graphs. Most of them are common words, which means that they are used frequently in our daily conversations. However, we can see some of them as special words, like Blake Griffin, Game of Thrones, Disney, etc. This indicates the usage of these phrases.

A normalized histogram of character count present in both Tweet1 and Tweet2 where considered to find the characters present in both sentences among a pair.

The peak of distributions both in Tweet1 and Tweet2 occurs at around 38 characters and has an average number of characters at 40. We observed that there are no tweets beyond 110 characters. The number of words that are present in both tweets of the tweet pair is observed. The number of words in tweets peak at 8 words. Therefore, we conclude that the maximum number of words in a tweet is 8 words, which indicates that most people tweets with approximately 8 words in an average. The threshold count of words can be set as 8. We conclude that the maximum number of words in a tweet is 18.

A word share ratio represents the number of words shared in common between two tweets in a tweet pair. The larger the ratio the more overlapped is the tweet pair. The number of words shared in common between tweets plays a major role in identifying similarity between sentences. The word share ratio graph indicated that the paraphrase tweets generally have more words shared between them.

Figure 1 shows the Term Frequency-Inverse Document Frequency (TF-IDF) word share ratio. TF-IDF is a numerical measurement that is planned to reflect how imperative a word is to a record in a corpus. TF-IDF word share ratio is an important feature in a document since it helps in identifying the occurrence of a particular term in a document. It is similar to word share ratio but here the terms that are having higher weights among the document would help in identifying the similarity between sentences.

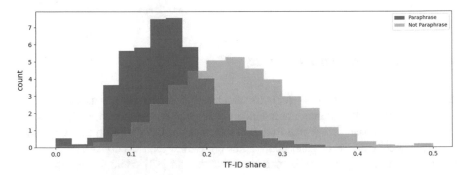

Fig. 1 TF-IDF share ratio in Twitter corpus

4.2 Quora Paraphrase Corpus

A Quora corpus consists of question pairs in which each question will be related to certain domain which is posted by the users [6]. The specialty of Quora dataset is that each question should be semantically correct; it means that the questions asked in Quora should be grammatically correct unlike Twitter because in Quora, the users ask questions to get appropriate answers from other users; this can be made sure only if the question asked is understandable to others. The Feature visualization of the Quora corpus is interpreted in the following. The section deals with visualization of the features extracted from both the questions in question pair. The most frequently occurring words and phrases present in question1 and question2 of the corpus will be represented using bigger fonts in word cloud for both questions in the question pair. The word cloud generated includes important informations showing the presence of words such as India, make and phrases such as "best way" in common to both questions in a pair which helps in analyzing the similarity. The total characters present in the Quora question pairs and the probability of the occurrence of characters in both duplicate and non-duplicate question pairs are found; the number of characters present is in an average of 40 in the Quora corpus. The character count also indicates that there are no questions having more than 200 characters. The word count that represents the total number of words in the questions is diagrammatically visualized. It finds the average word count per question for both duplicate and non-duplicate pairs. The average word count in a question is 8 and the maximum number of words in a question is 40.

The word share ratio represents the ratio of the number of the words shared in common between two questions among the duplicate and non-duplicate question pairs. The word share ratio was diagrammatically visualized and we observed that the word share ratio increases the chances for a sentence pairs to be a paraphrase or duplicate increases.

In TF-IDF, Term Frequency represents the number of times a token occurs in a document which makes it important to know how frequent a word is repeated in a document to understand the importance of that particular word in document. Inverse Document Frequency finds out the number of documents in which a word is occurring. Therefore, the TF-IDF share ratio finds the importance of words in the question pairs. The ratio is represented in the above graph. From Fig. 2, we can observe that, as the TF-IDF share ratio increases, the chances for a sentence pairs to be a paraphrase or duplicate increase.

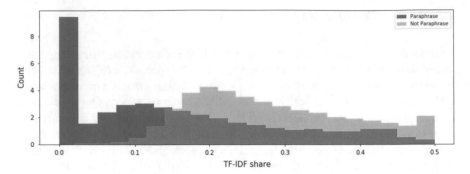

Fig. 2 TF-IDF share ratio in Quora corpus

5 Result

A total of six machine learning techniques were carried out on the Twitter and Quora corpus, and we have compared the performance of both the datasets on these algorithms separately. The results of classification for various models depicting the machine learning algorithms are shown in Tables 3 and 4.

The output statistics we obtained shows that while performing duplicate detection on both considered datasets, all six algorithms didn't result in equal performance. In Twitter corpus, Logistic Regression yielded better accuracy than all other techniques and this Logistic Regression method had achieved higher accuracy's compared to

Table 3 Results for Twitter dataset analysis

Model	Logistic Regression	Decision Tree	SVM	KNN	Naive Bayes	Random Forest
Precision	0.79	0.70	0.70	0.78	0.77	0.78
Recall	0.79	0.70	0.66	0.78	0.76	0.78
F1-score	0.78	0.70	0.67	0.77	0.74	0.77
Accuracy (%)	78.6	70.3	66.0	78.0	66.1	78.1

Table 4 Results for Quora dataset analysis

Model	Logistic Regression	Decision Tree	SVM	KNN	Naive Bayes	Random Forest
Precision	0.66	0.69	0.64	0.71	0.56	0.76
Recall	0.67	0.69	0.55	0.72	0.64	0.72
F1-score	0.66	0.69	0.55	0.72	0.50	0.73
Accuracy (%)	67.1	69.3	60.0	71.9	63.7	72.2

the existing approaches in this dataset. Logistic Regression is an algorithm which is highly suitable for correlated data. The dataset constitute classes that are linearly separable which makes them suitable to work efficiently with Logistic Regression. One of the major advantages of Logistic Regression is that it can be regularized with respect to data in order to avoid overfitting. The specialty of Twitter dataset is that the tweets may not convey a proper semantic meaning. Therefore, this dataset also includes sentences which don't have a proper grammatical meaning.

The Quora dataset yielded better result on Random Forest algorithm. The Quora dataset contains sentences with a proper semantic meaning. The advantages of Random Forest is that it can handle the missing values and since we have more trees in Random Forest the classifier will not overfit the model. SVM had yielded comparatively less results with respect to other algorithms in both datasets. Choosing a better kernel function for SVM was a difficult part and another disadvantage of SVM we incurred was that it took long training time on large datasets like Quora. Naive Bayes algorithm yielded a better result compared to SVM but it was the second worst performer among the six algorithms. KNN had a better performance equally in Quora and Twitter dataset which shows that distance measure-based mechanisms like KNN are also efficient in identifying the similarity between texts.

6 Conclusion

The duplicate detection carried out on the Quora and Twitter corpus suggests that machine learning algorithms work well in detecting the duplicate sentences among a sentence pair. The algorithms considered for this work included Logistic Regression, Decision Tree, Support Vector Machine, K-Nearest Neighbor, Naive Bayes, and Random Forest. Among these considered algorithms no algorithm failed for paraphrase detection in both corpora. The Random Forest and K-Nearest Neighbor algorithms performed equally well in both datasets. The Logistic Regression performed best among all algorithms for Twitter corpus; similarly, Random Forest provided best result for Quora corpus.

7 Future Work

The machine learning algorithms perform efficiently for duplicate detection tasks; however, the duplicate detection tasks can also be tackled using a neural network model. The duplicate detection tasks on these two corpora can be extended by implementing a model which performs better identification of the similarity of sentences [9]. A deep neural network model named Siamese network includes two LSTM networks in which each of them processes one sentence and hence finds the similarity between them. The Siamese network can be thus incorporated into the two corpora considered by us and the performance of sentence similarity identification can be validated.

References

1. Anand Kumar, M., Singh, S., Kavirajan, B., Soman, K.: Shared task on detecting paraphrases in indian languages (dpil): An overview. Lecture Notes in computer science (including subseries Lecture Notes in Artificial Intelligence and Lecture Notes in Bioinformatics) pp. 128–140 (2018)
2. Blacoe, W., Lapata, M.: A comparison of vector-based representations for semantic composition. In: Proceedings of the 2012 joint Conference on Empirical Methods in Natural Language Processing and Computational Natural Language Learning, pp. 546–556. Association for Computational Linguistics (2012)
3. Cordeiro, J., Dias, G., Brazdil, P.: A metric for paraphrase detection. In: International Multi-Conference on Computing in the Global Information Technology, 2007. ICCGI 2007, pp. 7–7. IEEE (2007)
4. Fernando, S., Stevenson, M.: A semantic similarity approach to paraphrase detection. In: Proceedings of the 11th Annual Research Colloquium of the UK Special Interest Group for Computational Linguistics, pp. 45–52 (2008)
5. Huang, C.H., Yin, J., Hou, F.: A text similarity measurement combining word semantic information with tf-idf method. Jisuanji Xuebao(Chin. J. Comput.) **34**(5), 856–864 (2011)
6. Iyer, S., Dandekar, N., Csernai, K.: First quora dataset release: Question pairs (2017)
7. Joao, C., Gaël, D., Pavel, B.: New functions for unsupervised asymmetrical paraphrase detection. J. Software **2**(4), 12–23 (2007)
8. Mahalakshmi, S., Anand Kumar, M., Soman, K.: Paraphrase detection for tamil language using deep learning algorithm. Int. J. of Appld. Engg. Res **10**(17), 13929–13934 (2015)
9. Mueller, J., Thyagarajan, A.: Siamese recurrent architectures for learning sentence similarity. In: AAAI, pp. 2786–2792 (2016)
10. Praveena, R., Kumar, M.A., Soman, K.P.: Chunking based malayalam paraphrase identification using unfolding recursive autoencoders. In: 2017 International Conference on Advances in Computing, Communications and Informatics (ICACCI) pp. 922–928 (2017)
11. Socher, R., Huang, E.H., Pennin, J., Manning, C.D., Ng, A.Y.: Dynamic pooling and unfolding recursive autoencoders for paraphrase detection. In: Advances in neural information processing systems, pp. 801–809 (2011)
12. Xu, W., Callison-Burch, C., Dolan, B.: Semeval-2015 task 1: Paraphrase and semantic similarity in twitter (pit). In: Proceedings of the 9th International Workshop on Semantic Evaluation (SemEval 2015), pp. 1–11 (2015)
13. Xu, W., Ritter, A., Callison-Burch, C., Dolan, W.B., Ji, Y.: Extracting lexically divergent paraphrases from twitter. Trans. Assoc. Comput. Linguist. **2**, 435–448 (2014)

Parallel Prediction Algorithms for Heterogeneous Data: A Case Study with Real-Time Big Datasets

Y. V. Lokeswari, Shomona Gracia Jacob and Rajavel Ramadoss

Abstract Parallel data mining algorithms are extensively used to mine and discover hidden knowledge from varied, unrelated data. Parallel data mining algorithms provide advantages such as reduced training time, less execution time, and less memory requirement. There are several issues in executing parallel data mining algorithms in a distributed environment. It is crucial to partition the data among processors such that there is minimal data dependency, proper synchronization, communication overhead, work load balancing among nodes in distributed processors and disk IO cost. Few of these issues can be resolved when parallel data mining algorithms are executed on Apache framework called Hadoop Map Reduce. Hadoop Map Reduce provides improved performance, reduced communication cost, reduced execution time, reduced training time, and reduced IO access. This paper proposes a novel framework that aims at enhancing the aforementioned advantages in terms of scalability by increasing the number of nodes in the Hadoop cluster and analyzing the performance of classification algorithms like K-Nearest Neighbor, Naïve Bayes and Decision Tree. This parallel framework could be extended to other fields of biotechnology where prediction on large datasets is essential.

Keywords Data mining algorithms · Hadoop map reduce · Scalability · Big data

Y. V. Lokeswari (✉) · S. G. Jacob
Department of CSE, Sri Sivasubramaniya Nadar College of Engineering,
Kalavakkam, Chennai, India
e-mail: lokeswariyv@ssn.edu.in

S. G. Jacob
e-mail: graciarun@gmail.com

R. Ramadoss
Department of ECE, Sri Sivasubramaniya Nadar College of Engineering,
Kalavakkam, Chennai, India
e-mail: rajavelr@ssn.edu.in

© Springer Nature Singapore Pte Ltd. 2019
J. D. Peter et al. (eds.), *Advances in Big Data and Cloud Computing*,
Advances in Intelligent Systems and Computing 750,
https://doi.org/10.1007/978-981-13-1882-5_46

1 Introduction

Our computerized world is moving towards "rich in data", but "poor in knowledge" scenario. Data mining and machine learning algorithms helps in discovering useful patterns and previously unknown relationships in a voluminous dataset [1–3]. Data mining algorithms like classification, regression, association rule mining, and clustering can be run in parallel to reduce the overall training and execution time [4, 5]. As mentioned above, parallel data mining algorithms have both advantages and disadvantages. Google has designed and developed an Apache framework called Hadoop Map Reduce that has alleviated the ill-effects of parallel data mining. Hadoop Map Reduce uses a "*map*" and "*reduce*" functions which work parallel on different subsets of data and store the results in the underlying Hadoop Distributed File System (HDFS) [1, 6].

Parallel Data mining algorithms have come to the rescue when predictions have to be made from massive datasets [7, 8]. Such algorithms when run on Hadoop Map Reduce generally result in improved performance, reduced execution time, reduced training time and reduced IO access [9]. However, investigations in execution of parallel data mining algorithms revealed the fact that when the algorithms were implemented on a single-node cluster, their performance in terms of training and execution time was low. Hence this motivated the authors to explore the options available to scale the existing system to incorporate huge volumes of data with Hadoop Map Reduce. Thus the authors proposed a novel framework by adjoining multiple nodes to the Hadoop Map Reduce cluster in order to further scale for massive datasets while ensuring prediction with very less training and execution time. Parallel K-Nearest Neighbor (K-NN), Naïve Bayes and Decision Tree classifiers were implemented in parallel using single-node Hadoop cluster and were extended to multi-node Hadoop cluster to investigate the proposed framework. Random Forest is an ensemble technique which improves the accuracy of prediction by generating multiple decision trees.

This research included the implementation of parallelized classification algorithms in Hadoop single-node- and multi-node cluster. The efficiency of the proposed framework was evaluated by analyzing the execution time for each of the above mentioned parallel classification algorithms. A concise description on parallel data mining algorithms is presented in the following sections.

1.1 Parallel K-NN Classifier on Hadoop Map Reduce

In K-NN classifier there is no training phase and so it is called as lazy learner classification algorithm. The test data is compared with each of the training dataset using distance metrics like Euclidean or Manhattan distance and is assigned to the K-most frequently occurring class. When parallel K-NN is implemented in a distributed environment, a master node splits the training dataset into N-subset and assigns to

N-slaves. Test data will be given to each slave. Slave processor computes the distance and reports to master. Once all slaves report distance values for its assigned subset, master sorts the distance in ascending order, selects top K distance and counts the number of classes in top K. Finally master assigns test data to the class having highest count [10]. This suffers from communication overhead between master and slaves. This can be eliminated by executing parallel K-NN on Hadoop Map Reduce where each Mapper will be assigned with data splits; Mapper finds the distance of test data with its assigned subset of training dataset and sorts in ascending order. Reducer will select top K distances and counts the number of classes occurring. Finally reducer assigns test data to the class which is frequently occurring in top K distances [9, 11, 12]. Parallel K-NN on Hadoop Map Reduce will reduce communication cost and improves performance. K-NN gives less accuracy compared to parallel Naïve Bayes and Decision tree classifiers due to the absence of the training phase. Speed of parallel K-NN is less compared to Naïve Bayes and Decision tree while the efficiency of parallel K-NN classifier can be improved by increasing the number of training instances [13].

1.2 Parallel Naïve Bayes Classifier on Hadoop Map Reduce

Naïve Bayes classifier works on the principle of Bayes theorem in which the values of one attribute will not influence on the values of other attributes in predicting the class. Test data is given to all slaves. When parallel Naïve Bayes classifier is executed in distributed environment, the master splits each attribute in the dataset and assigns to the slaves. Slaves will compute the class conditional probabilities in parallel and report to master. Master will find a class with highest probability and assigns test data to class with highest probability [14]. This suffers from communication overhead which can be eliminated when run on Hadoop Map Reduce. There are two phases such as training and prediction phase. In training phase, Mapper will find frequency of attributes for each class label. Reducer will count the frequency of attributes that belong to each class and outputs the <class-label and sum of frequency of attributes> as key value pairs. In Prediction phase, test data is given as input, for each attributes and each class in test data, calculate the count (class-label, attribute)/count (class-label) and assign test data to a class with maximum count [9, 15, 16]. Parallel Naïve Bayes on Hadoop Map Reduce will reduce training time, improves performance. Speed of parallel Naïve Bayes is faster compared to parallel Decision tree and K-NN classifiers. Efficiency of parallel Naïve Bayes can be improved by increasing the number of attributes in training dataset [13].

1.3 Parallel Decision Tree Classifier on Hadoop Map Reduce

The general process of building a decision tree is as follows. Given a set of training data, measurement function is applied onto all attributes to find a best splitting
attribute. Once the splitting attribute is determined, the instance space is partitioned
into several parts. Within each partition, if all training instances belong to one single
class, the algorithm terminates. Otherwise, the splitting process will be recursively
performed until the whole partition is assigned to the same class. Constructing a
decision tree can be done in parallel either using Synchronous tree construction
approach or partitioned tree construction approach [17, 18]. Attribute selection measures like Information gain or Gain Ratio or Gini Index are used to find the best
splitting attribute from the training dataset. Synchronous approach suffers from communication overhead and partitioned approach suffers from data movement among
processors. When decision tree is constructed in parallel on Hadoop Map Reduce,
both issues can be eliminated. Decision tree is constructed in breadth-first search
manner [19, 20]. Training dataset is split and given to each Mapper. Each Mapper
finds the class distribution information for attributes of its assigned training subset
of data. Mapper outputs attribute and its class distribution information to Reducer.
Reducer will collect class distribution information from all Mappers and aggregates
the count. Best attribute has to be selected using attribute selection measures and
training dataset is split according to splitting criteria. Now the split training dataset
based on best attribute are assigned to Mapper. The process repeats until all training
dataset in every split belongs to same class or no more training dataset is left to be
assigned to a class [13, 21, 22]. A map and reduce task is needed in each level of
the decision tree. Hence, constructing the decision tree model is an iterative process.
Once the training model is constructed, test data is checked along paths of decision
tree and class label of the test data will be found. Parallel Decision Tree on Hadoop
Map Reduce will reduce communication cost and execution time. It also improves
scalability [23]. Decision tree classifier has less error rate and easy to implement.
However the accuracy of decision tree classifier decreases with increase in number
of missing values in attributes [21].

2   Materials and Methods

2.1   Census Income Data

Dataset was extracted from UCI Machine Learning repository [24]. The *CarOwners*
dataset has 2500 instances. It contains information about the car buying patterns of
a person based on his/her personal details. This dataset was used to explore the parallel *K*-NN classifier on Hadoop Map Reduce. *Adult dataset/Census Income* dataset
was downloaded from UCI Machine Learning repository. The dataset has 48,842
instances and 14 attributes, including 6 continuous variables and eight categorical

Table 1 Description of real-time datasets utilized for Investigating parallel prediction

S. No.	Dataset	No. of instances	No. of attributes	No. of target classes	Distribution of instances for classes	Class names
1	CarOwners	2500	6	4	197 983 500 820	1. BMW5 2. Corsa 3. MX5 4. Zafira
2	Census income	48,842	14	2	7841 24,720	1. Income>50 K 2. Income ≤50 K
3	Chronic kidney disease	400	25	2	250 150	1. Chronic kidney disease 2. Not a chronic kidney disease

variables. There are two classes in this dataset denoting the set of instances where the individual income is greater than 50 K or not greater than 50 K. As there are missing values in this dataset, those missing value instances were not considered for training. This resulted in purging 16,281 instances from 48,842 giving out 32,561 instances. Census Income dataset was utilized to explore the parallel Naive Bayesian classifier on Hadoop Map Reduce.

2.2 Medical Data

Chronic Kidney Disease dataset was utilized to investigate the parallel formulation of Decision Tree and Random Forest. The dataset has 400 instances with two target classes in order to identify patients having chronic kidney disease and not having chronic kidney disease. The description of dataset is given in Table 1.

3 Methods

Apache Hadoop Map Reduce framework was used in order to execute classification algorithms in parallel and distributed environment. A single-node Hadoop cluster and multi-node Hadoop cluster need to be established in order to scale for data intensive applications like health care, retail marketing, Business, Stock market exchange, research, etc.

The following steps illustrate execution of parallel classification algorithms on Hadoop single-node- and multi-node cluster as depicted in Fig. 1.

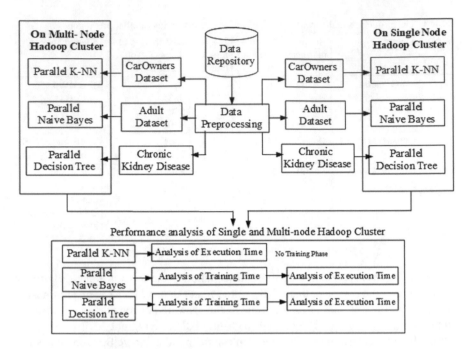

Fig. 1 Scalable parallel classifier on Hadoop cluster

3.1 Scalable and Efficient Parallel Classifiers on Hadoop Multi-node Cluster

1. Parallel K-Nearest Neighbor algorithm was implemented in single-node- and multi-node Hadoop cluster. The test data was given as input to K-NN classifier which classifies test data based on distance between training and test data.
2. Parallel Naïve Bayes algorithm was implemented in single-node- and multi-node Hadoop cluster. Test data was given as input to Naïve Bayes classifier which classifies test data based on probability of attribute values belonging to a class.
3. Parallel Decision Tree algorithm was implemented in single-node- and multi-node Hadoop cluster. Training dataset was used to construct the decision tree model based on Information gain as attribute selection measure to split training dataset. A Map and Reduce module was used at each level of the decision tree. Test data was fed to decision tree which outputs the class label of test data.
4. Performance of each parallel algorithm was evaluated based on training time and execution time parameter. Time to construct a model from training dataset is called training time. Time taken by the model to predict the class label of test data is known to be execution time. Since there is no training phase in K-NN classifier, only execution time was computed whereas for other parallel classifiers training time and execution time was computed. Time was measured for running parallel classification algorithms in single-node- and multi-node Hadoop cluster.

The following section describes the parallel classification algorithms that were executed on real-time datasets.

4   Results and Discussion

4.1   Performance Analysis in Hadoop Single-node- and Multi-node Cluster

The execution time is the time during which a program is running (executing), in contrast to other program lifecycle phases such as compile time, link time and load time. This is the sum of the time taken by the entire Map phase and Reduce phase. The parallel implementation of K-NN algorithm greatly increased the speed by reducing its time complexity from $O(D)$ to $O(D/p)$, where "D" is the number of training instances, "p" is the number of processors. Similarly executing Naive Bayes on Hadoop Map Reduce model across multiple nodes can save considerable amount of time compared to running the model on a single node. The implementation of parallel Naive Bayes classifier would significantly reduce the time complexity from $O(ND)$ to $O(N/p * D/p)$ where "N" is the number of attributes and "D" is the number of training instances.

The time taken for execution on a single-node cluster was calculated and it was compared with the time taken for execution on multi-node cluster. It was found that by executing on a multi-node cluster, the time taken was reduced to a great extent. Thus the time taken for executing the K-NN, Naive Bayes, and Decision Tree algorithms in both single and multi-node clusters were considered and compared. The number of Mappers and Reducers in single-node cluster was 2 and 1 respectively. The number of Mappers and Reducers in a multi-node cluster was 4 and 2 respectively. The time taken by Mapper and Reducer was called as training time. Total time given in Table 2 is the execution time which includes time taken to construct the model and to predict the class label for test data. The total time taken to execute parallel K-NN algorithm for 2500 instances on a single-node cluster was 1 min 29 s whereas on a multi-node cluster it took 28 s. The total time taken to execute parallel Naïve Bayes algorithm for 32,561 instances on a single-node cluster was 2 min 31 s whereas on a multi-node cluster it took 40 s. The decision tree algorithm for 11 instances was executed on a single-node cluster for 48 min 53 s whereas on a multi-node cluster it took 23 min 73. Decision tree was constructed with 11 training instances primarily to analyze execution time in single-node- and multi-node clusters. The same chronic kidney disease dataset with 400 training instances was run on single-node- and multi-node cluster which showed drastic decrease in execution time when executed on multi-node cluster.

Table 2 Comparison on execution time of parallel classification algorithms in Hadoop Map Reduce

Classification technique	No. of training instances	Single-node Hadoop cluster			Multi-node Hadoop cluster		
		Time taken by 2 mappers	Time taken by 1 reducer	Total time	Time taken by 4 mappers	Time taken by 2 reducers	Total time
Parallel K-NN	2500	8.89 s	8.34 s	1 min 29 s	4.07 s	4.53 s	28 s
Parallel naïve bayes	48,842	23.79 s	12.2 s	2 min 31 s	5.02 s	5.31 s	40 s
Parallel decision tree	For single tree: 11	29 min 50 s	12 min 48 s	48 min 53 s	11 min 07 s	6 min 90 s	23 min 73 s

Although it is a known fact that incorporating parallelism gains time, this research work has validated the same by computational methods and has alleviated the hurdles involved (synchronization/data consistency) in implementation. This research is believed to be the basis for computational advancements in the area of big data analytics that can be furthered to enhance clinical diagnostics/fraud detection, biomedical applications and finance predictions.

5 Conclusion

Parallel data mining algorithms in distributed environment suffers from communication cost, data movement among process and disk IO access. To process large amount of data, model construction takes much of the execution time. In order to reduce execution time Hadoop Map Reduce highly demonstrates the significance of using the Map Reduce programming model on top of the Hadoop distributed processing platform to process large volume of data. The authors have implemented the parallel K-NN classifier, parallel Naive Bayes classifier and parallel decision tree. Scaling to very large and massive datasets was made possible by extending single-node cluster to multi-node Hadoop cluster. Moreover, the authors have compared the execution time in single-node- and multi-node cluster and validated its efficiency. Decision tree was implemented using small set of training data due to absence of in-memory processing in Hadoop Map Reduce. Future work is aimed at implementing these algorithms on Machine Learning tools provided by Apache such as Mahout and Spark MLlib.

References

1. Grossman, L., Gou, Y.: Parallel Methods for Scaling Data Mining Algorithms to Large Data Sets, Handbook on Data Mining and Knowledge Discovery. Oxford University Press, Oxford (2001)
2. Talia, D.: Parallelism in knowledge discovery techniques. In: Applied Parallel Computing. Springer Berlin Heidelberg, pp. 127–136 (2002)
3. Wang, J., Chen, X., Zhou, K.: Research on a scalable parallel data mining algorithm, In: Fifth International Joint Conference on INC, IMS and IDC, 2009. NCM'09. IEEE, pp. 888–893 (2009)
4. Masih, S., Tanwani, S.: Data mining techniques in parallel and distributed environment-a comprehensive survey. Int. J. Emerging Technol. Adv. Eng. **4**(3), 453–461 (2014)
5. Zhou, L., Wang, H., Wang, W.: Parallel implementation of classification algorithms based on cloud computing environment, TELKOMNIKA Indonesian J. Electr. Eng. **10**(5), 1087–1092 (2012)
6. Xiao, H.: Towards parallel and distributed computing in large-scale data mining: a survey. Technical University of Munich, Technical Report (2010)
7. Hall, L.O., Chawla, N., Bowyer, K.W.: Combining decision trees learned in parallel. In: Working Notes of the KDD-97 Workshop on Distributed Data Mining, pp.10–15 (1998)
8. Joshi, M.V., Karypis, G., Kumar, V.: ScalParC: a new scalable and efficient parallel classification algorithm for mining large datasets. In: Parallel Processing Symposium, 1998, IPPS/SPDP 1998. Proceedings of the First Merged International and Symposium on Parallel and Distributed Processing 1998, IEEE pp. 573–579
9. Pakize, S.R., Gandomi, A.: Comparative study of classification algorithms based on MapReduce model. Int. J. Innovative Res. Adv. Eng. 2349–2163 (2014)
10. Parallel K-NN classifier: https://alitarhini.wordpress.com/2011/02/26/parallel-K-nearest-neighbor/ (2017)
11. Maha Lakshmi, N.V., Kanya Kumari, L., Rama Satish, A.: A study of classification algorithms using MapReduce framework. Int. J. Adv. Res. Comput. Sci. Software Eng. **5**(5), 885–891 (2015)
12. Anchalia, P.P., Roy, K.: The K-nearest neighbour algorithm using mapreduce paradigm. In: Fifth International Conference on Intelligent Systems. Modelling and Simulation (2014)
13. Wu, G., Haiguang, L.I., Hu, X., Bi, Y., Zhang, J., Wu, X.: MReC4.5: C4. 5 ensemble classification with MapReduce, In: Fourth IEEE ChinaGrid Annual Conference, pp. 249–255 (2009)
14. Parallel Naïve Bayesian Classifier: https://alitarhini.wordpress.com/2011/03/02/parallel-naive-bayesian-classifier/ (2017)
15. Katkar, V.D., Kulkarni, S.V.: A novel parallel implementation of naive bayesian classifier for big data. In: 2013 International Conference on Green Computing, Communication and Conservation of Energy (ICGCE), IEEE, pp. 847–852 (2013)
16. Zheng, S, Bayes, N.: Classifier: a mapreduce approach, a paper submitted to the graduate faculty of the North Dakota State University of agriculture and applied science (2014)
17. Srivastava, A., Han, E., Kumar, V., Singh, V.: Parallel Formulations of Decision-Tree Classification Algorithms, High Performance Data Mining. ISBN 978-0-7923-7745-0, Kluwer Academic Publishers. Manufactured in The Netherlands, pp. 237–261 (2002)
18. Paul, S.: Parallel and distributed data mining. In: Technical Report. ISBN: 978–953-307-547-1. Karunya University. Coimbatore, India (2011)
19. Ben-Haim, Y., Tom-Tov, E.: A streaming parallel decision tree algorithm. J. Machine Learning Res. **11**, 849–872 (2010)
20. Kubota, K., Nakase, A., Sakai, H., Oyanagi, S.: Parallelization of decision tree algorithm and its performance evaluation. In: The Fourth International Conference/Exhibition on IEEE High Performance Computing in the Asia-Pacific Region, 2000 Proceedings, vol. 2, pp. 574-579 (2000)
21. Chauhan, H., Chauhan, A.: Implementation of decision tree algorithm C4.5. Int. J. Sci. Res. Publications, **3**(10), 1–2 (2013)

22. Dai, W., Ji, W.: A map reduce implementation of C4.5 decision tree algorithm. Int. J. Database Theory Appl **4**, 49–60 (2014)
23. Shafer, J., Agrawal, R., Mehta, M.: SPRINT: a scalable parallel classifier for data mining. In: Proceeding of 1996 International Conference *on* Very Large Data Bases, pp. 544–555 (1996)
24. UCI Irvine Machine Learning Repository http://archive.ics.uci.edu/ml/datasets.html (2017)

Cloud-Based Scheme for Household Garbage Collection in Urban Areas

Y. Bevish Jinila, Md. Shahzad Alam and Prabhu Dayal Singh

Abstract Technical advancements have rapidly increased the population in urban areas. People residing in urban areas are scheduled with regular activities. The time-spend for managing the household cleanliness and garbage disposal on daily basis incurs extra overhead. A key challenge is to provide a proper garbage disposal for houses. In the present scenario, residents have to carefully monitor the garbage bins each day and dispose it when it is full. So, the problem is overflowing of garbage results in bad odour at home and surroundings. This limitation can be addressed by replacing the garbage bins with smart bins and connecting them to a cloud server. The processing entity in the cloud server further runs a shortest path route optimization algorithm and presents the shortest path for every 24 h basis. The employee can utilize this information for the collection of garbage from houses. The prototype of the proposed method of garbage collection is developed.

Keywords Garbage collection · Smart bin · Shortest path routing
Cloud computing · Waste management

1 Introduction

In developing countries like India, the growth of Urbanization is exponential. However, urbanization has brought more problems in areas like health, traffic, and security. The quality and the efficiency of the services provided in the urban region do not equip the needs of the public. The most important concern in cities is the cleanliness in public areas. Due to the migration of public towards cities, the quantity of waste disposal increases. If the garbage disposal is regulated in homes the cleanliness in the public areas can be maintained.

Y. Bevish Jinila (✉) · Md. Shahzad Alam · P. Dayal Singh
Department of Information Technology, Sathyabama Institute of Science
and Technology, Chennai 119, India
e-mail: bevishjinila.it@sathyabama.ac.in

© Springer Nature Singapore Pte Ltd. 2019
J. D. Peter et al. (eds.), *Advances in Big Data and Cloud Computing*,
Advances in Intelligent Systems and Computing 750,
https://doi.org/10.1007/978-981-13-1882-5_47

539

According to studies in 2016, the Chennai city in Tamil Nadu, India is dumped with more garbage as the number of residents in Chennai city increases exponentially for every year. According to the Waste-to-Energy Research and Technology Council (WTERT), the city generates 6404 tonnes of garbage every day with an average per capita of 0.71 kg. This is the highest in the country closely followed by Kolkata (0.66 kg), Delhi (0.65 kg), and Hyderabad (0.65 kg) and Bangalore (0.5 kg). There are around 8500 permanent employees in the Chennai Corporation, whose job is sweeping, collecting the garbage from streets and community bins.

The good quality of life in cities relies on the effective waste management schemes. There are two major problems in waste management. First, daily collection of garbage's may result in more number of empty garbage bins which results in wastage of time and man power. Secondly, weekly collection of garbage may result in the overflow of garbage bins which the increased rate of pests and spread of diseases. This requires a careful deployment of a smart garbage collection system.

Improper garbage disposal and the improper waste management end up with increase in the pests and spreading of diseases. This imposes bad odour in the household areas and on the roads.

At present in Chennai city, though the garbage is collected daily irrespective of its level, garbage overflows on roads which impose inconvenience to the public. The corporation employee is unaware of the filled up garbage bins and has to manually inspect and collect the garbage which is resultant of the garbage from houses. There is no smart application which helps the employee to collect the garbage based on its level.

The Ministry of Housing and Urban Affairs, Government of India has launched Swachhata—MoHUA to enable a citizen to post a civic related issue (e.g. garbage dump) which is further forwarded to the city corporation and thereafter assigned to sanitary inspector of the particular ward. This mobile application can be used by any citizen to capture the picture of a filled up garbage bin and the GPS location will be traced to identify the location of the dustbin. The officials notify the employees to clean the concerned garbage bin. Though this application helps citizens to keep the city clean, it requires manual intervention.

1.1 Problems to Be Addressed

The problems intended to be addressed includes the following,

- To collect the garbage on time based on the level of the garbage bins
- To generate a report on daily basis of filled up garbage bins for the corporation employee
- To integrate all the garbage bins and to develop a smart application that helps the employee to collect the garbage by shortest path.

The rest of the paper is organized as follows. Section 2 discusses on the existing studies in the area of waste management. Section 3 describes the architecture, materials and methods of the proposed model. Section 4 shows the prototype developed and the outputs generated. Section 5 concludes the work and gives a future direction.

2 Related Works

The existing studies and the new developments in garbage collection have been listed in this section. A study on the structure of the garbage collection system deployed with level detectors in the garbage bins is carried out by Vicentini et al. [1]. Most of the schemes in the literature are on the garbage collection in public areas. Fewer solutions have been given for household garbage collection.

The Radio Frequency Identification (RFID) technology is used in [2, 3] for smart waste management. The RFID tag is attached in all the garbage bins, the RFID reader and Geographic Information System (GIS) with camera in the GCV. When the GCV nears the garbage bin the RFID tag emits the information about the level of the bin. The RFID reader reads the information and the corporation employee decides to clean it or not. But, there are security issues, proximity issues, orientation problems and possibility of damage of tags. And, it also requires a complex process to detect the level of the bin. Hannan et al. [4, 5] has produced schemes for classification og garbages.

Hong et al. [6] has suggested the replacement of RFID with SGS (Smart Garbage Sensor) which helps to improve the energy efficiency up to 16%. The spatial inequality in garbage disposal in developing countries is analysed by Chen et al. [7]. A multiple regression model is proposed to quantify the relationship between various factors namely population density, level of education and unemployment rate. Fachmin Folianto et al. [8] have proposed the use of mesh networks for data collection. Whenever the bins are filled they need to be cleaned. The bin collector gives the route to collect the bins. However, this scheme results in complexity of the system.

Moreover, the development of Internet of Things (IoT) and their applications have been widely used in Smart City solutions. The data gathered by sensors can be sent to remote servers where it is stored, processed and used for tracking, monitoring and ultimately making intelligent decisions for infrastructure or service management [9].

Al-Fuqaha et al. [10] proposed that sketch of the IoT with a touch on technology, application and protocol concern. It explains about the differences between IoT and developing technologies like cloud computing and data analytics. It uses IoT for sensing the wastage level in the dustbins processes the data and sends it to the server for storing and process the data. The process is carried out by the Geographical Information system [11, 12].

To summarize, the existing schemes do not address the shortest path and visualization tools for household garbage collection. Such limitations are considered for implementing the proposed scheme.

3 Proposed System

This section addresses the proposed system on smart garbage collection. Based on the requirements like low cost, simplicity, and transparency, the proposed system is designed to be more transparent and simple.

The components of the system are listed below.

- Smart Bin
- Wi-fi access point
- Cloud Server
- Monitoring station
- Visualization tool

Figure 1 shows the schematic diagram of the proposed scheme.

In Fig. 1, the components in the proposed system are shown. The household smart bin generates an alert when it is full through the Wi-Fi to the registered cloud service. This is regularly updated in the database and monitored by a monitoring station. This information is accessible for the employee who collects garbage at houses in a mobile application. Figure 2 shows the flow diagram of the smart bin. Figure 3 shows the working of smart garbage processing unit.

The Ultrasonic Sensor unit is deployed in every bin. The sensor is linked to the Raspberry Pi circuit. When the garbage in the bins reaches the threshold value, the Raspberry pi generates a message with the entities namely the source IP address, bin ID, location and level. This message is transferred through the Wi-Fi to the cloud server. Then zone-wise split is done and is transferred to the local monitoring unit to find the optimized route. The details can be accessed through a mobile application.

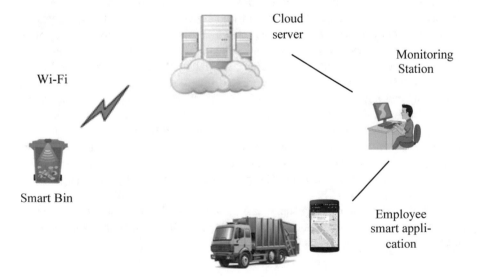

Fig. 1 Schematic diagram of the proposed system

Fig. 2 Flow diagram of
smart bin

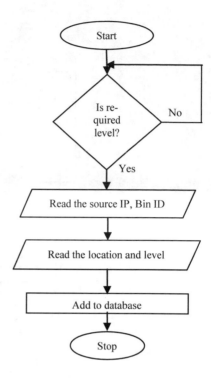

Ultrasonic Sensor

The ultrasonic sensor HC-SR04 is used to detect the level of the garbage in the bin. It measures the distance between the source and the target using ultrasonic waves. Based on the number of counts by the timer, the distance can be calculated between the object and transmitter. The TRD Measurement formula is expressed as: $D = R \times T$ which is known as the time/rate/distance measurement formula where D is the measured distance, and R is the propagation velocity (Rate) in air (speed of sound) and T represents time.

ZigBee

The ZigBee Alliance is an association of companies working together to define an open global standard for making low-power wireless networks.

The intended outcome of ZigBee Alliance is to create a specification defining how to build different network topologies with data security features and interoperable application profiles. Communication via zigbee network is interfaced with the Microcontroller such that the device can obtain the values.

IoT Gateway

IoT gateway acts as an interface between the IoT devices and the internet. They store and parse the information and send them to cloud server for processing and analytics. The cloud services are provided by major service providers like Amazon, Microsoft, etc. The IoT gateway transfers the data to the public cloud server.

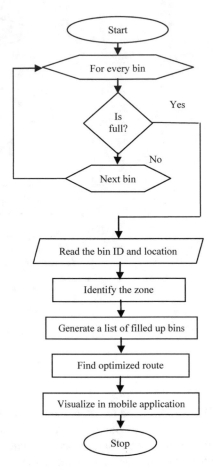

Fig. 3 Flow diagram of smart processing unit

4   Experimental Results

This section explains in brief the various experimental results. Table 1 show the sample cloud database tested with ten smart bins.

The system thus can maintain the entire details of waste management of the households in the city. The officials can easily monitor their area for cleanliness. Appropriate actions can be taken by the officials if the area is not cleaned. Figure 4 shows the prototype of the garbage collection.

Table 1 Cloud database

S. No.	Source IP Address	Bin ID	Location (Lat, Long)	Status (F/NF)	Level (%)	Timestamp
1	192.168.2.9	1	12°52′24″, 80°13′11″	F	80	2018-02-15:10:30:12
.	.	.	.	.		.
.	.	.	.	.		.
.	.	.	.	.		.
10	192.168.31.8	n	12°40′20″, 80°9′10″	F	75	2018-02-15:11:22:12

Fig. 4 Prototype of garbage collection

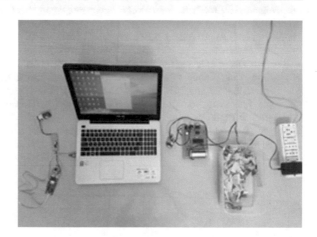

5 Conclusion

The merits of the proposed work includes the average volume of garbage collection per week, average route covered by each garbage vehicle per week, average time spent by each garbage vehicle per week, determination of cleanliness of cities, optimum garbage collection day wise and optimum location update of filled up garbage bins as map data. In future, the integration of public and household garbage collection can be incorporated.

References

1. Vicentini, F., Giusti, A., et al.: Sensorized waste collection container for content estimation and collection optimization. Waste Manag. **29**, 1467–1472 (2009)
2. Arebey, M., Hannan, M.A., Hassan, B., Begum, R.A., Huda, A.: RFID and integrated technologies for solid waste bin monitoring system. In: Proceedings of World Congress on Engineering (2010)

3. Hannan, M.A., Arebey, M., Begum, R.A.: Basri H (2011), Radio frequency identification (RFID) and communication technologies for solid waste bin and truck monitoring. Waste Manage. **31**, 2406–2413 (2011)
4. Hannan, M.A., Arebey, M., Begum, R.A., Basri, H.: An automated solid waste bin level detection system using a gray level aura matrix. Waste Manage. **32**, 2229–2238 (2012)
5. Islam, M.S., Hannan, M.A., Basri, H., Hussain, A., Areby, M.: Solid waste bin detection and classification using dynamic time warping and MLP classifier. Waste Manage. **34**, 281–290 (2014)
6. Hong, I., Park, S., Lee, B., Lee, J., Jeong, D., Park, S.: IoT-based smart garbage system for efficient food waste management. Scientific World J. (2014)
7. Chen, C.C.: Spatial inequality in municipal solid waste disposal across regions in developing countries. Int. J. Environ SciTechnol **7**(3), 447–456 (2010)
8. Folianto, F., Low, Y.S., Yeow, W.L.: Smart bin: smart waste management system-IEEE-April (2015)
9. Mitton, N., Papavassiliou, S., Puliafito, A., Trivedi, K.S.: Combining cloud and sensors in a smart city environment. EURASIP J. Wireless Commun. Networking, pp. 247 (2012)
10. Al-Fuqaha, A., Guizani, M., Mohammadi, M., Aledhari, M., Ayyash M.: Internet of Things: A Survey on Enabling Technologies, Protocols and Applications–IEEE (2015)
11. Sanjana, S.L., Jinila, Y.B.: An approach on automated rescue system with intelligent traffic lights for emergency services. In: International Conference on Innovations in Information, Embedded and Communication Systems 2017
12. Blessingson, J., Jinila, Y.B.: Multi utility/tracing kit for vehicles using RFID technology. RSTSCC pp. 273–276 (2010)

Protein Sequence Based Anomaly Detection for Neuro-Degenerative Disorders Through Deep Learning Techniques

R. Athilakshmi, Shomona Gracia Jacob and R. Rajavel

Abstract Exploring the effects of genetic information in causing potential brain disorders like Alzheimer's disease (AD) and Parkinson's disease (PD) is a relatively unexplored field. The aim of this investigation was to employ computational techniques at predicting anomalies that cause neuro-degenerative brain disorders with improved accuracy at an enhanced pace by analysis of gene and protein sequence data. The proposed methodology employed deep learning techniques to determine anomaly causing genes that played a significant role in causing potential brain disorders. The results revealed that deep learning models exhibit improved performance compared to conventional machine learning models, in identifying the optimal genes that cause neuro-degenerations.

Keywords Alzheimer's disease · Parkinson's disease · Autoencoders
Anomaly detection

1 Introduction

Alzheimer's disease (AD) and Parkinson's disease (PD) are neurological disorders affecting the older population in the world. Both the diseases are neuro-degenerative and typically begin late in life usually after the age of 60. AD is primarily a memory disorder and PD is a movement disorder [1, 2]. Earlier Diagnosis of AD and PD

R. Athilakshmi (✉) · S. G. Jacob
Department of CSE, Sri Sivasubramaniya Nadar College of Engineering,
Kalavakkam, Chennai, India
e-mail: athilakshmir@ssn.edu.in

S. G. Jacob
e-mail: graciarun@gmail.com

R. Rajavel
Department of ECE, Sri Sivasubramaniya Nadar College of Engineering,
Kalavakkam, Chennai, India
e-mail: rajavelr@ssn.edu.in

© Springer Nature Singapore Pte Ltd. 2019
J. D. Peter et al. (eds.), *Advances in Big Data and Cloud Computing*,
Advances in Intelligent Systems and Computing 750,
https://doi.org/10.1007/978-981-13-1882-5_48

reduces health cost, social care cost and betters our chances of delaying the onset of more debilitating symptoms, prolonging our independence, and maximizing our quality of life [3].

2 Literature Survey

The literature survey of the work related to our study is presented. A few researchers have worked on the AD/PD dataset, the details of which are given below. Javier et al. [4] gives an overview of statistical and machine-learning based method for personalized and cost effective detection of AD. Their work needs the support of new classifier for the enhancement of number and cost of biomarkers.

Guofeng et al. [5] have performed a comparative study on the performance of aging-related genes with normal genes and determine correlation cut off using Spearman's correlation and the study shows that age is one of the factor for Alzheimer's disease. In order to detect the Alzheimer disease early [6] extracted images from OASIS (Open Access Series of Imaging Studies) database by considering three parts of the brain—Hippocampus, Corpus Callosum, and Cortex. Their application used segmentation approach based on the Region of Interest (ROI) to extract images from the three sections of the brain. Support Vector Machine (SVM) was used for the classification of data and disease stage was set as the target class. Since images were used for prediction, the diagnosis consumed time and the disease had spread much by then. Sateesh et al. [7] performs Parkinson's disease Prediction using gene expression. They proposed a new classifier PBL-McRBFN which is used to predict PD using Micro-array gene expression data obtained from ParkDB database [8]. Their work results that the proposed classifier outperforms existing classifier in predicting PD. Yudong et al. [9] presented a Detection of Alzheimer's disease by displacement field estimation between a normal brain and an AD brain on OASIS dataset [10]. They have performed a comparative study on the performance of different classifier models in predicting Alzheimer disease. Their work results that the displacement field is effective in early detection of AD and related brain-regions. Most of the research work was based on AD or PD only and this research work take the two brain disorder into account for exploring and analyzing brain data. This paper aims at bringing the application of deep learning techniques in finding out the most contributed anomaly gene in brain disorders AD and PD.

3 Materials and Methods

3.1 Dataset

The following steps are used to acquire the dataset of Alzheimer's and Parkinson's disease.

Step 1:

The Gene Sets of Alzheimer's and Parkinson's were obtained from Kyoto Encyclopedia of Genes and Genomes (KEGG) which utilizes Gene Set Enrichment Analysis (GSEA) database [11]. KEGG is a knowledge database resource dealing with deciphering the genomes. A total of 169 genes were found to contribute to AD [12] and 133 genes were present in PD [13].

Step 2:

In order to find the level of similarity among the two types of brain disorders, the common genes were first identified and the gene set strength was found to be 95. Excluding the common genes, the strength of the Alzheimer's gene set reduced to 74 while the strength of the Parkinson's gene set was found to be 38.

Step 3:

The identified genes were submitted to Uniprot database [14] to obtain their associated protein sequence. The extracted sequence was saved as text file.

Step 4:

The next step is to submit the protein sequence to PROFEAT (Protein Features) server [15] which computes structural and physicochemical features of proteins and peptides for the given amino-acid sequence.

Step 5:

A Complete dataset of protein sequence based on structural and physicochemical properties associated with AD/PD is formed by mapping each protein descriptor value in excel sheet as shown in Fig. 1. A total of 1437 descriptor values were computed and represented as Gp, q, r, s where "s" represented the descriptor value and "r" indicated the descriptor while "q" signified the feature and "p" denoted the feature group. In this study, we are investigating the effect of Gene and Protein mutants in brain disorders (AD, PD) using bio-informatics and data-mining tools.

4 Deep Learning

Shallow structured architectures or conventional machine learning algorithms typically contain one or two layers of nonlinear feature transformations. It is also effective in solving many simple or well constrained problems but they have limitation in dealing with more complicated real world biomedical applications such as gene expression data, protein sequence data. Some of the conventional models are Gaussian mixture models (GMMs), Support Vector Machine (SVM), Logistic Regression and Multilayer Perceptrons with a single hidden layer. This leads to the need of deep

1	Genes	[G1.1.1.1]	[G1.1.1.2]	[G1.1.1.3]	[G1.1.1.4]	[G1.1.1.5]	[G1.1.1.6]	[G1.1.1.7]
2	ADAM10	4.946524	4.812834	5.748663	5.882353	4.411765	8.15508	3.877005
3	ADAM17	4.61165	4.247573	7.88835	7.038835	4.368932	5.461165	2.427184
4	APBB1	9.71831	1.830986	4.225352	9.577465	2.676056	7.605634	2.816901
5	APH1A	10.56604	1.132075	3.773585	2.641509	7.924528	8.679245	1.509434
6	APOE	12.30284	0.630915	3.470032	12.6183	1.26183	5.678233	0.630915
7	APP	8.181818	2.337662	6.493506	11.94805	2.727273	4.935065	3.246753
8	ATF6	5.820896	0.746269	4.328358	5.970149	2.089552	3.880597	1.791045
9	ATP2A1	9.090909	2.397602	5.294705	7.892108	3.596404	6.693307	0.899101
10	ATP2A2	8.06142	2.783109	4.702495	7.293666	4.126679	6.333973	1.247601
11	ATP2A3	9.87536	2.588686	4.506232	7.094919	3.451582	6.615532	1.438159
12	BACE1	6.786427	2.39521	5.389222	5.588822	4.391218	8.582834	1.996008
13	BACE2	10.03861	1.544402	3.861004	5.212355	4.826255	8.108108	0.579151

Fig. 1 Alzheimer Parkinson dataset based on structural and physicochemical properties of protein

learning architectures for extracting complex structure and building internal representation from inputs. Deep Learning is a branch of machine learning composed of multiple linear and nonlinear transformations. The essence of deep learning is to compute hierarchical features or representations of the observational data where the higher level features or factors are defined from lower level one. Deep structured architectures contain more layers of nonlinear transformations and it is originated from artificial neural network research. Some of them are Deep Neural Networks, Convolutional deep neural networks, Autoencoders, Deep Belief networks and recurrent neural network.

5 Proposed Methodology

5.1 Autoencoders

An Autoencoder (AE) is a special type of neural network (NN) where the output vector has the same dimensionality as the input vector. An AE tries to reconstruct the input in the output layer, passing data through the hidden layers. AEs were used for dimensionality reduction, pre-training deep NN, or as a features detector.

5.1.1 Autoencoder Training Algorithm

INPUT: Gene Dataset $x^{(1)}, \ldots, x^{(N)}$
OUTPUT: Encoder $e_\emptyset$, decoder d_θ
$\emptyset, \theta <$- Initialize parameters

repeat

$$E = \sum_{i=1}^{n} \|x^{(i)} - d_\theta(e_\phi(x^{(i)}))\|$$

Ø, θ <- Update parameters using gradients of E
until convergence of Ø, θ

An AE is a neural network that is trained by unsupervised learning, is used for reconstructions that are close to its original input. An AE is composed of two parts, an encoder and a decoder. A neural network with a single hidden layer has an encoder and decoder as in Eqs. (1) and (2), respectively.

W and b is the weight and bias of the neural network and σ is the nonlinear transformation function.

$$h = \sigma(W_{xh}x + b_{xh}) \tag{1}$$

$$z = \sigma(W_{hx}h + b_{hx}) \tag{2}$$

$$\|x - z\| \tag{3}$$

The encoder in Eq. (1) maps an input vector x to a hidden representation h by an affine mapping following a nonlinearity. The decoder in Eq. (2) maps the hidden representation h back to the original input space as a reconstruction by the same transformation as the encoder. The difference between the original input vector x and the reconstruction z is called the reconstruction error as in Eq. (3). An AE learns to minimize this reconstruction error.

Data passed into an AE experiences a reduction in dimensionality. With each reduction, the network summarizes the data as a set of features. With each dimensionality reduction, the features become increasingly abstract. The network reconstructs the original data from the abstract features and compares the reconstruction result against the original data. Based on the error between the reconstructed data and original data, the network uses back propagation to adjust its weights to minimize the reconstruction error. When reconstruction error is low, we can be confident that the particular gene found in the AE still carries important information that accurately represents the original data [16].

5.2 Autoencoder-Based Anomaly Detection

AE based anomaly detection is a deviation-based anomaly detection method using semi-supervised learning. It uses the reconstruction error as the anomaly score. Data points with high reconstruction error are considered to be anomalies. Only data with normal instances are used to train the AE. After training, the AE will reconstruct normal data very well, while failing to do so with anomaly data [17].

	reconstr_[G1.1.1.2].SE	reconstr_[G1.1.1.3].SE	reconstr_[G1.1.1.4].SE	reconstr_[G1.1.1.5].SE
1	0.05768397	0.105617179	1.520112e-04	0.01818498
2	0.05779121	0.017481791	2.412043e-04	0.01932255
3	0.08447466	0.013317244	1.981941e-03	0.01576301
4	0.09268716	0.021673529	1.062161e-05	0.01587752
5	0.08447466	0.004692875	6.350551e-04	0.00483001
6	0.03118956	0.068529801	1.274639e-05	0.05503218

	reconstr_[G1.1.1.6].SE	reconstr_[G1.1.1.7].SE	reconstr_[G1.1.1.8].SE	reconstr_[G1.1.1.9].SE
1	1.265576e-02	0.026319886	0.00231807	0.002130538
2	4.241184e-02	0.022689721	0.03445295	0.002850008
3	4.371389e-02	0.005610922	0.02767384	0.013506621
4	4.577804e-03	0.033141621	0.09927640	0.011095242
5	4.405602e-06	0.007672760	0.01743758	0.053338968
6	4.257920e-02	0.092484174	0.09522510	0.109671652

	reconstr_[G1.1.1.10].SE	reconstr_[G1.1.1.11].SE	reconstr_[G1.1.1.12].SE	reconstr_[G1.1.1.13].SE
1	0.0047866680	0.086488355	0.012809181	4.812778e-02
2	0.0002994447	0.043900111	0.003448707	2.781067e-02
3	0.0020898024	0.010078038	0.048243402	3.558784e-05
4	0.0042426715	0.003270587	0.002400973	5.929989e-02
5	0.0062005823	0.028463691	0.001077186	5.418362e-03
6	0.1105658424	0.101213380	0.006414315	1.354106e-01

Fig. 2   Reconstruction MSE in *R*

Fig. 3   Plotting of
reconstruction MSE in *R*

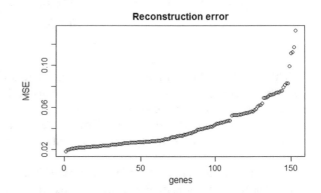

The Anomaly application computes the per-row reconstruction error for the test data set. It passes it through the AE model (built on the training data) and computes reconstruction mean square error (MSE) for each row in the test set. The quality of an estimator is measured by MSE, which is always non-negative or zero for better results.

Thus we train a Deep Learning AE based Anomaly detection technique to learn a nonlinear representation of the dataset, i.e.) learning the intrinsic structure of the training dataset. The AE model is then used to transform all test set data to their reconstructed data, by passing through the lower dimensional neural network and find outliers in a test dataset by comparing the reconstruction of each output data with its original data values. The reconstruction MSE of each gene and its plotting are shown in Figs. 2 and 3.

5.2.1 Autoencoder Based Anomaly Detection Algorithm

INPUT: Gene dataset X, threshold α
OUTPUT: Genes with high reconstruction error $\|x\text{-}\hat{x}\|$
$\emptyset$, θ <- train a autoencoder using the Gene dataset X
for i = 1 to N do

$$\text{reconstruction error (i)} = \sum_{i=1}^{n} \|x^{(i)} - d_\theta(e_\phi(x^{(i)}))\|$$

if reconstruction error (i) > α then
$x^{(i)}$ is an anomaly
else
$x^{(i)}$ is not an anomaly
end if
end for

6 Results and Discussions

In this study, an AE-Based Anomaly Detection method is used for detecting the anomaly gene in brain disorders AD and PD. In our study, AE was found to give the MSE of 0.04 The idea is that a high reconstruction error of a data indicates that the test set data does not conform to the structure of the training data and can hence be called an outlier and the genes such as PLCB1, ND4L, SLC18A1, UBE2G2, VDAC3, APAF1, ATP5A1, ATP5E, COX1, NDUFB7, NDUFB8, NDUFS1, NDUFS2 with high reconstruction MSE are taken as the anomaly genes based on structural and physicochemical properties of the protein sequences.

Acknowledgements This research work is a part of the Science and Engineering Research Board (SERB), Department of Science and Technology (DST) funded project under Young Scientist Scheme—Early Start-up Research Grant- titled "Investigation on the effect of Gene and Protein Mutants in the onset of Neuro-Degenerative Brain Disorders (Alzheimer's and Parkinson's disease): A Computational Study" with Reference No- SERB—YSS/2015/000737.

References

1. Deeg, D.J.H., Wahl, H.W., Litwin, H.: Ageing and transitions: looking back and looking forward. Eur. J. Ageing **15**, 1–3 (2018). https://doi.org/10.1007/s10433-018-0463-6,2018
2. Ganguli, M., Rodriguez, E.: Age, Alzheimer's disease, and the big picture. Int. Psychogeriatr. **23**(10), 1531–1534 (2011). https://doi.org/10.1017/S1041610211001906

3. Reeve, A., Simcox, E., Turnbull, D.: Ageing and Parkinson's disease: why is advancing age the biggest risk factor? Ageing Res. Rev. **14**, 19–30 (2014). https://doi.org/10.1016/j.arr.2014.01.004
4. Escudero, J., Ifeachor, E., Zajicek, J.P., Green, C., Shearer, J., Pearson, S.: Machine learning-based method for personalized and cost –effective detection of Alzheimer's disease. IEEE Trans. Biomed. Eng. **60**(1), 164–168 (2013). https://doi.org/10.1109/tbme.2012.2212278
5. Meng, G., Zhong, X., Mei, H.: A systematic investigation into aging related genes in brain and their relationship with alzheimer's disease. PLoS ONE (2016). https://doi.org/10.1371/journal.pone.0150624
6. Rabeh, A.B., Benzarti, F., Amiri, H.: Diagnosis of alzheimer diseases in early step using SVM (Support Vector Machine). In: 13th International Conference on Computer Graphics, Imaging and Visualization, pp. 364–367. IEEE computer society, Morocco (2016). https://doi.org/10.1109/cgiv.2016.76
7. Babu, G.S., Suresh, S.: Parkinson's disease prediction using gene expression—A projection based learning meta-cognitive neural classifier approach. Expert Syst. Appl. **40**, 1519–1529 (2013). https://doi.org/10.1016/j.eswa.2012.08.070
8. Taccioli, C., Tegnér, J., Maselli, V., et al.: ParkDB: a Parkinson's disease gene expression database. Database. Article ID bar007, 2011. https://doi.org/10.1093/database/bar007
9. Zhang, Y., Wang, S.: Detection of Alzheimer's disease by displacement field and machine learning. Peer J. **3**, e1251 (2015). https://doi.org/10.7717/peerj.1251
10. Marcus, D.S., Wang, T.H., Parker, J., et al.: Open access series of imaging studies (OASIS): cross-sectional mri data in young, middle aged, non demented, and demented older adults. J. Cogn. Neurosci. **19**, 1498–1507 (2007). https://doi.org/10.1162/jocn.2007.19.9.1498
11. Gene Card Database. Available: www.genecards.org
12. GeneSet Enrichment Analysis Data: Alzheimer GeneSet. Available: http://software.broadinstitute.org/gsea/msigdb/cards/KEGG_ALZHEIMERS_DISEASE.html
13. GeneSet Enrichment Analysis Data: Parkinson GeneSet. Available: http://software.broadinstitute.org/gsea/msigdb/cards/KEGG_PARKINSONS_DISEASE.html
14. Universal Protein Resource: Available: www.uniprot.org. Accessed 20 Jan 2018
15. Rao, H.B., Zhu, F., Yang, G.B., Li, Z.R., Chen, Y.Z.: Update of PROFEAT: a web server for computing structural and physicochemical features of proteins and peptides from amino acid sequence. Nucleic Acids Res. **39**(Web Server issue), W385–90 (2011).https://doi.org/10.1093/nar/gkr284
16. Lyudchik, O, Vlimant, J.R., Pierini, M.: Outlier detection using Autoencoders. CERN non-member state summer student report 2016 (2016)
17. Zhai, S., Cheng, Y., Lu, W., Zhang, Z.: Deep structured energy based models for anomaly detection. In: International Conference on Machine Learning, New York (2016). arXiv:1605.07717v2

Automated Intelligent Wireless Drip Irrigation Using ANN Techniques

M. S. P. Subathra, Chinta Joyson Blessing, S. Thomas George, Abel Thomas, A. Dhibak Raj and Vinodh Ewards

Abstract The aim of this work is to address the water scarcity prevalent in our country through smart irrigation practices. In this work we take certain environmental factors such as soil moisture content into consideration for creating a sustainable and smart irrigation system. The models proposed here for the day to day estimation of evapotranspiration are derived by using the daily data parameters such as temperature, solar radiation, wind speed and humidity for a period of 4 years (2009–2013) from Karunya University's meteorological station, at Coimbatore, Tamil Nadu, India. An Artificial Neural Network approach is adopted to run the software part using the environmental parameters and the output obtained from the ANN method with the least RMSE error is taken into account for the ETo value. The reliability of the computational models used are done based on the results achieved through two prominent empirical methods. These include Penman-Monteith equation and Hargreaves equation and comparing their respective Mean Square Errors (MSE) and also the Root Mean Square Errors (RMSE). Hargreaves method is suitable with the least RMSE error. In the hardware approach Hargreaves method has been implemented using Raspberry PI controller. The real-time data from the field controller is relayed to

M. S. P. Subathra (✉) · C. J. Blessing · S. Thomas George · A. Thomas · A. Dhibak Raj
Department of Electrical Sciences, Karunya Institute of Technology
and Sciences, Coimbatore 641114, Tamil Nadu, India
e-mail: subathra@karunya.edu

C. J. Blessing
e-mail: chintajoysonblessing@karunya.edu.in

S. Thomas George
e-mail: thomasgeorge@karunya.edu

A. Thomas
e-mail: abelt@karunya.edu.in

A. Dhibak Raj
e-mail: dhibakraj@karunya.edu.in

V. Ewards
Faculty of Computer Sciences and Technology, Karunya Institute
of Technology and Sciences, Coimbatore 641114, Tamil Nadu, India
e-mail: ewards@karunya.edu

© Springer Nature Singapore Pte Ltd. 2019
J. D. Peter et al. (eds.), *Advances in Big Data and Cloud Computing*,
Advances in Intelligent Systems and Computing 750,
https://doi.org/10.1007/978-981-13-1882-5_49

555

a hardware setup at the local base station. This is done through a wireless ZigBee protocol which eventually transmits the necessary data via a GPRS link to the remote station. The output volumetric water content was calculated using Crop coefficient ETC. Solenoid valves are remotely controlled to release a calculated value of water based on the data acquired at the local base station. This method of automated irrigation will mitigate the problems usually associated with farming and will finally result in generating greater yields of crop production.

Keywords Evapotranspiration ETo · Artificial neural network (ANN) · ZigBee

1   Introduction

Agriculture is man's oldest known profession. Even in the Indian Economy Agriculture output is an important component in calculating the GDP growth apart from providing sustenance to the entire population. In rural India, agriculture acts as a primary source of employment and trade. With globalization and exponential rise in population, the agriculture industry is facing a lot of more pressure than ever before. In the agricultural sector, according to statistics estimates that 60% of all the water taken for irrigation is taken from reservoirs, lakes and rivers alone. With the revolution in technology farmers and agriculturists across the globe have taken to more mechanized ways of irrigation to ensure judicious use of water. Even though we have many ways of irrigation the onus is always on to find the most efficient way to use water.

The primary goal of agriculture is to provide safe, fresh food and staples to the public at the most reasonable price possible. The day to day demand on the agricultural sector makes it necessary to automate the overall irrigation process. This type of automated irrigation considers factors such as temperature, humidity and also the topography of the surrounding environment are taken into consideration.

Therefore, to achieve a successful and high quality harvest it is necessary to design an irrigation system which takes all these factors into consideration. The main objective of irrigation is to provide plants with sufficient water to prevent stress that may reduce the yield. The frequency and quantity of water needed depends upon local climatic conditions, crop and stage of growth, and soil moisture plant characteristics.

The quantity of water required can be determined in different ways which do not require the knowledge of "Evapotranspiration (ET)" rates. By using the reference Evapotranspiration (ETo), the Evapotranspiration (ET) rates can be calculated. Though the number of equations available to estimate ETo are available in abundance, the FAO-56 Penman-Monteith (FAO-56 PM) [1] equation along with Hargreaves equation are used for this work. Soft computing methods have proved to be superior and reliable in forecasting and estimating ETo [2]. The primary aim of this study is to study and select from different evapotranspiration methods such as Penman Monteith and Hargreaves method, the best method to estimate the daily reference evapotranspiration by running it through Artificial Neural Network (ANN) model. The data used for these methods is sourced from a meteorological station situated in the southern part of India. The results achieved above are compared with the reference values calculated through the Penman Monteith method. The product achieved by multiplying the reference evapotranspiration (ETo) and the crop coefficient (Kc) forms the term crop evapotranspiration (ETc) [3]. In the hardware approach drip irrigation has been laid for 4 rows and sensors has been placed. Using wireless sensor network in a real-time field for tomato plant has been demonstrated using raspberry pi controller by considering the best technique (MLP) using Hargreaves method was used. The real-time data in order to calculate the water required for irrigation based on the respective crop's 'Evapotranspiration' coefficient.

2 Drip Irrigation

Drip irrigation is a process through which water and fertilizers used could be saved by allowing the aforementioned quantities to drip slowly, through a network of valves, pipes, tubes and emitters, thus reducing wastage of water and nutrients therefore improving the application efficiency in certain cases even up to 90%. The components involved include Pump, pipelines, fittings, lateral pipe, emitters, end cap, tee connector, straight connector.

3 Evapotranspiration

3.1 Evapotranspiration

Both the processes of Evaporation and transpiration tend to occur simultaneously and it isn't easy to differentiate between both. The amount of solar radiation, reaching the soil surface is primarily used to determine the evaporation from a cropped soil. As the crop grows, the amount of solar radiation received by the plant reduces due to the increase in canopy shade area. In the initial part of crop development, the primary loss of water is due to soil evaporation whereas in the later stages it is due to transpiration.

The FAO Penman-Monteith method based on expert consultation was recommended as the sole standard method for computing the reference evapotranspiration. This method requires certain parameter such as sunshine hours, solar radiation air humidity, air temperature and wind speed data.

3.2 Penman-Monteith Equation

The FAO Penman-Monteith equation determines the evapotranspiration from the hypothetical grass reference surface and [1] provides a standard to which evapotranspiration in different periods of the year or in other regions can be compared and to which the evapotranspiration from other crops can be related

$$\text{ETo} = \frac{0 \cdot 408\Delta(R_n - G) + \gamma \frac{900}{T+273} U_2(e_s - e_a)}{\Delta + \gamma(1 + 0 \cdot 34 U_2)} \tag{1}$$

where, ETo reference evapotranspiration [mm day^{-1}], R_n net radiation at the crop surface [MJ m^{-2} day^{-1}], G soil heat flux density [MJ m^{-2} day^{-1}], T air temperature at 2 m height [°C], u_2 wind speed at 2 m height [m s^{-1}], e_s saturation vapour pressure [kPa], e_a actual vapour pressure [kPa], $e_s–e_a$ saturation vapour pressure deficit [kPa], D slope vapour pressure curve [kPa °C^{-1}], g psychometric constant [kPa °C^{-1}].

3.3 Hargreaves Equation

As [4] an alternative when solar radiation data, relative humidity data and/or wind speed data are missing, reference evapotranspiration, ETo (mm d-1), can be estimated using the Hargreaves equation [5]. The FAO-56 Hargreaves equation [3] for daily computation is given by:

$$\text{ETo} = C_H(T_{max} - T_{min})^{\text{Eh}}(T_{mean} + 17.8)\text{Ra} \tag{2}$$

where T_{max} maximum day temperature T_{min} minimum day temperature, Ra extraterrestrial solar radiation, Environment constants $E_{h,\,\text{CH}}$

3.4 Crop Coefficient (ETc)

The reference evapotranspiration rates are multiplied by a crop coefficient to calculate the crop evapotranspiration. Here the crop coefficient expresses the difference in evapotranspiration between the cropped and reference grass surface. The differences achieved with respect to the evaporation and transpiration between both the surfaces can be combined into a single coefficient or described separately [6]. Certain factors which determine the selection approach are the climatic data available, the time at which the calculations are executed, also the accuracy required, play a pivotal role.

$$\text{ET}_{\text{Crop}} = \text{ETo} * \text{Kc} \quad \text{mm/day} \tag{3}$$

Table 1 Crop coefficient for tomato plant

Growth stage	Crop coefficient for tomato plant	
	No. of days	Crop coefficient (kc)
Initial	30	0.45
Development	40	0.75
Mid-season	40	1.15
Late season	25	0.80

where ETo—Evapotranspiration, Kc-Crop Coefficient. The crop coefficient for tomato plant is given in Table 1.

4 Artificial Neural Network (ANN)

ANN architecture is based on the structure and function of biological neural network. Similar to neurons in the brain ANN also consists of neurons which are arranged in various layers [7]. The popular neural network namely multilayer perceptron (MLP) consists of input layer to receive the external data to perform pattern recognition, output layer which gives the problem solution and hidden layer is an intermediate layer which separates the other layers. The adjacent neurons from input layer to output layer are connected through acyclic arcs. The MLP uses training algorithm to learn (Fig. 1).

Fig. 1 Multilayer perceptron

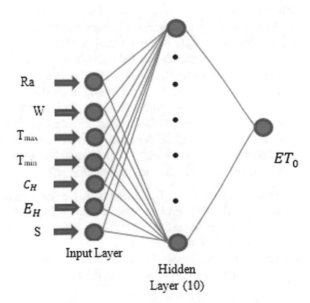

Fig. 2 Results for
hargreaves equation

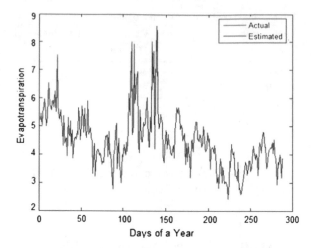

Table 2 Statistical error
achieved through penman and
hargreaves model

Methods	Penman monteith equation for Karunya data		Hargreaves equation for Karunya data	
	MSE	RMSE	MSE	RMSE
GRNN	35.5648	5.9636	22.5974	4.7537
PNN	23.6938	4.8676	12.9495	3.5985
RBF	34.9701	5.9136	22.5839	4.7523
MLP	0.3979	0.6306	0.0663	0.2575

The datasets which modifies the neuron weights depending on the error rate between target and actual output. First, for model establishment, various networks like generalized regression neural network, radial basis function neural network, probabilistic neural network and multilayer perceptron network were selected because of its wide application and high accuracy [8]. Total 4 years of data of Karunya (Karunya University weather station) was taken and 3 years were considered for training and 1 year was considered for testing using different models of Artificial Neural Network. The statistical error was calculated for a given area and equation to predict the better model for evapotranspiration calculations. For a given data, the neural network acquires the data and generates an ETo value which is referenced with the one calculated by equations such as Penman and Hargreaves. The model with the least RMSE error is chosen and implemented (Fig. 2; Table 2).

- From the above table it is evident that among Generalized Regression Neural Network (GRNN), Probabilistic Neural Network, Radial Basis Function (RBF), the Root Mean Square Error (RMSE) is the least for the Multi-layer Perceptron Neutral Network model.
- Generalized Regression Neural Network (GRNN) is generally used extensively for function approximation and consists of layers such as radial basis layer and

special linear layer. There is no need for an iterative training procedure here. It internally consists of four layers.

- Radial Basis Function Neural Network (RBF) has the abilities of quick training and generalization ability. The approximation of a non-linear object is done through RBF neural network.
- Multilayer Perceptron (MLP) is widely chosen for its ability ti solve tough and various issues. The output is influenced by several parameters to get the desired results.
- Probabilistic Neural Network (PNN) is a feed forward neural network, is used in pattern recognition problems and classification problems. This method ensures that mis-classification is minimized to a greater extent.

5 Study Area and Data

The study area involves an agricultural plot rectangular in size amounting to 11 square yards, with 30 tomato plants and 30 inline emitters. These plants are supplied by LDPE (Low Density Polyethylene) tubes which are 12 mm in diameter. The data such as temperature, solar radiation, wind speed and humidity are acquired for a period of 5 years from the meteorological station of Karunya University at Coimbatore (Latitude of $10.9397487°$ and Longitude of $76.7458484°$), Tamilnadu, India. It is located in the Western Ghats in southern part of India which is surrounded by high mountains.

6 Hardware Block Diagram

6.1 Hardware Implementation

The hardware requirements for this setup require a Solar panel. Through the concept of photovoltaic effect, solar panels are used to generate electricity from the sun. A Lead Acid battery is used in this setup, this is dependable and inexpensive when calculated on a cost-per-watt base. Most common types of Lead Acid battery used are the gel type also known as Valve-Regulated Lead Acid (VRLA). The battery used in this system is Sealed Lead Acid battery to prevent the batteries from overcharging, a charge controller is used. This is basically a voltage cum current regulator which regulates the voltage and current coming from the solar panels to the battery. Charge controller used here is a 3 A controller for a Lead-acid Battery. In the area of sensor networks, devices such as Tran's receivers play a very important role. It is necessary for sensor networks to use wireless modules to communicate and relay data. This setup uses one of the easiest modules named XBee wireless module. The Co-coordinator XBee is connected to the micro-controller board for transferring the data to processor. These are connected to the Transmitter and Receiver pins of the controller. An electronic component, model or subsystem whose primary aim is to act as a detector for detecting changes in the environment surrounding it and to further transmit the information to another processor or a data base is termed as a sensor. In Base station sensors used are Temperature, Humidity and Water Flow Sensor. The data of each sensor is logged in a coma separated value (csv) file in the system after a time interval. The first sensor used in this setup is a temperature and humidity sensor. This uses components such as a capacitive humidity sensor and thermistor to measure the surrounding air and transmits the digital signal on a data pin. One main disadvantage of this sensor is that the new data gets transmitted only once every 2 s. The temperature and humidity sensor used here works in the power range of 3–5 V and has maximum Input and Output current of 2.5 mA. The water flow sensor consists of parts such as a water rotor, valve body made out of plastic and a hall effect sensor. It works when the water flows through the rotor and the rotor rolls. The Hall Effect sensor further generates a consequent output based on the rotor rotation. This sensor was connected at the outlet-valve of the pump to measure the actual water applied to the plants each day. This sensor has a minimum dc working voltage of 4.5 V and a maximum working current of 15 mA. It has a DC working voltage of 5–24 V. Soil Moisture sensor is used to sense the moisture content in the soil at a particular place. It gives out the analog signal which is connected to the ADC of the XBee [9]. Their range ranges from 0 to 45% if volumetric water content in soil and has an accuracy of ±4% and power rating of 3 mA and 5 V DC supply. A relay circuit plays a crucial role in the hardware setup at the base station. A relay switch is used in this setup and the it has a coil which is driven by an NPN transistor switch. This switch acts in the cut-off region when base voltage of the transistor is zero. In such a condition no collector current flows into the base and no current flows through the relay coil too. Remote Sensor Node plays an important role in the whole system. This device is used to sense the soil moisture in the field and send the data to the base station. These are low power consuming unit with solar power and rechargeable Lithium ion battery for usage at time of inadequate solar radiation. These features include a pro-grammable Charge Current Up to 1000 mA/Pre-set 4.2 V Charge Voltage with 1.5% Accuracy. A Raspberry pi is used as here which has a quad core ARMv8 processor with 1.2 GHz clock speed. This is preferred for systems which require multitasking with high [10] speed and memory, for processing and storing the data obtained from remote sensor nodes. The programming language used in raspberry pi is Python. Python is versatile and provides multi-threading, hence two or more programs can run at same time using CPU time sharing. Both the equations are run through python and the results are noted (Figs. 3 and 4).

7   Working

The working of the system all starts with the deployment of remote sensor node in the field near to the tomato plant root at 20 mm safe distance. It reads the soil moisture content for every pre-set time interval and sends it to the base station, where

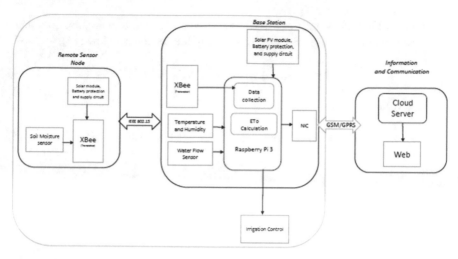

Fig. 3 Overall block diagram of system

Testbed

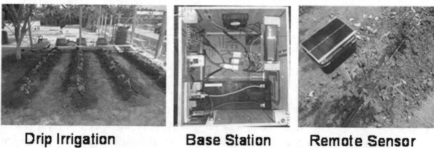

Drip Irrigation   **Base Station**   **Remote Sensor**

Fig. 4 Overall setup of the system

the packets of data received by the co-coordinator XBee are decrypted and data is obtained. The obtained values are Remote Node Battery Voltage and soil moisture with the node address. Upon receiving this, base station senses the environment temperature and Humidity and then logs the data into a CSV file and at the same time to the cloud server. The data upon reaching a certain count are then imported into a new program and the required parameters are calculated for finding the ETc required for the tomato plant. The calculations and crop parameters are given in Table 1. Then the water required is calculated and the pump is turned ON, post this operation the water flow from the laterals are continuously monitored using water flow sensor, and upon reaching the calculated amount of water needed the pump will be turned OFF. All these process details are updated to the server, upon which a user can view in their personal devices in the form of a webpage. The volumetric analysis performed allows an average of 9.1(L)/day of water for the tomato crop on a daily basis. The calculations are achieved by applying the formulas below [11].

$$ETo = C_H(T_{max} - T_{min})^{Eh}(T_{mean} + 17.8)Ra \qquad (4)$$

Final Volume

$$v = (KcETo - rm)(1 \div 1 - Lf(1 - LR) \div Lf(1 - LR) \qquad (5)$$

where

rm   The average monthly rain volume (mm)
Lf   Leaching efficiency coefficient as a function of the irrigation water applied
LR   Leaching fraction given by the Humidity that remains in the soil

$$LR = ECw \div (5ECe - ECw) \qquad (6)$$

ECw: the electrical conductivity of the irrigation water (ds . $m^{(-1)}$) and ECe: the crop salt tolerance (ds . $m^{(-1)}$).

7.1 Database Server

A data base server is generally a computer program which is tasked with providing database services for other computer programs and computers. Several DBMS softwares provide database-server functionality and some exclusively rely on a client-server model in order to access the database (Fig. 5).

7.2 Cloud Computing

Cloud computing is a recently developed type of computing which is based on the concept of sharing computer re sources rather than having other network related hardware to handle applications. Cloud computing is offered by various companies either as open source or a licensed version, for example IBM's Blue Cloud. Cloud computing is used in this setup to make the data acquired by the local base station accessible for anyone anywhere (Fig. 6).

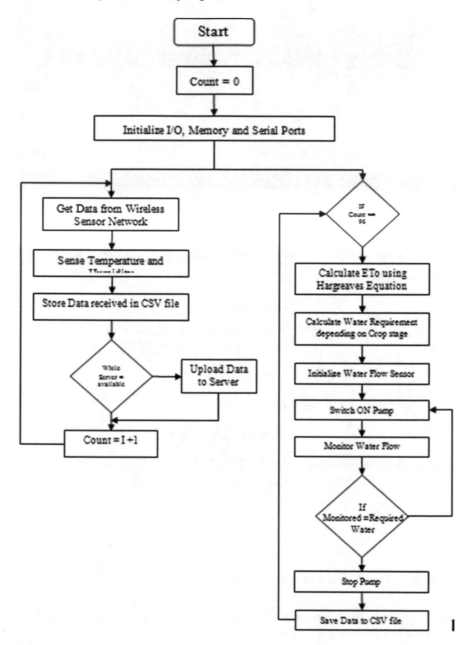

Fig. 5 Flowchart representing the working of the system

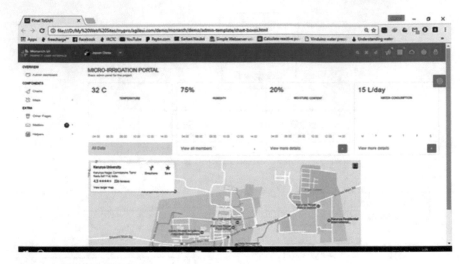

Fig. 6 Cloud connected system

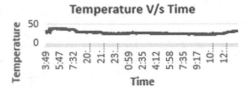

Fig. 7 Temperature variation for a day

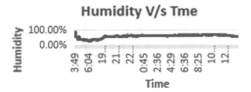

Fig. 8 Humidity variation for a day

8 Results and Discussion

Results shown in the below figures are plotted from the data obtained or stored in the database after testing it for 24 h.

Figure 7 shows the temperature variation for a day. Figure 8 Humidity variation for a day and Fig. 9. Soil Moisture variation for a day.

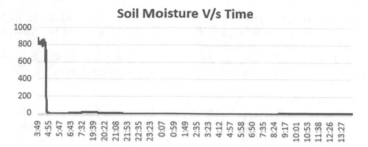

Fig. 9 Soil moisture variation for a day

9 Conclusion

In this work Information and Communication Technologies (ICT) are used extensively. The foremost concern in the Indian Agricultural sector is the lack of efficient irrigation techniques. Drip irrigation to a certain extent solves this issue. By considering cum monitoring environmental parameters such as temperature, relative humidity, sunshine hours, the water required can be calculated. This enables us to efficiently use the natural resources available to us ensures that a high yield is achieved. The work done here highlights the need for a Multi-layer Perceptron (MLP) enables irrigation network. The most relevant input variables for predicting the evapotranspiration are found to be Minimum Temperature, Maximum Temperature, and Minimum Humidity, Maximum Humidity, Wind Speed and Sunshine hours. The Penman and Hargreaves equation are used to calculate the Evapotranspiration rates and these models are run in a python program which is fed into the Raspberry pi. The Raspberry pi acts as an interface between the soil and the computations involved. It is found that wind speed has minimum effect on evapotranspiration prediction. From the trained models for P-M and Hargreaves equation, Hargreaves equation was more accurate. For Karunya data, Hargreaves equations the root mean square error (RMSE) and Mean square error (MSE) obtained were 6 and 2.5%, showing, high accuracy for MLP which utilizes the most relevant input variables. The developed MLP model can be used for prediction of evapotranspiration rates for particular location depending on the environmental factors. Also, the hardware model was tested in the real time under an experimental setup in university premises. Thus, in conclusion, the proposed ANN technique increases the irrigation efficiency, which ultimately results in reducing the labor cost thus saving water and electricity.

References

1. Pandey, P.K., Dabral, P.P., Pandey, V.: Evaluation of reference evapotranspiration methods for the northeastern region of India. Elsevier J. Int. Soil Water Conservation Res. **4**, 52–63 (2016)

2. Shamshirband, S., Amirmojahedi, M., Goci, M., Akib, S., Petkovi, D., Piri, J., Trajkovic, S.: Estimation of reference evapotranspiration using neural networks and cuckoo search algorithm. Elsevier, J. Irrigation Drainage Eng. (2015)
3. Allen, R.G., Pereira, L.S. Raes, D., Smith, M., et al.: Crop evapotranspiration-guidelines for computing crop water requirements-FAO irrigation and drainage paper 56, vol. 300. FAO, Rome (1998)
4. www.rasberrypi.org
5. Hargreaves, G.H., Samani, Z.A.: Reference crop evapotranspiration from temperature. Applied Engineering in Agriculture, American Society of Agricultural and Biological Engineers (1985)
6. Kim, S., Hung, S.K.: Neural networks and genetic algorithm approach for nonlinear evaporation and evapotranspiration modeling. J. Hydrol. **351**, 299–317 (2008)
7. Antonopoulos, V.Z., Antonopoulos, A.V.: Daily reference evapotranspiration estimates by artificial neural networks technique and empirical equations using limited input climate variables. Elsevier J. Comput. Electron. Agric. **132**, 86–96 (2017)
8. Huo, Z., Feng, S., Kang, S., Dai, X.: Artificial neural network models for reference evapotranspiration in an arid area of northwest China. Elsevier, J. Arid Env. (2012)
9. Gutiérrez, J., Villa-Medina, J.F. Nieto-Garibay, A., Porta, M.: Automated irrigation system using a wireless sensor network and GPRS module. IEEE Transactions on Instrumentation and Measurement (2014)
10. Cobaner, M.: Evapotranspiration estimation by two different neuro-fuzzy inference systems. Elsevier J. Hydrol. **398**, 292–302 (2011)
11. Yahyaoui, I., Tadeo, F., Segatto, M.V.: Energy and water management for drip-irrigation of tomatoes in a semi-arid district. Elsevier, Agric. Water Manage (2017)

Certain Analysis on Attention-Deficit Hyperactivity Disorder Among Elementary Level School Children in Indian Scenario

R. Catherine Joy, T. Mercy Prathyusha, K. Tejaswini, K. Rose Mary,
M. Mounika, S. Thomas George, Anuja S. Panicker and M. S. P. Subathra

Abstract Attention-Deficit/Hyperactivity Disorder (ADHD) is a neurodevelopmental condition encompassing symptoms of inattention, hyperactivity, and impulsivity that interfere with a child's daily functioning prevailing around the world. In this work, the children of 6–10 years were chosen from six different schools in Coimbatore under Thondamuthur union. It helps us to identify the prevalence of ADHD in primary school children relating to gender difference, socioeconomic status, and also the presence of co-morbid factors that have an effect on their academic performance. The Vanderbilt ADHD diagnosis rating scale questionnaire was given to teachers and 950 elementary school children in Coimbatore to assess the ADHD symptoms from the samples (526 boys and 424 girls) were employed in this study. The IBM Statistical Package for the Social Sciences (SPSS) software version 20 was accustomed to categorize the prevalence of ADHD. There were 58.17% bright learners, 27.61% average learners, and 14.23% slow learners. Concerning attention symptoms 12.5% of students have attention symptoms to school works, 11.91% has difficulty sustaining attention to tasks or activities. The results shows a high prevalence of ADHD among primary school children mostly among males than females. This study indicates the importance of early identification and intervention by doctors. Information assortment and storing the big data hubs pertains to such disorders might be the new challenge of the long run health applications.

Keywords ADHD · Vanderbilt questionnaire · SPSS software

R. Catherine Joy · T. Mercy Prathyusha · K. Tejaswini · K. Rose Mary · M. Mounika
S. Thomas George (✉) · M. S. P. Subathra
Department of Electrical Sciences, Karunya Institute
of Technology and Sciences, Coimbatore, India
e-mail: thomasgeorge@karunya.edu

A. S. Panicker
Department of Psychiatry, P. S. G. Institute of Medical Sciences
and Research, Coimbatore, India

© Springer Nature Singapore Pte Ltd. 2019
J. D. Peter et al. (eds.), *Advances in Big Data and Cloud Computing*,
Advances in Intelligent Systems and Computing 750,
https://doi.org/10.1007/978-981-13-1882-5_50

1 Introduction

Recently with the availability of information and communication technology-based devices around the globe and its advancements, children are more addicted to Internet, virtual and animated gaming and electronic gadgets like smart phones, tablets, and televisions. They tend to do multitasking like doing homework, eating, watching TV, etc., all at the same time. Some children fail to do any two works at a time, as they struggle to concentrate on any of these activities completely. This lack of attention tends to affect the children with problems like Attention-Deficit and Hyperactivity Disorder (ADHD), Internet addiction and other addictions.

ADHD is a complex neurodevelopmental disorder [1] and the most chronic mental disturbance predominantly seen in preschool and early school years [3]. Attention-deficit hyperactivity disorder could also be a sturdy genetic basis multifunctional disorder caused during pregnancy [2]. Features of ADHD are hyperactivity, reduced cognitive process, poor educational performance, behavioral difficulty, emotional, impulsivity and social functioning, etc. Symptoms of ADHD in children continue as they enter adolescence and adulthood [2, 3].

According to studies over 8.33% of ADHD has psychiatric disorder in their families [4]. Among the factors which will increase a child's risk for developing ADHD are: (a) A mother's intake of drugs, alcohol and tobacco during pregnancy, (b) Birth difficulties or terribly low birth weight: Babies born before their due date or premature infant, (c) Environmental exposure: Disclosure to lead or alternative virulent substances, (d) Extremely neglected, abused or socially deprived, (e) Food supplements like artificial food colors and (f) Brain injury.

The knowledge concerning the ADHD affected children shows the high risk of negative outcomes like learning disabilities, depressions, school failure, and dropout. Especially in primary school children speech and language delay with normal hearing reflect ADHD. In addition to it there is a high probability of losing the ability to learn which influence their behavior negatively [5, 6].

In adults we have a tendency to see failing relationships, troubled in workplace, substance abuse, low shallowness, irritability, and mood swings [7, 8]. Individuals with ADHD classically bother about being organized, keeping targeted, making sensible plans, and thinking before acting. They are tense, noisy, and unable to urge themselves to the reality of dynamical situations which requires behavioral adaptability. This disorder could be a behavioral condition that produces challenges in concentrating on everyday requests and routines challenges [9, 10].

According to Centre for Disease Control and prevention (CDC), 5.5 Million children of age group 4–17 years are diagnosed by ADHD. Mild ADHD may be diagnosed at the age of 7, moderate ADHD at the age of 6 and severe ADHD in 4 years of age. In 2009, the statistics showed 7.3%, in 2011 it showed 9.5%. The yearly proportion rise of ADHD throughout the years 2003–2011 is 5% [11]. A survey on attention symptoms in and around Coimbatore was performed to relate with the global statistics.

2 Methods

2.1 Participants

A recent survey to know the percentage of children having ADHD related symptoms was conducted by visiting two Elementary schools and three Primary schools in Coimbatore district under Thondamuthur union, in which one thousand students were chosen who are in the age group of 6–10. Among 1000 students, 950 volunteered to participate in the survey with the help of teachers. There are 526 male students and 424 female students and they were assured that the data collected will be confidential. The number of students participated according to the classes from I to V are 207, 192, 158, 186, 207 respectively. The distribution of gender based on age is shown in the Table 1.

2.2 Survey Scale

A psychological assessment tool for attention-deficit hyperactivity disorder (ADHD) symptoms and their effects on behavior and academic performance in children ages 6–12 is to be identified. The Vanderbilt ADHD diagnosis Teacher Rating Scale suggested by National Institute for Children's Health Quality (NICHQ) is used considering its standards. The questionnaire employed in this survey is useful for screening the students stricken by ADHD. This scale includes 43 items which measures the total range of symptoms, classroom behavior, and academic performance. Every symptom was scored on four choices of "never", "occasionally", "often" and "very often" and the academic and classroom behavioral performance was scored on five choices of "excellent", "above average", "average", "somewhat of a problem" and "problematic". Four purpose response for symptoms is scored from {0 = never, 1 = often, 2 = Often, 3 = very often}. Five purpose response for classroom behavior and academic performance is scored from one to five {1 = excellent, 2 = above average, 3 = Average, 4 = somewhat of a problem, 5 = Problematic}. The collected information was tabulated and it was fed in the statistical software for analysis. The IBM Statistical Package for the Social Sciences (SPSS) Software version 20 was used to categorize the prevalence of ADHD.

Table 1 Distribution of male and female students in different age groups

Age	Male	Female
6	104 (51.23%)	99 (48.77%)
7	115 (59.59%)	78 (40.41%)
8	94 (58.75%)	66 (41.25%)
9	111(59.04%)	77 (40.96%)
10	102 (49.52%)	104 (50.48%)

3   Procedure

The teachers were asked to fill the Vanderbilt ADHD checklist by observing every individual child. Administrative clearance was obtained from the school assuring safety of the data collected. They answered the questionnaire and it was collected back within one-month time from the date of handing over. These rating scores were entered into SPSS software version 20 which is used to analyze the percentage of students with ADHD symptoms. The overall score of each rating of every item among the participants can be viewed statistically and graphically.

4   Results

By the study in the primary level school children, the prevalence of Attention-Deficit Hyperactivity Disorder (ADHD) symptoms is seemingly higher in boys than in girls in the primary level school children.

4.1   Assessment of ADHD in Children

A total of 950 students [out of which 526 are boys (52.39%), 424 are girls (47.61%)] were recruited in the present study. Median range age (in years) of the children was found to be 8 (6–10). A total of 135 of 950 children were found to meet the criteria of ADHD symptoms based on the response from the teachers. There were 58.17% bright students, 27.61% average performing students and 14.23% slow learning children (Slow learning children are considered as ADHD children). Of the 135 ADHD children with symptoms, 97 (71.85%) were boys and 38 (28.15%) were girls and this was statistically significant as around. The schematic representation of sample collection is shown in Fig. 1.

The result for three categories of Vanderbilt ADHD rating is (a) Symptom of ADHD (b) Academic Performance (c) Class room behavior are as follows:

Fig. 1 Schematic representation of sample collection

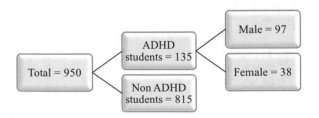

Table 2 Distribution for "fails to give attention to details or makes careless mistakes"

Score	Percentage (%)	No. of students
Never	55.11	523
Occasionally	17.28	164
Often	15.07	143
Very often	12.54	120

Fig. 2 Graphical distribution for "Fails to give attention to details or makes careless mistakes"

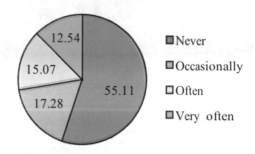

Table 3 Distribution for Mathematical performance

Score	Percentage (%)	No. of students
Excellent	48.37	459
Above average	7.38	70
Average	29.82	283
Somewhat of a problem	3.27	31
Problematic	11.17	107

(a) **Symptom-Fails to give attention to details or makes careless mistakes in schoolwork**

Some children cannot focus or pay attention to class for long time and they get distracted or do some daydreaming, disturbing class, or making noise. Few students make careless mistakes in doing their homework and this is also one of the symptoms of ADHD. The distribution for this symptom is shown statistically in Table 2 and graphically in Fig. 2.

(b) **Academic Performance—Mathematics**:

Depending on the academic performance, children with ADHD find difficult to finish math homework and are problematic in performing their studies. According to the statistics 107 (11.17%) children are having this difficulty. The distribution for this performance is shown statistically in Table 3 and graphically in Fig. 3.

(c) **Classroom behavior—Relationship with peers**

Due to their hyperactivity ADHD students have problem in mingling with peers. The distribution for this behavior is shown statistically in Table 4 and graphically in Fig. 4.

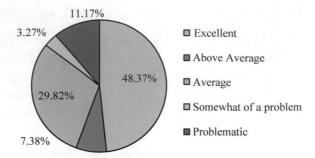

Fig. 3 Graphical distribution for mathematical performance

Table 4 Distribution for relationship with peer

Score	Percentage (%)	No. of students
Excellent	48.89	473
Above average	6.65	63
Average	30.59	291
Somewhat of a problem	2.74	26
Problematic	10.13	97

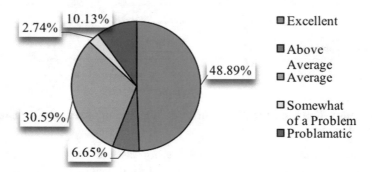

Fig. 4 Graphical distribution for relationship with peer

From this study, the symptom "Has difficulty in sustaining attention in tasks and activities" was 57.64% for "never", 17.28% for "occasionally", 13.17% for "often", 11.91% for "very often" seen in children.

For the performance of Reading, among 950 students, 48.63% were excellent, 7.17% were above average, 27.74% were average, 5.17% had somewhat of a problem, 11.29% were problematic.

For the classroom behavior "Following directions", among 950 students, 49.32% were excellent, 7.27% were above average, 27.50% were average, 5.90% had somewhat of a problem, 10.01% were problematic.

In addition to the above results, the other 37 items were also calculated using SPSS software version 20 for an extensive analysis which is to be a part of future studies.

5 Conclusion

We conclude that ADHD is one of the neurodevelopmental psychiatric disorders most prevalent in children. As it is very difficult to recognize ADHD students visually, standardized rating scale can be used as a first step to identify ADHD. According to this survey, percentage of children with ADHD symptoms were more in male children than the female children (male—71.85% and females—28.15%). This study helps in early identification and treating them in advance which helps in proper diagnosis. Researchers are suggesting that through neuro-feedback mechanism and Virtual reality (VR) games, one can improve their cognitive performance skills in ADHD children. Information assortment and storing the big data hubs pertains to such disorders might be the new challenge of the long run health applications and devising future strategies.

References

1. Sridhar, C., Bhat, S., Acharya, U.R., Adeli, H., Bairy, G.M.: Diagnosis of attention deficit hyperactivity disorder using imaging and signal processing techniques. Comput. Biol. Med. **88**, 93–99 (2017)
2. Biederman, J.: Attention-deficit/hyperactivity disorder: a selective overview. Biol. Psychiat. **57**(11), 1215–1220 (2005)
3. Feiz, P., Emamipour, S.: A survey on prevalence rate of attention-deficit hyperactivity disorder among elementary school students (6–7 years old) in Tehran. Procedia-Social Behavioral Sci. **84**, 1732–1735 (2013)
4. Venkata, J.A., Panicker, A.S.: Prevalence of attention deficit hyperactivity disorder in primary school children. Indian J. Psychiatry **55**(4), 338 (2013)
5. DA, A.N.S.P.B.: Prevalence of ADHD in a rural Indian population (2016)
6. Talaei, A., Mokhber, N., Abdollahian, E., Bordbar, M.R.F., Salari, Elham: Attention deficit/hyperactivity disorder: a survey on prevalence rate among male subjects in elementary school (7–9 years old) in Iran. J. Attention Disorders **13**(4), 386–390 (2010)
7. Sayal, K., Prasad, V., Daley, D., Ford, T., Coghill, D.: ADHD in children and young people: prevalence, care pathways, and service provision. The Lancet Psychiatry (2017)
8. Marcano, J.L., Bell, M.A., Beex, A.A.L.: Classification of ADHD and non-ADHD subjects using a universal background model. Biomed. Signal Process. Control **39**, 204–212 (2018)

9. Albatti, T.H., Alhedyan, Z., Alnaeim, N., Almuhareb, A., Alabdulkarim, J., Albadia, R., Alshahrani, K.: Prevalence of attention deficit hyperactivity disorder among primary school-children in Riyadh, Saudi Arabia; 2015–2016. Int. J. Pediatrics Adolescent Med. **4**(3), 91–94 (2017)
10. Venkatesh, C., Ravikumar, T., Andal, A., Virudhagirinathan, B.S.: Attention-deficit/hyperactivity disorder in children: clinical profile and co-morbidity. Indian J. Psychological Med. **34**(1), 34 (2012)
11. Choudhury, H.A., Ghosh, P., Victor, R.: Prevalence of attention deficit hyperactivity disorder among primary school children in cachar, assam, north-eastz. Indian J. Psychiatry **59**(6), S199–S199

An IoT-Enabled Hadoop-Based Data Analytics and Prediction Framework for a Pollution-Free Smart-Township and an Asthma-Free Generation

Sherin Tresa Paul, Kumudha Raimond and Grace Mary Kanaga

Abstract With the ever-growing infrastructure developments, social and economic standards, the rate of emergency department visits with respiratory illnesses has been on an all-time high. A wide range of factors are the cause to the debilitating ailments like Asthma. Among them atmospheric pollution is the most uncontrollable and dangerous contributor to this alarming development which inevitably comes along with the comforts we enjoy in this modern era. Unsurprisingly, children are the most affected, being the most vulnerable category of any community. This paper portrays an up-close picture of the wide range of pollutants, its hazardous impact on the pediatric population and the various prevalent techniques to monitor the pollution. The aim of this paper is to propose a Continuous Air Pollution Data Analytics Framework for a Pollution-Free Smart-Township thereby bringing forth an Asthma-Free generation.

Keywords Indoor and outdoor air pollution · Asthma
TRAP (Traffic Related Air Pollutants) · WBE (Wastewater-based Epidemiology)
Mobile air pollution monitors · Hadoop · IoT · Citizen science · Data analytics

1 Introduction

With the increase in comforts we, as a generation are inviting the destruction of the natural environmental resources we are blessed with from the beginning of time. Air, being one of the most basic need to live, has become one of the most polluted one

S. T. Paul (✉) · K. Raimond · G. M. Kanaga
Department of Computer Science and Technology, Karunya Institute
of Technology and Sciences, Coimbatore, India
e-mail: sherintresapaul777@gmail.com

K. Raimond
e-mail: kraimond@karunya.edu

G. M. Kanaga
e-mail: grace@karunya.edu

© Springer Nature Singapore Pte Ltd. 2019
J. D. Peter et al. (eds.), *Advances in Big Data and Cloud Computing*,
Advances in Intelligent Systems and Computing 750,
https://doi.org/10.1007/978-981-13-1882-5_51

among the lot. Since the Clean Air Act of 1970, the top criteria of pollutants involving Nitrogen Oxides, CO, Lead, Sulfur Oxides, ground level Ozone, and particulate matter are being vigilantly monitored and regulated individually which is proven to be insufficient to determine the possible combinatory effects of the individual chemicals [1]. According to the most recent WHO statistics, at least 235 million people are victims of this enervating illness and a disturbing rate of deaths estimated up to 383,000 were reported in the year 2015 and these estimates seems to be increasing every year at a troubling rate [2]. The existence of asthma from childhood has been observed to be linked to many factors. Among this plethora of triggers, air pollution seems to be the most unstable, uncontrollable, and dangerous one so far [3]. Children are considered to be the most susceptible part of the population affected by asthma owing to their growing and sensitive respiratory system, higher breathing rate, outdoor exposure, etc. It is also an interesting factor that the children belonging to the 3–12 age range spend a lot of their spare time in the children's parks, playgrounds, etc. which is very rich in usually ignored pollutants and triggers of asthma called pollen grains and ground level ozone [4, 5]. With the emerging technologies like IoT, Big Data Analytics, Cloud and Fog Computing, the research pertaining to monitoring the atmospheric pollution, and its quantification has become extensive. The Royal College of Physicians has put forth a paper [6] which reveals the reality of air pollution, particularly due to the transportation fuelled by fossil fuels and its hazardous health effects. Moreover, this paper has noted the urgent need of a new Clean Air Act for the new-age population.

The objective of this paper is to depict a clearer picture of an assortment of air pollutants and its effects on children's respiratory health, particularly pediatric asthma. An IoT-Enabled Hadoop-Based Continuous Air Pollution Data Analytics Framework for a Pollution-Free Smart-Township and an Asthma-Free Generation is proposed. The review of recent and relevant papers on pollutant exposures is given in Sect. 2. Section 3 gives a review on pollution monitoring techniques. Section 4 explains the proposed framework and Sect. 5 concludes the paper followed by future works and references.

2 Related Works on Pollutant Exposures

A detailed review on indoor air pollution focusing on usage of biomass fuel and its effects on respiratory health has been done. The link between respiratory diseases and biomass fuels are evident from the review and it points to another significant economic factor in choosing the cheaper alternative as fuel even when cleaner options are available in developing nations [7]. With the aim to find the relevance of indoor air pollution, hospital readmissions were correlated to childhood asthma, focusing on the child's bedroom by using logistic regression method and additionally with random forest method for better prediction from the enormous dataset of risk dynamics. Strong indicators were found supporting the association between frequent vacuuming (with bagged cleaners), usage of synthetic duvets, usage of carpeting, presence of airborne yeast, cladosporium, etc., and hospital readmissions with OR = 15.7, 14.6, 4.07, 1.52, and 1.68 respectively [8]. The children in Greater New Orleans with an eligibility criteria of moderate-sense scientifically diagnosed asthma aged 5–17 were included in a study to investigate the association between development of asthma and its correlation to cockroach exposure [9]. An inter-generational study was organized for observing association between the grandmothers' smoking when pregnant with the mother and the possibility of occurrence of asthma in the grandchild using the log binomial regression method which pointed towards a positive correlation for the risk of asthma development in the grandchild, irrespective of the mother's smoking habits [10]. A meticulous study was conducted on the smoking habits of parents and its correspondence with the likelihood of offspring developing asthma with the use of logistic regression. Occurrence of asthma in the offspring increased by 1.7-folds than non-smoking subjects when the mother is a smoker. When father presented smoking habit during the gestational period, the risk was higher, a 2.9-fold surge and when both parents sustained smoking, the risk amounted to an overwhelming rate of 3.7-fold. The quitting of smoking on mother's side and its effect on the decreased risk for asthma development in the offspring was estimated as a minimal change from 3.7 to 2.8 while quitting on the father's side amounted to a significant decrease in the risk [11]. A study was primed to find the effects of mother's smoking and/or passive smoking at home on the children's risk of developing asthma by employing Cox proportional and discrete-time hazard endurance analyses on a cohort constituting 7-year old children who belonged to Toronto. The outcome of the study established that even if the mother does not belong to the active smoking criteria, the presence of passive smoking at home at gestational period will contribute to the development of asthma in children at a rapid pace [12]. Incidences of asthma, allergic rhinitis, and bronchitis in children living in the proximities of a petrochemical complex is monitored and relevant SO2 pollution levels from concerned monitoring stations were analyzed for inferring short-term and long-term impact of SO2 pollution [13]. A study with an objective to estimate the collective effects of pollutants by inspecting the data of CO, ground level Ozone, NO_2 and $PM_{2.5}$ exposure and ED visits of children with asthma in Atlanta, Georgia was analyzed with CART. The results pointed towards the highest risk tallied to the days in which $PM_{2.5}$ was in the highest section and NO_2 belonging to the lowest two sections [14].

A 10-year emergency call records from 2002 to 2011 has been subjected to multiple regression analysis and the results implicated that despite the various asthma care and prevention techniques, the fact that pollens (especially Parietaria), rains, wind velocity, potential pollutants (SO_2, O_3, NO_x) and seasons (mainly spring, autumn) are the key contributing environmental factors concerning asthma attacks in children is the bitter truth which needs to be dealt with for the betterment of quality of life led by asthmatics [15]. A study employing Association rule mining, a case-crossover study with training and testing technique has been performed to find the multiple pollutant effect on the respiratory health of children. The significant role of O_3 in the development of asthma in pediatric subjects has been evident throughout the studies [16]. A case-crossover design was done to investigate the relationship between asthma and atmospheric pollutants like PM_{10}, $PM_{2.5}$, SO_2, NO_2, O_3, and CO. The effect modification of parameters is also assessed through time-stratified analysis with sex, age, and seasons [17]. A paper pertaining to a study of child and adolescent hospitalizations due to asthma exacerbation over 5 years to find association between asthma and outdoor fungi was proposed. A bi-directional time-stratified case-crossover study was conducted proved to be suitable for short-term exposures and its related events [18]. A study utilizing the land-use regression model and IDW model was conducted in clusters of neighborhoods which are prone to high TRAP exposures and by using spatial analytical methods like geocoding, the study reflected a much realistic picture of atopic asthma development with respect to TRAP exposure at birth place [19]. The emergency care facility data in Korea, particularly asthma-related data was analyzed for the purpose of finding hourly effects of pollutants and asthma attacks. A case-crossover analysis was done with the day of the visit to emergency healthcare facility as case and the same day for every other weeks were considered as controls. It is evident from the study that the Korean air quality policies need a thorough revision taking SES and hourly analyzed data into consideration as the pollutant exposure at different time of the same day can cause very different allergic reactions [20]. Studies published from 1999 to 2016 were analyzed for the detection of association between TRAP exposure and its correlation to development of childhood Asthma. Overall and age-specific meta-analysis were conducted and statistically significant risk assessment rates with BC, NO_2, $PM_{2.5}$ and PM_{10} exposures were found [21].

The ground level Ozone is known for its irritant nature to the respiratory system and thereby has a major role in atmospheric pollution [22]. A study in this regard which assembled the Linear regression models and time series analysis on the patient data of ED visits corresponding to asthma attacks, and ozone (ground level) concentrations resulted in a clearer C-R Curve and the nonlinear curve indicated the susceptibility of younger population to the negative respiratory impact of ozone exposure. In short, it is high time the role of ozone in our Healthcare regime to be highlighted [23]. The middle-aged TRAP exposure is analyzed to find its role associated with reduced pulmonary function, asthma, etc. The TRAP exposures were assessed relative to annual mean NO_2, which is detected by a satellite-based Land-Use Regression model [24]. A study observed that CO and NO_x exposure are major contributors to premature births among asthmatics and O_3 exposure posed the same threat for non-asthmatics. It has been concluded that the premature birth risk is much more elevated in asthmatic mothers who came into contact with pollutants when compared to non-asthmatic mothers [25]. The correlation between maternal exposure to air pollution in consecutive trimesters and the occurrences of Asthma, Eczema, and Allergic Rhinitis in 2598 pre-school going children of age 3–6 years has been analyzed and evaluated in a comprehensive study [26]. A population-based birth cohort study which included four locations with repeated questionnaires, meta-analysis, pooled analysis and a Spatial Land-Use Regression model was proposed to find the relationship between early-life contact and its potential to be the cause of development of asthma and its prevalence throughout childhood and adolescence. The outcomes concluded high probability of development and prevalence of asthma in childhood and adolescence [27]. Scientifically diagnosed asthma in pre-school going children of age 3–6 years were closely examined via questionnaires related to the exposure of critical air pollutants SO_2, NO_2, and PM_{10}. Logistic regression analysis was applied to identify an association between early-life exposure (in utero and first year of life) and childhood asthma by using Odds Ratio and 95% CI [26]. The hospitalization data along with pollutant and aeroallergen exposure data was analyzed while focusing on the combinatorial effects of synoptic weather changes and has procured a reliable notion that, the synoptic combination of weather types can significantly vary the effects of pollutants and aeroallergens on the allergy-prone population. A pooled dataset of 10 cities was used along with combined single-pollutant-specific regression coefficients for analysis [28].

3 Review on Pollution Monitoring Techniques

There are several air pollution monitoring hubs available these days but most of them are expensive and bulky. In other words our era of portability is not properly equipped with a portable and economical air quality monitoring system. Recently a research [29] has been done by using mobile-microscopy and machine learning to assess the impact of air pollution and it is tested in the areas around Los Angeles International Airport (LAX) and the results proved an increase in PM in areas surrounding the airport for more than 7 kms, when monitored for 24 h. A crowdsourcing tool called Smart Citizen Kit (SCK) is developed by HabitatMap which effectively monitors the atmospheric components like CO_2, humidity, temperature, light, pressure, etc. The SCK can collect and send the pollutant data and atmospheric data over the Internet (Wi-Fi) and can be visualized through an online portal as well [30]. A Multi-Agent Framework for a Hadoop-Based Air Quality Decision Support System was developed [31] which is motivated by the need to do real-time analysis of high-volume data pertaining to atmospheric pollution. The Framework is based on Hadoop and the database used is HBase. This agent enabled framework used K-means algorithm, MapReduce and Artificial Neural Network for Clustering, Projection and Prediction respectively. A cloud based IoT framework for the monitoring of healthcare data at a home-based environment was proposed [32] termed as Healthcare Industrial IoT (HealthIIoT). This framework may help improvise the healthcare monitoring of asthmatics on a home-based environment with respect to the real-time analytics of air pollution. A low-cost sensor-based air pollution monitoring system was proposed [33] using LabVIEW and ZigBee-enabled Wireless Sensor Network and though this is a low-complexity low-cost system, it lacks the intelligence of advanced machine learning algorithms. A recent research [34] implicated the need for a community-based health monitoring system and has developed a prototype with a remodeled version of the monitor, Dylos DC1700 (Dylos Corporation) equipped with Internet connectivity and was able to deploy in a community based on the Citizen Science Concept [35]. The model lacked a 3-Dimensional pollutant monitoring because the monitor was placed on a stationary location along the roadside. The proposed system has both stationary monitors on the selected locations and it also uses wearable devices equipped with sensors for active pollutant monitoring which will be stored in the Cloud for real-time data analytics using machine learning algorithms. The Smart

Citizen Kit [30] can be mounted on a drone or unmanned vehicle for continuous and dynamic data monitoring. Extensive recent review articles [36–38] on low-cost pollution monitoring devices and wireless sensor networks indicate an increase in the public concern regarding the uncontrollable and exponentially growing dilemma of air pollution. The reviews also show the availability of good quality, low-cost, solar-powered pollution monitors. Personal Ozone Monitor (POM) is another promising one on the market which can be used in the proposed system [39].

The WBE (Wastewater-based Epidemiology) is a method which has no personal information ties and it can be analyzed for the presence and quantity of respiratory medicines like Salbutamol, Deryphilline, and certain steroids like Prednisolone, Hydrocortisone, etc. These pharmaceutical concentrations can be detected and analyzed for any particular target area like a pre-school, a maternity ward or even a rehabilitation center. It is important to collect data before it goes into the water processing plant. Simple to moderate and moderate to severe asthma can also be detected by assessing more steroid usage for frequent and chronic asthma attacks which brings us to a study based on wastewater-based epidemiology (WBE) approach to render the association between the increase in use of salbutamol and increase in air pollution in order to estimate the prevalence of asthma in the sub-population under study which is analyzed by log-linear Poisson-regression model [40]. IoT (Internet of Things) is a disruptive technology [41] which basically enables the things we use in day to day life to connect to the internet and share data. IoT is widely implemented as a wearable technology with sensors to monitor heart rate, temperature, etc. The steep surge in the health hazards related to the upsurge of air pollution has been a pressing concern and the need for a framework utilizing the IoT and sensors for collecting the pollutant and atmospheric data is needed now more than ever. To enable the data collection and the real-time analysis of the said collected data, a collective sense of responsibility should be built in the community. Citizen Science [35] is one such policy which enables the community members to be an active part of an ongoing research. The citizens which become an active part of the ongoing research are called a Citizen Scientists. A recent research [34] implicated the need for a community-based health monitoring system and has developed a prototype with a remodeled version of the monitor, Dylos DC1700 (Dylos Corporation) equipped with Internet connectivity and was able to deploy in a community based on the Citizen Science Concept [35]. The model lacked a 3-Dimensional pollutant monitoring in which the monitor was placed on a stationary location along the roadside. The proposed system has both stationary monitors on the selected locations and it also uses wearable devices equipped with sensors for active pollutant monitoring which will be stored in the Cloud for real-time data analytics using machine learning algorithms. The Smart Citizen Kit [30] can be mounted on a drone or unmanned vehicle for continuous and dynamic data monitoring. Extensive recent review articles [36–38] on low-cost pollution monitoring devices and wireless sensor networks indicate an increase in the public concern regarding the uncontrollable and exponentially growing dilemma of air pollution. The reviews also show the availability of good quality, low-cost solar-powered pollution monitors. Personal Ozone Monitor (POM) is another promising one in the market which can be used in the proposed system [39].

4 Proposed Data Analytics and Prediction Framework

A Citizen Science [35] based concept is used to propose a Continuous Air Pollution Data Analytics Framework. The Framework collects the data through IoT-Enabled Citizen Scientists which will be stored in the Cloud on a real-time basis. An enhanced Hadoop foundation with Hbase equips the proposed framework with Big Data Analytics and MapReduce. As and when needed the framework accesses the air pollution data stored in the Cloud and the data is subjected to analysis by machine learning algorithms to detect sudden atmospheric changes and alerts the parents, teachers and authorities about the same. This unit is also responsible for predictive analytics using historical air pollution and epidemiological data analysis and informs the Smart Citizens about the same (Fig. 1).

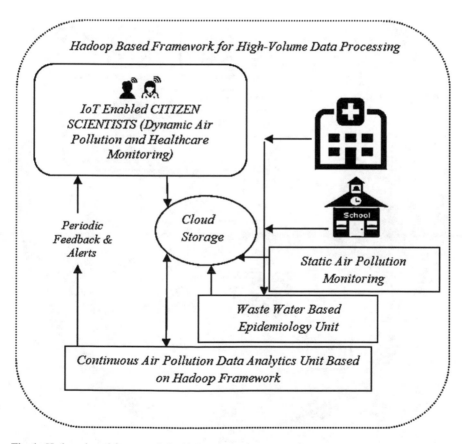

Fig. 1 Hadoop-based framework for high-volume data processing

The proposed framework is a novel one for the development of a Smart-Township which is pollution-free and thereby asthma-free in the long run with the help of continuous air pollution data analytics. The proposed framework is designed based on the following methodologies and concepts. Figure 1 portrays the IoT-Enabled Hadoop-Based Continuous Air Pollution Data Analytics Framework and Fig. 2 depicts the proposed Smart-Township model based on the Proposed Framework.

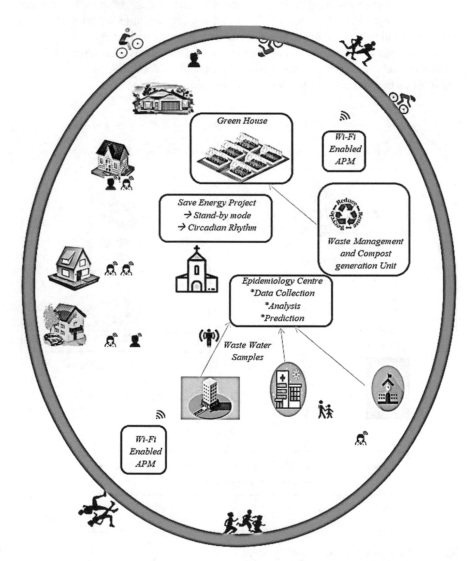

Fig. 2 The proposed Smart-Township model based on the proposed IoT-enabled Hadoop-based continuous air pollution data analytics and prediction framework for a pollution-free and asthma-free generation

WBE [40] is used in the proposed system to collect, analyze, and correlate the presence and increase of medicines for the respiratory diseases in the wastewater samples obtained from the schools and hospitals in the area. This data is used for further analysis with the air pollution data acquired over a week and the cause for the increase in respiratory attacks can be easily determined and precautions can be communicated to parents, teachers, and caregivers through SCK.

5 Conclusion and Future Work

The magnitude of danger atmospheric pollution poses to the younger generation is thoroughly understood and analyzed by surveying several studies and methodologies. The prominent role of air pollution in childhood asthma exacerbation is properly investigated by referring to the most relevant set of literature and it is duly documented. A framework for IoT-Enabled Hadoop-Based Continuous Air Pollution Data Analytics for developing a Pollution-Free Smart-Township and an Asthma-Free Generation is proposed and the required literature is reviewed. The proposed framework mitigates the unmet need for a static as well as dynamic air pollution monitoring. The framework works on the community level which is primarily based on the Citizen Science and Citizen Scientist concepts. The proposed framework can lead to certain revolutionary changes in the respiratory health of our younger generation. The Smart-Township Phase 2 includes an Action plan including but not limited to a Solar Power Station, Solar-Powered Vehicles and a Community-Oriented Air Quality Policy Renewal with active inputs from the laymen perspective through crowdsourcing tools. The Solar Power utilization will ultimately lead to the usage of cleanest energy and thereby reducing the fossil fuel usage which in turn brings down the Carbon Footprint contribution on a Global Level.

References

1. EPA. http://www.airnow.gov. Last Accessed May 2017
2. World Health Organization (WHO). World Health Organization Fact Sheet: Asthma. Last Updated April 2017. http://www.who.int/mediacentre/factsheets/fs307/en/. Last Accessed June 2017
3. CDC. Asthma in the US Vital Signs May, 2011
4. Bateson, T.F., Schwartz, J.: Children's response to air pollutants. J. Toxicol Environ. Health-A 71(3), 238–243 (2008)
5. EPA: Ozone—Good and Bad. https://www.airnow.gov/index.cfm?action=aqibasics.ozone Last Accessed May 2017
6. Holgate, S.T.: Every breath we take: the lifelong impact of air pollution'—a call for action. Clin. Med. J. R. Coll. Physicians London. 17(1), 8–12 (2017)
7. Bruce, N., Perez-Padilla, R., Albalak, R.: Indoor air pollution in developing countries: a major environmental and public health challenge. Bull. World Health Organ. 78(9), 1078–1092 (2000)
8. Vicendese, D., et al.: Quantitative assessments of indoor air pollution and respiratory health in a population-based sample of French dwellings. Environ. Res. 111(3), 425–434 (2011)

9. Rabito, F.A., Carlson, J.C., He, H., Werthmann, D., Schal, C.: PhDc New Orleans, La, and Raleigh, NC (2016)
10. Magnus, M.C., et al.: Grandmother's smoking when pregnant with the mother and asthma in the grandchild: the Norwegian mother and child cohort study. Thorax **70**, 237–243 (2015)
11. Harju, M., Keski-Nisula, L., Georgiadis, L., Heinonen, S.: BMC Public Health (2016) 16:428, https://doi.org/10.1186/s12889-016-3029-6, published online - 24 May 2016
12. Simons, E., To, T., Moineddin R., Stieb, D., Dell, S.D.: Journal of Allergy and Clinical Immunology: In Practice, 2014-03-01, Volume 2, Issue 2, Pages 201-207.e3, Copyright © 2014 American Academy of Allergy, Asthma & Immunology
13. Chiang, T.Y., Yuan, T.H., Shie, R.H., Chen, C.F., Chan, C.C.: Increased incidence of allergic rhinitis, bronchitis and asthma, in children living near a petrochemical complex with SO2 pollution. Environ. Int. **96**, 1–7 (2016)
14. Gass, K., Klein, M., Chang, H.H., Flanders, W.D., Strickland, M.: Classification and regression trees for epidemiologic research: an air pollution example. Environ Health **13**(17) 2014
15. Tosca, M.A., et al.: Asthma exacerbation in children: relationship among pollens, weather, and air pollution. Allergol Immunopathol (Madr). (2013)
16. Toti, G., et al.: Analysis of correlation between pediatric asthma exacerbation and exposure to pollutant mixtures with association rule mining. Artif. Intell. Med. **74**, 44–52
17. Ding, L., et al.: Airpollution and asthma attacks in children: a case cross over analysis in the city of Chongqing, China. Environ. Pollut. (2016). https://doi.org/10.1016/j.envpol.2016.09.070
18. Tham, R., Erbas, B.: School of Public Health, La Trobe University, Rm 129, Health Sciences 1, Bundoora 3086, Victoria, Australia. Asthma & Immunology (2016)
19. Shankardass, K., Jerrett, M., Dell, S.D., Foty, R., Stieb, D.: Spatial analysis of exposure to traffic-related air pollution at birth and childhood atopic asthma in Toronto. Ontario. Health Place **34**, 287–295 (2015)
20. Kim, J., Kim, H., Kweon, J.: Hourly differences in air pollution on the risk of asthma exacerbation, Environ. Pollut. 203:15e21 (2015)
21. Khreis, H., et al.: Exposure to traffic-related air pollution and risk of development of childhood asthma: a systematic review and meta-analysis, Environ Int (2016)
22. EPA guidelines about Air Quality Index., https://www3.epa.gov/airnow/aqi_brochure_02_14.pdf. Last Accessed May 2017
23. Zu, K.: Environ. Int. (2017). https://doi.org/10.1016/j.envint.2017.04.006
24. Bowatte, G., et al.: Traffic-related air pollution exposure is associated with allergic sensitization, asthma and poor lung function in middle age. J. Allergy Clin. Immunol. (2016). https://doi.org/10.1016/j.jaci.2016.05.008
25. Mendola, P., Wallace, M., Hwang, B.S., Liu, D., Robledo, C., Mnnist, T., et al.: Preterm birth and air pollution: critical windows of exposure for women with asthma. J. Allergy Clin. Immunol (2016)
26. Deng, Q., Lu, C., Li, Y., Sundell, J., Norback, D.: Environ. Res. **150**, 119–127 (2016)
27. Gehring, U., et al.: Exposure to air pollution and development of asthma and rhinoconjunctivitis throughout childhood and adolescence: a population-based birth cohort study, Lancet Respir Med 2015, Published online November 10 2015. http://dx.doi.org/10.1016/S2213-2600(15)00426-9
28. Hebbern, C., Cakmak, S.: Synoptic weather types and aeroallergens modify the effect of air pollution on hospitalizations for asthma hospitalizations in Canadian cities. Environ. Pollut. **204**, 9–16 (2015)
29. Wu, Y.-C., et al.: Air quality monitoring using mobile microscopy and machine learning. Light Sci. Appl. **6**(9), e17046 (2017)
30. Smart Citizen: Smart Citizen documentation (2016). http://docs.smartcitizen.me/
31. Fazziki, A.E., Sadiq, A., Ouarzazi, J., Sadgal, M.: A multi-agent framework for a hadoop based air quality decision support system, pp. 45–59
32. Hossain, M.S., Muhammad, G.: Cloud-assisted industrial internet of things (IIoT)—Enabled framework for health monitoring **101**, 192–202 (2016)

33. Telagam, N., Kandasamy, N., Nanjundan, M.: Smart Sensor network based high quality air pollution monitoring system using labview 13(8), 79–87 (2017)
34. English, P.B., et al.: The imperial county community air monitoring network: a model for community-based environmental monitoring for public health action. Environ. Health Perspect. 125(7), 1–5 (2016)
35. Broeder, D.L., Devilee, J., Van Oers, H., Schuit, A.J., Wagemakers, A.: Citizen science for public health. Health Promot Int Dec 23. pii: daw086, https://doi.org/10.1093/heapro/daw086. (2016)
36. Spinelle, L. et al.: Review of portable and low-cost sensors for the ambient air monitoring of benzene and other volatile organic compounds. Sensors (Switzerland). 17(7) (2017)
37. Mckercher, G.R., et al.: Characteristics and applications of small, portable gaseous air pollution monitors. Environ. Pollut. 223, 102–110 (2017)
38. Kumar, P., et al.: The rise of low-cost sensing for managing air pollution in cities. Environ. Int. 75, 199–205 (2015)
39. POM, Personal Ozone MonitorTM, https://www.twobtech.com/pom-personal-ozone-monitor. html
40. Fattore, E., Davoli, E., Castiglioni, S., Bosetti, C., Re Depaolini, A., Marzona, I., Zuccato, E., Fanelli, R.: Environ. Res. ISSN: 0013-9351, (2016). http://dx.doi.org/10.1016/j.envres.2016. 05.051
41. Ashton, Kevin: That 'Internet of Things' Thing. RFiD J. 22, 97–114 (2009)

Printed in the United States
By Bookmasters